Nanomedicines
Design, Delivery and Detection

RSC Drug Discovery Series

Editor-in-Chief:
Professor David Thurston, *King's College, London, UK*

Series Editors:
Professor David Rotella, *Montclair State University, USA*
Professor Ana Martinez, *Centro de Investigaciones Biologicas-CSIC, Madrid, Spain*
Dr David Fox, *Vulpine Science and Learning, UK*

Advisor to the Board:
Professor Robin Ganellin, *University College London, UK*

Titles in the Series:
 1: Metabolism, Pharmacokinetics and Toxicity of Functional Groups
 2: Emerging Drugs and Targets for Alzheimer's Disease; Volume 1
 3: Emerging Drugs and Targets for Alzheimer's Disease; Volume 2
 4: Accounts in Drug Discovery
 5: New Frontiers in Chemical Biology
 6: Animal Models for Neurodegenerative Disease
 7: Neurodegeneration
 8: G Protein-Coupled Receptors
 9: Pharmaceutical Process Development
10: Extracellular and Intracellular Signaling
11: New Synthetic Technologies in Medicinal Chemistry
12: New Horizons in Predictive Toxicology
13: Drug Design Strategies: Quantitative Approaches
14: Neglected Diseases and Drug Discovery
15: Biomedical Imaging
16: Pharmaceutical Salts and Cocrystals
17: Polyamine Drug Discovery
18: Proteinases as Drug Targets
19: Kinase Drug Discovery
20: Drug Design Strategies: Computational Techniques and Applications
21: Designing Multi-Target Drugs
22: Nanostructured Biomaterials for Overcoming Biological Barriers
23: Physico-Chemical and Computational Approaches to Drug Discovery
24: Biomarkers for Traumatic Brain Injury
25: Drug Discovery from Natural Products
26: Anti-Inflammatory Drug Discovery
27: New Therapeutic Strategies for Type 2 Diabetes: Small Molecules
28: Drug Discovery for Psychiatric Disorders
29: Organic Chemistry of Drug Degradation
30: Computational Approaches to Nuclear Receptors

31: Traditional Chinese Medicine

32: Successful Strategies for the Discovery of Antiviral Drugs

33: Comprehensive Biomarker Discovery and Validation for Clinical Application

34: Emerging Drugs and Targets for Parkinson's Disease

35: Pain Therapeutics; Current and Future Treatment Paradigms

36: Biotherapeutics: Recent Developments using Chemical and Molecular Biology

37: Inhibitors of Molecular Chaperones as Therapeutic Agents

38: Orphan Drugs and Rare Diseases

39: Ion Channel Drug Discovery

40: Macrocycles in Drug Discovery

41: Human-based Systems for Translational Research

42: Venoms to Drugs: Venom as a Source for the Development of Human Therapeutics

43: Carbohydrates in Drug Design and Discovery

44: Drug Discovery for Schizophrenia

45: Cardiovascular and Metabolic Disease: Scientific Discoveries and New Therapies

46: Green Chemistry Strategies for Drug Discovery

47: Fragment-Based Drug Discovery

48: Epigenetics for Drug Discovery

49: New Horizons in Predictive Drug Metabolism and Pharmacokinetics

50: Privileged Scaffolds in Medicinal Chemistry: Design, Synthesis, Evaluation

51: Nanomedicines: Design, Delivery and Detection

How to obtain future titles on publication:
A standing order plan is available for this series. A standing order will bring delivery of each new volume immediately on publication.

For further information please contact:
Book Sales Department, Royal Society of Chemistry, Thomas Graham House, Science Park, Milton Road, Cambridge, CB4 0WF, UK
Telephone: +44 (0)1223 420066, Fax: +44 (0)1223 420247,
Email: booksales@rsc.org
Visit our website at www.rsc.org/books

Nanomedicines
Design, Delivery and Detection

Edited by

Martin Braddock
AstraZeneca Research and Development, Macclesfield, UK
Email: martin.braddock@astrazeneca.com

THE QUEEN'S AWARDS
FOR ENTERPRISE:
INTERNATIONAL TRADE
2013

RSC Drug Discovery Series No. 51

Print ISBN: 978-1-84973-947-4
PDF eISBN: 978-1-78262-253-6
EPUB eISBN: 978-1-78262-751-7
ISSN: 2041-3203

A catalogue record for this book is available from the British Library

Published by The Royal Society of Chemistry,
Thomas Graham House, Science Park, Milton Road,
Cambridge CB4 0WF, UK

Registered Charity Number 207890

For further information see our web site at www.rsc.org

Printed in the United Kingdom by CPI Group (UK) Ltd, Croydon, CR0 4YY, UK

Preface

The science, or rather multiple sciences that support nanotechnology and specifically nanomedicine, has exploded in breadth and complexity over the past several decades. It is regularly cited that the renowned physicist Richard Feynman's 1959 lecture "There's plenty of room at the bottom" later published in 1960,[1] sowed the conceptual scientific seeds for the nanotechnology that has spawned such rapid technological development over the past 55 years. We have witnessed an explosive growth in the design, delivery and detection of nanomedicines,[2] especially in the field of oncology, largely as a result of necessity to limit the undesirable side effects of drugs amenable for cancer treatment by targeting the therapy to the tumour site and limiting exposure of the drug to otherwise healthy tissue.[3]

By the very nature of the field, this volume cannot do complete justice to the many talented individuals, teams and institutes who are intensively devoting their time and lives to making new scientific discoveries daily and attempting to harness new knowledge for the development of candidate novel medicines and potential products. For this, I apologise to those whom I have omitted. In this book, we aim to take the reader on our journey of "design, delivery and detection" and illustrate the voyage with an understanding of some design considerations of nanocarriers that are applicable to the cargo, whether it be a drug or a gene. We further illustrate delivery to a series of well-defined molecular targets relevant to oncology and illustrate the potential of siRNA and mesencyhmal stem cells. We next concentrate on aspects of technology which allow us to visualize the location and fate of nanomedicines *in vitro* and *in vivo* with specific emphasis on translation of non-clinical studies into the clinical trial environment. The next section illustrates the progress made in oncology with a variety of agents and constructs describing both non-clinical and clinical studies and their outcomes.

RSC Drug Discovery Series No. 51
Nanomedicines: Design, Delivery and Detection
Edited by Martin Braddock
© The Royal Society of Chemistry 2016
Published by the Royal Society of Chemistry, www.rsc.org

Notwithstanding the apparently exponential growth in scientific understanding and clinical development of new and potential new medicines, as with other medical modalities, there is no shortage of obstacles and hurdles to overcome before we see widespread use of nanomedicines across multiple therapeutic indications.[4] The last chapters of this book discuss aspects of the regulatory path and some of the promises and challenges which face us today, together with an overview of the development of Doxil®, the lessons we have learnt from an integrated understanding of many disciplines and the continued science in discovery today.

Looking further to the future; what may we envisage? It is inevitable that as science progresses and more clinical trials complete, the true potential of nanomedicine will be better understood. Advances in tumour targeting and imaging continue to be made[5] and the first human clinical trial assessing the safety of "C dots", referred to in Chapter 5, has been completed[6] with further studies under way. Improvements in construct design continue to be made and the potential utility of functionalized nanoparticles for drug delivery to the brain and neurodegenerative diseases is an emerging area.[7,8]

At the end of this journey, if the reader has acquired a good working knowledge of the field, and is able to use this book as a reference and is inquisitive enough to explore the field further, we will have achieved our ambition.

Martin Braddock

References

1. R. P. Feynman, *Eng. Sci.*, 1960, **23**, 22.
2. M. S. Ventola, *Pharm. Ther.*, 2012, **37**, 582.
3. G. Pillai, *SOJ Pharm. Pharm. Sci.*, 2014, **1**, 13.
4. A. J. Balakrishnan, *Nanomed. Res.*, 2014, **1**, 12.
5. G. Lin, X. Wang, F. Yin and K.-T. Yong, *Int. J. Nanomed.*, 2015, **10**, 335.
6. E. Phillips, O. Penate-Medina, P. B. Zanzonica, R. D. Carvajal, P. Mohan, Y. Ye, J. Humm, M. Gönen, H. Kalaigian, H. Schöder, W. H. Strauss, S. M. Larson, U. Wiesner and M. S. Bradbury, *Sci. Transl. Med.*, 2014, **6**, 260ra149.
7. C.-T. Lu, Y.-Z. Zhao, H. L. Wong, J. Cai, L. Peng and X.-Q. Tian, *Int. J. Nanomed.*, 2014, **9**, 2241.
8. G. Tosi, M. A. Vandelli, F. Forni and B. Ruozi, *Exp. Opin. Drug Delivery*, 2015, **12**, 1041.

Contents

Chapter 1   **Design Considerations for Properties of Nanocarriers
on Disposition and Efficiency of Drug and
Gene Delivery**   **1**
Jose Manuel Ageitos, Jo-Ann Chuah and Keiji Numata

1.1   Introduction   1
1.2   Types of Nanocarriers/Nanoparticles   2
    1.2.1   Viral Nanoparticles (VNPs)   2
    1.2.2   Micelles and Liposomes   2
    1.2.3   Polymeric Nanoparticles   4
    1.2.4   Dendritic Nanoparticles   5
    1.2.5   Peptidic Nanoparticles   6
    1.2.6   Nanocrystals and Nanosuspensions   7
    1.2.7   Metallic Nanoparticles   7
    1.2.8   Silica Nanoparticles   8
    1.2.9   Carbon-based Nanoparticles   8
1.3   Physicochemical Factors that Affect Nanoparticle
Efficiency   8
    1.3.1   Size   10
    1.3.2   Shape   11
    1.3.3   Surface Charge   11
    1.3.4   Ligands   12
1.4   Conclusions   13
References   14

RSC Drug Discovery Series No. 51
Nanomedicines: Design, Delivery and Detection
Edited by Martin Braddock
© The Royal Society of Chemistry 2016
Published by the Royal Society of Chemistry, www.rsc.org

**Chapter 2 Targeting Cyclins and Cyclin-dependent Kinases Involved
in Cell Cycle Regulation by RNAi as a Potential Cancer
Therapy**   23
Manoj B. Parmar and Hasan Uludağ

2.1   Introduction   23
2.2   The Cell Cycle   24
      2.2.1   An Overview   24
      2.2.2   Restriction Points and Checkpoints   26
      2.2.3   Regulation of Cell Cycle   27
2.3   Deregulation of the Cell Cycle in Cancer   28
      2.3.1   Conventional Drug Therapy Against Cyclins
              and CDK Inhibitors   29
2.4   RNA Interference   29
      2.4.1   Mechanism of Action   29
      2.4.2   RNAi and Cell Cycle Proteins   30
      2.4.3   Overcoming RNAi Intended for Cell Cycle
              Regulation   36
2.5   Concluding Remarks   38
Acknowledgements   39
References   39

**Chapter 3 Nanoparticle Carriers to Overcome Biological Barriers
to siRNA Delivery**   46
Hamidreza Montazeri Aliabadi and Hasan Uludağ

3.1   Introduction   46
      3.1.1   *In Vitro* siRNA Delivery   48
      3.1.2   *In Vivo* siRNA Delivery   66
3.2   Extracellular Barriers   66
      3.2.1   Surviving Degradation in the Systemic
              Circulation   68
      3.2.2   Escaping the Immune System   69
      3.2.3   Exiting the Systemic Circulation at the
              Site of Action   70
      3.2.4   Extracellular Matrix   73
3.3   Cellular Barriers   74
      3.3.1   Binding to the Cell Membrane   74
      3.3.2   Entering the Cell   77
      3.3.3   Permeating the Lipid Bilayer of
              Endosomes   78
      3.3.4   Intra-cytoplasmic Trafficking   79

3.4   siRNA Delivery Systems   80
     3.4.1   Lipid-based Delivery Systems   81
     3.4.2   Polymer-based Delivery Systems   82
     3.4.3   Peptides   84
     3.4.4   Recent Accomplishments in siRNA
          Delivery   85
References   90

Chapter 4   **Magnetic Targeting as a Vehicle for the Delivery
of Nanomedicines**   **106**
Mitsuo Ochi and Goki Kamei

4.1   Introduction   106
4.2   External Magnetic Device   107
4.3   Labeling Mesenchymal Stem Cells with
Supraparamagnetic Iron Oxide   107
     4.3.1   Cell Proliferation   109
     4.3.2   Multipotential Differentiation Capacity   110
4.4   Adhesion of m-MSCs to the Tissue Injured Site   112
     4.4.1   Cell Adhesion Rate in *Ex vivo* Studies   112
     4.4.2   Cell Distribution (Bioluminescence Imaging)   112
4.5   Animal Studies   114
     4.5.1   Cartilage Regeneration   114
     4.5.2   Bone Regeneration   115
     4.5.3   Muscle Regeneration   116
4.6   Conclusion   118
Acknowledgements   118
References   118

Chapter 5   **The Development of Theranostics – Imaging
Considerations and Targeted Drug Delivery**   **120**
*Wa'el Al Rawashdeh, Siem Wouters and
Fabian Kießling*

5.1   Introduction   120
5.2   Theranostic Carrier Materials   124
     5.2.1   Polymeric Nanoparticles   124
     5.2.2   Liposomes   129
     5.2.3   Antibodies   129
     5.2.4   Metal Nanoparticles   130
     5.2.5   Nanocarbons   131
     5.2.6   Microbubbles   132

5.3   Theranostics and Imaging   133
       5.3.1   Nuclear Imaging   133
       5.3.2   Computed Tomography   134
       5.3.3   Magnetic Resonance Imaging   136
       5.3.4   Ultrasound   139
       5.3.5   Optical Imaging   141
5.4   Conclusions   146
References   147

Chapter 6   **The Role of Imaging in Nanomedicine Development and**
             Clinical Translation   **151**
             Jinzi Zheng, Raquel De Souza, Manuela Ventura,
             Christine Allen and David Jaffray

6.1   Introduction   151
6.2   Imaging for *In vivo* Evaluation of the Spatio-temporal
       Distribution Characteristics of Nanomedicines   152
       6.2.1   Rationale for Spatio-temporal
                Biodistribution Assessment   152
       6.2.2   Imaging as a Non-invasive Method for
                Nanoparticle Biodistribution Assessment   153
6.3   Use of Imaging to Understand and Optimize
       Nanomedicine Performance   157
       6.3.1   Investigation of Size-dependence and Lesion
                Targeting Ability   157
       6.3.2   Investigation of the Effectiveness of Active
                Versus Passive Targeting   163
       6.3.3   Stroma Modification to Enhance
                Nanomedicine Delivery and Efficacy   170
       6.3.4   Assessment of the Performance of
                Activatable Nanomedicines   171
6.4   Clinical Experience and Future Considerations   175
References   178

Chapter 7   **Anticancer Agent-Incorporating Polymeric Micelles:**
             from Bench to Bedside   **182**
             Yasuhiro Matsumura

7.1   Introduction   182
7.2   Anticancer Agents Incorporating Micelles under
       Clinical Evaluation   184
       7.2.1   NK105, a Paclitaxel-incorporating Micelle   184
       7.2.2   NC-6004, Cisplatin-incorporating Micelle   187
       7.2.3   NC-6300, Epirubicin-incorporating Micelle   189

7.3 Verification of the EPR Effect using Imaging Mass
     Spectrometry   193
7.4 Discussion and Conclusion   194
References   195

**Chapter 8 Polymeric Nanoparticles and Cancer: Lessons Learnt from
            CRLX101**   **199**
*Ismael Gritli, Edward Garmey, Scott Eliasof, Andres Tellez,
Mark E. Davis and Yen Yun*

8.1 Introduction   199
8.2 Topoisomerase 1 Inhibitors   200
     8.2.1 Carbohydrate-based Polymeric
            Nanoparticles   203
     8.2.2 Polyamine Polymeric Nanoparticles   204
     8.2.3 HPMA Copolymeric Nanoparticles   205
     8.2.4 PEG Polymeric Nanoparticles   206
     8.2.5 Amphiphilic Polymeric Nanoparticles   209
     8.2.6 Bioconjugates   210
     8.2.7 Non-polymeric Nanoparticles   211
8.3 Hypoxia Inducible Factor-1 Inhibitors   211
     8.3.1 2ME2   211
     8.3.2 Camptothecins   212
     8.3.3 siRNA Technologies   213
     8.3.4 Endogenous HIF-1α Inhibitors   213
     8.3.5 Other HIF-1α Inhibitors   213
8.4 Cancer Stem Cells   214
8.5 Combination Therapy   215
8.6 CRLX101   216
     8.6.1 CRLX101 Chemistry   217
     8.6.2 CRLX101 Preclinical Results   217
     8.6.3 CRLX101 Clinical Results   220
8.7 Conclusion   221
References   224

Chapter 9 Nanodelivery Strategies in Breast Cancer Chemotherapy **233**
Vuong Trieu, Osmond J. D'Cruz and Larn Hwang

9.1 Introduction   233
9.2 Nanocarriers for Drug Delivery to Solid Tumors   234
     9.2.1 Liposomal Nanoparticles   235
9.3 Doxil®—The First FDA-approved Nano-drug   236
9.4 Taxane-based Nanodelivery (Abraxane® and
     Genexol-PM®)   238

9.5   CrEL-free Formulations of PTX   240
9.6   Albumin-bound PTX (*nab*-PTX/Abraxane®)   240
        9.6.1   Clinical Efficacy and Safety of Abraxane®   242
9.7   Polymeric Paclitaxel Micelles (Genexol-PM®)   244
        9.7.1   Clinical Studies of Genexol-PM®   246
        9.7.2   Clinical Efficacy of Genexol-PM®   246
9.8   Conclusion   247
References   248

**Chapter 10 Developing a Predictable Regulatory Path for Nanomedicines
                 by Accurate and Objective Particle Measurement   253**
                 Amy J. Phillips

10.1   Introduction   253
10.2   Regulation of Nanomedicines   254
          10.2.1   The Need to Develop Regulatory
                        Pathways for Nanomedicines   254
          10.2.2   Current Status of Nanomedicine
                        Regulation   256
10.3   Accurate and Objective Particle Measurement   259
          10.3.1   Specific Challenges of Nanoscale
                        Measurements   260
          10.3.2   Important Parameters of Nanoscale
                        Materials Used in Bioapplications   261
10.4   Current Techniques for Characterising
           Nanomaterials   265
          10.4.1   Ensemble Measurement Techniques   267
          10.4.2   Single-particle Techniques   268
          10.4.3   TRPS for Accurate Particle-by-particle
                        Measurement   271
10.5   Conclusions and Outlook   276
References   276

Chapter 11 Nanomedicine: Promises and Challenges   281
                 R. L. Juliano

11.1   The Evolution of Nanomedicine   281
11.2   Unique Capabilities of Nanomaterials:
           The Promise of Nanomedicine   282
11.3   Conceptual Issues in Nanomedicine   283
11.4   Challenges to the Implementation of Nanomedicine   285
11.5   Conclusion   286
References   287

Chapter 12 The Challenge of Regulating Nanomedicine: Key Issues 290
Raj Bawa, Yechezkel Barenholz and Andrew Owen

12.1 Introduction 290
12.2 Defining "Nano": A Problem for Regulators? 291
12.3 Lessons Learned from Doxil®: The First
      FDA-approved Nanodrug 295
12.4 Baby Steps Lead to Regulatory Uncertainty:
      The FDA as an Example 298
12.5 Importance of Understanding Pharmacokinetics
      and Distribution in Development and Regulatory
      Submission 306
12.6 Conclusions 310
References 312

Chapter 13 Doxil® – the First FDA-approved Nano-drug: from Basics *via*
CMC, Cell Culture and Animal Studies to Clinical Use 315
Yechezkel (Chezy) Barenholz

13.1 Introduction 315
13.2 Introducing Doxil® 316
13.3 Doxil®: Historical Perspectives in Short 324
13.4 Clinical Indications for Doxil® 326
13.5 Doxil®-related Intellectual Property 327
13.6 From the Failure of DOX-OLV to the Success
      of Doxil® 327
13.7 The Obligatory Need for Animal Studies and the
      Issue of the Relevance of Studies Using Cells in
      Culture (*in vitro*) to Doxil® Development 330
    13.7.1 The Issue of Animal Studies 330
    13.7.2 *In vitro–In vivo* Correlation 332
    13.7.3 Studies Based on Cells in Culture 332
    13.7.4 Lessons Learned from *In vitro* Cell in
            Culture Studies During Doxil®
            Development 333
13.8 The Time is Ripe for Generic Doxil®-like PLDs 335
13.9 Doxil®: New Findings (2012–2015) 336
13.10 Doxil® Still Keeps Some Secrets 341
Special Acknowledgements 341
References 342

Subject Index 346

CHAPTER 1

Design Considerations for Properties of Nanocarriers on Disposition and Efficiency of Drug and Gene Delivery

JOSE MANUEL AGEITOS, JO-ANN CHUAH AND KEIJI NUMATA*

Enzyme Research Team, Biomass Engineering Program Cooperation Division, RIKEN Center for Sustainable Resource Science, 2-1 Hirosawa, Wako-shi, Saitama 351-0198, Japan
*Email: keiji.numata@riken.jp

1.1 Introduction

Delivery of drugs or genetic material into cells is one of the emerging areas of biotechnology.[1] Nanoparticle (NP)-based drug carriers are especially interesting, due to their ability to deliver drugs inside target cells, thereby reducing the side-effects of non-specific treatments.[2] Among the various cargoes that are transportable by NPs, genetic material allows the reprogramming of cells both temporarily as well as permanently.[3] Several approaches are available for gene delivery, such as the use of natural vectors, like modified viruses, or artificial vectors, in the form of liposomes or peptidic complexes. Among the key considerations for designing NPs to effectively overcome biological barriers are their physicochemical properties and effects they produce in the living organism. This chapter discusses how the

RSC Drug Discovery Series No. 51
Nanomedicines: Design, Delivery and Detection
Edited by Martin Braddock
Published by the Royal Society of Chemistry, www.rsc.org

properties of nanocarriers can affect biological responses as well as the functionality of drug and gene delivery systems.

1.2   Types of Nanocarriers/Nanoparticles

NPs can be classified following several parameters, size being one of the first established. In this way, NPs are defined as particles with a diameter <100 nm,[1,4,5] but in practical applications[6] it is common for larger particles to be used (up to 1000 nm), especially for drug[4] and gene delivery. NPs can be differentiated based on composition, structure or properties; however, given their heterogeneity it is difficult to talk about pure types. Cargoes can interact with NPs either covalently, non-covalently *via* weak forces such as electrostatic or hydrophobic interactions, by hydrogen bonding, or can be physically entrapped in the matrix. The various types of NPs described in this chapter are summarized in Figure 1.1.

1.2.1   Viral Nanoparticles (VNPs)

Viruses are infectious nano-sized pathogens (10–200 nm) that naturally deliver their genetic material into a determinate cell line or organ. Viruses are mainly composed of proteinaceous materials (capsid) that have provided the basis for the development of drug and gene nanocarriers.[7–11] Viral-like NPs (VLPs) differ from VNPs by the absence of endogenous genetic material and their inability to replicate or to alter the host genome.[7,12,13] VLPs (Figure 1.1A) can be formed by self-assembly of capsid proteins over a functionalized inorganic NP core[13] into well-characterized monodisperse structures. Capsid proteins can be engineered and modified with different chemicals and proteins to promote specific targeting and improve penetration properties in VNPs/VLPs.[8,9,14,15] Gene delivery is one of the most studied fields of VNPs, since viruses are natural agents that transfect their genetic material[3,16–19] into cells. Although RNA and DNA viruses are potential candidates for gene delivery,[3] they usually present problems of toxicity and immunogenicity, with limitations in the types of transportable cargo.[20,21] VNP/VLPs have been used to improve vaccine effectiveness[7,12] due to the strong immune response that is produced in the host. The development of artificial gene vectors such as polymeric NPs or liposomes[22] has attracted the attention of researchers as an alternative to natural viruses, which can produce harmful side-effects when introduced into living organisms.[3,17,18]

1.2.2   Micelles and Liposomes

In general, transport of hydrophilic compounds is more favorable in biological aqueous conditions.[23] Hydrophobic cargoes can be transported by amphipathic NPs forming the classical core–shell carriers.[5,24–28] For example, in micelles (Figure 1.1B), the hydrophilic part is exposed to medium

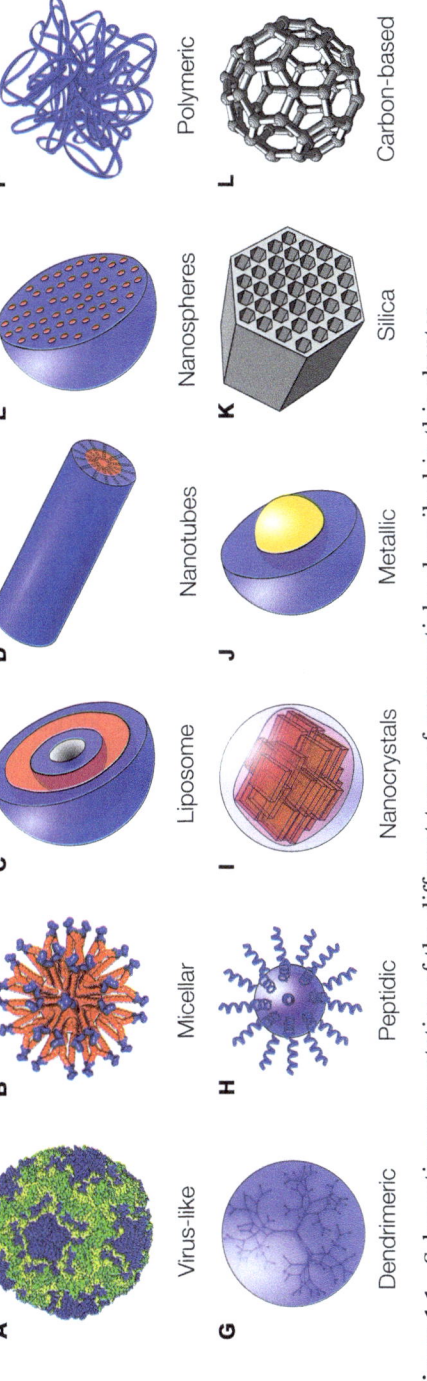

Figure 1.1 Schematic representation of the different types of nanoparticles described in this chapter.

while forming a hydrophobic core.[29] Amphipathic interactions can produce NPs with several layers, the transport of hydrophilic cargoes in liposomes being possible[24,30] (Figure 1.1C). Liposomes can easily penetrate the cytoplasmic membrane[6] and promote cellular uptake.[31] Micellization of chemotherapeutic agents can reduce their cytotoxic effect and increase their effectiveness.[28,32–35] Liposomes formed with cationic lipids[36] are extensively used in gene delivery, being one of the most efficient strategies.[37] Cationic lipids can spontaneously combine and condense negatively charged DNA molecules, thus allowing penetration of DNA into cells while being protected from nuclease attack.[38] Although liposomes have high loading capacity, their low stability and strong interaction with cargo can produce a non-stable release.[39,40] *In vivo* employment of cationic lipids is limited by their high cytotoxicity and their unspecific absorption by phagocytic cells of the reticulo-endothelial system.[20] A drawback of liposomes is their low stability, which can result in fusion, aggregation, or leakage of the encapsulated drug substance during storage.[41] An alternative to increase the stability of lipid NPs is the use of solid lipid NPs that mainly consist of solid lipid stabilized with surfactants.[5,24,42] These NPs show good physical stability and bio-compatibility,[25,42,43] although, similar to other hydrophobic NPs, they have a short half-life *in vivo* and are removed from the blood circulation by the reticulo-endothelial system, particularly in the liver and the spleen.[40]

1.2.3   Polymeric Nanoparticles

Polymeric NPs consist of non-biodegradable and/or biodegradable polymers.[42] Biodegradable polymers, which are widely used for drug delivery,[25,44,45] can be divided into two groups, namely biopolymers (protein, peptide and polysaccharide) and synthetic polymers [poly(lactic acid) (PLA) and poly(ε-carpolactone) (PCL)]. Polymer composition and physical properties are factors that influence the effectiveness of these NPs.[24] The high versatility of these NPs, produced by the precipitation of linear polymers into colloidal nanoparticles solution, is due to the countless numbers of available polymers and their combinations[6] (Figure 1.1F).

The more commonly used non-degradable synthetic polymers are N-(2-hydroxypropyl)-methacrylamide copolymer (HPMA),[24,46] poly(vinylpirrolidone) (PVP)[39] and polyethylene glycol (PEG),[25] since they do not induce a significant cytotoxicity within biological systems.[5,24] Poly(N-isopropylacrylamide) [poly(NIPAAm)] is a thermosensitive polymer and exhibits a low critical solution temperature at body temperature.[47] These characteristics allow poly(NIPAAm) to be employed extensively as a drug carrier[21,48,49] for thermally controlled release. Natural polymers including albumin, silk, chitosan, and heparin have been employed for the delivery of oligonucleotides, proteins, and drugs.[45,46,50] A substantial part of the studies on polymeric NPs focus on the encapsulation of larger molecules like DNA or proteins rather than drugs.[23] Nanospheres (Figure 1.1E) are NPs in which the cargo is dispersed through the polymeric matrix.[25] Biodegradable nanospheres are especially suitable for the controlled release of drugs,[4,51] because the choice of polymer for NP formation promotes a different degradation rate under selected conditions.[23] Biodegradable polymeric NPs[52] comprise PLA, poly(glycolic acid), poly(lactic-glycolic acid) (PLGA),[4] poly(methyl methacrylate) (PMMA), or poly(L-glutamic acid) (PGA).[25,53] These NPs are advantageous because they can undergo hydrolysis to form biodegradable metabolites in biological systems. For applications that require a long-term biocompatible stay in the host organism, PCL can be used for NP synthesis due to its slower degradation rate in comparison with other biodegradable NPs.[52] Polycationic polymers such as poly(L-lysine)[54] (PLL) or linear poly(ethylenimine) (PEI) have been employed for the condensation of DNA to form poly-ion complexes. PLL has been shown to mediate gene transfer by compacting pDNA into a tight toroid structure of ~ 100 nm and rendering it resistant to DNase digestion.[54] Even so, the high cationic nature of PLL and PEI produces cytotoxicity and triggered an immune response by the activation of complement.[55] Another cationic polymer is the natural polysaccharide chitosan,[56] which has shown positive attributes of biocompatibility and degradability.[21] This linear polymer, which is a soluble derivative of chitin (the main compound of arthropod shells), is composed of randomly distributed D-glucosamine and *N*-acetyl-D-glucosamine.[56] Chitosan NPs have the ability to adhere to mucosal surfaces and penetrate into cells;[20,57] the presence of hydroxyl and amine groups allows chemical modification to increase its bioactivity.[20,21,50,57–60] For all the above, chitosan derivatives have been studied as non-viral vectors, since its cationic charges allows complexation with DNA or RNA.[21,56] Cyclodextrins[61–63] are cyclic oligosaccharides with a lipophilic inner cavity and a hydrophilic outer surface. Their amphipathic nature allows formation of non-covalent inclusion complexes with drugs,[63] although inorganic compounds are not generally suitable for complexation.[62] These versatile molecules have been employed to increase the loading capacity of NPs, since they are able to enhance the number of hydrophobic sites in NPs structure.[62,64] Cyclodextrins can mask drug cytotoxicity[65] and even form the backbone of more complex structures for the transport of genetic materials.[21,55,66]

1.2.4 Dendritic Nanoparticles

Dendrimers are radially hyperbranched polymers with regular repeat units.[6] They are attractive systems for drug delivery due to their highly defined dispersity, nanometer size range, spheroid-like shape and multifunctionality.[47,67,68] Aside from their ease of preparation, dendrimers have multiple copies of functional groups on the molecular surface which enables derivatization for biological recognition processes.[65] Even so, dendrimers are reported to cause hematological toxicity,[69] especially in the case of non-functionalization. Examples of typical dendrimers are poly(propyleneimine) (PPI), poly(amido amine) (PAMAM),[70,71] poly(2,2-bis(hydroxymethyl)propionic acid (bis-MPA), poly(glycerol-succinic acid) (PGLSA-OH)[72] or epsilon derivatives of PLL.[73] Although their high charge density allows easy insertion into membranes, and can facilitate endosomal escape,[74] dendrimers have low water solubility and exhibit elevated cytotoxicity.[75] For example, PAMAM NPs have limited applications in medicine due to their original toxicity.[55,69,70] PAMAM is known to induce nephrotoxicity as well as hepatotoxicity, and its cationic charge can cause platelet aggregation by disruption of membrane integrity.[76] Nevertheless, it is possible to reduce the cytotoxicity by chemical modification and the combination of different polymeric ends.[47,55,69,75,77,78] Branched PEI is one of the most studied and commonly used branched polycationic polymers[6,21,70,79–83] for gene delivery, due to its cationic charges and lower cytotoxicity compared to PAMAM, although its cytotoxic effect is higher than other cationic polymers.[84] In general, the buffering capacity of polyamines promotes endosomal escape of NPs, since osmotic swelling occurs by the accumulation of chloride ions in the endosome.[85,86]

1.2.5   Peptidic Nanoparticles

Peptidic NPs are based on the use of peptide sequences that promote cellular internalization known as cell-penetrating peptides (CPPs) or protein transduction domains.[29,87] CPPs are short peptides (6–30 aa) that are able to cross the cellular membrane for intracellular trafficking of cargoes.[6,29,50,87] It has been postulated that the ability of CPPs to be internalized by cells is related to their strong affinity for lipid bilayers.[88] CPP sequences are based on natural protein-transduction domains[87] such as transactivator of transcription of human immunodeficiency virus (HIV-1 TAT peptide),[6,26,29] or penetratin (pAnt), which is derived from the third helix of the *Drosophila* antennapedia homeodomain.[26,89] Low molecular weight protamine (LMWP) is derived from the natural protein protamide, an arginine-rich nuclear protein that replaces histones during spermatogenesis.[90,91] LMWP has demonstrated comparable performance to TAT peptide for cellular translocation while being neither as antigenic, mutagenic nor cytotoxic as other cationic peptides.[90] Similar to other NPs, CPPs can be covalently linked to the cargo, forming a conjugate that promotes transport and internalization of the complex[25,29,87] *via* cellular pathways, but this covalent modification may alter the biological activity of cargoes. To circumvent this limitation, a non-covalent strategy for the attachment of cargo[29,87,92,93] without the necessity of chemical cross-linking or modification is preferred. The presence of cationic amino acids such as lysine or arginine in CPPs seems to be one of the factors that help to improve their transfection efficiency.[88,94] Meanwhile histidine-rich regions can enhance endosomal escape through the pH buffering, or proton sponge effect.[45] In general, cationic CPPs can form complexes with negatively charged DNA molecules based on electrostatic interactions.[50,94] Besides CPPs, cationic peptides have been used for gene delivery, and consist of consecutive basic amino acid sequences, which compact DNA into spherical complexes, or chromatin-like components such as histones or protamine, which compact DNA in a structured manner.[3] In this way, oligo-arginine[95–97] has demonstrated similar characteristics to CPP in cell translocation, being superior to other polycationic homopolymers.[98,99] The peptide motive Arg-Gly-Asp (RGD) is able to recognize and bind to the $\alpha_\nu\beta_3/\alpha_\nu\beta_5$ integrins that are expressed in certain cell types such as endothelial cells, osteoclasts, macrophages, and platelets.[44,100] Integrins are transmembrane glycoproteins that interact with the cellular matrix and promote receptor-mediated endocytosis. $\alpha_\nu\beta_3/\alpha_\nu\beta_5$ integrins are overexpressed in angiogenic endothelial cells, also being suitable markers for neoplasms.[38] Modifications to increase the specificity of NPs [97,101] for targeted delivery to a specific organelle within a cell[6] include the use of signal peptides[87,101] such as nuclear localization signals[102] or mitochondrial-targeting peptides.[103] NPs composed of amphipathic peptides[6,87,93,104] can also be used for the delivery of hydrophobic drugs.[27]

1.2.6 Nanocrystals and Nanosuspensions

Nanocrystals are associations of molecules in a crystalline form,[105] composed of pure drug with only a thin coating of surfactants[25] (Figure 1.1I). Drug nanocrystals can be generated by "bottom-up" (intermolecular association) or "top-down" (milling of crystals) technologies.[64] Nanocrystals NPs are composed of 100% drug without the addition of carrier materials such as polymeric NPs.[64] Nanocrystals have been more studied for material science than for drug delivery, given that not all therapeutic compounds can be easily crystallized.[105] However, they are the usual choice for the oral administration of drugs,[25,64,106] since their nano-scale size improves drug solubility and dissolution rate as well as increasing adhesion to the intestinal wall and capillary uptake.[64,107] Nanocrystals have also been successfully employed in the parenteral delivery of compounds such as antibiotics or insulin.[5]

1.2.7 Metallic Nanoparticles

Metallic NPs (Figure 1.1J) are heavily utilized in biomedical sciences because they can be prepared and surface-functionalized in many different ways.[108] These NPs can be used in diagnostics as well as for drug and gene delivery.[24] Metallic NPs can be easily synthesized over a broad range of sizes and shapes, and are usually composed of gold, platinum, titanium dioxide, copper, iron oxides [as magnetite ($M_xFe_{3-x}O_4$, M = Mn, Ni, Co, Fe) or maghemite (Fe_2O_3)][105] and can be functionalized *via* thiol-metal chemistry.[6] Metallic NPs can combine properties as surface plasmon resonance, magnetism, or anti-oxidant capabilities.[109] Among the most used are the colloidal gold NPs,[4,24,109,110] given that cells can intake gold NPs without apparent cytotoxicity. Shelling NPs with gold will reduce their cytotoxicity while increasing their stability.[110] Gold NPs are one of the most successful inorganic carriers in oncology, where they have shown applicability as drug carriers and in the thermal ablation of tumors.[4,111] However, the gold NPs with quantum sizes (1.5 nm diameter) can be toxic, because of their ability to penetrate into the cellular nucleus and bind irreversibly to DNA.[112] Moreover, metallic NPs have shown the production of reactive oxygen species (ROS) and oxidative stress,[40] although this effect has been shown ubiquitously in several types of NPs.[69] The safety of the use of metallic NPs *in vivo* is under debate, considering that divalent cations and heavy metals are toxic.[40,69]

1.2.8 Silica Nanoparticles

Silicon dioxide NPs (Figure 1.1K) are multifunctional structures available in micro- or mesoporous forms suitable for the encapsulation of various cargoes.[6,105,113,114] Their low cytotoxicity and easy derivatization with different surface chemistries make them versatile tools for cargo delivery[25,105,111,113–119] with excellent physicochemical stability. Silica NPs can be prepared as hollow or multichannel structures, which can be used for drug/gene cargoes. Mesoporous silica NPs present a high surface area and large pore volumes for functionalization.[118,111] For example, by using complexes that limit the release of the cargo (nanovalves), it is possible to induce drug release following an intracellular event or external stimulus.[111,114] Silica NPs have shown cytotoxicity caused by ROS generation, which could be correlated with size (20 nm being more toxic than 100 nm) and surface charge.[120]

1.2.9 Carbon-based Nanoparticles

Fullerenes and nanotubes are hollow, carbon-based cage-like NPs (Figure 1.1L) that have been employed for drug and gene delivery. These materials are constructed within a sheet of graphene arranged into small cylindrical or spherical structures.[25,69] Carbon nanotubes can be formed using single or multi-walled graphene sheets and have different sizes and diameters depending on the synthesis conditions.[121–123] Unmodified carbon NPs are insoluble, therefore requiring modification of their surface to improve solubility and reduce cytotoxicity.[122,123] Carbon NPs can be functionalized by the addition of compatible functional groups for the delivery of various biomolecular cargoes.[121] The efficiency of their transport across cells is related to their hydrophobic nature, which allows penetration without damaging the cellular membranes. However, these carriers have been reported to cause cellular apoptosis due to ROS production in the mitochondria[69] and, similar to other fullerene compounds, they reduce the systemic immune response.[12]

1.3 Physicochemical Factors that Affect Nanoparticle Efficiency

NPs have a high surface area to mass ratio, which allows maximization of the functional surface while minimizing the biological response to the amount of exogenous compounds in the organism. The biological properties and interactions of nano-scale materials dramatically change in comparison with bulk material.[5] Thereby, NP design must take into account some critical parameters[65] such as a cytotoxic effect, transport and release of NPs into cells[6,124] or target tissues[4,26] (Figures 1.2 and 1.3). With the current advances

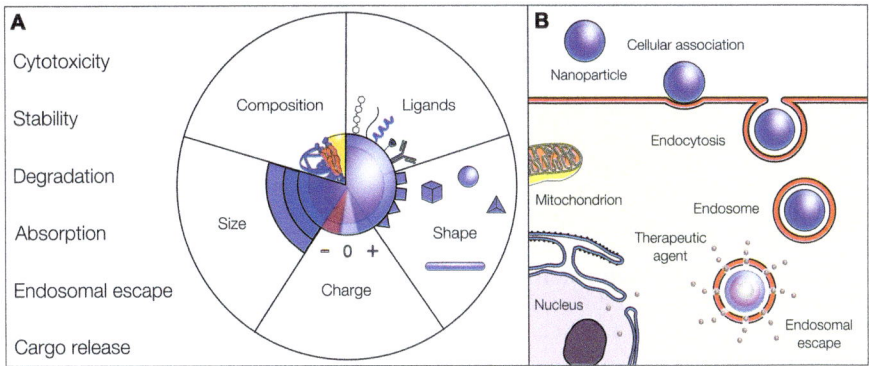

Figure 1.2 A. Critical parameters in nanoparticle design. **B.** Endocytic mechanism of cellular uptake.

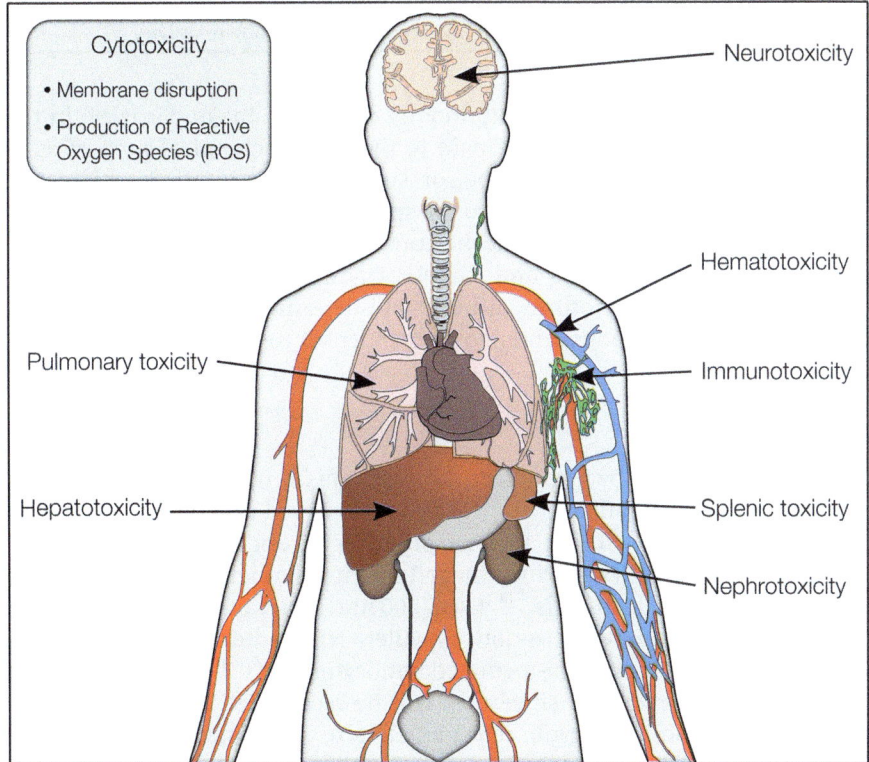

Figure 1.3 Major forms of *in vivo* toxicity of nanoparticles.
Based on Mikael Häggström human body diagram and Aillon *et al.*, 2009.[69]

Table 1.1   General advantages and drawbacks of the different types of nanoparticles.

Nanoparticle	Advantages	Disadvantages
Virus	High efficiency	Trigger immune response
		Poor drug-loading capacity
Micellar	Efficient loading of	Poor stability
	hydrophobic drugs	Liver accumulation
	Excellent cell penetration	Cytotoxic
Polymeric	Excellent functionalization	Polydisperse
	High versatility	Cytotoxic
Dendrimers	Facilitate endosomal escape	Cytotoxic
	Controllable generation	
	Easy functionalization	
Peptidic	Increased cell penetration	Poor stability
	High specificity	Self-aggregation
Nanocrystals	High loading capacity	Limited applicability
	Good dissolution	
Metallic	Tunable shape and size	Cytotoxic
		Not biodegradable
Sillica	High loading capacity	Poor solubility
	Easy functionalization	
	Good biocompatibility	
Carbon-based	Stable	Cytotoxic
	Good cell penetration	Immune suppressors
	Tunable size and shape	Poor reactivity

in material design it is possible to modify or custom-make particles; in the design of NPs attention must be paid to the physicochemical properties of nanomaterials such as size, shape, or surface charge, and the biological response they induce. In addition to the specific advantages that a type of NP may present (Table 1.1), it should be considered that, due to their optimal properties, the applications and routes of administration will depend on whether the cargo is cytotoxic and this will determine if they are administered *via* oral, intravenous, cutaneous or mucosal routes.[3,4,6,26,64]

1.3.1   Size

The size of NPs is known to affect the mechanism of absorption and residence time in the organism.[105,125,126] It was observed that NPs of 20 nm can penetrate into cells without the contribution of endocytic mechanisms.[4] However, it is generally described that NPs with a diameter <100 nm present a favored entrance into cells.[124] Even 500 nm NPs can enter the cells *via* caveolae-mediated internalization, smaller NPs enter by phagocytosis, macropinocytosis or clathrin-mediated endocytosis.[124,126] The absorption of large NPs can be reduced, since they will be recognized by neutrophils.[105] However, for non-phagocytic cells, small size correlates with increased cytotoxicity.[85] In the case of gold and silica NPs, ROS formation as well as cytotoxicity increases with a decrease in particle size.[112,120] The NPs with smaller sizes have a higher tendency to aggregate and interact with biological components, thereby increasing their cytotoxicity.[69] The size of NPs also influences their biodistribution, as demonstrated in a study using gold NPs. Upon administration of different-sized particles in rats, NPs with a diameter of 10 nm were localized in the main part of the organs, while those with diameters of 50 and 250 nm were only present in the liver and the spleen.[107] Relatively small particles with diameters in the range 1–20 nm have a long circulatory residence.[105] The mobility of NPs inside the cell depends on their size and interaction with the uptake pathway,[6] for example NPs with a diameter of 40 nm can enter the cellular nucleus.[85] The effect of size on the internalization of NPs varies depending on cell type, and it is also influenced by the shape of NP, varying between nanospheres and nanotubes.[124]

1.3.2 Shape

NP shape is a parameter that can significantly influence time of residence and mechanisms of cellular uptake.[127–129] The cellular immune system respond to the shape of exogenous particles;[125] macrophages phagocytose rod-shaped NPs faster than spherical ones, and internalization velocity depends on the orientation of the NP.[127] Rod-like shape (aspect ratio) is one of the factors that intracellular pathogens use to enter into cells, even in non-phagocytic cells; for instance, rod-like NPs showed the highest uptake in HeLa cells, followed by spheres, cylinders, and cubes.[125] Even though the overall process of phagocytosis is a result of the complex interplay between shape and size, the effect of shape becomes less evident when the size of NPs is <100 nm.[124] Engineering the shape of NPs can tailor them to either avoid or promote phagocytosis.[127] Conversely, an increase in phagocytosis will help to induce immune response[112] and enhance cellular uptake, which will consequently reduce the bio-availability of NPs by accumulation in the reticulo-endothelial system.

1.3.3 Surface Charge

The surface charge of NPs in suspension is usually determined by zeta (ζ)-potential analysis. When charged NPs are in suspension, surface electric charge attracts the surrounding counter-ions to form an electrical double layer, with an inner region (Stern layer) where the ions are strongly bound. ζ-potential is the electric potential that is produced at the surface of hydrodynamic shear between the ions of a solvent and the surface of a particle. Higher ζ-potential contributes greater resistance to flocculation of NPs due to their high self-repulsion.[130,131] The surface charge of NPs plays a crucial role in cell association,[40] where, generally, positively charged NPs show a higher internalization rate than neutral or negatively charged NPs.[43,85,124,125] Positively charged carriers are frequently employed for gene transport, as they are able to form complexes with the negatively charged DNA molecule.[29,87,92,93] Polarity of surface charge can promote the specificity of cellular uptake, given that phagocytic cells tend to preferably interact with negatively charged NPs.[85] In addition to cellular uptake and stability of NPs, polarity of surface charge affects toxicity; for example, polycations are generally recognized as cytotoxic and hemolytic and can activate the innate immune system,[55] whereas polyanions can cause anticoagulant activity and stimulate cytokine release.[39] Positively charged NPs were generally found to be more apt to cause inflammation compared to a neutral or negatively charged particle.[24] Cationic charged NPs produce toxicity in non-phagocytic cells by plasma-membrane disruption, while anionic NPs produce toxicity in phagocytic cells by apoptosis.[85] When NPs are introduced into a living organism they readily associate with serum proteins and lipids, forming an external corona.[40,43,132,133] The protein corona can affect the efficiency of cargo delivery, by affecting circulation lifetime, signaling, kinetics, and transport, and triggering the immune response[133] (Figure 1.3). Formation of the corona can influence the size and shape of NPs, and in the case of strongly charged NPs proteins can be denatured, increasing the aggregation of NPs. The nature of the proteins forming the corona will depend on the NP's composition and surface charge, in this way, cationically charged NPs absorb proteins with an isoelectric point <5.5, such as albumin, while anionic NPs promote the absorption of proteins with an isoelectric point of >5.5, such as immunoglobulin G.[133] Interaction with serum protein can completely alter the ζ-potential of NPs,[134] changing the theoretical values predicted in protein-free solutions and alter or suppress the intended function of NPs.[135]

1.3.4   Ligands

The surface properties of NPs are known to be fundamental in determining NP–cell interaction, hence the conjugation of various chemical or biological ligands to NP surfaces has been explored. PEGylation (modification of the surface with PEG) is the predominant method used to reduce toxicity[6,26,35,38,39,65,136–138] as the resultant "stealth" brushes mimic the cell's glycocalyx.[139,140] PEGylation increases a colloidal carrier's stability *in vivo* by its steric effect, which acts as a barrier to aggregation[24,105] and the formation of the protein corona.[43] Nevertheless, PEGylation reduces uptake[38,124,141,142] and increases the circulation time of NPs.[4] In an attempt to mimic viral particles, stearylation of peptidic NPs has dramatically increased cell transfection efficiency for gene delivery,[97,143,144] albeit with a mild increase in cytotoxicity.[145] Molecules attached to the NP surface can act as a homing sequence and change the biological distribution of the NP. For instance, it is known that sigma receptors are overexpressed in many human cancers and benzamides exhibit a high affinity for these sigma receptors.[35] Hence, modification of the NP surface by the addition of benzamide (*i.e.* anisamide) enables the targeting of cancerous cells.[35,146,147] Another method, folate receptor-mediated targeting, is extensively employed in the targeting of NPs into tumor cells.[34,57,83,118,148–151] The folate receptor is overexpressed in many types of cancers (breast, ovary, endometrium, kidney, lung, head and neck, brain, and myeloid) and is internalized into cells after ligand binding.[34] Another common strategy for the modification of NPs is the attachment of peptidic sequences, which can help with the improvement of translocation and absorption properties such as gastric mucoadhesivity[152,153] and blood-vessel specificity.[154] For example, LMWP,[90] TAT peptide,[26,155] and neurotensin-targeting peptide[142] have been successfully employed for NP functionalization to enable the delivery of specific cargoes into the brain, while the RGD sequence can be employed for selectively targeting angiogenesis.[50,156] Lactoferrin,[157] lectins,[158,159] or transferrin[30,160,161] have also been employed for delivery of NPs into the brain. Arginine-based or derivatized NPs have shown to increase cellular uptake without increasing cytotoxicity,[95,137,143,162,163] while the addition of histidine residues to the NP improved intracellular release of cargo by the proton sponge effect.[58,70,164] Functionalization of NPs with tumor-homing peptides has shown that is possible to deliver NPs specifically to some types of tumors.[136,165]

Specific delivery of NPs is a crucial point when they transport chemotherapeutic drugs for treatment, for instance to cancerous cells. The first molecules that specifically recognized tumor antigens were monoclonal antibodies.[83] Antibodies are able to specifically recognize target molecules (antigens) by multiple weak interactions between complementary three-dimensional surfaces.[135] However, despite their high specificity, direct ligation of antibodies to drugs resulted in non-significant improvements in treatment,[46] since their recognition properties are reduced after drugs are loaded. Even so, antibody modification can be employed for liposomes,[30,70,157,166] cationic peptides,[167,168] polymeric NPs,[169,170] magnetic NPs,[171] or mesoporous silica.[113,172] Despite the demonstration of antibody application *in vitro*, its *in vivo* application is limited owing to their high cost, limited shelf-life, potential immunogenicity and retention in the reticuloendothelial system.[34,108]

1.4 Conclusions

The nano-technological approach to improve gene/drug delivery has been a dynamic and promising research field for a few decades. In the design of nanocarriers, the intrinsic properties of NPs to select their composition for controlled release of cargo. Therefore, one of the main functions of NPs must be to serve as a transport system without increasing endogenous toxicity of the transported substance (Figure 1.3). The different NPs described in this chapter present advantages and drawbacks depending on the application and route of administration (Table 1.1). Since NPs represent the maximization of surface by unit of mass, improvement of the transporting capacities of NPs can be performed by functionalization of their surface (Table 1.2). NPs administered systemically tend to accumulate in tumors due to the enhanced permeability and retention effect.[28,173] Nevertheless, NPs

Table 1.2   General advantages and drawbacks of the surface modifications of nanoparticles.

Modification	Advantages	Disadvantages
PEG	Reduced cytotoxicity	Increased circulation time
		Reduced cellular uptake
Cell-penetrating peptides	Increase cellular uptake	Cytotoxic
Folate	Tumor homing tag	Limited application
Anisamide	Tumor homing tag	Limited application
Tumor-homing peptides	Tumor homing tag	Less transfection efficiency
Arginine	Increase cellular uptake	
Histidine	Increase endosomal escape	
Stearyl (group)	Increase transfection	
Gold shelling	Reduce cytotoxicity	Non-biodegradable
Antibodies	Increase specificity	Increases immune response

also tend to accumulate in the reticulo-endothelial system. In order to reduce the side-effects of a systemic treatment,[83] design of chemotherapeutical drug nanocarriers must be focused into the specific release of cargo into the target cells/organs. One of the most promising fields of research in NPs is based on their synergic effect with cytotoxic drugs in the treatment of cancer.[174] The studies carried out with NPs incorporating signaling molecules, such as benzamides, folate, or tumor-homing sequences showed their potential to improve target delivery functions. Furthermore, the large bioactive surface of NPs becomes fundamental to improve their efficiency and reduce their toxicity by tailoring signaling molecules.

References

1. G. M. Whitesides, *Small*, 2005, **1**, 172–179.
2. C.-S. Lee, W. Park, S.-J. Park and K. Na, *Biomaterials*, 2013, **34**, 9227–9236.
3. D. J. Glover, H. J. Lipps and D. A. Jans, *Nat. Rev. Genet.*, 2005, **6**, 299–310.
4. W. H. De Jong and P. J. A. Borm, *Int. J. Nanomed.*, 2008, **3**, 133–149.
5. J. E. Kipp, *Int. J. Pharm.*, 2004, **284**, 109–122.
6. L. Y. T. Chou, K. Ming and W. C. W. Chan, *Chem. Soc. Rev.*, 2011, **40**, 233–245.
7. I. Yildiz, S. Shukla and N. F. Steinmetz, *Curr. Opin. Biotechnol.*, 2011, **22**, 901–908.
8. J. K. Pokorski and N. F. Steinmetz, *Mol. Pharm.*, 2011, **8**, 29–43.
9. Z. Wu, K. Chen, I. Yildiz, A. Dirksen, R. Fischer, P. E. Dawson and N. F. Steinmetz, *Nanoscale*, 2012, **4**, 3567.
10. I. Yacoby, H. Bar and I. Benhar, *Antimicrob. Agents Chemother.*, 2007, **51**, 2156–2163.

11. M. L. Flenniken, L. O. Liepold, B. E. Crowley, D. A. Willits, M. J. Young and T. Douglas, *Chem. Commun.*, 2005, **2**, 447–449.

12. D. M. Smith, J. K. Simon and J. R. Baker, *Nat. Rev. Immunol.*, 2013, **13**, 592–605.

13. J. Sun, C. DuFort, M.-C. Daniel, A. Murali, C. Chen, K. Gopinath, B. Stein, M. De, V. M. Rotello, A. Holzenburg, C. C. Kao and B. Dragnea, *Proc. Natl. Acad. Sci. U. S. A.*, 2007, **104**, 1354–1359.

14. M. Manchester and P. Singh, *Adv. Drug Delivery Rev.*, 2006, **58**, 1505–1522.

15. Y. Ren, S. M. Wong and L.-Y. Lim, *Bioconjugate Chem.*, 2007, **18**, 836–843.

16. A. Izembart, E. Aguado, O. Gauthier, D. Aubert, P. Moullier and N. Ferry, *Hum. Gene Ther.*, 1999, **10**, 2917–2925.

17. E. Cevher, A. D. Sezer and E. S. Çaglar, in *Recent Advances in Novel Drug Carrier Systems*, ed. A. D. Sezer, InTech, 2012, pp. 437–469.

18. C. Kaeppel, S. G. Beattie, R. Fronza, R. van Logtenstein, F. Salmon, S. Schmidt, S. Wolf, A. Nowrouzi, H. Glimm, C. von Kalle, H. Petry, D. Gaudet and M. Schmidt, *Nat. Med.*, 2013, **19**, 889–891.

19. T. Azzam and A. Domb, *Curr. Drug Delivery*, 2004, **1**, 165–193.

20. H. Lv, S. Zhang, B. Wang, S. Cui and J. Yan, *J. Controlled Release*, 2006, **114**, 100–109.

21. T.-H. Kim, H.-L. Jiang, D. Jere, I.-K. Park, M.-H. Cho, J.-W. Nah, Y.-J. Choi, T. Akaike and C.-S. Cho, *Prog. Polym. Sci.*, 2007, **32**, 726–753.

22. E. V. B. van Gaal, R. van Eijk, R. S. Oosting, R. J. Kok, W. E. Hennink, D. J. A. Crommelin and E. Mastrobattista, *J. Controlled Release*, 2011, **154**, 218–232.

23. C. Wischke and S. P. Schwendeman, *Int. J. Pharm.*, 2008, **364**, 298–327.

24. S. Naahidi, M. Jafari, F. Edalat, K. Raymond, A. Khademhosseini and P. Chen, *J. Controlled Release*, 2013, **166**, 182–194.

25. M. Rawat, D. Singh, S. Saraf and S. Saraf, *Biol. Pharm. Bull.*, 2006, **29**, 1790–1798.

26. T. Kanazawa, F. Akiyama, S. Kakizaki, Y. Takashima and Y. Seta, *Biomaterials*, 2013, **34**, 9220–9226.

27. J. Li, J. Li, S. Xu, D. Zhang and D. Liu, *Colloids Surf. B. Biointerfaces*, 2013, **110**, 183–190.

28. A. D. Miller, *J. Drug Delivery*, 2013, **165981**, 1–9.

29. M. C. Morris, S. Deshayes, F. Heitz and G. Divita, *Biol. Cell*, 2008, **100**, 201–217.

30. V. P. Torchilin, *Nat. Rev. Drug Discovery*, 2005, **4**, 145–160.

31. H. Yang, H. Mao, Z. Wan, A. Zhu, M. Guo, Y. Li, X. Li, J. Wan, X. Yang, X. Shuai and H. Chen, *Biomaterials*, 2013, **34**, 9124–9133.

32. T. Nakanishi, S. Fukushima, K. Okamoto, M. Suzuki, Y. Matsumura, M. Yokoyama, T. Okano, Y. Sakurai and K. Kataoka, *J. Controlled Release*, 2001, **74**, 295–302.

33. M. Baba, Y. Matsumoto, A. Kashio, H. Cabral, N. Nishiyama, K. Kataoka and T. Yamasoba, *J. Controlled Release*, 2012, **157**, 112–117.

34. J. F. Kukowska-Latallo, K. A. Candido, Z. Cao, S. S. Nigavekar, I. J. Majoros, T. P. Thomas, L. P. Balogh, M. K. Khan and J. R. Baker, *Cancer Res.*, 2005, **65**, 5317–5324.
35. R. Banerjee, P. Tyagi, S. Li and L. Huang, *Int. J. Cancer*, 2004, **112**, 693–700.
36. P. L. Felgner, T. R. Gadek, M. Holm, R. Roman, H. W. Chan, M. Wenz, J. P. Northrop, G. M. Ringold and M. Danielsen, *Proc. Natl. Acad. Sci. U. S. A.*, 1987, **84**, 7413–7417.
37. P. Hawley-Nelson, V. Ciccarone and M. L. Moore, *Curr. Protoc. Mol. Biol.*, 2008, Chapter 9, Unit 9.4, pp. 9.4.1–9.4.17.
38. X. Guo and L. Huang, *Acc. Chem. Res.*, 2012, **45**, 971–979.
39. R. Duncan, *Nat. Rev. Drug Discovery*, 2003, **2**, 347–360.
40. A. E. Nel, L. Mädler, D. Velegol, T. Xia, E. M. V. Hoek, P. Somasundaran, F. Klaessig, V. Castranova and M. Thompson, *Nat. Mater.*, 2009, **8**, 543–557.
41. C. Chen, D. Han, C. Cai and X. Tang, *J. Controlled Release*, 2010, **142**, 299–311.
42. S. A. Wissing, O. Kayser and R. H. Müller, *Adv. Drug Delivery Rev.*, 2004, **56**, 1257–1272.
43. D. Liu, E. Mäkilä, H. Zhang, B. Herranz, M. Kaasalainen, P. Kinnari, J. Salonen, J. Hirvonen and H. A. Santos, *Adv. Funct. Mater.*, 2013, **23**, 1893–1902.
44. K. Numata, J. Hamasaki, B. Subramanian and D. L. Kaplan, *J. Controlled Release*, 2010, **146**, 136–143.
45. K. Numata and D. L. Kaplan, *Biomacromolecules*, 2010, **11**, 3189–3195.
46. K. Cho, X. Wang, S. Nie, Z. G. Chen and D. M. Shin, *Clin. Cancer Res.*, 2008, **14**, 1310–1316.
47. K. Kono, *Polym. J.*, 2012, **44**, 531–540.
48. J. E. Chung, M. Yokoyama, T. Aoyagi, Y. Sakurai and T. Okano, *J. Controlled Release*, 1998, **53**, 119–130.
49. J. E. Chung, M. Yokoyama, M. Yamato, T. Aoyagi, Y. Sakurai and T. Okano, *J. Controlled Release*, 1999, **62**, 115–127.
50. K. Numata and D. L. Kaplan, *Adv. Drug Delivery Rev.*, 2010, **62**, 1497–1508.
51. C. Buzea, I. I. Pacheco and K. Robbie, *Biointerphases*, 2007, **2**, MR17–MR71.
52. A. Mahapatro and D. K. Singh, *J. Nanobiotechnol.*, 2011, **9**, 55.
53. Y. Fu and W. Kao, *Expert Opin. Drug Delivery*, 2010, 7, 429–444.
54. C.-K. Chan and D. A. Jans, *Immunol. Cell Biol.*, 2002, **80**, 119–130.
55. R. Duncan and L. Izzo, *Adv. Drug Delivery Rev.*, 2005, **57**, 2215–2237.
56. R. Riva, H. Ragelle, A. des Rieux, N. Duhem, C. Jerome and V. Preat, *Adv. Polym. Sci.*, 2011, **244**, 19–44.
57. Z. Liu, Z. Zhang, C. Zhou and Y. Jiao, *Prog. Polym. Sci.*, 2010, **35**, 1144–1162.
58. V. M. Gaspar, J. G. Marques, F. Sousa, R. O. Louro, J. A. Queiroz and I. J. Correia, *Nanotechnology*, 2013, **24**, 275101.

59. T. Kean, S. Roth and M. Thanou, *J. Controlled Release*, 2005, **103**, 643–653.
60. L. Liu, Y. Bai, C. Song, D. Zhu, L. Song, H. Zhang, X. Dong and X. Leng, *J. Nanoparticle Res.*, 2009, **12**, 1637–1644.
61. R. Challa, A. Ahuja, J. Ali and R. K. Khar, *AAPS PharmSciTech*, 2005, **6**, E329–E357.
62. A. Vyas, S. Saraf and S. Saraf, *J. Incl. Phenom. Macrocycl. Chem.*, 2008, **62**, 23–42.
63. T. Loftsson and M. E. Brewster, *J. Pharm. Sci.*, 1996, **85**, 1017–1025.
64. J.-U. A. H. Junghanns and R. H. Müller, *Int. J. Nanomed.*, 2008, **3**, 295–309.
65. S. K. Sahoo, F. Dilnawaz and S. Krishnakumar, *Drug Discovery Today*, 2008, **13**, 144–151.
66. S. Srinivasachari, K. M. Fichter and T. M. Reineke, *J. Am. Chem. Soc.*, 2008, **130**, 4618–4627.
67. H. Liu, Y. Chen, D. Zhu, Z. Shen and S.-E. Stiriba, *React. Funct. Polym.*, 2007, **67**, 383–395.
68. E. Burakowska, J. R. Quinn, S. C. Zimmerman and R. Haag, *J. Am. Chem. Soc.*, 2010, **131**, 10574–10580.
69. K. L. Aillon, Y. Xie, N. El-Gendy, C. J. Berkland and M. L. Forrest, *Adv. Drug Delivery Rev.*, 2009, **61**, 457–466.
70. C.-X. He, Y. Tabata and J.-Q. Gao, *Int. J. Pharm.*, 2010, **386**, 232–242.
71. K. Esumi, H. Houdatsu and T. Yoshimura, *Langmuir*, 2004, **20**, 2536–2538.
72. M. T. Morgan, M. a Carnahan, S. Finkelstein, C. a H. Prata, L. Degoricija, S. J. Lee and M. W. Grinstaff, *Chem. Commun.*, 2005, 4309–4311.
73. M. A. Mintzer and M. W. Grinstaff, *Chem. Soc. Rev.*, 2011, **40**, 173–190.
74. W. Tian and Y. Ma, *Soft Matter*, 2012, **8**, 6378–6384.
75. T.-I. Kim, H. J. Seo, J. S. Choi, H.-S. Jang, J.-U. Baek, K. Kim and J.-S. Park, *Biomacromolecules*, 2004, **5**, 2487–2492.
76. M. A. Dobrovolskaia, A. K. Patri, J. Simak, J. B. Hall, J. Semberova, S. H. De Paoli Lacerda and S. E. McNeil, *Mol. Pharm.*, 2012, **9**, 382–393.
77. L. Albertazzi, F. M. Mickler, G. M. Pavan, F. Salomone, G. Bardi, M. Panniello, E. Amir, T. Kang, K. L. Killops, C. Bräuchle, R. J. Amir and C. J. Hawker, *Biomacromolecules*, 2012, **13**, 4089–4097.
78. Y. Choi, T. Thomas, A. Kotlyar, M. T. Islam and J. R. Baker, *Chem. Biol.*, 2005, **12**, 35–43.
79. Y.-Q. Wang, J. Su, F. Wu, P. Lu, L.-F. Yuan, W.-E. Yuan, J. Sheng and T. Jin, *Int. J. Nanomed.*, 2012, 7, 693–704.
80. A. Swami, R. Goyal, S. K. Tripathi, N. Singh, N. Katiyar, A. K. Mishra and K. C. Gupta, *Int. J. Pharm.*, 2009, **374**, 125–138.
81. F. Meyer, V. Ball, P. Schaaf, J. C. Voegel and J. Ogier, *Biochim. Biophys. Acta*, 2006, **1758**, 419–422.
82. J. Ziebarth and Y. Wang, *Biophys. J.*, 2009, **97**, 1971–1983.

83. L. Brannon-Peppas and J. O. Blanchette, *Adv. Drug Delivery Rev.*, 2012, **64**, 206–212.
84. O. Veiseh, F. M. Kievit, V. Liu, C. Fang, Z. R. Stephen, R. G. Ellenbogen and M. Zhang, *Mol. Pharm.*, 2013, **10**, 4099–4106.
85. E. Fröhlich, *Int. J. Nanomed.*, 2012, 7, 5577–5591.
86. N. D. Sonawane, F. C. Szoka and a S. Verkman, *J. Biol. Chem.*, 2003, **278**, 44826–44831.
87. L. Crombez, M. C. Morris, S. Deshayes, F. Heitz and G. Divita, *Curr. Pharm. Des.*, 2008, **14**, 3656–3665.
88. Y. Su, T. Doherty, A. J. Waring, P. Ruchala and M. Hong, *Biochemistry*, 2009, **48**, 4587–4595.
89. P. E. G. Thorén, D. Persson, P. Isakson, M. Goksör, A. Önfelt and B. Nordén, *Biochem. Biophys. Res. Commun.*, 2003, **307**, 100–107.
90. H. Xia, X. Gao, G. Gu, Z. Liu, N. Zeng, Q. Hu, Q. Song, L. Yao, Z. Pang, X. Jiang, J. Chen and H. Chen, *Biomaterials*, 2011, **32**, 9888–9898.
91. D. Lochmann, V. Vogel, J. Weyermann, N. Dinauer, H. von Briesen, J. Kreuter, D. Schubert and A. Zimmer, *J. Microencapsul.*, 2004, **21**, 625–641.
92. S. Deshayes, M. Morris, F. Heitz and G. Divita, *Adv. Drug Delivery Rev.*, 2008, **60**, 537–547.
93. E. Gros, S. Deshayes, M. C. Morris, G. Aldrian-Herrada, J. Depollier, F. Heitz and G. Divita, *Biochim. Biophys. Acta*, 2006, **1758**, 384–393.
94. N. A. Alhakamy and C. J. Berkland, *Mol. Pharm.*, 2013, **10**, 1940–1948.
95. M. Furuhata, H. Kawakami, K. Toma, Y. Hattori and Y. Maitani, *Int. J. Pharm.*, 2006, **316**, 109–116.
96. I. Nakase, T. Takeuchi, G. Tanaka and S. Futaki, *Adv. Drug Delivery Rev.*, 2008, **60**, 598–607.
97. K. Kogure, H. Akita, Y. Yamada and H. Harashima, *Adv. Drug Delivery Rev.*, 2008, **60**, 559–571.
98. D. J. Mitchell, D. T. Kim, L. Steinman, C. G. Fathman and J. B. Rothbard, *J. Pept. Res.*, 2000, **56**, 318–325.
99. A. Mann, G. Thakur, V. Shukla, A. K. Singh, R. Khanduri, R. Naik, Y. Jiang, N. Kalra, B. S. Dwarakanath, U. Langel and M. Ganguli, *Mol. Pharm.*, 2011, **8**, 1729–1741.
100. T. G. Park, J. H. Jeong and S. W. Kim, *Adv. Drug Delivery Rev.*, 2006, **58**, 467–486.
101. D. V. Schaffer and D. A. Lauffenburger, *J. Biol. Chem.*, 1998, **273**, 28004–28009.
102. R. Cartier and R. Reszka, *Gene Ther.*, 2002, **9**, 157–167.
103. N. Bolender, A. Sickmann, R. Wagner, C. Meisinger and N. Pfanner, *EMBO Rep.*, 2008, **9**, 42–49.
104. N. R. Lee, C. J. Bowerman and B. L. Nilsson, *Biomacromolecules*, 2013, **14**, 3267–3277.
105. A. H. Faraji and P. Wipf, *Bioorg. Med. Chem.*, 2009, **17**, 2950–2962.
106. L. Gao, G. Liu, J. Ma, X. Wang, L. Zhou, X. Li and F. Wang, *Pharm. Res.*, 2013, **30**, 307–324.

107. B. S. Zolnik and N. Sadrieh, *Adv. Drug Delivery Rev.*, 2009, **61**, 422–427.
108. R. Weissleder, K. Kelly, E. Y. Sun, T. Shtatland and L. Josephson, *Nat. Biotechnol.*, 2005, **23**, 1418–1423.
109. S. A. Durazo and U. B. Kompella, *Mitochondrion*, 2012, **12**, 190–201.
110. T. A. Erickson and J. W. Tunnell, in *Nanomaterials for the life Science Vol. 3: Mixed Metal Nanomaterials*, ed. C. S. S. R. Kumar, WILEY-VCH Verlag GmbH & Co., Weinheim, 2009, vol. 3, pp. 1–44.
111. W. X. Mai and H. Meng, *Integr. Biol.*, 2013, **5**, 19–28.
112. A. E. Gregory, R. Titball and D. Williamson, *Front. Cell. Infect. Microbiol*, 2013, **3**, 1–13.
113. C.-P. Tsai, C.-Y. Chen, Y. Hung, F.-H. Chang and C.-Y. Mou, *J. Mater. Chem.*, 2009, **19**, 5737–5743.
114. J. Zheng, X. Tian, Y. Sun, D. Lu and W. Yang, *Int. J. Pharm.*, 2013, **450**, 296–303.
115. L. Zhang, F. Gu and J. Chan, *Clin. Pharmacol. Ther.*, 2007, **83**, 761–769.
116. H. Zhang, M.-A. Shahbazi, E. M. Mäkilä, T. H. da Silva, R. L. Reis, J. J. Salonen, J. T. Hirvonen and H. A. Santos, *Biomaterials*, 2013, **34**, 9210–9219.
117. I.-T. Teng, Y.-J. Chang, L.-S. Wang, H.-Y. Lu, L.-C. Wu, C.-M. Yang, C.-C. Chiu, C.-H. Yang, S.-L. Hsu and J. A. Ho, *Biomaterials*, 2013, **34**, 7462–7470.
118. M. Xie, H. Shi, Z. Li, H. Shen, K. Ma, B. Li, S. Shen and Y. Jin, *Colloids Surf. B. Biointerfaces*, 2013, **110**, 138–147.
119. M. Xie, H. Shi, K. Ma, H. Shen, B. Li, S. Shen, X. Wang and Y. Jin, *J. Colloid Interface Sci.*, 2013, **395**, 306–314.
120. Y.-H. Park, H. C. Bae, Y. Jang, S. H. Jeong, H. N. Lee, W.-I. Ryu, M. G. Yoo, Y.-R. Kim, M.-K. Kim, J. K. Lee, J. Jeong and S. W. Son, *Mol. Cell. Toxicol.*, 2013, **9**, 67–74.
121. W. Cheung, F. Pontoriero, O. Taratula, A. M. Chen and H. He, *Adv. Drug Delivery Rev.*, 2010, **62**, 633–649.
122. L. Lacerda, A. Bianco, M. Prato and K. Kostarelos, *Adv. Drug Delivery Rev.*, 2006, **58**, 1460–1470.
123. C. Klumpp, K. Kostarelos, M. Prato and A. Bianco, *Biochim. Biophys. Acta*, 2006, **1758**, 404–412.
124. J. Rauch, W. Kolch, S. Laurent and M. Mahmoudi, *Chem. Rev.*, 2013, **113**, 3391–3406.
125. S. E. A. Gratton, P. A. Ropp, P. D. Pohlhaus, J. C. Luft, V. J. Madden, M. E. Napier and J. M. DeSimone, *Proc. Natl. Acad. Sci. U. S. A.*, 2008, **105**, 11613–11618.
126. J. Rejman, V. Oberle, I. S. Zuhorn and D. Hoekstra, *Biochem. J.*, 2004, **377**, 159–169.
127. J. A. Champion and S. Mitragotri, *Proc. Natl. Acad. Sci. U. S. A.*, 2006, **103**, 4930–4934.
128. J. A. Champion, Y. K. Katare and S. Mitragotri, *Proc. Natl. Acad. Sci. U. S. A.*, 2007, **104**, 11901–11904.

129. B. D. Chithrani and W. C. W. Chan, *Nano Lett.*, 2007, 7, 1542–1550.
130. M. Kaszuba, J. Corbett, F. M. Watson and A. Jones, *Philos. Trans. A. Math. Phys. Eng. Sci.*, 2010, **368**, 4439–4451.
131. Y. Zhang, M. Yang, N. G. Portney, D. Cui, G. Budak, E. Ozbay, M. Ozkan and C. S. Ozkan, *Biomed. Microdevices*, 2008, **10**, 321–328.
132. M. P. Monopoli, C. Aberg, A. Salvati and K. A. Dawson, *Nat. Nanotechnol.*, 2012, 7, 779–786.
133. M. Rahman, S. Laurent, N. Tawil, L. Yahia and M. Mahmoudi, in *Protein-Nanoparticle Interactions*, Springer Berlin Heidelberg, Berlin, Heidelberg, 2013, vol. 15, pp. 21–45.
134. A. H. van Asbeck, A. Beyerle, H. McNeill, P. H. M. Bovee-Geurts, S. Lindberg, W. P. R. Verdurmen, M. Hällbrink, U. Langel, O. Heidenreich and R. Brock, *ACS Nano*, 2013, 7, 3797–3807.
135. Y. Hoshino, H. Koide, T. Urakami, H. Kanazawa, T. Kodama, N. Oku and K. J. Shea, *J. Am. Chem. Soc.*, 2010, **132**, 6644–6645.
136. Z. Li, C. Wang, L. Cheng, H. Gong, S. Yin, Q. Gong, Y. Li and Z. Liu, *Biomaterials*, 2013, **34**, 9160–9170.
137. Y. Maitani and Y. Hattori, *Expert Opin. Drug Delivery*, 2009, **6**, 1065–1077.
138. N. Arnida, N. Nishiyama, W.-D. Kanayama, Y. Jang, Yamasaki and K. Kataoka, *J. Controlled Release*, 2006, **115**, 208–215.
139. P. L. Rodriguez, T. Harada, D. A. Christian, D. A. Pantano, R. K. Tsai and D. E. Discher, *Science*, 2013, **339**, 971–975.
140. R. Hong, C. Huang, Y. Tseng, V. Pang, S.-T. Chen, J.-J. Liu and F.-H. Chang, *Clin. Cancer Res.*, 1999, **5**, 3645–3652.
141. Z. Amoozgar and Y. Yeo, *Wiley Interdiscip. Rev. Nanomed. Nanobiotechnol*, 2012, **4**, 219–233.
142. G. D. Kenny, A. S. Bienemann, A. D. Tagalakis, J. a Pugh, K. Welser, F. Campbell, A. B. Tabor, H. C. Hailes, S. S. Gill, M. F. Lythgoe, C. W. McLeod, E. a White and S. L. Hart, *Biomaterials*, 2013, **34**, 9190–9200.
143. S. Futaki, W. Ohashi, T. Suzuki, M. Niwa, S. Tanaka, K. Ueda, H. Harashima and Y. Sugiura, *Bioconjugate Chem.*, 2001, **12**, 1005–1011.
144. A. El-Sayed, T. Masuda, I. Khalil, H. Akita and H. Harashima, *J. Controlled Release*, 2009, **138**, 160–167.
145. H. Wang, J. Chen, Y. Sun, J. Deng, C. Li, X. Zhang and R. Zhuo, *J. Controlled Release*, 2011, **155**, 26–33.
146. J. Guo, J. R. Ogier, S. Desgranges, R. Darcy and C. O'Driscoll, *Biomaterials*, 2012, **33**, 7775–7784.
147. Y. Chen, S. R. Bathula, Q. Yang and L. Huang, *J. Invest. Dermatol.*, 2010, **130**, 2790–2798.
148. H. Yao, S. S. Ng, W. O. Tucker, Y.-K.-T. Tsang, K. Man, X.-M. Wang, B. K. C. Chow, H.-F. Kung, G.-P. Tang and M. C. Lin, *Biomaterials*, 2009, **30**, 5793–5803.
149. J. Sudimack and R. J. Lee, *Adv. Drug Delivery Rev.*, 2000, **41**, 147–162.

150. P. S. Low, W. A. Henne and D. D. Doorneweerd, *Acc. Chem. Res.*, 2008, **41**, 120–129.
151. Y. Bae and K. Kataoka, *J. Controlled Release*, 2006, **116**, 49–50.
152. M. P. Sarparanta, L. M. Bimbo, E. M. Mäkilä, J. J. Salonen, P. H. Laaksonen, a M. K. Helariutta, M. B. Linder, J. T. Hirvonen, T. C. Laaksonen, H. a Santos and A. J. Airaksinen, *Biomaterials*, 2012, **33**, 3353–3362.
153. H. Valo, M. Kovalainen, P. Laaksonen, M. Häkkinen, S. Auriola, L. Peltonen, M. Linder, K. Järvinen, J. Hirvonen and T. Laaksonen, *J. Controlled Release*, 2011, **156**, 390–397.
154. M. E. Akerman, W. C. W. Chan, P. Laakkonen, S. N. Bhatia and E. Ruoslahti, *Proc. Natl. Acad. Sci. U. S. A.*, 2002, **99**, 12617–12621.
155. H. Xia, X. Gao, G. Gu, Z. Liu, Q. Hu, Y. Tu, Q. Song, L. Yao, Z. Pang, X. Jiang, J. Chen and H. Chen, *Int. J. Pharm.*, 2012, **436**, 840–850.
156. F. Tang, L. Li and D. Chen, *Adv. Mater.*, 2012, **24**, 1504–1534.
157. K. Hu, Y. Shi, W. Jiang, J. Han, S. Huang and X. Jiang, *Int. J. Pharm.*, 2011, **415**, 273–283.
158. X. Gao, W. Tao, W. Lu, Q. Zhang, Y. Zhang, X. Jiang and S. Fu, *Biomaterials*, 2006, **27**, 3482–3490.
159. Z. Wen, Z. Yan, K. Hu, Z. Pang, X. Cheng, L. Guo, Q. Zhang, X. Jiang, L. Fang and R. Lai, *J. Controlled Release*, 2011, **151**, 131–138.
160. Z. Pang, H. Gao, Y. Yu, J. Chen, L. Guo, J. Ren, Z. Wen, J. Su and X. Jiang, *Int. J. Pharm.*, 2011, **415**, 284–292.
161. H. Hatakeyama, H. Akita, K. Maruyama, T. Suhara and H. Harashima, *Int. J. Pharm.*, 2004, **281**, 25–33.
162. J. Wu, D. Yamanouchi, B. Liu and C.-C. Chu, *J. Mater. Chem.*, 2012, **22**, 18983–18991.
163. W. J. Kim, L. V. Christensen, S. Jo, J. W. Yockman, J. H. Jeong, Y.-H. Kim and S. W. Kim, *Mol. Ther.*, 2006, **14**, 343–350.
164. T. Merdan, J. Kopecek and T. Kissel, *Adv. Drug Delivery Rev.*, 2002, **54**, 715–758.
165. K. Numata, A. J. Mieszawska-Czajkowska, L. A. Kvenvold and D. L. Kaplan, *Macromol. Biosci.*, 2012, **12**, 75–82.
166. D. D. Spragg, D. R. Alford, R. Greferath, C. E. Larsen, K. D. Lee, G. C. Gurtner, M. I. Cybulsky, P. F. Tosi, C. Nicolau and M. a Gimbrone, *Proc. Natl. Acad. Sci. U. S. A.*, 1997, **94**, 8795–8800.
167. E. Song, P. Zhu, S.-K. Lee, D. Chowdhury, S. Kussman, D. M. Dykxhoorn, Y. Feng, D. Palliser, D. B. Weiner, P. Shankar, W. a Marasco and J. Lieberman, *Nat. Biotechnol.*, 2005, **23**, 709–717.
168. X. Li, P. Stuckert, I. Bosch, J. D. Marks and W. A. Marasco, *Cancer Gene Ther.*, 2001, **8**, 555–565.
169. H. Han and M. E. Davis, *Mol. Pharm.*, 2013, **10**, 2558–2567.
170. P. Kocbek, N. Obermajer, M. Cegnar, J. Kos and J. Kristl, *J. Controlled Release*, 2007, **120**, 18–26.

171. R. Rezaeipoor, R. John, S. G. Adie, E. J. Chaney, M. Marjanovic, A. L. Oldenburg, S. A. Rinne and S. A. Boppart, *J. Innov. Opt. Health Sci.*, 2009, **2**, 387–396.
172. E. Secret, K. Smith, V. Dubljevic, E. Moore, P. Macardle, B. Delalat, M.-L. Rogers, T. G. Johns, J.-O. Durand, F. Cunin and N. H. Voelcker, *Adv. Healthcare Mater.*, 2013, **2**, 626.
173. Y. Matsumura and H. Maeda, *Cancer Res.*, 1986, **46**, 6387–6392.
174. S. Aryal, C.-M. Jack, Hu, V. Fu and L. Zhang, *J. Mater. Chem.*, 2012, **22**, 994–999.

CHAPTER 2

Targeting Cyclins and Cyclin-dependent Kinases Involved in Cell Cycle Regulation by RNAi as a Potential Cancer Therapy

MANOJ B. PARMAR[a] AND HASAN ULUDAĞ*[a,b,c]

[a] Faculty of Pharmacy and Pharmaceutical Sciences, University of Alberta, Edmonton, AB, Canada; [b] Department of Chemical & Materials Engineering, Faculty of Engineering, University of Alberta, Edmonton, AB, Canada; [c] Department of Biomedical Engineering, Faculty of Medicine, University of Alberta, Edmonton, AB, Canada
*Email: hasan.uludag@ualberta.ca

2.1 Introduction

The cell cycle is the complex, tightly regulated process of cell division.[1] When the cell cycle is deregulated, normal cells could transform into cancer cells with enhanced potential to migrate and proliferate. In such cases, proteins involved in cell cycle progression may no longer appropriately regulate different stages of events critical for cell division, leading to decoupling of various integrated processes. Transformed cells typically proliferate at a rate faster than the normally tightly regulated reproduction of normal cells without any (or less) quality control, which could lead to the formation of a tumor.[2,3] The cell cycle is usually arrested upon detecting the presence of damaged DNA in normal cells, where the DNA repair process

RSC Drug Discovery Series No. 51
Nanomedicines: Design, Delivery and Detection
Edited by Martin Braddock
© The Royal Society of Chemistry 2016
Published by the Royal Society of Chemistry, www.rsc.org

kick-starts the repair of damaged genetic information. In contrast, transformed cells can proceed with the synthesis of damaged DNA followed by
their division without arresting the cell cycle, resulting in an unregulated
proliferation of cells. The altered genetic information that appears in the
progeny is sometimes benign with no functional consequences, but often
imparts undesired properties to daughter cells.[3-5] The essential cell cycle
proteins that are up-regulated and are capable of imparting undesired
properties to transformed cells would be potential targets for cancer therapy;
the inhibition of their expression could be pursued in order to arrest cells at
quality-control checkpoints and/or to induce the apoptosis of transformed
cells.[2,6]

Two important classes of mediators involved in the progression of cell
cycle are cyclins and cyclin-dependent kinases (CDKs), and deregulation of
the cell cycle is often mediated by alterations in the activity of cyclins and
CDKs in hyper-proliferative cancer cells. The deregulation of cyclins and
CDKs induces unscheduled proliferation due to genomic instability that
generally leads to chromosomal instability.[7-10] Therapeutic strategies that
block the activities of cyclins and CDKs may decrease the tumor development. Although several chemical inhibitors have been developed to target
cyclins and CDKs and tested clinically, none of these inhibitors has been
approved for routine clinical use.[11,12] Therefore, an effective strategy that
targets cell cycle proteins to control malignant cell growth is yet to be established. RNA interference (RNAi) could be a potential therapeutic strategy
for this purpose, and here we focus on RNAi attempts to target and silence
(*i.e.* down-regulate) the expression of cyclins and CDKs in order to realize a
benefit in cancer therapy. In this chapter, we review the regulation of the cell
cycle by cyclins and CDKs and their crucial role in the unregulated cell cycle
of transformed cells, as well as current studies on the silencing of cyclins
and CDKs with special emphasis on the use of synthetic carriers and delivery
systems to achieve effective silencing. Critical factors that contribute to the
efficiency of RNAi have been highlighted to bring this therapeutic approach
to a clinical realm.

2.2 The Cell Cycle

2.2.1 An Overview

The cell cycle constitutes a series of complex events that lead to the physical
division and multiplication of a cell.[13] The cell cycle is broadly defined by
two phases: *interphase*, where the cell prepares itself for a division, and
mitosis, where the cell divides its nuclear materials into two separate
daughter cells.[13,14] Several checkpoints exist in the cell cycle to ensure that
the cell is ready to divide. If severe damage is detected at the checkpoints,
cell division is arrested at that point and the apoptotic pathway is immediately activated to eradicate the cell. The cell cycle is a continuous process

for certain types of cells, and cells are continually entering and exiting this process during their lifetime.[13,14]

The interphase includes G1, S and G2 phases (Figure 2.1), where the first gap phase (G1) occupies the longest time of the cell cycle. G1 phase comes immediately after the end of the previous cell division during the active stage of cell proliferation. It is the phase where the cell prepares itself for the next cell division by synthesizing new proteins and organelles needed for daughter cells, resulting in the growth of cell size. During the G1 phase, the cell has to decide whether to stay in this phase and commit to a division, or leave this phase and wait in the G0 phase. If the cell is not ready for division, it enters into the latent G0 phase, which can last for a longer period of time until sufficient growth factors and other essential elements become available. Once the required amount of nutrients is available, cell re-enters the G1 phase and starts synthesizing the proteins needed for DNA replication. The duration of G1 is different depending on the cell type, and for human somatic cells, it may take approximately 40% of a complete cell cycle.[15,16]

After the necessary proteins are synthesized in the G1 phase, cell proceeds into the S phase, where the DNA replication is the major event. However, proteins and enzymes that are required for DNA replication are also synthesized in this phase. DNA replication is a very tightly regulated process.

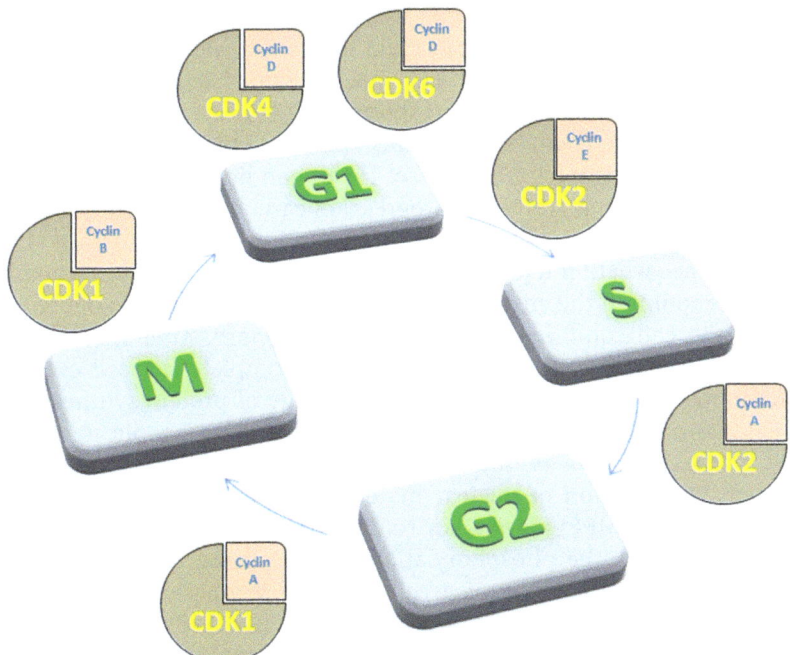

Figure 2.1 Schematic representation of the cell cycle. Cell cycle progression is regulated by various cyclins and cyclin–CDK complexes at different phases, which may serve as potential targets for RNAi-mediated cancer therapy.

Only a single replication is ensured by loading pre-initiation complex onto the DNA, which duplicates the entire genome for the two daughter cells. If DNA damage occurs in this phase, it is repaired by initiating several DNA damage strategies, depending on the type of damage. In case of un-repairable DNA damage, expression of apoptotic proteins increases to eliminate the mutated cell.[15,17,18]

The last and the shortest gap phase is G2, where the proteins that are required for mitosis are synthesized. The G2 phase ends with the prophase of mitosis, where cellular chromatin condenses into chromosomes. Some transformed cells can directly enter into mitosis from S phase, as the G2 phase is least important in cell cycle.[19] Since this is the last phase before entering into mitosis, it also allows a checkpoint to ensure the correctness of DNA replication and synthesis of organelles. G2 phase ends with the mitosis (M phase), that is the division of chromosomes into two identical nuclei followed by division of cytoplasm and cell membrane. The stages of M phase include prophase, metaphase, anaphase, telophase and cytokinesis. At the end of a 'normal' cell cycle, two identical daughter cells are expected to be formed.[20]

2.2.2   Restriction Points and Checkpoints

The concept of checkpoints in cell cycle regulation was first introduced by Hartwell and Weinert in 1989.[21] The mechanism of sending an inhibitory signal to later cell cycle events in response to a current incomplete event is called a checkpoint. There are three checkpoints in the eukaryotic cell cycle: G1/S, G2/M and metaphase/anaphase. If the conditions of each phase are not met at these points, cell cycle progression can be halted. The point of no return during G1 phase of cell cycle is called the restriction point.[22,23] The cell determines whether to exit cell cycle and remain in quiescent (G0) phase at this checkpoint. If the environmental factors favor replication of the entire genome, the cell commits to enter into the S phase. After the restriction point is cleared, the cell has the capacity to complete the cell cycle with limited nutrient availability following a slow rate of protein synthesis.[24]

The G2/M checkpoint ensures that the whole genome is replicated and the newly synthesized DNA is undamaged before going for chromosome con-densation and division of nuclear material in the M phase. The M-phase promoting factor (MPF) plays a key role at this checkpoint.[25] If the cell de-tects DNA damage, the phosphorylation of MPF at the tyrosine and threo-nine residues halt the entry of a cell into the M phase. The metaphase/anaphase checkpoint is also known as spindle checkpoint, which prevents separation of sister chromatids until all chromosomes are aligned at the mitotic plate and attached to the bipolar spindle through their kinetochores. Drugs that depolymerize the microtubules can arrest the cell cycle at the M phase. The activation of anaphase-promoting complex through MPF signals the entry of the cell into anaphase, which leads to the destruction of securin, an inhibitor of a protease separase.[26,27] Once the cell passes spindle checkpoint, completion of the cell cycle is assured.

2.2.3 Regulation of Cell Cycle

2.2.3.1 Cyclins and CDKs

Cyclins are so-called because their synthesis and degradation during cell cycle are cyclical.[28,29] Several cyclins are active in different stages of the cell cycle. The main role of cyclins is to bind CDKs and activate them to facilitate cell cycle progression.[16,29] The CDKs, as the name suggests, have the ability to phosphorylate proteins at serine and threonine residues. The CDKs were identified before their known regulatory function in cell cycle, and therefore the nomenclature of CDKs has not been uniform. However, the nomenclature was harmonized by consecutive numeric naming (*e.g.* CDK1, CDK2, *etc.*) based on a Cold Spring Harbor meeting in 1991.[30] Cyclin–CDK complexes are positive modulators of cell cycle progression as their activity leads to another phase of the cycle.[31,32] CDKs are deactivated when cyclins are degraded, resulting in a cell cycle arrest.[16,32]

The progression from one phase to another occurs in an orderly and tightly regulated fashion. Of the 13 major CDKs identified, five are active during cell cycle: CDK2, CDK4 and CDK6 during G1, CDK2 in S, CDK1 in G2 and M and CDK7 as the CDK-activating kinase.[16,33] The levels of CDKs are constant during cell cycle, while the concentrations of cyclins fluctuate according to cell cycle transition. Different cyclins are required at different phases and specific cyclins bind to specific CDKs (Figure 2.1). For the entry of a cell into G1, cyclins D1, D2 and D3 bind to CDK4 and CDK6.[34] Cyclin E associates with CDK2 to facilitate the entry of cell from G1 to S phase.[35] The complex of CDK2 and cyclin A is essential during the S phase. Cyclin A binds to CDK1 in late G2 for the progression of a cell from G2 to M phase.[36,37] CDK1 and cyclin B complex activates the M phase.[38] Each cyclin undergoes proteolysis at the end of each phase because of the presence of destruction box motif in cyclin A and cyclin B, and the PEST sequence (rich with amino acids proline, glutamic acid, serine and threonine) in cyclin D and cyclin E.[39,40]

2.2.3.2 CDK Inhibitors

The negative regulators of the cell cycle are CDK inhibitors, whose functions are to arrest cell cycle progression by inhibiting the activity of CDKs. The CDK inhibitors prevent the ability of cyclin–CDK complexes to phosphorylate their targets. Two families of CDK inhibitors are the INK4 family and Cip/Kip family.[41] The members of INK4 family are p15 (INK4b), p16 (INK4a), p18 (INK4c) and p19 (INK4d), which inhibit the activity of CDK4 and CDK6 in the G1 phase.[42] The Cip/Kip family includes p21 (Cip1), p27 (Cip2) and p57 (Kip2). These inhibitors inactivate the CDK–cyclin complexes of cyclins A, D and E, and to a lesser extent, that of cyclin B.[43] The p21 was identified by two separate research groups in 1993 and reported as a negative regulator of G1 phase.[44,45] The inhibition of cell cycle by p21 is thought to occur by its important role in the p53-dependent repression of several genes that are necessary in the cell cycle.[44] The p21 suppresses the activity of proliferating cell nuclear antigen (PCNA), an important protein in DNA synthesis and thus inhibits the DNA synthesis in the S phase.[46,47] Polyak *et al.* (1994) first reported the binding of p27 to cyclin E–CDK2 complex following the arrest of the cell cycle at the G1/S checkpoint.[48] Since then, many important roles of p27 in negative regulation of the cell cycle have been established. Cyclin A–CDK2, cyclin D–CDK4 and cyclin E–CDK2 complexes were shown to be inactivated by the binding of p27. The p57 interacts with cyclin–CDK complexes of the G1 phase and has the ability to arrest the cell cycle at the G1 phase. The p57 binds PCNA and plays a role in inhibiting DNA synthesis.[49]

2.3   Deregulation of the Cell Cycle in Cancer

Unscheduled proliferation is the main defect in cancer cells due to constitutive mitogenic signals and constant response to those signals by cancer cells. The genetic alterations in tumors are due to genomic instability that leads to chromosomal instability, which together contribute to the progression of the tumor.[7-10] These three main defects in the cell cycle may be mediated by deregulation of cyclins and CDKs. As mentioned earlier, checkpoints prevent the cell cycle progression until the previous phase has been accomplished and deregulations of checkpoints due to mutations ultimately affect the activities of cyclins and CDKs. Generally, DNA damage checkpoints can detect oncogene-induced cell cycling, especially the high demand of DNA replication stress, and failure to do so by this checkpoint increases the genomic instability.[9,50,51] DNA damage checkpoints are mostly regulated by CDK activity, which can be inhibited by higher expression of CDK inhibitors upon detecting DNA damage. CDK1 is the master regulator during mitosis for equal distribution of genetic material into two daughter cells.[52] CDK1 with cyclin A or cyclin B is responsible for the activation of >70 substrates (by phosphorylation) that are essential for mitosis.[52] Spindle checkpoint ensures proper segregation of chromosomes in two daughter cells.[53] CDK1 activity may be deregulated due to defects in spindle checkpoint, resulting in abnormal chromosome segregation. All these defects are directly or indirectly associated with deregulation of cyclins and CDKs.

The cyclins and the inhibitors (INK4 or Cip/Kip) mainly regulate CDKs. Structurally, CDKs have two domains, the *N*-terminal and *C*-terminal, and two regulatory sequences, a PSTAIRE sequence and a T-loop.[54] Upon binding to cyclins, conformational changes in CDK allow phosphorylation of the threonine residue of a PSTAIRE sequence and activate the complex. Binding of inhibitors to CDK prevents the interaction with cyclin by allosteric changes at the cyclin-interacting sequence of CDK.[54,55] Any small deregulation of CDK activation and deactivation by cyclins and their inhibitors could lead to unregulated cell cycle progression. In many cancer types, deregulation of CDK4 and CDK6 has been found along with the binding of cyclin D or their INK4 inhibitors.[7,55] The mutation in the coding amino acid R24C in CDK4 blocks the binding of INK4 in melanoma, while translocation of CDK6 is the main cause of its overexpression in leukemia.[55,56] Cyclin D overexpression and inability to bind INK4 inhibitors to CDK4 and CDK6 because of a mutation in the INK4 family or changes in the catalytic subunit of CDK seem to be the common feature of certain cancer types. The high expression of cyclin E and inactivation of p21 and p27 inhibitors have also been observed during tumor development, resulting in the activation and deregulation of CDK2.[57,58] Some primary tumors such as breast, colon, lung and prostate activate CDK1 aberrantly due to overexpression of cyclin B.

2.3.1 Conventional Drug Therapy Against Cyclins and CDK Inhibitors

The therapeutic value of cyclin and CDK interference has been studied, and various inhibitors of these proteins have been screened at the clinical stage in the past two decades. The first-generation compounds that target a broad spectrum of CDKs are pan-CDK inhibitors such as flavopiridol.[59] These inhibitors did not meet outcome expectations or acceptable pharmacokinetic profiles in clinical studies.[60] The second-generation CDK inhibitors, purvalanol A,[61] NU6140[62] and the 2,6,9-trisubstituted purine analog, *R*-roscovitine,[63] which mainly target CDK2, CDK7 and CDK9, are being studied to identify their novel synthetic analogs with superior activity, which requires further investigation in terms of their antitumor activity and pharmacokinetic profiles. The thiazole urea CDKi-277[64] and the acyl-substituted triazole diamine JNJ-7706621[65] are other broad-spectrum CDK inhibitors. JNJ-7706621 targets CDK1 and CDK2 and shows a unique inhibitory profile. However, drug resistance has been reported for these inhibitors with increasing doses. The CDK1 and cyclin B inhibitor thiazolinone RO-3306 mainly arrests G2/M transition and has demonstrated efficiency to induce apoptosis in transformed cells compared to normal cells.[66] Unfortunately, none of these cyclins and CDKs inhibitors have been approved for routine clinical use. Therefore, a new strategy to decrease the expression of cyclins and CDKs will be a fruitful pursuit to decrease the growth of malignant cells.

2.4 RNA Interference

2.4.1 Mechanism of Action

The post-transcriptional gene silencing mediated by double-stranded RNA (dsRNA) is described as RNA interference (RNAi). Fire *et al.* (1998)[67] first discovered RNAi in *Caenorhabditis elegans* as homology-dependent gene silencing. They were able to show that the introduction of dsRNA was effective and specific to knock down a targeted gene, whereas the effect of single-stranded RNA was negligible to decrease the copies of the targeted mRNA. In recent years, RNAi has been established unequivocally as a promising approach to silence sequence-specific targets, especially for targets that are up-regulated in abnormal conditions, such as cancer. The translation arrest or degradation of specific mRNA is mediated by RNAi as a regulation of post-transcription of a gene. The RNAi can be mediated by three means: (i) synthetic small interfering RNA (siRNA)[68] that is 20–25 nucleotides long with few overhang bases can interfere with the expression of specific genes with a complementary nucleotide sequence in the mRNA transcript; (ii) microRNA (miRNA),[69] an endogenously expressed non-coding nuclear transcript, is processed by intracellular enzymes to yield 20–25-nucleotide sequences for mRNA silencing; and (iii) short hairpin RNA (shRNA),[70] which is expressed from vectors that are artificially transfected into cells.

The endogenous RNAi can be broadly divided into two steps: cleavage of primary (pri-)miRNA into mature miRNA and the transport of mature miRNA to its intracellular target.[71,72] A complex hairpin structure of endogenously expressed pri-miRNA is cleaved to small hairpin structures, the precursor (pre-)miRNA, in the nucleus. The ribonuclease enzyme, dicer, cleaves the hairpin structure once the pre-miRNA is transported to the cytoplasm, which, in turn, forms mature miRNA 20–25 nucleotides long with two to three overhang nucleotides. An ectopically expressed shRNA from a vector and synthetically designed siRNA can be introduced directly to cells since the nuclear processing can be avoided with the appropriate design. The siRNA is incorporated into the RNA-inducing silencing complex (RISC) following the release of the passenger strand and leaving the guide strand of siRNA to target the mRNA of interest, which is followed by cleavage of the target mRNA *via* endonuclease activity or translational arrest depending on the nature of base pairing between the mRNA and the guide strand.[73,74] Because of high specificity and minimal side-effects, RNAi using siRNA is gaining the upper hand as a useful approach for the treatment of cancer.[75] In this case, siRNA could be visualized as a 'drug', but one that is distinct from conventional drugs in the sense that its delivery requires special attention due to its hydrophilic and anionic nature that makes it practically non-permeable to cell membrane (see later).[76]

2.4.2 RNAi and Cell Cycle Proteins

2.4.2.1 *siRNA and its Delivery*

For cancer therapy, cyclins and CDKs have been targeted extensively using siRNA in different models of cancer (Table 2.1). Among the 13 identified CDKs, CDK4 and CDK6 are the most studied targets, including their activators cyclin D1, cyclin D2 and cyclin D3 in order to impede cell proliferation, as these are essential for the G1 progression which takes the maximum amount of time within a complete cell cycle. As CDK1 plays

Table 2.1 Cyclins and CDKs that have been targeted by the siRNA in different cell lines.

Target	Cell lines	Transfection agent (supplier)	Ref.
Lipid-based carriers			
Cyclin E	SK-BR3, MDA-MB157, MDA-MB436, MDA-MB453, T47D	Oligofectamine (Invitrogen)	Liang et al. (2010)[77]
CDK7	PC3	Lipofectamine (Invitrogen)	Manzo et al. (2012)[78]
CDK2, CDK4, CDK6, Cyclin D1, Cyclin E	HCC1954, SK-BR-3, MDA-MB-231, BT474, MCF-7	Lipofectamine (Invitrogen)	Sahin et al. (2009)[79]
Cyclin A, Cyclin B	HeLa	Gene Silencer (Genlantis)	Gong et al. (2007)[80]
CDK1 to CDK9	MDA-MB-435	siPORT NeoFX (Ambion)	Ovcharenko et al. (2007)[81]
CDK5	E18 rat hippocampal neuronal cells	Lipofectamine (Invitrogen)	Zheng et al. (2007)[82]
CDK7, CDK9	MDA-MB-231	Lipofectamine (Invitrogen)	Johnstone et al. (2008)[83]
CDK4, Cyclin D1	MCF-7	Lipofectamine (Invitrogen)	Grillo et al. (2006)[84]
CDK4	Tca8113	Oligofectamine (Invitrogen)	Zhou et al. (2009)[85]
CDK1, CDK2, CDK5	LNCaP	Lipofectamine (Invitrogen)	Lin et al. (2004)[86]
Cyclin A	HeLa, U2-OS, HEK293T	Lipofectamine (Invitrogen)	De Boer et al. (2008)[87]
CDK2, Cyclin C	Cortical cells from E18 rat	HiPerFect (Qiagen)	Tomashevski et al. (2010)[88]
CDK6, CDK8	HeLa	Oligofectamine (Invitrogen)	MacKeigan et al. (2005)[89]
Cyclin D1	Pancreatic tumor cell line	Lipofectamine (Invitrogen)	Biliran et al. (2005)[90]
CDK5	SH-SY5Y-N	TransMessenger (Qiagen)	Wang et al. (2006)[51]
CDK5	CAL51	Oligofectamine (Invitrogen)	Turner et al. (2008)[92]
CDK7, CDK9	PC3, DU154, T98G, U87MG	DharmaFECT (Thermo Scientific)	Caracciolo et al. (2012)[93]
CDK4	DanG	Oligofectamine (Invitrogen)	Retzer-Lidl et al. (2007)[94]
CDK4, CDK6	GH3, GH4, MCF7	Oligofectamine (Invitrogen)	Jirawatnotai et al. (2004)[95]
CDK2	IMR32, NGP, LAN-5, SK-N-FI, F2112	Lipofectamine (Invitrogen)	Molenaar et al. (2009)[96]
CDK1	HeLa	Lipofectamine (Invitrogen)	Xiao et al. (2009a)[97]
CDK1, CDK2	HeLa	Lipofectamine (Invitrogen)	Xiao et al. (2009b)[98]
CDK8	HCT116	Lipofectamine (Invitrogen)	He et al. (2011)[99]
CDK5	INS-1	Lipofectamine (Invitrogen)	Ubeda et al. (2006)[100]
CDK6	A375, MelJUSO, Skmel 28, 518A2, 607B	Oligofectamine (Invitrogen)	Okamoto et al. (2005)[101]
CDK1, CDK2, CDK7, CDK9	PC3	Oligofectamine (Invitrogen)	Mohapatra et al. (2009)[102]
CDK7	MEFs	Lipofectamine (Invitrogen)	Helenius et al. (2009)[103]

Table 2.1 (*Continued*)

Target	Cell lines	Transfection agent (supplier)	Ref.
Cyclin A, Cyclin B	H157, H596	siPORT Amine (Ambion)	Cho et al. (2006)[104]
CDK3, Cyclin C	T98G	Oligofectamine (Invitrogen)	Ren and Rollins (2004)[105]
Cyclin D1	HeLa and H2009	Lipofectamine (Invitrogen)	Jirawatnotai et al. (2011)[106]
Polymeric carriers			
CDK2, CDK5, CDK7	A549	TransIT-siQUEST (Mirus)	Choudhary et al. (2011)[107]
CDK4	SW480, HepG2	TransIT-TKO (Mirus)	Osabe et al. (2008)[108]
CDK2, CDK4	HepG2, HuH6, Caco2, MCF7	TransIT-siQUEST (Mirus)	Sugatani et al. (2010)[109]
Cyclin D1	K562	ExGen 500 (Fermentas)	Peer et al. (2007)[110]
Without carriers			
CDK6	hESCs	Nucleofection	Zhang et al. (2009)[111]
CDK4	HUVEC	Nucleofection	Zumbansen et al. (2010)[112]
Cyclin D1, Cyclin D2	TK-1	Recombinant Protamine (Abnova)	Peer et al. (2008)[113]
CDK2, CDK4, CDK6, Cyclin D1, Cyclin E	LY18	Nucleofection	Gumina et al. (2010)[114]

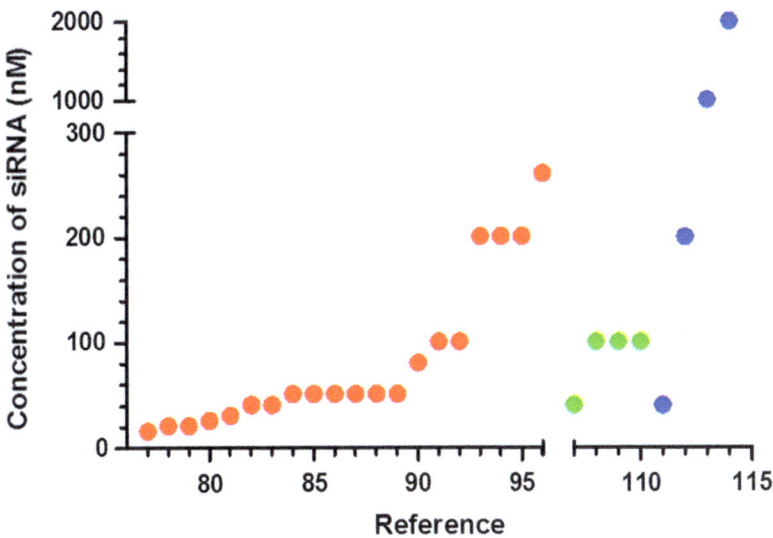

Figure 2.2 Concentrations of siRNA used to target cyclins and CDKs in different cell lines. Red, green and blue dots represent the use of lipid-based, polymeric and no carriers, respectively, for transfection of siRNA. See also Table 2.1.

crucial role during mitosis by phosphorylating other substrates essential for progression during mitosis, it also serves as a vital target in order to arrest the cell cycle. Various amounts of siRNA targeting cyclins and CDKs have been used (Figure 2.2). The amount required to transfect siRNA by nucleofection or electrofection is usually higher, as shown in Figure 2.2. The higher amount of siRNA could lead to non-specific effects, resulting in poor specificity for a 'therapeutic' siRNA. Lipid-based and polymeric carriers have been developed recently for the efficient transfection of siRNA, and a smaller amount of siRNA (20–200 nM) and specific to the target has been used for cyclins and CDKs (Figure 2.2). The concentration ratio between the cationic carrier and siRNA should also be taken into account. siRNA–to-carrier ratios from 1 : 1 to 1 : 12 have been implemented based on the efficacy of siRNA and the type of cell-lines. Gong *et al.* (2007)[80] have targeted cyclin A and cyclin B using a new unconventional dicer-substrate siRNAs (DsiRNAs) that are synthetically designed siRNAs to be optimally processed by dicer before engaging the RISC complex. The potency of DsiRNA seems to be higher than traditional siRNA as it goes through natural processing pathway in the cell.

The binding of siRNA to cell membranes and subsequent uptake is nearly impossible due to its highly labile and anionic nature, as well as large (~ 14 kDa) molecular weight.[76] siRNA can be transported into the cells using cationic carriers which form cationic (or neutral) nanoparticles as the carrier interacts with the siRNA through ionic interactions to form the nanoparticles.[76,115] Biodegradable and biocompatible carriers are desirable so as not to display any adverse effects on the cells while leaving no trace of the 'passive' carrier in the cells.[116] Among the carriers used for siRNA delivery, commercially available lipid-based carriers are more popular for delivery against CDKs and cyclins (Table 2.1). Lipid-based carriers form liposomal structures with a lipid bilayer envelope and an aqueous core.[76,116] Cationic lipids in the lipid bilayer structure interact with negatively charged siRNA to form an ionic complex, and immobilize siRNA on the envelope structure (either internal or external). The siRNA could also be entrapped in the core of the liposome. Some carriers form 'solid' lipidic nanoparticles of homogenous siRNA–lipid complexes. Cationic lipids display more disruptive activity on cell membranes (*i.e.* more toxic) compared to neutral lipids and may have higher immunogenicity due to enhanced liposomal uptake by macrophages.[117] Certain cationic polymers such as polyethylenimine (PEI) have also been used to deliver siRNA against CDKs and cyclins. Polymers condense siRNA into positively charged polymeric micelles (believed to be homogenous structures) that are subsequently internalized through the anionic cell membrane by endocytosis. In the case of PEI, osmotic imbalance bursts the endosomal vesicles (the so-called proton-sponge effect of PEI), releasing polymer–siRNA complexes into the cytoplasm.[118] Cationic polymers are also toxic due to their ability to disrupt cell membrane and mitochondrial membrane and their toxicity usually increases with their molecular weight. The efficacy of siRNA critically depends on the release of siRNA from the carrier in the cytoplasm so that it is available in free form for RISC incorporation, and its efficient intracellular transport to the site(s) of silencing.

2.4.2.2   shRNA Expression for RNAi

The cyclins and CDKs have also been successfully silenced using shRNA and non-viral vectors (Table 2.2). The amount of plasmid DNA ligated with shRNA used to target cyclins and CDKs is in the range of 0.4–10 µg in preclinical models. Klier *et al.* (2008)[130] have achieved stable expression of cyclin D shRNA from plasmid DNA for 2 weeks in mantle cell lymphoma cell lines, while the knockdown of cyclin D1 was shown for 3 weeks in mice using shRNA.[131] The data shown in Table 2.2 represent the silencing of specific cyclins or CDKs by shRNA for 3–7 days *in vitro*. The main concern with shRNA is its limited duration of expression in the eukaryotic cells resulting in the silencing of a target for short period of time, which is the same drawback as that facing the siRNA itself.

Although non-viral vectors are more popular than viral vectors for shRNA delivery because of their superior safety profile, non-viral vectors (plasmid DNA) have certain limitations and disadvantages. As in siRNA, the major issue is the delivery of plasmid DNA into the cells, which requires a carrier as well. Though many commercially available lipid-based carriers and polymers have been used for successful transfection of plasmid DNA *in vitro*, the carrier and/or plasmid DNA complexes may induce an immune response

Table 2.2 Cyclins and CDKs that have been silenced by shRNA.

Target	Vector	Cell lines	Transfection agent (supplier)	Ref.
Lipid-based carriers				
CDK2	BS/U6	HeLa, U2-OS	Lipofectamine (Invitrogen)	Long et al. (2010)[119]
CDK2	pTER	IMR32	Lipofectamine (Invitrogen)	Molenaar et al. (2009)[96]
CDK2	pSilencer 2.0-U6	95D	Lipofectamine (Invitrogen)	Xu et al. (2010)[120]
CDK4	pSIREN	PC12	Lipofectamine (Invitrogen)	Biswas et al. (2005)[121]
CDK5	pLKO.1	HEK293T	Lipofectamine (Invitrogen)	Choi et al. (2010)[122]
CDK5	pCAG–MCS2	COS7	Lipofectamine (Invitrogen)	Kawauchi et al. (2006)[123]
CDK5	MISSION Lentiviral construct	A549	Lipofectamine (Invitrogen)	Zhu et al. (2011)[124]
CDK5	Lentiviral construct	SH-SY5Y	Lipofectamine (Invitrogen)	Chen et al. (2007)[125]
CDK7	pENTR-H1	U2-OS	FuGENE (Promega)	Helenius et al. (2009)[103]
Cyclin C	pSilencer 2.0-U6	SaoS-2	Lipofectamine (Invitrogen)	Ren and Rollins (2004)[105]
Cyclin D1	pLKO.1	Tca8113	Oligofectamine (Invitrogen)	Zhou et al. (2009)[85]
Polymeric carriers				
CDK3	pU6pro	HEK293, T98G	jetPEI (Polyplus)	Zheng et al. (2008)[126]
CDK3	pU6pro	SaoS-2, JB6	jetPEI (Polyplus)	Cho et al. (2009)[127]
Without carriers				
CDK1 and CDK2	pSuperior.puro	NCI-H1299, U2-OS	CalPhos mammalian transfection kit (Clontech)	Cai et al. (2006)[128]
CDK7	pCMV-GFP	HEK293	Calcium phosphate precipitation	D'Alessio et al. (2007)[129]
Cyclin D1, Cyclin D2, Cyclin D3	pSuper	Granta 519, Jeko-1, JVM-2, Hbl-2, Rec-1, UPN-1, Z-138	Lentiviral-mediated	Klier et al. (2008)[130]
CDK4, Cyclin D1	pMKO.1	V720, A249	Not specified	Yu et al. (2006)[131]
CDK5	pBS/U6	SaoS-2	Not specified	Yang and Hinds (2006)[132]
CDK6	pMKO.1	LG	Not specified	Fujimoto et al. (2007)[133]
CDK8	pMKO.1	DLD-1, HCT116	Not specified	Firestein et al. (2008)[134]
Cyclin D1	pRetroSuper	U2-OS, BTR	Not specified	Rowland et al. (2005)[135]

in vivo, mainly the innate immune response by causing tissue damage.[136-138] Intracellular trafficking affects the expression of shRNA from plasmid DNA. Plasmid DNA must dissociate from its carrier in the cytoplasm and must be transported to the nucleus for transcription (unlike siRNA which needs to remain in the cytoplasm), and exposes itself to the transcriptional machinery by decondensation of dsDNA.[139] Being a foreign DNA, activities such as inhibition of transcription, nuclease-mediated degradation of plasmid DNA, induced mutation in the plasmid DNA or induction of apoptosis may be associated with the delivery of plasmid DNA.[140,141] Moreover, adaptive immune response may be activated because of the sensitivity of the pathogen recognition *in vivo*.[142,143] As the plasmid lacks the mechanism of maintaining its copy number in the eukaryotic cells, plasmid concentration decreases in the daughter cells, and eventually shRNA expression diminishes.[144] Some of these limitations can be avoided by designing vectors that can complement the mammalian DNA sequence, such as the mammalian tissue-specific promoters and enhancer sequences. The positive charge of the carriers can be reduced by substitution of the amine groups, so that plasmid DNA can easily dissociate after transfection into cells.

2.4.2.3   miRNA

As the synthesis of several cyclins and CDKs is endogenously controlled by miRNA, its delivery can be used to target these proteins in order to decrease the proliferation rate of cancer cells.[153] Silencing of cyclins and CDKs by miRNA was shown using plasmid DNA as a vector, while naked miRNA using cationic carriers was introduced into cells for the same purpose (Table 2.3). The concentration of miRNA used for a targeted effect was in the range of 30–50 nM (similar to that of siRNA), while the amount of plasmid DNA ligated with miRNA was 0.3–10 µg. Lin *et al.* (2010)[150] have shown the knockdown of CDK2, CDK4, CDK6, cyclin D and cyclin E by miR-302 for 3 weeks *in vivo* using plasmid DNA, which is again the limitation of a vector for the expression of specific miRNA. However, the advantage of using miRNA over siRNA or shRNA is the wide range of potential targets that can be silenced for a specific miRNA. The siRNA or shRNA would be very specific to silence just single target, while miRNA can silence more than one target based on its complementary sequence to different genes.

2.4.3   Overcoming RNAi Intended for Cell Cycle Regulation

Although RNAi seems to be a powerful approach to regulate cell cycle progression, there might be specific mechanisms employed by malignant cells in order to overcome its inhibitory activities. If mutated copies of a specific cyclin or CDK are amplified in the genome or overexpressed, these mutated transcripts might not be recognized by siRNA and thus escape processing by the RISC complex. Chromosomal instability may also lead to mutations in essential survival pathways and malignant cells may nullify the RNAi

Table 2.3 Cyclins and CDKs that have been silenced by miRNA. The studies include miRNA delivery by an expression vector as well as delivery of miRNA itself.

Target	miRNA	Vector	Cell lines	Transfection agent (supplier)	Ref.
Lipid-based carriers					
Cyclin D1	miR-17-5p, miR-20a	pMSCV	MCF7, NAFA	Lipofectamine (Invitrogen)	Yu et al. (2008)[145]
CDK6, Cyclin D1	miR-34a	pGL3	A549	Lipofectamine (Invitrogen)	Sun et al. (2008)[146]
Cyclin D1, Cyclin E	miR-15b		U87, U118	X-tremeGENE (Roche)	Xia et al. (2009)[147]
CDK8, Cyclin A, Cyclin B, Cyclin E	let-7		HeLa, A549, HepG2	NeoFx (Ambion)	Johnson et al. (2007)[148]
Cyclin D1	miR-193b		Malme-3M, SKMEL-28, SKMEL-5	Lipofectamine (Invitrogen)	Chen et al. (2010)[149]
Without carriers					
CDK2, CDK4, CDK6, Cyclin D, Cyclin E	miR-302	pTet-tTS	MCF7, HepG2, Tera-2	Electroporation	Lin et al. (2010)[150]
CDK6	pre-miR-9		TOM-1, NALM-20, MY, LOUCY, JOURKAT, TANOUE	Nucleofection	Rodriguez-Otero et al. (2011)[151]
CDK6	pre-miR-124a		TOM-1, NALM-20	Nucleofection	Agirre et al. (2009)[152]

treatment by utilizing the mutated survival pathway. Usually, the miRNAs silence more than one target based on their complementary sequence to different mRNAs. The miRNAs targeting cyclins and CDKs may have complementary sequences to CDK inhibitors, which may decrease the expression of inhibitors and amplify cell cycle progression. In such instances, these miRNAs will not be appropriate for RNAi treatment. Sometimes, targeting just one cyclin or CDK may not be enough to stop unscheduled cell proliferation, and targeting multiple cyclins or CDKs might be required to inhibit the cell growth drastically. The choice of the RNAi mediator (siRNA *vs.* shRNA *vs.* microRNA) will be critical for this purpose. The transcript copy numbers of specific cyclins or CDKs are different among various cancer types. However, the genes with low transcript copy numbers, crucial for cell cycle progression, would be more effective to target than those with high copy numbers. Any dynamic regulation of transcript copy number by malignant cells might also provide a means to overcome therapeutic RNAi. These are unresolved issues at this stage and careful studies are required to reveal their impact in targeting cell cycle regulators.

2.5   Concluding Remarks

Several other proteins play a crucial role during cell cycle events in addition to CDKs and cyclins, such as the kinesin protein family, cell division cycle proteins, checkpoint kinases, centromere proteins and RAD homolog proteins. These proteins are sometimes up-regulated in transformed cells and therefore should be taken into account as potential targets for RNAi treatment. Kinesin proteins are essential mitotic motor proteins that are required early in mitosis to form a bipolar spindle by separating the emerging spindle poles.[154,155] Kinesin spindle protein (KSP) is the most effective target from this family and its efficacy is currently being tested at clinical stage after successful preclinical testing in a cancer model.[156] Cell division cycle (CDC) proteins are widely distributed among almost all phases of the cell cycle, and serve important roles at particular stages; for example, CDC2 and CDC7 serve as kinases during the G2 and M phases, while CDC20 activates the anaphase promoting complex during mitosis.[157,158] Checkpoint kinases have an important function of cell cycle arrest upon receiving confirmation of DNA damage.[159] Centromere proteins mainly facilitate the formation of centromeres when nuclear materials condense into chromosomes.[160] RAD homolog proteins involve in the repair of DNA double-strand break during homologous recombination.[161]

Other potential targets that might act in conjunction with cell cycle proteins are anti-apoptosis proteins, which are often up-regulated in transformed cells and help the cells to withstand intracellular pressures due to accelerated cell cycle progression.[162] Silencing a cell cycle protein may increase the expression of anti-apoptosis proteins based on the conjunction of cell cycle proteins and anti-apoptosis proteins,[163,164] and therefore, anti-apoptosis proteins such as B-cell lymphoma (BCL-2), B-cell lymphoma extra-large (BCL-XL), myeloid cell leukemia (MCL-1) and survivin could be silenced by RNAi with cell cycle proteins. The dual silencing of a cell cycle protein and an anti-apoptosis protein might be exceptionally effective and specific for the treatment of cancer by RNAi.

Finally, carrier design to deliver effective anti-cancer agents is an ongoing activity as more effective carriers is constantly needed for nucleic acid-based therapies. Aliabadi *et al.* (2013)[165] have reported silencing of MCL-1 and P-glycoprotein using engineered PEIs which are substituted with lipid molecules. Such tailored carriers are expected to be preferable to commercially available alternatives, since they can offer superior delivery kinetics and can be tailored depending on the cell type and chosen target. The siRNA against KSP was delivered using a lipid-based carrier in the clinical studies, which also seems to be effective for the delivery of nucleic acids.[156] The delivery of nucleic acids is not readily possible without carriers at the clinical stage, and electrofection and nucleofection is, therefore, most likely remain an experimental tool. As cationic carriers derived from polymers typically possess a high density of positive charges, it is difficult for nucleic acid–cationic carrier complexes to easily dissociate in the cytoplasm. The efficiency of nucleic acid release can be improved by decreasing the charge density of a carrier by substituting some positively-charged moieties with suitable ligands. Cationic lipids can be advantageous in this regard due to lower charge densities. Studies for such systematic investigations remain to be conducted for deploying siRNA-based RNAi for cyclin and CDK inhibition.

Acknowledgements

The siRNA research in the authors' laboratory was financially supported by operating grants from Canadian Breast Cancer Foundation (CBCF), Alberta Innovates Health Sciences (AIHS) and Natural Sciences and Engineering Council of Canada (NSERC), and equipment support from Alberta Heritage Foundation for Medical Research (AHFMR). The authors declare that there are no competing financial interests in relation to the work described. We are grateful to numerous past trainees in the Uludag laboratory who contributed to the development of non-viral delivery systems, in particular Dr Vanessa Incani, Dr Meysam Abbasi, Dr Hamidreza M. Aliabadi, Dr Charlie Hsu, Dr Aws Alshamsan and Dr Remant K. C. We also acknowledge our long-term collaborators Dr Afsaneh Lavasanifar, Dr Tian Tang, Dr Michael Weinfeld and Dr Xiaoyan Jiang for their valuable contributions to our RNAi research.

References

1. K. A. Schafer, *Vet. Pathol.*, 1998, **35**, 461.
2. K. Vermeulen, D. R. Van Bockstaele and Z. N. Berneman, *Cell Proliferation*, 2003, **36**, 131.
3. A. Ahmad, Z. Wang, R. Ali, B. Bitar, F. T. Logna, M. Y. Maitah, B. Bao, S. Ali, D. Kong, Y. Li and F. H. Sarkar, *Targeting New Pathways and Cell*

Death in Breast Cancer, ed. R. L. Aft, InTech, Rijeka, Croatia, 2012, ch. 8, p. 113.

4. Z. A. Stewart, M. D. Westfall and J. A. Pietenpol, *Trends Pharmacol. Sci.*, 2003, **24**, 139.
5. A. Maya-Mendoza, C. W. Tang, A. Pombo and D. A. Jackson, *Front. Biosci.*, 2009, **14**, 4199.
6. B. Novák, J. C. Sible, J. J. Tyson, *Encyclopedia of Life Sciences*, John Wiley & Sons Ltd, Chichester, 2003.
7. M. Malumbres and M. Barbacid, *Nat. Rev. Cancer*, 2001, **1**, 222.
8. J. Massague, *Nature*, 2004, **432**, 298.
9. M. B. Kastan and J. Bartek, *Nature*, 2004, **432**, 316.
10. G. J. Kops, B. A. Weaver and D. W. Cleveland, *Nat. Rev. Cancer*, 2005, **5**, 773.
11. S. Lapenna and A. Giordano, *Nat. Rev. Drug Discovery*, 2009, **8**, 547.
12. M. Malumbres and M. Barbacid, *Nat. Rev. Cancer*, 2009, **9**, 153.
13. G. M. Cooper, *The Cell: A Molecular Approach*, ASM Press, Washington, DC, 2nd edn, 2000, ch. 14.
14. C. Norbury and P. Nurse, *Annu. Rev. Biochem.*, 1992, **61**, 441.
15. H. Lodish, A. Berk, C. A. Kaiser, M. Krieger, M. P. Scott, A. Bretscher, H. Ploegh, P. Matsudaira, *Molecular Cell Biology*, W.H. Freeman, New York, 6th edn, 2008, ch. 20.
16. D. Morgan, *Cell Cycle: Principles of Control*, New Science Press, London, 2007.
17. S. P. Bell and A. Dutta, *Annu. Rev. Biochem.*, 2002, **71**, 333.
18. D. Branzei and M. Foiani, *Curr. Opin. Cell Biol.*, 2005, **17**, 568.
19. R. M. Liskay, *Proc. Natl. Acad. Sci. U. S. A.*, 1977, **74**, 1622.
20. A. Maton, *Cells: Building Blocks of Life*, Prentice Hall, New Jersey, 3rd edn, 1997.
21. L. H. Hartwell and T. A. Weinert, *Science*, 1989, **246**, 629.
22. A. B. Pardee, *Science*, 1989, **246**, 603.
23. A. Zetterberg, O. Larsson and K. G. Wiman, *Curr. Opin. Cell Biol.*, 7, 835.
24. A. B. Pardee, *Proc. Natl. Acad. Sci. U. S. A.*, 1974, **71**, 1286.
25. J. Gautier, J. l. Maller, T. A. Langan, M. J. Lohka, S. Shenoy, D. Shalloway and P. Nurse, *J. Cell Sci., Suppl.*, 1989, **12**, 53.
26. J. M. Peters, *Curr. Opin. Cell Biol.*, 1998, **10**, 759.
27. R. Ciosk, W. Zachariae, C. Michaelis, A. Shevchenko, M. Mann and K. Nasmyth, *Cell*, 1998, **93**, 1067.
28. T. Evans, E. T. Rosenthal, J. Youngblom, D. Distel and T. Hunt, *Cell*, 1983, **33**, 389.
29. J. Pines, *Cell Growth Differ.*, 1991, **2**, 305.
30. A. M. Abukhdeir and B. H. Park, *Expert Rev. Mol. Med.*, 2008, **10**, e19.
31. J. Pines, *Adv. Cancer Res.*, 1995, **66**, 181.
32. D. O. Morgan, *Nature*, 1995, **374**, 131.
33. R. P. Fisher and D. O. Morgan, *Cell*, 1994, **78**, 713.
34. C. J. Sherr, *Cell*, 1994, **79**, 551.
35. M. Ohtsubo, A. M. Theodoras, J. Schumacher, J. M. Roberts and M. Pagano, *Mol. Cell Biol.*, 1995, **15**, 2612.

36. F. Girard, U. Strausfeld, A. Fernandez and N. J. Lamb, *Cell*, 1991, **67**, 1169.
37. D. H. Walker and J. L. Maller, *Nature*, 1991, **354**, 314.
38. R. W. King, P. K. Jackson and M. W. Kirschner, *Cell*, 1994, **79**, 563.
39. M. Glotzer, A. W. Murray and M. W. Kirschner, *Nature*, 1991, **349**, 132.
40. M. Rechsteiner and S. W. Rogers, *Trends Biochem. Sci.*, 1996, **21**, 267.
41. C. J. Sherr and J. M. Roberts, *Genes Dev.*, 1995, **9**, 1149.
42. A. Carnero and G. J. Hannon, *Curr. Top. Microbiol. Immunol.*, 1998, **227**, 43.
43. L. Hengst and S. I. Reed, *Curr. Top. Microbiol. Immunol.*, 1998, **227**, 25.
44. W. S. el-Deiry, T. Tokino, V. E. Velculescu, D. B. Levy, R. Parsons, J. M. Trent, D. Lin, W. E. Mercer, K. W. Kinzler and B. Vogelstein, *Cell*, 1993, **75**, 817.
45. J. W. Harper, G. R. Adami, N. Wei, K. Keyomarsi and S. J. Elledge, *Cell*, 1993, **75**, 805.
46. Z. Q. Pan, J. T. Reardon, L. Li, R. H. Flores, R. Legerski, A. Sancar and J. Hurwitz, *J. Biol. Chem.*, 1995, **270**, 22008.
47. S. Waga, R. Li and B. Stillman, *Leukemia*, 1997, **11**, 321.
48. K. Polyak, M. H. Lee, B. H. Erdjument, A. Koff, J. M. Roberts, P. Tempst and J. Massague, *Cell*, 1994, **78**, 59.
49. H. Watanabe, Z. Q. Pan, N. Schreiber-Agus, R. A. DePinho, J. Hurwitz and Y. Xiong, *Proc. Natl. Acad. Sci. U. S. A.*, 1998, **95**, 1392.
50. J. Bartek, C. Lukas and J. Lukas, *Nat. Rev. Mol. Cell Biol.*, 2004, **5**, 792.
51. A. Aguilera and B. Gómez-González, *Nat. Rev. Genet.*, 2008, **9**, 204.
52. M. Malumbres and M. Barbacid, *Trends Biochem. Sci.*, 2005, **30**, 630.
53. A. Musacchio and E. D. Salmon, *Nat. Rev. Mol. Cell Biol.*, 2007, **8**, 379.
54. N. P. Pavletich, *J. Mol. Biol.*, 1999, **287**, 821.
55. S. Ortega, M. Malumbres and M. Barbacid, *Biochim. Biophys. Acta*, 2002, **1602**, 73.
56. S. G. Rane, S. C. Cosenza, R. V. Mettus and E. P. Reddy, *Mol. Cell Biol.*, 2002, **22**, 644.
57. A. Martín, J. Odajima, S. L. Hunt, P. Dubus, S. Ortega, M. Malumbres and M. Barbacid, *Cancer Cell*, 2005, 7, 591.
58. E. Aleem, H. Kiyokawa and P. Kaldis, *Nat. Cell Biol.*, 2005, 7, 831.
59. G. I. Shapiro, *J. Clin. Oncol.*, 2006, **24**, 1770.
60. A. M. Senderowicz, *Invest. New Drugs*, 1999, **17**, 313.
61. N. Villerbu, A. M. Gaben, G. Redeuilh and J. Mester, *Int. J. Cancer*, 2002, **97**, 761.
62. M. Pennati *et al.*, *Mol. Cancer Ther.*, 2005, **4**, 1328.
63. S. J. McClue *et al.*, *Int. J. Cancer*, 2002, **102**, 463.
64. M. Payton *et al.*, *Cancer Res.*, 2006, **66**, 4299.
65. S. Emanuel *et al.*, *Cancer Res.*, 2005, **65**, 9038.
66. L. T. Vassilev *et al.*, *Proc. Natl. Acad. Sci. U. S. A.*, 2006, **103**, 10660.
67. A. Fire, S. Xu, M. K. Montgomery, S. A. Kostas, S. E. Driver and C. C. Mello, *Nature*, 1998, **391**, 806.
68. A. J. Hamilton and D. C. Baulcombe, *Science*, 1999, **286**, 950.

69. I. Bentwich *et al.*, *Nat. Genet.*, 2005, **37**, 766.
70. C. B. Moore, E. H. Guthrie, M. T. Huang and D. J. Taxman, *Methods Mol. Biol.*, 2010, **629**, 141.
71. D. H. Kim and J. J. Rossi, *Nat. Rev. Genet.*, 2007, **8**, 173.
72. R. C. Wilson and J. A. Doudna, *Annu. Rev. Biophys.*, 2013, **42**, 217.
73. P. J. Leuschner, S. L. Ameres, S. Kueng and J. Martinez, *EMBO Rep.*, 2006, 7, 314.
74. J. B. Preall and E. J. Sontheimer, *Cell*, 2005, **123**, 543.
75. Z. Wang, D. D. Rao, N. Senzer and J. Nemunaitis, *Pharm. Res.*, 2011, **28**, 2983.
76. H. M. Aliabadi, B. Landry, C. Sun, T. Tang and H. Uludağ, *Biomaterials*, 2012, **33**, 2546.
77. Y. Liang, H. Gao, S. Y. Lin, J. A. Goss, F. C. Brunicardi and K. Li, *PLoS One*, 2010, **5**, e12860.
78. S. G. Manzo *et al.*, *Cancer Res.*, 2012, **72**, 5363.
79. O. Sahin *et al.*, *BMC Syst. Biol.*, 2009, **3**, 1.
80. D. Gong *et al.*, *Curr. Biol.*, 2007, **17**, 85.
81. D. Ovcharenko, K. Kelnar, C. Johnson, N. Leng and D. Brown, *Cancer Res.*, 2007, **67**, 10782.
82. Y. L. Zheng, B. S. Li, J. Kanungo, S. Kesavapany, N. Amin, P. Grant and H. C. Pant, *Mol. Biol. Cell*, 2007, **18**, 404.
83. C. N. Johnstone *et al.*, *Mol. Cell. Biol.*, 2008, **28**, 687.
84. M. Grillo *et al.*, *Breast Cancer Res. Treat.*, 2006, **95**, 185.
85. X. Zhou, Z. Zhang, X. Yang, W. Chen and P. Zhang, *Int. J. Cancer*, 2009, **124**, 483.
86. H. Lin, J. L. Juang and P. S. Wang, *J. Biol. Chem.*, 2004, **279**, 29302.
87. L. De Boer *et al.*, *Oncogene*, 2008, **27**, 4261.
88. A. Tomashevski, D. R. Webster, P. Grammas, M. Gorospe and I. I. Kruman, *Cell Death Differ.*, 2010, **17**, 1189.
89. J. P. MacKeigan, L. O. Murphy and J. Blenis, *Nat. Cell Biol.*, 2005, 7, 591.
90. H. Biliran Jr *et al.*, *Clin. Cancer Res.*, 2005, **11**, 6075.
91. C. X. Wang, J. H. Song, D. K. Song, V. W. Yong, A. Shuaib and C. Hao, *Cell Death Differ.*, 2006, **13**, 1203.
92. N. C. Turner *et al.*, *EMBO J.*, 2008, **27**, 1368.
93. V. Caracciolo *et al.*, *Cell Cycle*, 2012, **11**, 1202.
94. M. Retzer-Lidl, R. M. Schmid and G. Schneider, *Int. J. Cancer*, 2007, **121**, 66.
95. S. Jirawatnotai, A. Aziyu, E. C. Osmundson, D. S. Moons, X. Zou, R. D. Kineman and H. Kiyokawa, *J. Biol. Chem.*, 2004, **279**, 51100.
96. J. J. Molenaar *et al.*, *Proc. Natl. Acad. Sci. U. S. A.*, 2009, **106**, 12968.
97. H. Xiao *et al.*, *Front. Med. China*, 2009a, **3**, 384.
98. H. Xiao *et al.*, *Chin.-Ger. J. Clin. Oncol.*, 2009b, **8**, 371.
99. S. B. He, Y. Yuan, L. Wang, M. J. Yu, Y. B. Zhu and X. G. Zhu, *J. Exp. Clin. Cancer Res.*, 2011, **30**, 109.
100. M. Ubeda, J. M. Rukstalis and J. F. Habener, *J. Biol. Chem.*, 2006, **281**, 28858.

101. I. Okamoto *et al.*, *Neoplasia*, 2005, **7**, 303.
102. S. Mohapatra, B. Chu, X. Zhao, J. Djeu, J. Q. Cheng and W. J. Pledger, *Int. J. Biochem. Cell Biol.*, 2009, **41**, 595.
103. K. Helenius, Y. Yang, J. Alasaari and T. P. Mäkelä, *Mol. Cell. Biol.*, 2009, **29**, 315.
104. N. H. Cho *et al.*, *Cancer Sci.*, 2006, **97**, 1082.
105. S. Ren and B. J. Rollins, *Cell*, 2004, **117**, 239.
106. S. Jirawatnotai *et al.*, *Nature*, 2011, **474**, 230.
107. S. Choudhary, K. P. Rosenblatt, L. Fang, B. Tian, Z. H. Wu and A. R. Brasier, *J. Biol. Chem.*, 2011, **286**, 37187.
108. M. Osabe, J. Sugatani, A. Takemura, Y. Yamazaki, A. Ikari, N. Kitamura, M. Negishi and M. Miwa, *Biochem. Biophys. Res. Commun.*, 2008, **369**, 1027.
109. J. Sugatani, M. Osabe, M. Kurosawa, N. Kitamura, A. Ikari and M. Miwa, *Drug Metab. Dispos.*, 2010, **38**, 177.
110. D. Peer, P. Zhu, C. V. Carman, J. Lieberman and M. Shimaoka, *Proc. Natl. Acad. Sci. U. S. A.*, 2007, **104**, 4095.
111. X. Zhang *et al.*, *J. Cell Biol.*, 2009, **184**, 67.
112. M. Zumbansen *et al.*, *J. RNAi Gene Silencing*, 2010, **6**, 354.
113. D. Peer, E. J. Park, Y. Morishita, C. V. Carman and M. Shimaoka, *Science*, 2008, **319**, 627.
114. M. R. Gumina, C. Xu and T. C. Chiles, *Cell Cycle*, 2010, **9**, 820.
115. H. M. Aliabadi, B. Landry, R. K. Bahadur, A. Neamnark, O. Suwantong and H. Uludağ, *Macromol. Biosci.*, 2011, **11**, 662.
116. J. Wang, Z. Lu, M. G. Wientjes and J. L. Au, *AAPS J.*, 2010, **12**, 492.
117. M. Hashida, S. Kawakami and F. Yamashita, *Chem. Pharm. Bull.*, 2005, **53**, 871.
118. O. Boussif, F. Lezoualc'h, M. A. Zanta, M. D. Mergny, D. Scherman, B. Demeneix and J. P. Behr, *Proc. Natl. Acad. Sci. U. S. A.*, 1995, **92**, 7297.
119. X. E. Long, Z. H. Gong, L. Pan, Z. W. Zhong, Y. P. Le, Q. Liu, J. M. Guo and J. C. Zhong, *BMB Rep.*, 2010, **43**, 291.
120. L. Xu, C. Wang, Z. Wen, X. Yao, Z. Liu, Q. Li, Z. Wu, Z. Xu, Y. Liang and T. Ren, *Immunol. Lett.*, 2010, **127**, 93.
121. S. C. Biswas, D. X. Liu and L. A. Greene, *J. Neurosci.*, 2005, **25**, 8349.
122. J. H. Choi *et al.*, *Nature*, 2010, **466**, 451.
123. T. Kawauchi, K. Chihama, Y. Nabeshima and M. Hoshino, *Nat. Cell Biol.*, 2006, **8**, 17.
124. Y. X. Zhu *et al.*, *Blood*, 2011, **117**, 3847.
125. T. C. Chen, Y. K. Lai, C. K. Yu and J. L. Juang, *Cell Microbiol.*, 2007, **9**, 2676.
126. D. Zheng, Y. Y. Cho, A. T. Lau, J. Zhang, W. Y. Ma, A. M. Bode and Z. Dong, *Cancer Res.*, 2008, **68**, 7650.
127. Y. Y. Cho, F. Tang, K. Yao, C. Lu, F. Zhu, D. Zheng, A. Pugliese, A. M. Bode and Z. Dong, *Cancer Res.*, 2009, **69**, 272.
128. D. Cai, V. M. Latham Jr, X. Zhang and G. I. Shapiro, *Cancer Res.*, 2006, **66**, 9270.

129. A. C. D'Alessio, I. C. Weaver and M. Szyf, *Mol. Cell. Biol.*, 2007, **27**, 7462.
130. M. Klier *et al.*, *Leukemia*, 2008, **22**, 2097.
131. Q. Yu *et al.*, *Cancer Cell*, 2006, **9**, 23.
132. H. S. Yang and P. W. Hinds, *Cancer Res.*, 2006, **66**, 2708.
133. T. Fujimoto, K. Anderson, S. E. Jacobsen, S. I. Nishikawa and C. Nerlov, *EMBO J.*, 2007, **26**, 2361.
134. R. Firestein *et al.*, *Nature*, 2008, **455**, 547.
135. B. D. Rowland, R. Bernards and D. S. Peeper, *Nat. Cell Biol.*, 2005, **7**, 1074.
136. H. Sakurai *et al.*, *J. Controlled Release*, 2007, **117**, 430.
137. A. Gautam, C. L. Densmore and J. C. Waldrep, *Gene Ther.*, 2001, **8**, 254.
138. H. Sakurai, K. Kawabata, F. Sakurai, S. Nakagawa and H. Mizuguchi, *Int. J. Pharm.*, 2008, **354**, 9.
139. D. Lechardeur and G. L. Lukacs, *Hum. Gene Ther.*, 2006, **17**, 882.
140. L. Zang, M. Nishikawa, K. Machida, M. Ando, Y. Takahashi, Y. Watanabe and Y. Takakura, *Gene Ther.*, 2011, **18**, 891.
141. S. C. Ribeiro, G. A. Monteiro and D. M. Prazeres, *J. Gene Med.*, 2004, **6**, 565.
142. K. A. Whitehead, J. E. Dahlman, R. S. Langer and D. G. Anderson, *Annu. Rev. Chem. Biomol. Eng.*, 2011, **2**, 77.
143. N. Bessis, F. J. GarciaCozar and M. C. Boissier, *Gene Ther.*, 2004, **11**, S10.
144. E. Riu, Z. Y. Chen, H. Xu, C. Y. He and M. A. Kay, *Mol. Ther.*, 2007, **15**, 1348.
145. Z. Yu *et al.*, *J. Cell Biol.*, 2008, **182**, 509.
146. F. Sun, H. Fu, Q. Liu, Y. Tie, J. Zhu, R. Xing, Z. Sun and X. Zheng, *FEBS Lett.*, 2008, **582**, 1564.
147. H. Xia *et al.*, *Biochem. Biophys. Res. Commun.*, 2009, **380**, 205.
148. C. D. Johnson *et al.*, *Cancer Res.*, 2007, **67**, 7713.
149. J. Chen *et al.*, *Am. J. Pathol.*, 2010, **176**, 2520.
150. S. L. Lin, D. C. Chang, S. Y. Ying, D. Leu and D. T. Wu, *Cancer Res.*, 2010, **70**, 9473.
151. P. Rodriguez-Otero *et al.*, *Br. J. Haematol.*, 2011, **155**, 73.
152. X. Agirre *et al.*, *Cancer Res.*, 2009, **69**, 4443.
153. Z. Yu, R. Baserga, L. Chen, C. Wang, M. P. Lisanti and R. G. Pestell, *Am. J. Pathol.*, 2010, **176**, 1058.
154. R. D. Vale, T. S. Reese and M. P. Sheetz, *Cell*, 1985, **42**, 39.
155. R. D. Vale, *Cell*, 2003, **112**, 467.
156. J. Tabernero *et al.*, *Cancer Discovery*, 2013, **3**, 406.
157. J. M. Kim, M. Yamada and H. Masai, *Mutat. Res.*, 2003, **532**, 29.
158. E. R. Kramer, C. Gieffers, G. Hölzl, M. Hengstschläger and J. M. Peters, *Curr. Biol.*, 1998, **8**, 1207.
159. Y. Sanchez, C. Wong, R. S. Thoma, R. Richman, Z. Wu, H. Piwnica-Worms and S. J. Elledge, *Science*, 1997, **277**, 1497.
160. S. Westermann and A. Schleiffer, *Trends Cell Biol.*, 2013, **23**, 260.
161. E. L. Ivanov and J. E. Haber, *Curr. Biol.*, 1997, 7, R492.

162. O. Méndez, Y. Fernández, M. A. Peinado, V. Moreno and A. Sierra, *Clin. Exp. Metastasis*, 2005, **22**, 297.

163. M. E. Harley, L. A. Allan, H. S. Sanderson and P. R. Clarke, *EMBO J.*, 2010, **29**, 2407.

164. Y. Saintigny, A. Dumay, S. Lambert and B. S. Lopez, *EMBO J.*, 2001, **20**, 2596.

165. H. M. Aliabadi, P. Mahdipoor and H. Uludağ, *Cancer Gene Ther.*, 2013, **20**, 169.

CHAPTER 3

Nanoparticle Carriers to Overcome Biological Barriers to siRNA Delivery

HAMIDREZA MONTAZERI ALIABADI* AND HASAN ULUDAĞ*

Department of Chemical & Material Engineering, Electrical & Computer Engineering Research Facility (ECERF), Faculty of Engineering, University of Alberta, Edmonton, AB, Canada
*Email: montazer@chapman.edu; hasan.uludag@ualberta.ca

3.1 Introduction

A different approach to traditional medicine, which commonly aims to control and/or eliminate the symptoms of a disorder, or foreign organisms in the body, is to down-regulate or delete the intracellular factors responsible for the problem. The majority of cellular functions and characteristics are controlled by the expression of different proteins with different categories of functions, such as metabolic enzymes, transcription factors and cell cycle proteins. Therefore, controlling the expression of specific proteins could potentially change the fate of a cell, and ultimately, the fate of a patient. Although the possibility of inhibiting protein expression has been recognized for some time, the possibility of an endogenous pathway for inhibiting protein expression was established in the late 1990s, when an attempt to overexpress chalcone synthase in pigmented petunia petals by introducing a chimeric petunia CHS gene unexpectedly resulted in a block in anthocyanin biosynthesis, due to a 50-fold decrease in the mRNA levels of the protein.[1] In 1998, Fire et al. showed that the introduction of an exogenous RNA into cells

RSC Drug Discovery Series No. 51
Nanomedicines: Design, Delivery and Detection
Edited by Martin Braddock
© The Royal Society of Chemistry 2016
Published by the Royal Society of Chemistry, www.rsc.org

could interfere with the expression of a protein at the mRNA level, and even showed that a double-stranded RNA (dsRNA) was more effective for this purpose.[2] Since then, RNA interference (RNAi) has come a long way, and "silencing" a protein is now a powerful investigational tool and a potential therapeutic strategy for a wide range of clinical disorders.

RNAi is a general term that describes various strategies used to interfere with expression of a target protein at the post-transcriptional stage, either by degrading the mRNA responsible for the synthesis of the protein or blocking the translational access to the mRNA. However, different RNAi approaches all depend on an intracellular macromolecular assembly known as the RNA-induced silencing complex (RISC). A dsRNA, either synthesized endogenously or exogenously introduced, is recruited into this complex, which contains the Argonaute 2 (Ago2) enzyme which is capable of "slicing" the mRNA target. Only one strand of the dsRNA is retained in the complex, which is called "guide strand" as it matches the target mRNA sequence. The guide strand then binds to the complementary region of the target mRNA, which in turn triggers the Ago2-directed cleavage of the mRNA transcript. The fragmented mRNA is further degraded by the intracellular RNases.[3] Inhibiting protein synthesis by blocking translation, rather than mRNA degradation, can also occur, depending on the degree of complementarity between the guide strand and the mRNA sequence.[4]

Different RNAi strategies can be broadly categorized into three, based on the use of microRNAs (miRNA), short hairpin RNAs (shRNA) or small interfering RNA (siRNA) for silencing. miRNAs are an endogenous group of non-coding RNAs designed for post-transcriptional regulation of protein expression. The functional miRNAs sequences are encoded (integrated) individually in the primary miRNA (pri-miRNA) sequence that is hundreds to a few thousand base pairs in length with a hairpin structure.[5] The synthesized pri-miRNA is processed in the nucleus to the shorter pre-miRNA, which is then exported out of the nucleus and into the cytoplasm, where it is further processed by the enzyme Dicer (a member of RNase III family) to a 20–24 nucleotide dsRNA that enters the RISC.[6] Binding of miRNA to RISC could rarely and paradoxically promote degradation, as shown for micro-ribonucleoproteins (microRNP) in different cell cycle phases.[7] Since miRNAs do not require a perfect match with target mRNA for silencing, they are capable of silencing a group of proteins (rather than individual proteins). miRNAs could become a therapeutic strategy after viral delivery for incorporation into the cellular genome for a long-term effect,[8] or after non-viral delivery of the final functional form of the miRNA.[9] The shRNAs are another hairpin structure that could be involved in RNAi. Transgenes encoding shRNA are delivered to a cell by expression vectors, where they could be integrated into the cell genome and lead to a stable expression of shRNA, which is transferred to the cytoplasm for silencing.[10] The loop-shaped portion of shRNAs is chopped off by Dicer, and the resulting dsRNA is bound to the RISC. Although beneficial for long-term effects, the need to design a high-expression and/or integrating system for hard-to-transfect patient (or primary) cells impedes the practical use of shRNA in a clinical setting.

Delivering synthetic siRNAs is the third RNAi strategy which essentially provides the cell with the final mediator of miRNA- or shRNA-based silencing, which is a short (20–25 pairs of nucleotides) dsRNA that matches the target mRNA, and is capable of degrading it after being incorporated into RISC.[11] The significant disadvantage of this approach is that the siRNA-mediated silencing is temporary and limited by the molecules delivered with each dose; that is until the provided siRNA is consumed or degraded. However, this approach has several advantages: (i) once in the cell, siRNA only requires an efficient RISC mechanism and therefore does not rely on the expression of a transgene and subsequent processing systems; (ii) the target for the siRNA is in the cytoplasm and uptake into nucleus is not required, which eliminates a critical barrier (nuclear envelope) in the delivery of therapeutic nucleic acids; (iii) while the fact that one miRNA could silence several proteins could be ideal for certain therapeutic purposes, siRNA (if designed correctly and used within a reasonable dose) is highly specific, which is ideal for research purposes, as well as certain therapeutic strategies involving a single target (this is also true for shRNA); (iv) while transfection with "incorporating" transgenes encoding hairpin structures creates a more permanent effect, and eliminates the need for multiple doses, it also requires selection of the transfected cells for an efficient silencing. This usually means transfecting the cells *in vitro*, selecting the transduced cells, and then transferring the cells to the animal model or patient for efficient engraftment (and even expansion), which could limit the clinical use of this technique to some extent; and (v) synthetic siRNA can be viewed as a "drug" and the biopharmaceutical industry has developed ample experience for formulation development to bring such a drug into the clinical setting. Figure 3.1 summarizes the three approaches to RNAi at a glance.

The focus of this chapter is on siRNA delivery, the study of which has displayed significant growth in the last decade to answer some of the basic scientific questions about the mechanisms involved in silencing and the potential of this approach for clinical therapy. Figure 3.2 illustrates the steady increase in the number of scientific publications on siRNA silencing, and the efforts put into place to optimize carriers for delivering these molecules in an efficacious manner. In this chapter, we examine different applications of protein silencing *via* siRNAs, review the intra- and extra-cellular barriers for effective siRNA delivery, present (and analyze) various delivery systems used to overcome these barriers, and assess some of the recent achievements in the field.

3.1.1 *In Vitro* siRNA Delivery

Figure 3.3a (data from Table 3.1) categorizes the types of siRNA delivery studies reported in the literature between 1 January 2013 and 30 November 2013. Approximately 74% of the reports were restricted to *in vitro* studies,

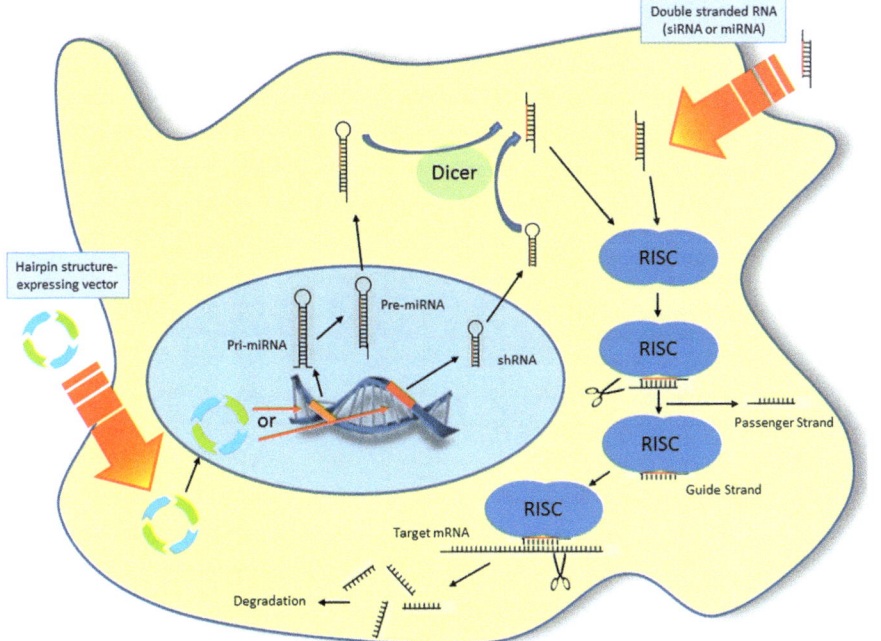

Figure 3.1 Mechanisms of RNAi. Different RNAi strategies are illustrated in a single cell. The permanent expression of hairpin structures could be accomplished by transfecting the cells with a plasmid, which will be incorporated in the chromosome, and transcribed as either shRNAs or pri-miRNAs. The transcribed hairpin structures will then exit the nucleus and are sliced by dicer to form a double stranded RNA structure, which is then incorporated into the RISC complex. After removal of the passenger strand, the remaining guide strand will identify and bind to the target mRNA, which triggers the slicing of mRNA by the Argonaute 2 enzyme in the RISC. The mRNA fragments are then degraded in the cytoplasm. The second strategy is based on delivering the double stranded RNA structures directly to the cytoplasm, which will continue on the same path described for hairpin structures. These strategies, however, are temporary, and the effect "wears off" after degradation of the delivered RNA molecules.

with an additional ∼17% undertaking an *in vivo* component. This is an indication of the importance of *in vitro* studies as well as challenges in *in vivo* studies. Delivering siRNA to different cell lines and primary cells *in vitro* has been performed for three main purposes: (i) as an investigational tool to reveal the operational mechanism behind the action of a target protein; (ii) to develop and characterize particular delivery system to enable effective silencing; and (iii) to assess the potential of a target for a clinical therapy. More often, these three distinct purposes have merged in individual scientific investigations and they become an integral (*i.e.* inter-dependent) part of the "structure–property–function" studies. Below, we provide a summary of these three operational paradigms.

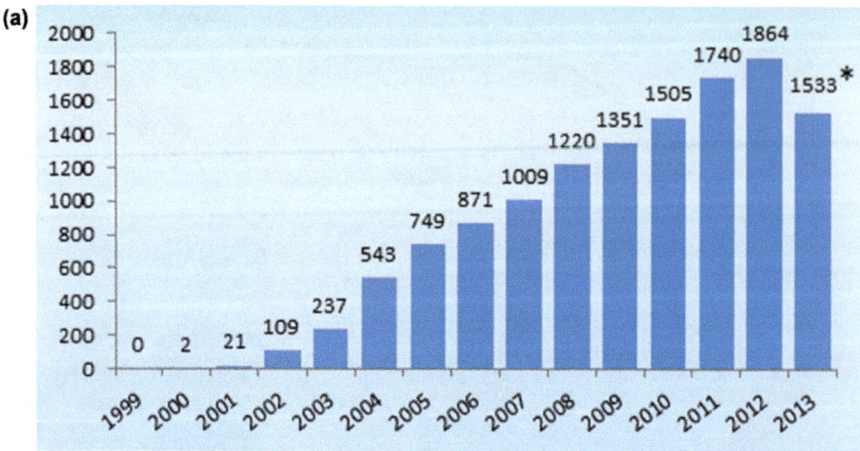

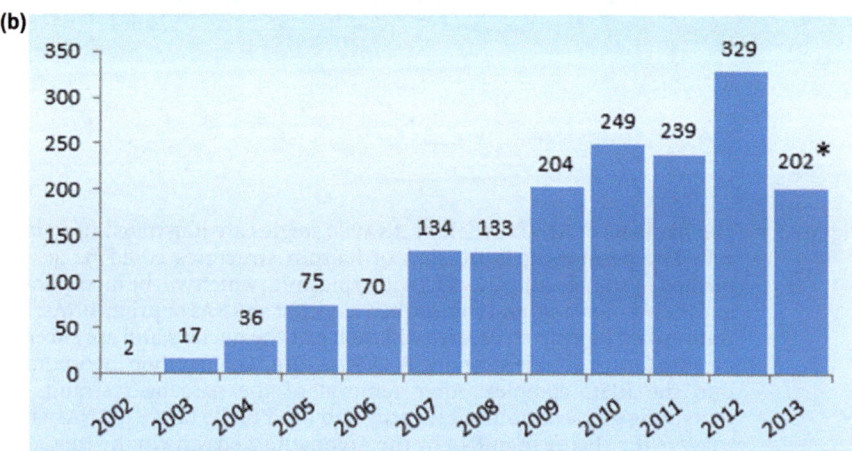

Figure 3.2 siRNA research over the years. A PubMed search with keywords "siRNA AND silencing" (a) or "siRNA AND (Delivery systems OR Carriers)" was performed on November 20th 2013, and the number of papers (including review articles and non-English language manuscripts) are separated for each calendar year. This database search did not identify any research published previous to 1999. *The last bar representing the year 2013 only demonstrates papers that entered the PubMed database before the search date.

3.1.1.1   siRNA as an Investigational Tool

In vitro studies involving silencing a protein (or a library of proteins) are not always focused on the silenced protein. Many studies are designed to investigate the effect of silencing a particular protein on the activity or the abundance of another protein, a cellular mechanism or a signalling/metabolic pathway. To re-examine a few examples: (i) it has been shown that Wnt5a plays an active role in metastasis of pancreatic cancer. In an article on the role of Wnt5a in epithelial-to-mesenchymal transition and subsequent metastasis of pancreatic adenocarcinoma cells, Bo *et al.* reported the effect of siRNA silencing of β-catenin on Wnt5a and therefore, the invasiveness in pancreatic cancer cells;[12] (ii) in a study investigating the role of SIRT1 in p53 activation and cell survival, Zhao *et al.* reported silencing a protein called deleted in breast cancer 1 (DBC1) and the resultant effect on PUMA and BAX, two major transcriptional targets of p53, which demonstrated the utility of siRNA to investigate the mechanism of activation of another protein (p53 activation through SIRT1);[13] (iii) interleukin (IL)-17 is overexpressed in some tumors but its role in tumor growth, angiogenesis, and other components of cancer biology has been a topic of debate, and connections to multiple intracellular pathways have been suggested. Using siRNAs targeting STAT3 and AKT proteins, potential connections of IL-17 to JAK2/STAT3 and/or AKT pathways have been investigated.[14] In this study, targeting STAT3 did not affect IL-17-induced AKT phosphorylation, while silencing AKT significantly reduced IL-17-induced STAT3 phosphorylation. In contrast, a siRNA targeting AKT blocked IL-17-induced up-regulation of IL-6, but STAT3 silencing did not affect the IL-6 levels. These results indicated that IL-17 mediated tumor promotion through IL-6 induction by activating the AKT pathway, and IL-6 in turn activates JAK2/STAT3 signalling;[14] and (iv) while tyrosine kinase inhibitors of epidermal growth factor receptor (EGFR) (such as gefitinib and erlotinib) are quite effective in the treatment of non-small cell lung cancer with mutations activating EGFR, resistance to drugs is often developed within 6–12 months. Harada *et al.* reported a potential role of the JAK2 pathway in the development of such a resistance, which was confirmed by a combination of erlotinib and siRNA silencing of JAK2 that restored the sensitivity to erlotinib in resistant cells.[15] This type of study inherently assumes that the siRNA delivery has been undertaken efficaciously (often validated at the mRNA level of target proteins) and, more importantly, the details of delivery (*i.e*, intracellular pharmacokinetics and pharmacodynamics of the effect, as well as the physicochemical nature of the delivery system) do not affect the observed outcomes. Based on the studies in Table 3.1 and as summarized in Figure 3.3b, ∼15% of the studies could be categorized as "investigational".

In a more comprehensive approach to elucidate physiological functions, screening siRNA libraries (*i.e.* delivering a library of siRNAs) have also been pursued to analyze intracellular events and reveal unknown mediators or inter-connections among signalling molecules. These libraries typically contain numerous siRNAs against a family of proteins (*e.g.* kinases or proteases), proteins involved in a cellular event (*e.g.* apoptosis or cell cycle), or the entire set of proteins in the genome. The studies by Izrailit *et al.* with a kinase siRNA library in a breast cancer cell line engineered to express Notch protein is an example of this strategy. Activation of Notch and over-expression of Notch ligand Jagged 1 (JAG1) have been considered as indicators of poor prognosis in breast cancer and, in this study, a kinase siRNA library was used to identify proteins involved in Notch activation. Tribbles

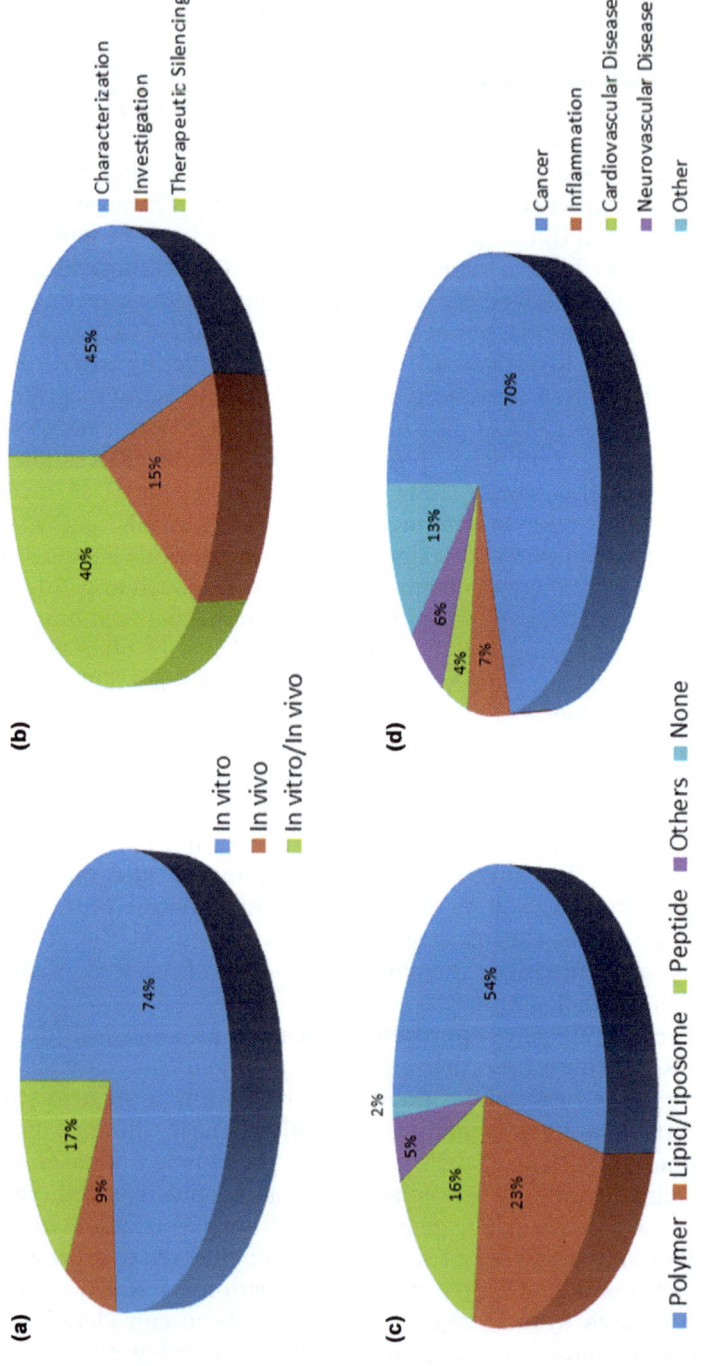

homolog 3 (TRB3), a known regulator of MAPK-ERK pathway was one of strong 'hits' in this investigation. Further experiments indicated the TRB3 to be a "master" regulator of Notch through MAPK-ERK and TGF-β pathways, to play an important role in tumor growth, and to be a potential therapeutic target in breast cancer treatment.[16]

3.1.1.2 Characterization and Development of a Delivery System

The second category of studies with siRNAs involves explorations of designing and characterizing novel carriers for siRNA delivery. In these studies, the majority of outcomes probe the chemical structure of the delivery system, the physical characteristics of the carrier, and cellular uptake and cytotoxicity of the delivery system. Silencing efficiency is usually evaluated by targeting a reporter gene, including fluorescent proteins such as green fluorescent protein (GFP)[17-19] or red fluorescent protein (RFP),[20,21] enzymes creating bioluminescence (*e.g.* firefly luciferase that creates light from oxidizing luciferin),[22-24] or a housekeeping gene such as glyceraldehyde 3-phosphate dehydrogenase (GAPDH).[25,26] While convenient to assay, such marker genes are not normally expressed in human cells, and therefore these studies employ cells that are transfected to permanently express the reporter gene and it is assumed that such a modification does not interfere with silencing efficiency. Whether silencing is effective in normal

---

Figure 3.3 Trends in siRNA research in 2013. A PubMed search with keywords "siRNA AND (Delivery systems OR Carriers)" was performed on November 30th 2013 with the time frame filter of "From 2013/01/01 to 2013/11/30". After excluding the review articles, non-English manuscripts, reports on siRNA research without actual silencing, and two papers that were not accessible to authors, the 116 remaining reports (summarized in Table 3.1) were analyzed statistically to identify different recent trends in siRNA research: (a) only ~26% of the selected literature included an *in vivo* component, which could be interpreted as an indication of the importance of the *in vitro* studies and/or the challenges that researchers face for *in vivo* experiments. No report on human studies was identified; (b) about half of the selected literature only reported silencing of a reporter protein as a tool for characterization of their proposed carrier (*in vitro* or *in vivo*), while ~40% of them reported silencing a protein with a crucial intracellular role, which could potentially be used for therapeutic purposes. The rest of the selected literature reported silencing of a protein to investigate a different cellular mechanism or pathway; (c) more than half of the selected literature used a polymer-based carrier for siRNA delivery. Lipid-based carriers and peptides were ranked second and third, respectively. Only 2% of the manuscripts reported siRNA silencing without using a carrier ("naked" chemically-modified siRNA), which is an indication of the importance of the carriers for siRNA delivery; (d) an overwhelming majority of the selected literature (~70%) categorized as "therapeutic silencing" in (b) focused on silencing a cancer-related protein, which could be an indication of the significant potential of this strategy in cancer treatment.

Table 3.1 Summary of studies performed on siRNA silencing in 2013 (up to 30 November)[a,ff]

	Delivery system/type	Type	Cell line/animal model	Target protein	Objective	Dose (nM)	Ref.
1	Silica-PEI nanoparticles/polymer	In vitro	Osteosarcoma U2OS	Luciferase	Characterization	10–100	253
2	PEI and OEI derivatives/polymer	In vitro	K562 erythroleukemic cells	Luciferase	Characterization	~700	23
3	Branched histidine/peptide	In vitro	MDA-MB-435 cells	Luciferase	Characterization	~140	24
4	Lipofectamine® 2000/liposome	siRNA screen	HeLa cells	Rab GTPases	Investigation[b]	10	264
5	PDMAEMA-PCL/polymer	In vitro	HEK 293T and SKNO-1 cells	Luciferase	Characterization	500	265
6	CA-NC VLPs/peptide	In vitro	HEK 293T cells	Aldoketoreductase	Characterization[c]	50	266
7	P[MAA-co-CMA]&P(DMAEMA-co-CMA)/polymer	In vitro	Neuroblastoma SHEP cells	GFP	Characterization	50	17
8	Modified BSA/peptide	In vitro/in vivo	B16 lung cancer cells/C57BL/6 mice	Bcl-2	Lung cancer therapy	100/80 Kg^{-1}	267
9	LPEI(dimeric siRNA)/polymer	In vitro	MDA-MB-435 (breast cancer)	GFP	Characterization	~140	19
10	Silicon-PEI nanoporous particles/polymer	In vitro/in vivo	MDA-MB-435/FBV mice	ATM	Cancer therapy/biocompatibility	30–70/15–75 µg Kg^{-1}	254
11	Crosslinked HPD-siRNA/polymer	In vitro/in vivo	B16F10 melanoma cells/Balb/C nude mice	RFP	Characterization	50/10 µg Kg^{-1}	20
12	tHSA-polymerized siRNA/peptide	In vitro/in vivo	B16F10 melanoma cells/PC-3 bearing Balb/C nude mice	RFP/VEGF	Characterization/cancer therapy	50/1.5 mg Kg^{-1}	21
13	GAPMA-s-APMA/polymer	In vitro	Liver carcinoma (HepG2)	hTERT	Cancer therapy	~140	268
14	PFNBr/polymer	In vitro	PANC-1 Cells	PLK1	Cancer therapy	100–200	269
15	Thiolated gelatin (polysiRNA)/peptide	In vitro/in vivo	B16F10 melanoma cells/athymic nude mice	RFP	Characterization	100/40 µg/mice	239

#	Carrier		Cell/Model	Target	Application	Dose/Conc.	Ref.
16	Functionalized mesoporous silica nanoparticles/polymer	*In vitro/in vivo*	HeLa cells/Zebrafish larvae	Bcl-2/GFP	Cancer therapy/characterization	~140/0.04 μg/larvae	255
17	DOPE-PEI and DPPE-PEI/polymer	*In vitro*	c 166 cells	GFP	Characterization	100	212
18	CSVs and PCSVs viroplexes	*In vitro*	Jurkat and CEM leukemia cells	Vimentin	Characterization[d]	0.2	270
19	PEI-C2DRBD/polymer	*In vitro*	HeLa cells	GFP	Characterization	50	271
20	PF6/PF6-NP/peptide	*In vitro*	B16-F1 cells	Luciferase	Characterization	~40	272
21	Lipofectamine® 2000/liposome	*In vitro*	293T cells	DSTYK	Investigation[e]	5–100	273
22	Cationic liposomes	*In vivo*	CD 1 mice	Intersectin-1	Cancer therapy[f]	100 μg/mice	205
23	PAMAM dendrimer/polymer	*In vitro*	Multiple cell lines	Hsp27 and TCTP	Characterization	20–50	226
24	PPM/polymer	*In vitro/ex vivo*	Raw 264.7 macrophages/FVB mice colon	TNF-α	IBD Treatment	100 nM	274
25	Chitosan powder/polymer	*In vivo*	BALB/c mice	Luciferase	Characterization	60 μg/mice	44
26	Amphiphilic cyclodextrin	*In vitro*	*mHypoE N41* cells	Luciferase	Characterization	50	275
27	KPAE/polymer	*In vitro*	Raw 264.7 macrophages	TNF-α	Inflammation	Not specified	276
28	Cationic lipids	*In vitro*	CHO cells/HT29 cells	GFP/Luciferase	Characterization	100	199
29	Oligofectamine®/lipid	*In vitro*	K562 erythroleukemic cells	$U2AF^{65}$ and Jmjd6	Investigation[g]	10–100	277
30	Library of dynamic amphiphiles	*In vitro*	HeLa cells	GFP & GAPDH	Characterization	~70	26
31	FA-HP-β-CD-PEI/polymer	*In vitro/in vivo*	HeLa cells/nude mice	VEGF	Cancer therapy	100/100–300 mg Kg^{-1}	278
32	F127-CaP/polymer	*In vitro*	H1299 lung cancer cells	GFP	Characterization	~56	279
33	β-CD-P(HMA-co-DMAEMA-co-TMAEMA)/polymer	*In vitro*	UM-SCC-17B cancer cells	Bcl-2	Cancer therapy[h]	100	280
34	PEI-LA/polymer	*In vitro/in vivo*	MDA-MB-435 cells/nude mice	Mcl-1/RPS6KA5	Cancer therapy	18–54/1.5–10 μg/mice	29
35	Lipid modified PEI/polymer	*In vitro*	MDA-MB-435	STAT3	Cancer therapy	9–72	281

Table 3.1 (Continued)

	Delivery system/type	Type	Cell line/animal model	Target protein	Objective	Dose (nM)	Ref.
36	PEI-PA/polymer	In vitro	K562 erythroleukemic cells	BCR-ABL	Cancer therapy	100	138
37	X-tremeGENE/lipid	In vitro	HEK293T cells	Stx5	Investigation[i]	~1400	282
38	H5-S4$_{13}$-PV/peptide DOTAP:DOPE/liposome	In vitro	HT1080, A549, and HeLa cells	GFP/Survivin	Characterization/ cancer therapy	50	204
39	F127 poloxamer/polymer	In vivo	C57BL/6 mice	Smad3	Wound healing	10 000/mice	283
40	PEI/Lap or PEI/DS on PrS/ CaP/PrS/DSL LBL film	In vitro	NIH-3T3, MDA-MB-435, and M4A4 cells	GFP	Characterization	0.5×0.5 cm^{j}	263
41	Gu-PAMAM-MWCNTs/ polymer	In vitro	A549 lung cells	PLK1	Cancer therapy	80	246
42	DOTAP-DSPE-PEG/lipid	In vitro/ in vivo	A549 lung cells/ athymic nu/nu mice	MRP1/Bcl-2	Cancer therapy	170/170 µg Kg^{-1}	284
43	Ac-X3-gT/peptide	In vitro	Hepatocellular Huh7 cells	GAPDH	Characterization	~280	285
44	PEG-PCL-ss-PDMAEMA/ polymer	In vitro/ in vivo	HeLa cells/BALB/C nude mice	Luciferase/PLK1	Characterization/ cancer therapy	1 µg/well /2 mg Kg^{-1}	286
45	PGS/polymer	In vitro	HeLa cells	NFκB p65	Cancer therapy	100	287
46	TPP/OTMC/PVBLG-8/OPM/ OPC/polymer	In vitro/ in vivo	RAW 264.7 cells/ C57BL/6 mice	TNF-α	Control systemic inflammation	~10/0.2 mg Kg^{-1}	288
47	Library of CPPs/peptide	In vitro	SKNO-1 cells	Luciferase	Characterization	50–200	235
48	PCL-ss-p(GMA-TEPA)/polymer	In vitro	HeLa cells	GAPDH	Characterization	100	289
49	Chitosan-TPP/polymer	In vitro	Jurkat/MCF7	HSP70	Cancer therapy	50–200	220
50	EGDMA-DMAEMA/polymer	In vitro	Drosophila Schneider 2 cells	Luciferase	Characterization	90 µM	290
51	PAEA-GALA/polymer	In vitro	Multiple cell lines	iNOS	Lung injury	100	291
52	PHD-PPF/polymer	In vitro/ in vivo	HeLa cells/athymic nude Balb/c mice	GFP	Characterization	150/100Kg^{-1}	152
53	CT/D-D/polymer/liposome	In vitro	HEK 293 cells	GFP	Characterization^{k}	100	222

#	Carrier/type	In vitro/in vivo	Cell line / model	Target	Application	Concentration / size	Ref.
54	DOTAP/liposome	*In vivo*	Nude nu/nu mice	MRP1/Bcl-2	Cancer therapy	0.125 mg Kg^{-1}	206
55	Lipofectamine®/liposome	*In vitro*	HUVEC cells	NPR-C	Investigation^l	50	292
56	M-MSN-PEI/polymer	*In vitro*	A549 lung cells	GFP	Characterization	40	171
57	PEI-DA/polymer	*In vitro/ in vivo*	Rat heart H9C2 cells/ Sprague–Dawley rats	SHP-1	Cardiac ischemia treatment	50/10 μM/rat	213
58	PAA derivatives/polymer	*In vitro*	Caco-2 cells	Luciferase	Characterization	50	293
59	HTCC/polymer	*In vitro/ in vivo*	Caco-2 cells/LLC-bearing C57BL/6 mice	TERT	Cancer therapy	$3/0.2$ μg Kg^{-1}	294
60	PEI/LA-gelatin/polymer	*In vitro*	MDA-MB-435 cells	KSP	Cancer therapy^m	36	295
61	Lipofectamine® 2000/ liposome	*In vitro*	RAW 264.7 macrophage cells	Rab6	Investigation^n	Not specified	296
62	PEI-PPHPG/polymer	*In vitro*	Recombinant HT-1080 cells	Luciferase/GFP	Characterization	~70	214
63	PEI-LA/polymer	*In vitro*	MDA-MB-435/MDA-MB-231	Mcl-1/P-gp	Cancer therapy	18–54	28
64	PAMAM-PEG-CPP/polymer	*In vitro/ in vivo*	Rat cardiomyocytes/ Sprague–Dawley rats	AT1R	Cardiovascular disease therapy	$50/5$ μg Kg^{-1}	231
65	CPLA Thiolene/polymer	*In vitro*	PANC-1 pancreatic cells	k-Ras	Cancer therapy	200	297
66	THP/polymer	*In vitro*	HEK293, HeLa and CHO cells	GFP	Characterization	~60	298
67	MP-g-OEI/polymer	*In vitro*	CT26 and Huh-7 cells	Luciferase	Characterization	~70	299
68	PLL, PLH, and PLA/peptides	*In vitro*	OHS and SK-MEL-28 cells	S100A4/MEK1/ MEK2	Characterization	50	173
69	Lipofectamine RNAiMAX®/ lipid	*In vitro*	C3H10T1/2 cells	Notch1	Investigation^o	20	300
70	PEG-PLL-PLLeu/peptide	*In vitro/ in vivo*	MCF7 breast cancer cells/BALB/c nude mice	Bcl-2	Cancer therapy	$50–150/0.2$ mg Kg^{-1}	301
71	Vitamin E TPGS/polymer	*In vitro*	NIH3T3/MCF7/SK-BR-3 cells	PLK1	Cancer therapy	125	262
72	Chitosan/polymer	*In vitro*	Rat RBE4 cells	P-glycoprotein	Investigation^p	50–300	221

Table 3.1 (*Continued*)

	Delivery system/type	Type	Cell line/animal model	Target protein	Objective	Dose (nM)	Ref.
73	None (modified siRNA)	*In vivo*	Sprague–Dawley rats	ASPP/PUMA/Fas	Neuron survival	10 µg/rat	302
74	GeneTrans® II/lipid	*In vitro*	HUVEC cells	Scribble	Investigation[q]	Not specified	303
75	P[(HPMA-co-PDSMA)-b-(PAA-co-DMAEMA-co-BMA)]/polymer	*In vitro*	HeLa cells	GAPDH	Characterization	12.5–50	304
76	Linear and branched peptides	*In vitro*	Neuro-2A-Luc cells	Luciferase	Characterization	50–100	236
77	Oligofectamine® lipid	*In vitro*	A549 lung cells	TFRC	Investigation[r]	30	115
78	P(BPD)-P(D)-P(Az)-man/polymer	*In vitro*	THP-1/MDA-MB-231/BMDM	PPIB	Characterization	50	153
79	CAP-dopa-Chi/polymer	*In vitro*	MDA-MB-435 cells	GFP	Characterization	~70	305
80	PEGMA-PDMA-(PDMA-POMA)	*In vitro*	HeLa cells	GFP	Characterization	80	306
81	TTMC-TPP/polymer	*In vitro*	Raw 264.7 macrophages	IL-6	Characterization[s]	~80	307
82	PEI-g-THA/polymer	*In vitro*	CT26 cells	Luciferase	Characterization	~70	215
83	DOPC-TX100/lipid	*In vitro*	U87 glioblastoma cells	Luciferase	Characterization	~15	308
84	MSN-PEI-PEG/polymer	*In vitro*/in vivo	Multidrug resistant MCF7 breast cancer cells	P-glycoprotein	Cancer therapy (overcome MDR)	1–70/1.2 mg Kg^{-1}	309
85	DOGS-DOPE-C/lipid	*In vitro*	MAGI-CCR5 cells	CXCR4	Control HIV1	100	131
86	Cholesteryl diamine/lipid	*In vitro*	FL-SiHa cells	Luciferase	Characterization	50	310
87	P19-YSA/peptide	*In vitro*	SKOV3 ovary cancer cells	RFP	Characterization	200	311
88	CS-TPP-HA/polymer	*In vitro*	A549 lung cells	Luciferase	Characterization	~150	259
89	Transfection reagent[t]	*In vitro*	INS1 pancreatic β cells	c-Kit & Fas	Investigation[u]	Not specified	312
90	Optifect/lipid	*In vitro*	OP9 adipocyte cells	HCA1	Investigation[v]	30	313

No.	Carrier/material	In vitro/in vivo	Cell line/animal model	Target gene	Application	Dose	Ref.
91	Gal-functionalized PEG-b-PCL and PEG-PPEEA micelles/polymer	In vitro/in vivo	Primary hepatocytes (mice)/BALB/c mice	Apolipoprotein B	Treatment of high cholesterol	400/4–12 mg Kg^{-1}	314
92	Oligofectamine®/lipid	In vitro	HeLa cells	PKD1 & PKD2	Investigation^w	Not specified	315
93	PKAA/polymer	In vitro/in vivo	Raw 264.7 macrophages/ICR mice	TNF-α	Acute liver failure	~90/35 µg Kg^{-1}	175
94	STR-PEG-LNP/liposome	In vitro	HeLa cells	GAPDH	Characterization	30	161
95	Oligofectamine®/lipid	In vitro	HaCaT cells	P63	Investigation^x	100	316
96	RNAiMAX®/lipid	In vitro	Hec1A, HCT116, and SW48 cells	K-Ras	Investigation^y	300	317
97	Library of PBAEs/polymer	In vitro	GB 319 cells	GFP	Characterization	60	318
98	CG-DOPE-Ch-DC-DSPE-PEG/lipid	In vitro	A549 lung cells	Bcl-2	Cancer therapy	~70	319
99	PEG-PLGA-PLL-RGD/polymer	In vitro	PC-3 prostate cancer cells	VEGF	Cancer therapy	NA^z	320
100	Lipid-modified β-cyclodextrin	In vitro/in vivo	Rat ST14A and human GM04691 cells/R6/2 mice	Huntingtin Proein	Huntington's disease therapy	100/2.5 µg/mice	321
101	Capsomer-transferrin/peptide	In vitro	HeLa cells	Bcl-2	Cancer therapy	5–25	322
102	Atelocollagen/polymer	In vivo	C.B17 SCID/SCID mice	NEDD1	Cancer therapy	50 µg/mice	323
103	PEG-PEI(25000 Dalton)/polymer	In vitro	C17.2 neural stem cells	ROCK II	Alzheimer's	100	324
104	[Mg$_n$Al(OH)$_{2n+2}$]$^+$(A$^-$)·mH$_2$O	In vitro	HEK 293T cells	DCC	Characterization^{aa}	250	325
105	PTDMs/mimic peptides	In vivo	Jurkat T cells	NOTCH1	Cancer therapy	80–100	326
106	Oligofectamine®/lipid	In vitro	HeLa	COG6	Investigation^{bb}	Not specified	327
107	PAAC(Agm or Gal)/polymer	In vitro	MDA-MB-231 cells	Luciferase	Characterization	~20–180	328
108	SWNT-PEI-NGR/polymer	In vitro/in vivo	PC-3 prostate cancer cells/BALB/c nude mice	hTERT	Cancer therapy	150/1 mg Kg^{-1}	247
109	Exosomes	In vitro	HCC70 cells	Luciferase/EGFR	Characterization^{cc}	50	251

Table 3.1 (Continued)

	Delivery system/type	Type	Cell line/animal model	Target protein	Objective	Dose (nM)	Ref.
110	Dextran-C6 lipid-PEG/polymer	In vitro	KHOS osteosarcoma cells	P-glycoprotein	Overcome MDR	100	329
111	MO-OA-PEI/polymer	In vivo	HRS/J hairless mice	GAPDH	Characterization	10 µM/mice	216
112	TAT-A1/peptide	In vitro	HUVEC cells	GAPDH	Characterization	50	330
113	None (modified siRNA)	In vivo	Lewis rats	Interleukin – 23	EAE therapy	50 µg(×9)/rat	331
114	DPPE-Ch-HA/lipid	In vitro	A549 and NAR cells	Luciferase/P-gp	Characterization[dd]	300	196
115	RVG/peptide	In vivo	CAFI mice	Calpain 2	Investigation[ee]	50 µg/mice	238
116	Chimeric capsid/peptide	In vivo	Balb/c nude mice	RFP	Characterization	33 µg/mice	332

[a]The studies were selected from 214 papers retrieved by a PubMed search on 30 November 2013, by the following criteria. Keywords: "siRNA" AND ("Delivery systems" OR "Carriers"); filters: "from 2013/01/01 to 2013/11/30". The following types of studies were excluded: (i) manuscripts in a non-English language; (ii) manuscripts reporting siRNA studies without any actual silencing component; (iii) review papers (due to lack of specific experimental data); and (iv) two manuscripts that were not accessible to us through the University of Alberta library, and the required data were not included in the abstract. These two papers were titled: *Developing lipid nanoparticle-based siRNA therapeutics for hepatocellular carcinoma using an integrated approach and Fabrication of 14 different RNA nanoparticles for specific tumor targeting without accumulation in normal organs.*

[b]To identify the main Rab (family of G-proteins) involved in Herpes simplex virus 1 infection.

[c]The authors' only justification provided in the manuscript is "… because silencing of its expression is well established in our laboratory". However, involvement of this protein in cell proliferation and apoptosis has been reported in other publications.

[d]Although vimentin is a member of intermediate filament family and plays a significant role in supporting and anchoring organelles in the cytosol, and also a sarcoma tumor marker, it seems that in this project it was only used as a proof-of-principle model protein.

[e]To investigate the role of mutations in DSTYK gene in congenital abnormalities of the kidney or urinary tract, and its connections with fibroblast growth factor (FGF) signaling. The results show that DSTYK is downstream of FGF signaling.

[f]Intersectin-1 (ITSN-1) is shown to regulate vascular permeability and endothelial cells (ECs) survival, with a role in neuroblastoma tumorigenesis, silencing of which has induced apoptotic cell death. In this study, however, authors investigated the role of ITSN-1s in the ECs function and lung homeostasis in healthy mice using electron microscopy.

[g]U2AF65 and Jmjd6 silencing created a similar effect on ferrochelatase (FECH) splicing as iron deficiency, which indicates the involvement of these splicing factors in modulation of FECH splicing, and therefore, erythropoietic protoporphyria.

[h]Bcl-2 silencing was also studied in combination with AT-101 (a Bcl-2 small molecule inhibitor).

[i]Syntaxin 5 is a protein involved in intracellular vesicle trafficking and an interaction partner of the very low density lipoprotein (VLDL) receptor. The effect of Syntaxin 5 silencing on the VLDL-receptor was investigated in this study.

[j]A 0.5 ×0.5 cm of the siRNA-coated dressing was introduced to each well with the cells.

[k]This study was designed to compare chitosan-based nanoparticles to DOTAP/DOPE cationic liposomes in their efficiency in delivering plasmids and siRNA to the selected cell line. The results indicated that cationic liposomes were more efficient (∼70% vs. ∼57%) in delivering siRNA.

[l]C2238 atrial natriuretic peptide (ANP) minor allele (substitution of thymidine with cytosine in position 2238) is often associated with cardiovascular events. This study showed that silencing NPR-C would rescue the cells from apoptosis induced by C2238-αANP, which indicates that C2238-αANP reduces endothelial cell survival and impairs endothelial function through NPR-C signaling.

[m]While kinesin spindle protein (KSP) has been studied as a potential target for siRNA silencing in curbing tumor cell growth with considerable success, this study was performed mainly to evaluate the role of gelatin in stabilizing siRNA complexes, and therefore could be categorized as a "characterization" study.

[n]This study was designed to investigate the pathways involved in TNF secretion (a pro-inflammatory cytokine). Silencing Rab6 *via* siRNA reduced the delivery of TNF to the cell surface and subsequent release into the media, which confirms the role of Rab6 in TNF transport.

[o]This study was designed to investigate the effect of silk proteins on osteoblast differentiation. Silencing Notch1 signalling induced alkaline phosphatase (ALP) activity and the authors concluded that silk proteins stimulate osteoblast differentiation of mesenchymal progenitor cells by regulating the Notch signaling pathway.

[p]While P-glycoprotein (P-gp) is a well-known membrane protein involved in development of multidrug resistance in cancer cells, this study was designed to investigate the effect of P-gp silencing on the drug delivery across blood–brain barrier (where P-gp is overexpressed).

[q]This study investigates the function of polarity proteins in mesenchymal cells. Silencing of scribble *via* siRNA blocked directed but not random migration of human umbilical vein endothelial cells. This study indicates that scribble is a novel regulator of integrin α5 turnover and sorting, which is required for oriented cell migration and sprouting angiogenesis.

[r]Transferrin is one of the targeting moieties used for active targeting. This study speculates that transferrin could be shielded in the biological environment, which could result in loss of specificity in targeting. Transferrin receptor was silenced to evaluate this hypothesis.

[s]Interleukin 6 (IL-6) is a multifunctional cytokine that plays important roles in host defense, acute phase reactions, immune response, nerve cell function, and hematopoiesis. However, since the manuscript does not elaborate on any specific reason for selecting this target, we assumed that IL-6 was used as a reporter gene (*in vivo* experiments were also performed with TAMRA-labeled siRNA for evaluating the biodistribution with no silencing data).

[t]The only data available in manuscript reads: "sc-270241; Santa Cruz Biotechnology" siRNA transfection reagent. The company's website offers no information about this carrier.

[u]c-Kit is shown to be important for β-cell survival and maturation, and interactions between the Fas receptor and ligand are capable of triggering β-cell apoptosis. This study investigates the interrelationship between c-Kit and Fas activation that mediates β-cell survival and function. While down-regulation of Fas by siRNA silencing might have therapeutic value in prevention of β-cell apoptosis in onset of diabetes, this study undertakes a pure investigational approach.

[v]The hydroxy-carboxylic acid receptor (HCA1) is a G protein-coupled receptor highly expressed on adipocytes and naturally occurring missense mutation in this receptor could impact lipid disposition. In this study, siRNA silencing of HCA1 resulted in an increase in lipid accumulation.

[w]PKD is known to be involved in Golgi-to-cell-surface transport by controlling the biogenesis of specific transport carriers. In this paper the effect of silencing PKD vi siRNA on cell cycle, Raf-1 and mitogen-activated protein kinase kinase (MEK) activation, and cleavage of the non-compact zones of Golgi membranes in G2 phase is studied.

Table 3.1 (*Continued*)

[x]Clouston syndrome is a rare autosomal dominant condition and is caused by mutations in GJB6 gene, which encodes a gap junction protein connexin 30 (Cx30). P63 protein was knocked-down by siRNA to investigate a potential connection with GJB6. In fact the GJB6-transcript was also significantly decreased after siRNA silencing of p63, which indicates that GJB6 might be a potential target gene for p63.

[y]This study was designed to investigate the impact of mutation in K-Ras signaling and biological properties of a few isogenic cell lines. The authors found that single copy mutant K-Ras causes surprisingly modest activation of downstream signaling to ERK and Akt. Also, knock-out of the wild type K-ras allele consistently increased growth in soft agar.

[z]Unfortunately, the full text of this paper was not available to the authors. Other information presented was extracted from the abstract.

[aa]DCC is known as a tumor suppressor gene, and silencing this protein is not beneficial in cancer treatment. In this study the authors examined both transfecting the cells to express the protein and silencing the expression as a characterization process for the carrier.

[bb]The conserved oligomeric Golgi (COG) complex has been specifically implicated in the tethering of retrograde intra-Golgi vesicles, and the COG6 subunit of the COG complex is capable of interacting with a subset of Golgi SNAREs. In this study, siRNA silencing of COG6 disrupted the integrity of Golgi complex. These results show that COG6–SNARE interactions are important for both COG6 localization and Golgi integrity.

[cc]This study was designed to investigate siRNA and miRNA delivery to breast cancer cells. The siRNA was only used for luciferase knockdown *in vitro* as a characterization test, and the *in vivo* studies were performed with miRNA (which is not included in the table).

[dd]While P-glycoprotein (P-gp) is a well-known membrane transporter involved in MDR, it was used as a reporter gene to characterize the silencing efficiency of the carrier.

[ee]Calpain 1 and 2 are two isoforms of calpain highly expressed in central nervous system. This study was designed to investigate the role of calpain 2 in hippocampal synaptic plasticity and in learning and memory. Calpain2 gene silencing eliminated long-term potentiation and impaired learning and memory.

[ff]Abbreviations: **Agm:** Agmatine; **AT1R:** angiotensin II type 1 receptor; **ATM:** ataxia telangiectasia mutated; **Bcl-2:** B-cell lymphoma 2; **BCR-ABL:** breakpoint cluster region – Abelson; **β-CD-P(HMA-co-DMAEMA-co-TMAEMA):** β-cyclodextrin – poly(hydrophobic hexyl methacrylate-co-2-(dimethylamino)ethyl methacrylate – co-N,N,N-trimethylaminoethyl methacrylate iodide); **BMDM:** bone marrow-derived macrophages; **BSA:** bovine serum albumin; **C2DRBD:** immunoglobulin G (IgG) Fc binding domain (C2)-double stranded RNA binding domain; **CA:** capsid; **CAP-dopa-Chi:** calcium phosphate-3,4-dihydroxy-L-phenylalanine-chitosan; **CG-DOPE-Ch-DC-DSPE-PEG:** cholesteryl oleate/glycerol trioleate-(1,2-dioleoyl-*sn*-glycero-3-phosphoethanolamine)-cholesterol-(3β-[N-(N',N'-dimethylaminoethane)-carbamoyl]-cholesterol hydrochloride)-(1,2-distearoyl-sn-glycero-3-phosphoethanolamine)-poly(ethylene glycol); **COG:** conserved oligomeric Golgi; **CPPs:** cell penetrating peptides; **CPLA:** cationic poly(lactic acid); **CSV:** cationic Sendai virosomes; **CT:** chitosan – tripolyphosphate (TPP); **DA:** deoxycholic acid; **DCC:** deleted in colorectal carcinoma; **D-D:** DOTAP – DOPE; **DOGS-DOPE-C:** dioctadecylamidoglycylspermine-DOPE-cardiolipine; **DOPC-TX100:** 1,2-dioleoylsn-glycero-3-phosphocholine – Triton X-100; **DOPE:** 1,2-dioleoyl-*sn*-glycero-3-phosphoethanolamine; **DOTAP:** 1,2-dioleoyl-3-trimethylammonium-propane; **DPPE:** 1,2-dipalmitoyl-*sn*-glycero-3-phosphoethanolamine; **DPPE-Ch-HA:** 1,2-dipalmitoyl-*sn*-glycero-3-phosphoethanolamine-cholesterol-hyaluronan; **DSPE:** 1,2-distearoyl-*sn*-glycero-3-phosphoethanolamine; **DSTYK:** dual serine–threonine and tyrosine protein kinase; **EAE:** experimental autoimmune encephalomyelitis; **EGDMA-DMAEMA:** ethylene glycol dimethacrylate-2-(dimethylamino)ethyl methacrylate; **F127-CaP:** pluronic F127/calcium phosphate hybrid nanoparticles; **FA-HP-β-CD-PEI:** folic acid-2-hydroxypropyl-β-cyclodextrin – polyethyleneimine; **Gal:** galactosamine; **GAPDH:** glyceraldehyde 3-phosphate dehydrogenase; **GAPMA-s-APMA:**

guanidinylated 3-gluconamidopropyl methacrylamide-s-3-aminopropyl methacrylamide copolymers; **GFP:** green fluorescence protein; **Gu-PAMAM-MWCNTs:** guanidinium – polyamidoamine – multi-walled carbon nanotubes; **H5-S4₁₃-PV:** S413-PV cell penetrating peptide modified with 5 histidines; **HCA1:** hydroxy-carboxylic acid receptor 1; **HPD:** hyaluronic acid-*graft*-poly(dimethylaminoethyl methacrylate); **HSP70:** heat shock protein 70; **HTCC:** *N*-((2-hydroxy-3-trimethyl-lammonium) propyl] chitosan chloride; **hTERT:** human telomerase reverse transcriptase; **HUVEC:** human umbilical vein endothelial cell; **IBD:** inflammatory bowel disease; **IL-6:** interleukin-6; **Jmjd6:** 2-oxoglutarate-dependent dioxygenase Jumonji domain-containing protein 6; **KPAE:** ketal containing poly(β-amino ester); **k-Ras:** V-Ki-ras2 Kirsten rat sarcoma viral oncogene homolog; **KSP:** Kinesin spindle protein; **LPEI:** linear polyethyleneimine; **Mcl-1:** myeloid leukemia cell differentiation protein; **M-MSN:** magnetic mesoporous; **MO-OA-PEI:** monoolein-oleic acid-polyethyleneimine; **MP-g-OEI:** poly(aspartate-graft-oligoethylenimine); **MRP1:** multidrug resistance protein 1; **MSN:** mesoporous silica nanoparticles; **NC:** nucleocapsid; **NEDD1:** neural precursor cell expressed, developmentally down-regulated 1; **NFκB:** Nuclear factor kappa-light-chain-enhancer of activated B-cells; **NPR-C:** natriuretic peptide receptor-C; **OEI:** oligoethyleneimine; **PAAC:** poly(acrylic acid); **PAA:** poly(allylamine); **PAEA-GALA:** poly(acrylamidoethylamine)-*b*-poly(styrene) – GALA peptide; **PAMAM:** polyamidoamine; **PBAE:** poly(beta-amino esters); **P(BPD)-P(D)-P(Az)-Man:** poly(butyl methacrylate-*co-2*-propylacrylic acid-*co-2*-dimethylaminoethyl methacrylate)-poly(dimethylaminoethyl metha-crylate)-poly(2-azidoethyl methacrylate)-mannose; **PCL:** poly(caprolactone); **PCL-ss-p(GMA-TEPA):** poly(caprolactone) – disulfide bond – poly(glycidyl methacry-late-tetraethylenepentamine); **PCSV:** protamine sulfate (PS)-condensed cationic Sendai virosomes; **PDMAEMA:** poly((dimethylamino)ethylene methacrylate); **P(DMAEMA-co-CMA):** poly(dimethylamino ethyl methacrylate-cocholesteryl methacrylate); **PEG:** poly(ethylene glycol); **PEGMA-*b*-PDMA-*b*-(PDMA-*co*-POMA):** poly[poly(ethylene glycol) methyl ether methacrylate]-blockpoly[2-dimethylaminoethyl methacrylate]-*block*-poly[2-dimethylaminoethyl methacrylate-*co*-octyl methacrylate]; **PEG-PLL-PLLeu:** poly[(ethylene glycol)-*b*-poly(L-lysine)-bpoly(L-leucine); **PEI:** polyethyleneimine; **PEI/DS:** layers of polyethyleneimine and dextran sulfate; **PEI-g-THA:** polyethyleneimine – grafted – tris(hydroxymethyl) acrylamidomethane; **PEI-LA:** polyethyleneimine – linoleic acid; **PEI/Lap:** layers of poly-ethyleneimine and laponite silicate; **PEI-PA:** polyethyleneimine – palmitic acid; **PF6/PF6-NP:** PepFect6/DOPE-covered PepFect6/siRNA nanoparticles; **PFNBr:** cationic conjugated polyelectrolyte brushes; **P-gp:** P-glycoprotein; **PGS:** poly(L-glutamic acid) – spermine; **PHD-PPE:** PEI-hydrazone-doxorubicin – PEI-PEG-folate; **PI[HPMA-*co*-PDSMA)-*b*-(PAA-*co*-DMAEMA-*co*-BMA)]:** poly[*N*-(2-hydroxypropyl)methacrylamide-*co-N*-(2-(pyridin-2-yldisulfanyl)ethyl)methacrylamide]-*b*-propy-lacrylic acid-*co*-dimethylaminoethyl methacrylate-*co*-butyl methacrylate; **PKAA:** poly(ketal amidoamine); **PKD:** protein kinase D; **PLA:** poly-L-arginine; **PLH:** poly-L-histidine; **PLK1:** polo-like kinase 1; **PLL:** poly-L-lysine; **P(MAA-co-CMA):** poly(methacrylic acid-*co*-cholesteryl methacrylate; **PPHPG:** P[PEG-histidine-PEG-glutamic acid]; **PPM:** poly(cystamine bisacrylamide-branched PEI) – polyethylene glycol – mannose (p(CBA-bPEI)-PEG-Man); **Prs/CaP/PrS/DSL LBL Film:** protamine sulfate/ calcium phosphate/protamine sufate/dextrose sulfate light layer-by-layer film; **PTDM:** protein transduction domain mimics; **RFP:** red fluorescence protein; **RPS6KA5:** ribosomal protein S6 kinase polypeptide 5; **RVG:** rabies virus glycoprotein-chimeric; **SHP-1:** Src homology phosphatase-1; **Smad3:** mothers against decapentaplegic homolog 3; **STAT3:** signal transducer and activator of transcription 3; **STR-PEG-LNP:** strophanthidin-poly(ethylene glycol)-lipid nanoparticles; **Stx5:** syntaxin 5; **TERT:** telomerase reverse transcriptase; **TFRC:** transferrin receptor; **THP:** 6-(*N*,*N*,*N′*,*N′*-tetramethylguanidinium chloride)-hexanoyl-poly-ethylenimine; **tHSA:** thiolated human serum albumin; **TNF-α:** tumor necrosis factor alpha; **TPGS:** D-α-tocopheryl polyethylene glycol succinate; **TPP:** tripoly-phosphate; **TPP/OTMC/PVBLG-8/OPM/OPC:** sodium tripolyphosphate – oleyl trimethyl chitosan – poly(g-(4-(((2-(piperidin-1-yl)ethyl)-amino)methyl)benzyl-L-glutamate) – oleyl-PEGmannose – oleyl-PEG-cysteamine; **TTMC-TPP:** thiolated trimethyl chitosan-tripolyphosphate; **VEGF:** vascular endothelial growth factor; **VLPs:** virus-like particles.

(unmodified and reporter gene-missing) cells and whether silencing of physiologically relevant targets are undertaken with similar efficiency are inherent unknowns in such studies. An inspection of Table 3.1 and Figure 3.3a suggests that a majority of studies ($\sim$45%) up to 2013 were undertaken to characterize siRNA delivery systems.

3.1.1.3   Evaluating Silencing Potential of Specific Targets

The focus of these investigations is on silencing a specific protein (or a library of proteins) in order to study the potential of the resultant effects for therapy. Silencing the targeted protein (unlike the first category where silencing was mainly undertaken to study a different protein) is pursued with a more practical approach to explore the potential of siRNA silencing as a therapeutic strategy. These studies are often performed to confirm the potency of silencing and its effect on the targeted disorder as a prelude to preclinical *in vivo* (animal) models and eventually clinical trials. Approximately 40% of the studies in 2013 (Figure 3.3b) were dedicated to this type of study, where a specific target was silenced for a therapeutic purpose. Categorizing these studies based on therapeutic indication (Figure 3.3d), $\sim$70% of the studies were directed towards treatment of different types of cancer or overcoming multidrug resistance (MDR). An overview of the research in this field (PubMed search with "siRNA" and "silencing" as keywords with no time limitations; Figure 3.4) shows that almost half of the

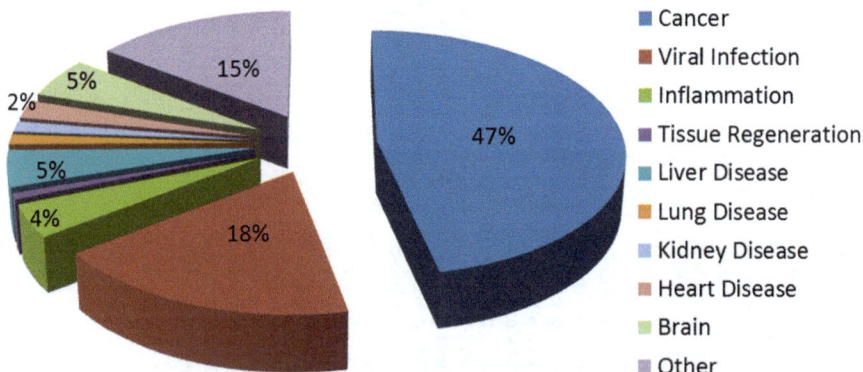

Figure 3.4   Historical therapeutic trends in siRNA research. The literature retrieved by PubMed search "siRNA AND silencing" (presented in Figure 3.2a) were further categorized based on their therapeutic interest to provide a more general picture of the practical aspect of siRNA-based research. The general search was categorized based on the following additional keywords: "cancer OR metastasis OR tumor OR leukemia" for Cancer, "virus OR viral OR viral infection" for Viral Infection, "liver disease OR hepatitis" for Liver Disease, "lung disease NOT lung cancer" for Lung disease, "heart OR heart disease" for Heart Disease, "CNS OR brain" for Brain, and "inflammation", "tissue regeneration" and "kidney" for respective categories. The dominance of cancer-related siRNA research was observed in this analysis as well.

studies in the literature were devoted to cancer treatment, in parallel with the dominant role of this disease in clinical therapy. Our own work has identified several targets that could be silenced to retard breast cancer cell line growth *in vitro*, as well as identifying synergistic combinations that were very effective in controlling cell growth both *in vitro* and *in vivo* models of breast cancer.[25,27–30]

Cancer treatment *via* siRNA silencing has taken different directions, specifically to prevent invasion and/or metastasis, to impede tumor growth, to trigger apoptosis, or to overcome MDR to chemotherapeutic agents (for the latter, siRNA is to be used as a part of a therapeutic regiment along with anticancer drugs). However, these apparently separate phenomena significantly overlap in many cases. Metastasis is the leading cause of death by cancer,[31] and even though it is basically an *"in vivo"* phenomenon (*i.e.* execution of its full mechanism of action requires many factors that are hard to replicate in *in vitro* models), many *in vitro* studies focused on studying different strategies in controlling identified mechanism(s) involved. Receptor of activated protein kinase C1 (RACK1) is an anchoring or adaptor protein, expression of which was shown to be correlated with metastasis and poor clinical outcome. *In vitro* RACK1 siRNA silencing showed a significant inhibitory effect on cell adhesion, migration and invasion in human oral squamous cell carcinoma HSC-3 and CAL-27 cell lines,[32] which could be a promising sign for later efforts to prevent metastasis *in vivo*. In another study on non-small cell lung carcinoma (NSCLC) cell lines, as well as primary NSCLC cells, SPARC/osteonectin, cwcv and kazal-like domains proteogly-can 1 (SPOCK1; a protein shown to regulate epithelial-to-mesenchymal transition that is believed to play a key role in metastasis) were expressed significantly higher in lung cancer cells, and in metastasizing tumor tissue compared to non-metastasizing tumor tissue.[31] In this study, *In vitro* SPOCK1 silencing inhibited lung cancer cell growth, colony formation and invasion. Angiogenesis (formation of new vasculature to supply rapidly growing tumors) is another key factor for tumor growth and a possible path for migration of metastatic cells. Vascular endothelial growth factor (VEGF) is considered the most potent factor involved in this phenomenon,[33] and therefore an ideal target for siRNA silencing. *In vitro* studies on this protein are reported in the literature, with promising effect on cell growth; for example, silencing in RKO human colorectal carcinoma cells showed a 67% decrease in cell proliferation evaluated with WST-1 proliferation assay.[34]

Another popular approach to cancer treatment is to inhibit the intracellular proteins involved in anti-apoptotic response and mediators of MDR, or combinations of both. This approach has been studied extensively in our laboratory, and the targets we employed for *in vitro* siRNA silencing included survivin,[30] myeloid cell leukemia sequence 1 (Mcl-1),[28] signal transducer and activator of transcription 3 (STAT3),[35] P-glycoprotein (P-gp),[25,36] and breast cancer resistance protein (BCRP).[27] Needless to say, co-silencing both anti-apoptotic proteins and drug transporters such as P-gp resulted in potent therapeutic effect when both siRNA and conventional drug therapy were employed in *in vitro* models.

Despite the overwhelming volume of data generated *in vitro*, delivering siRNA effectively into the cells is not an easy task. The negative charge of the nucleotides (due to phosphate groups), their hydrophilic nature, and their size (which is too large for passive diffusion) minimize the chances of siRNA passing the cell membrane on its own. Therefore, an effective and non-toxic carrier to enhance cellular uptake is crucial for reliable and reproducible results. The characteristics and structures of the proposed delivery systems are reviewed here.

3.1.2   *In Vivo* siRNA Delivery

The siRNA delivery studies in preclinical models are the natural follow-up to promising *in vitro* studies and can translate the therapeutic potential of silencing a specific protein to a more realistic setting. These are studies that usually start in small animal models such as rodents, especially in cancer models, and in the case of positive results, might move on to larger models (if available) or a clinical setting. However, there are several additional obstacles that a siRNA molecule has to overcome in order to reach its site of action in the preclinical models. The unprotected and unmodified siRNA molecule is extremely unstable in the physiological milieu. The serum half-life of siRNA is <30 min[3] due to rapid degradation of the molecule by RNase A type nucleases, among other enzymes.[37] The relatively small size of the siRNA ($\sim$14 kDa) means that it is readily cleared from the blood circulation through kidneys by glomerular filtration,[38] which contributes to the short half-life. Even if siRNA is able to reach its target tissue, the cellular uptake is another obstacle, as explained in the previous section. However, this seemingly discouraging picture has inspired many attempts to tailor efficient carriers for siRNA delivery. We first review the barriers that siRNA has to overcome in order to reach the cytoplasm (and more accurately, the RISC complex in cytoplasm) of target cells, before reviewing the different categories of carriers designed for siRNA delivery and the most recent accomplishments in this field.

3.2   Extracellular Barriers

The extracellular barriers to be overcome *in vivo* depend mostly on the accessibility of the site of action and the route of administration of siRNA (which might be also selected based on the site of action). The routes of administration could be categorized as local (*i.e.* applying the delivery system directly to the site of action) or systemic (requires the circulatory system to reach the target). While local routes of administration offer the opportunity to maximize the accumulation of siRNA in the target tissue and minimize non-specific distribution to other tissues (which introduces the risk of off-target effects and toxicity), it is not always feasible for clinical practice. Intratumoral injection, for instance, is a straightforward method of administration of siRNAs for solid tumors;[29,39] however, this method is not practical in a clinical setting and is limited to preclinical studies (except for cases where tumors are accessible, such as skin lesions). Intravitreal injections of siRNAs targeting VEGF[40] or VEGF receptor 1 (VEGFR 1)[41] have been reported in the preclinical setting, as well as in clinical trials.[42,43] In these studies, designed to curb the uncontrolled ocular neovascularization in different types of age-related macular degeneration, siRNA was injected directly into the vitreous cavity. In another example of local administration of siRNA, Okuda *et al.* recently reported a "powder for inhalation" formulation for delivering siRNA to the lungs.[44] A powder mixture of siRNA targeting luciferase, chitosan, and manitol was prepared by supercritical CO_2, and delivered directly to the lungs of a lung metastasis model through the trachea using polyethylene tubing. The promising results of this study (silencing luciferase expression in colon26 cells metastasizing to the lung) might be the first chapter of a novel local route of administration for siRNA. Other examples of local administration of siRNAs include intravaginal administration of siRNA targeting viral UL27 and UL29 genes (encoding an envelope glycoprotein and a DNA binding protein, respectively) to battle in herpes simplex virus 2 (HSV-2) infection,[45] intra-testicular injection of human fibroblast growth factor (FGF)-4 siRNA to inhibit growth of NEC8 cells (derived from human testicular tumors),[46] subcutaneous injection near tumor sites to silence bcl-2,[47] intranasal administration of siRNA for the treatment of pulmonary conditions,[48] and intrathecal injection of siRNA that silences pain-related cation-channel P2 × 3 for the treatment of chronic neuropathic pain[49] or siRNA targeting delta opioid receptor (DOR) to treat nociception induced by the DOR selective agonist, [D-Ala(2), Glu(4)] deltorphin II (DELT).[50] Oral administration for a local effect in the gastrointestinal tract has also been reported;[51] however, depurination of nucleic acids in the low-pH stomach, as well as the endonucleases in gut lumen pose additional physiological barriers for this type of administration.[52,53]

Systemic administration of siRNAs, though generally more challenging for tissue-specific delivery, seems to be the more practical route for many disorders, including cancer, where the malignant growth is not readily accessible for direct application of the siRNA in the majority of cases. While oral administration of siRNA for systemic silencing has been reported (most recently, for instance, by a chitosan-based nanoparticle administered to C57BL/6J mice where the siRNA was detected in spleen, kidney, and liver[54]), the parenteral route remains the major route of administration for siRNA-0delivery systems. Tang *et al.* have reported efficient GFP silencing in both transient adenovirus-infected C57BL/6 mice and GFP-transgenic mice using intramuscularly injected methoxypoly(ethylene glycol) (mPEG)-modified poly(amidoamine) (PAMAM) dendrimers delivering GFP siRNA.[55] Intravenous (IV) injection seems to be the prominent route of administration, and most practical for the clinical treatment of systemic conditions. Several studies have reported higher siRNA accumulation in the liver, kidney or lungs with IV delivery, which could be utilized for efficient delivery to these organs, but also may pose significant risk of toxicities due to unwanted exposure of these vital organs. Intraperitoneal (IP) injection could an alternative to IV injection for systemic delivery, since comparable silencing efficiency to IV injection has been reported for a liposomal delivery system.[56] The IP injection could lead to a slower release into blood circulation (known as a "depot effect") compared to the immediate availability associated with IV injection,[57] which could potentially lower the risk of toxicity. While IP injection offers ease of administration, less volume restriction, and better reproducibility in preclinical animal models, it is less practical in a clinical setting. In reviewing the extracellular barriers for siRNA delivery, we focus on IV injection as the route of administration and omit the barriers in order to reach systemic blood circulation when administered *via* other routes of administration. We also focus on barriers for reaching a solid tumor as the most common application of siRNA silencing.

3.2.1 Surviving Degradation in the Systemic Circulation

The siRNA in the circulation is immediately exposed to serum nucleases.[58–60] While double-stranded nucleic acids, including siRNAs, are more stable than their single-stranded counterparts, degradation by nuclease attack eliminates free siRNA fairly quickly in the body.[61] The primary nuclease in serum is a processive 3′-exonuclease which may be ERI-1 (THEX1),[62,63] which makes the single-stranded 3′-overhangs in traditional siRNAs particularly susceptible to degradation. Nuclease attack could initiate at the 3′-overhangs and proceeds in a 3′ → 5′ direction.[64] RNase A-like activity has also been implicated in siRNA degradation in serum,[65,66] which is surprising, since RNase A cleaves at pyrimidine bases in single-stranded RNA. It is possible that transient transitions to single-stranded forms ("breathing") in short duplexes could create a suitable substrate for this class of nuclease.[61] Longer RNA duplexes appear to have slightly greater stability in serum.[67,68]

One approach to prevent, or at least delay, this almost instant enzymatic attack is modification of siRNA without jeopardizing its silencing efficiency. Nuclease stability could be achieved by phosphorothioate (PS) backbones at the 3′-end for exonuclease resistance and a 2′ modification (2′-O-methyl) for endonuclease resistance,[69] without significantly affecting the silencing capability.[70] The 2′-fluoro (2′-F) modification has been also shown to stabilize the siRNA against nucleases.[61] Incorporation of 2′-F at pyrimidine positions does not affect the siRNA activity *in vitro*[71] or *in vivo*.[58] The combined use of 2′-F modified pyrimidine with 2′-O-methyl purines can result in siRNAs that are extreme stable in serum for improved *in vivo* efficiency.[72] Modification at the 5′-end of the antisense strand is reported to decrease the silencing activity more than the modification at the 3′-end. The 2′-O-methoxyethyl modification on either strand also leads to loss of activity.[43] Grafting PEG to siRNA has also been reported for enhanced its serum stability,[73] presumably due to the steric protection of the siRNA against nucleases, as well as other molecules responsible for siRNA clearance from circulation. PEGylation was reported to increase the serum half-life of siRNA from <35 min to >30 h after adding PEG at the 5′-end of scrambled siRNA.[74] The other strategy is modifying the siRNA with cholesterol. Modifying the sense strand of siRNA with 3′ cholesterol has been reported to increase the activity of free nuclease-resistant siRNA *in vitro* and *in vivo*.[75] In another study comparing the chemical modification of siRNA and nano-particle formulations revealed that cholesterol modification increased the circulatory half-life to >30 min (compared to <5 min for unmodified dsRNAs); however, the protective effect of nanoparticles seemed to be more significant.[76]

However, the chemical modification of siRNA, while effective in delaying enzymatic degradation to some extent, may not affect other characteristics of siRNA that play a role in their rapid clearance from systemic circulation. The relatively small size of siRNA, which facilitates its renal filtration, for in-stance, does not change significantly with this type of modification. Carriers designed for systemic delivery of siRNA are usually in the nanometer range and siRNA could undergo minimal renal filtration while being associated with such delivery systems.

3.2.2 Escaping the Immune System

The interaction of exogenous siRNAs with the immune system has two as-pects. The mononuclear phagocyte system (MPS), previously known as the reticuloendothelial system, is defined as a family of cells that include bone marrow progenitors, blood monocytes and tissue macrophages,[77] such as the Kupffer cells in liver, intraglomerular mesangial cells in the kidneys, alveolar macrophages in the lungs and red pulp scavenger macrophages in the spleen. The main function of these phagocytic cells is to eliminate for-eign intruders and affected autologous cells.[78] While siRNA on its own is too small to trigger uptake by the MPS cells, aggregation with serum proteins, such as complement proteins and immunoglobulins (*i.e.* 'opsonins'), could initiate their removal from the circulation. The free extrinsic nucleotides could be easily recognized by scavenger receptors, leading to non-parenchymal cell internalization (*i.e.* the Kupffer cells in the liver).[79] While chemical modification of siRNAs has virtually no effect on this uptake, PEGylation could be an effective strategy to minimize the MPS uptake. Delivery systems for siRNA are larger in size and may actually increase the risk of uptake by the MPS. However, this risk could be also minimized by optimizing the size and surface characteristics of delivery systems.

Secondly, triggering an immune response has been a concern with siRNAs where certain motifs are more prone to stimulate an immunogenic response. However, such a risk is significantly less than the plasmid DNA, and 2′-*O*-methyl modifications have been reported to further decrease this risk.[80] In special cases, such as cancer therapy, stimulation of immune system (as an off-target effect) could be advantageous, and could be exploited as an adjuvant to RNAi-mediated silencing.[81] Association of siRNA with a delivery system may conceal the reactive motifs in the siRNA structure; however, immunogenic reactions have been reported for some of the carriers (or components of the carrier) commonly used in siRNA delivery, including immunogenicity of monoclonal antibodies (used as targeting moiety in siRNA carriers),[82] cell penetrating peptides (interacting with Toll-like receptors and therefore affecting the innate immune system),[83] cationic lipoplexes[84] and negatively charged anionic lipids integrated into liposomal structures.[85] Even PEGylation, which has been used with considerable success in preventing opsonization and interaction with the immune system could be immunogenic as a result of repeated administration and promote antibody (especially IgM) responses.[86]

3.2.3 Exiting the Systemic Circulation at the Site of Action

"Naked" siRNA, if not degraded by nucleases or associated with serum proteins, should be rapidly accumulated in urine and excreted. In fact, Hatanaka *et al.* have reported a novel positron emitter-labeling technique for tracking unmodified double-stranded siRNA *in vivo*, where ~60% of IV-administered siRNA was accumulated in the bladder within 45 min of injection.[87] No sign of siRNA was observed in any of the analyzed organs after 60 min, except traces in the kidney, which probably represents the remainder of the siRNA destined for excretion. Chemically modified siRNA should have a better chance of distributing intact in the body, and the intact molecules could easily leave the circulation in a random manner through the pores in the vasculature that allow the passage of particles <5 nm.[39] This random pattern was reported by Gao *et al.*, where modified radio-labeled siRNAs where monitored after IV injection.[76] In this study, the circulatory half-life of unmodified siRNA was <5 min, with degradation observed within 1 min of injection. Conversely, modified siRNA was detected (with comparable quantities) in all selected organs after 30 min and disappeared after 24 h. Therefore, egressing the circulation does not seem to be a barrier, but the challenge for siRNA is exiting at a pre-programmed site, which is known as "targeting". While chemical modifications intended to improve serum stability have no significant effect on targeting siRNA, several delivery system have achieved significant levels of targeting by undertaking a "passive" and an "active" approach to targeting.

3.2.3.1 Passive Targeting

Delivery systems undertaking passive targeting are designed to alter the pharmacokinetic profile and biodistribution of siRNA without an active mechanism to "seek" a target. Passive targeting, a concept often applied for targeting solid tumors, takes advantage of the characteristics of the delivery system and the unique physiology of target tissues to accumulate the carrier and the siRNA associated with it at a specific site. The particle size and surface characteristics are keys in this process. A delivery system intended for passive targeting needs to be large enough to avoid extravasation from continuous capillaries, so as to minimize distribution throughout the body.

Generally speaking, molecules >5 nm in diameter do not readily cross the tight capillary endothelium wall and will remain in the circulation for a longer period.[39] The main pathway by which siRNA leaves the bloodstream is through renal elimination. The pore size of the glomerular filtration barrier is ~8 nm.[88] Most delivery systems are >20 nm and will avoid renal clearance.[89] Additionally, particles in the 10–100 nm range (especially particles <70 nm) would easily perfuse through fenestrated capillaries in the liver and will therefore accumulate in this organ.[90] Particles >100 nm are easily recognized and removed from blood circulation by the MPS, which might increase the accumulation of the particles in the spleen. Therefore, it appears that ~100 nm particles could avoid renal elimination and removal by liver, and still be small enough to escape uptake by the MPS, which means the particles would remain in circulation until they reach the discontinuous endothelium of "leaky" vasculature of the tumor (see below). Therefore, it is obvious that the fate of a particle could be manipulated by its size or, in other words, the size of the particles could be tailored for delivery to a specific tissue.

The other crucial characteristic for an efficient carrier is "stealth" properties, *i.e.* properties that help the delivery system evade the MPS recognition and uptake and enable its residence in circulation for a longer period of time. The key for a "stealth" carrier is to possess "biomimetic" properties including steric shielding, decreased unspecific binding and reduced complement activation.[91] PEGylation seems to be a popular strategy to minimize particle aggregation and opsonization due to the hydrophilic PEG shell. It has been shown that PEGylation could increase the mean residence time in circulation and minimize MPS uptake,[52] and the area under the curve (AUC) of the concentration in plasma *vs.* time (*i.e.* decreased clearance and/or the volume of distribution).[92] Polysaccharide coating of particles has been also considered as an alternative to PEG grafting.[93]

Passive accumulation of delivery systems in solid tumors relies on prolonged circulation time of stealth particles and a phenomenon known as the enhanced permeation and retention effect, which consists of two aspects. The "enhanced" permeation for a solid tumor is usually due to abnormal and "leaky" vasculature of a tumor. These vasculature, which form rapidly during the angiogenesis procedure are often defective and contain large gaps in the endothelial wall of the capillaries responsible for nurturing tumor cells.[94] The resulting gaps could be an ideal escape route for carriers in circulation to accumulate in the tumor. Formation of fenestrated (with endothelial openings of 40–80 nm) and discontinuous capillaries (with openings of 100–1000 nm) in tumor capillaries is the main reason for the enhanced permeation of tumor vasculature.[95] The "retention" effect is mainly due to a lack of lymphatic drainage in solid tumors that normally could drain the extravasated macromolecules from normal tissue.[11] The carriers could be trapped in leaky tissue in this way. Although originally recognized for tumorigenic tissues, other physiological processes involving *de novo* vascularization (*e.g.* tissue re-growth), vascular leakage/damage (*e.g.* sites of inflammation[96]) or bacterial infections[97] could be primed for passive targeting.

3.2.3.2 Active Targeting

The presence of a moiety that specifically binds to a component on target cells or tissues underlies active targeting, which is a strategy that could be added to delivery systems already tailored for passive targeting. Two assumptions are fundamental for this approach: (i) the targeting moiety association with siRNA (or the delivery system carrying it) should remain stable up to the destination; and (ii) the targeted component should exclusively present on targeted cells or tissues or significantly overexpressed compared to non-targeted tissue. Different categories of targeting moieties have been employed for direct conjugation to siRNA and/or carrier. These categories include peptides, antibodies, carbohydrates other endogenous molecules and synthetic aptamers.[98] Aptamers could be identified from randomly synthesized nucleic acid libraries using a process termed SELEX (systematic evolution of ligands by exponential enrichment).[99,100] They have been used in siRNA targeted delivery in three different methods:

(i) Non-covalent siRNA binding through a connector. This method of binding has been investigated through different linkers including: (a) streptavidin, which is a tetrameric protein that was first identified in the culture supernatants of *Streptomyces avidinii* and *Streptomyces lavendulae*;[101,102] and (b) packaging RNAs (pRNAs) that are derived from bacteriophage phi29 small RNAs that normally package DNA into procapsids.[98]

(ii) Aptamer–siRNA chimeras. This is a complete nucleotide-based approach where the aptamer can be linked to siRNA by transcription. As an example, an aptamer specific for prostate-specific membrane antigen (PSMA) was covalently linked to the passenger strand of siRNA targeting polo-like kinase 1 (PLK1) or Bcl-2 with no negative impact on the aptamer binding or siRNA silencing efficiency *in vitro* or *in vivo*.[103] A second generation of aptamer–siRNA chimeras was also prepared by adding a U–U overhang to the guide strand of siRNA, and a smaller aptamer specific for PSMA was covalently attached to the guide strand.[74] These changes resulted in regression of PSMA-expressing tumors in athymic mice by IP injection of the siRNA, while the unmodified chimeras had no significant effect on the tumor growth.

(iii) Integration into delivery systems. As an example, RNA aptamers have been included in a polyethyleneimine (PEI)-sodium citrate nanocore carrier designed for delivering siRNA targeting anaplastic lymphoma kinase (ALK).[104]

Other targeting moieties used for siRNA conjugation include a cyclic RGD,[105] a glycoprotein extracted from rabies virus (RVG),[106] a prostate-specific membrane antigen,[107] a TAT peptide[108] and a monoclonal antibody against transferrin receptor.[109] Different antibodies or antibody fragments have been successfully used as targeting moieties integrated in delivery systems. Immunoliposomes (liposomes incorporating antibodies) have been reported for targeting siRNA to xenografts in animal models.[110] The folic acid receptor has been recognized as a marker for ovarian carcinomas and frequently overexpressed in a wide range of tumors,[111] which makes folate a useful ligand for tumor targeting.[112] For hepatocytes that overexpress asialoglyco-protein receptors, galactose has been evaluated for targeted delivery of cationic liposomes.[113] Other ligands studied for active targeting of siRNAs include hyaluronic acid and anisamide. A list of ligands used for active targeting of siRNA can be found in one of our recent review papers.[11]

Active targeting *via* cell-surface binding ligands not only increases accumulation of siRNA in specific tissues, but may also increase cellular uptake of the siRNA. In NOD/SCID mice bearing luciferase-expressing Neuro2A xenografts, a similar biodistribution pattern and tumor accumulation was reported for non-targeted particles and particles with transferrin as the targeting moiety.[114] However, siRNA delivered by targeted particles reduced the luciferase activity by approximately two–fold compared to silencing achieved by the non-targeted formulation. Transferrin could improve the association of the particles with target cells, and therefore, increase cell uptake and silencing efficiency. A more recent study with transferrin-conjugated nanoparticles demonstrated that in a serum-rich medium, proteins could shield transferrin from binding to its receptors on cells and in solution.[115] Therefore, the authors concluded that in a complex biological environment, interaction with other proteins could mask the targeting molecules and cause loss of specificity.

3.2.4 Extracellular Matrix

Extravasated carriers delivering siRNA need to navigate their way through the extracellular matrix (ECM) before making contact with cell membranes. ECM is a non-cellular tissue component that forms a scaffold around the cells, mainly for physical support, but also for initiation of crucial biochemical and biomechanical cues required for tissue morphogenesis, differentiation and homeostasis.[116] The composition of the ECM might be different for each organ.[117] The main ingredients of the ECM are usually proteoglycans and fibrous proteins. Proteoglygans are structural proteins that also play a major role in signal transduction with regulatory functions in various cellular processes. These molecules are composed of a specific core protein substituted with covalently-linked glycosaminoglycan chains. Hyaluronan is an exception, since it lacks a protein core.[118] The main fibrous proteins are collagens, elastins, fibronectins and laminins.[116] Some ECM ingredients, such as sulphated glycosaminoglycans are negatively charged, and could be an obstacle for anionic siRNA molecules. Conversely, ECM could also pose a challenge for cationic delivery systems, by sequestering cationic carriers, or competing for anionic siRNA binding, which could de-stabilize the carrier/siRNA assembly. In both cases, the ECM could decrease the concentration of siRNA available for cell uptake. This negative impact of ECM on siRNA carriers has been shown for PEI-based carriers even after PEGylation.[119] Phagocytosis by MPS components is another challenge in the ECM that could eliminate siRNAs (or siRNA carriers) before reaching their destination.[39]

3.3   Cellular Barriers

After reaching the target cells, siRNA needs to interact strongly with the cell surface for a stable association, become internalized (the nature of which might vary quite significantly in different cell types) and reach its cytoplasmic destination to integrate into RISC complexes (which, along with other factors such as mRNA copy number and turnover rate, determines the silencing efficiency). The native siRNA has little chance for a significant uptake on its own, and mainly relies on a carrier's characteristics to overcome these obstacles. The size of particles formed by siRNA/carriers is an important characteristic for cellular internalization. As described for the extravasation of siRNA particles, a size of ~ 100 nm seems to be optimal for overcoming the obstacles to reach the target cells. However, effective *in vitro* uptake has been reported for particles with sizes as large as 200 nm, or greater.[11] The total surface charge of the particles, usually characterized by its ζ-potential, is another important feature, and a cationic charge density is desirable for interaction with anionic cell surface. The pathway of internalization and the fate of siRNA inside the cell could significantly impact the number of siRNAs available and intracellular siRNA pharmacokinetics, which eventually determines the silencing efficiency.[120,121] Depending on the uptake pathway, rate limiting step and the silencing efficiency might vary among binding efficiency, internalization rate, endosomal escape kinetics, carrier dissociation or intracellular transport.[121] Any significant discrepancy between the total cellular exposure to siRNA and the intact siRNA at the cytoplasmic target would cause an inconsistency between the degree of internalization and silencing efficiency.[122] In the next few sections, we focus on the steps involved in cell internalization and trafficking.

3.3.1   Binding to the Cell Membrane

Strong interactions with the cell surface are crucial for the effective internalization of siRNA, which is a challenge due to anionic phosphate groups on the nucleotide backbone. Numerous studies have indicated negligible cellular uptake of "naked" siRNA. However, different categories of carriers (to be reviewed later) have accomplished significant improvements in delivering their cargo intracellularly. Interaction of particles with cell surface as a whole (and not individual components) is still the key factor for efficiency, and sufficient intra-particle binding strength is necessary to prevent dissociation of particles at the cell surface due to higher concentrations of competing polyelectrolytes.[123] Involvement of different cell surface proteins in this interaction has been suggested, and these include anionic glycosaminoglycans,[124] and in particular, β1-integrin receptors[125,126] that are known to internalize bacteria and viruses[127] *via* endocytosis.[128] Successful carriers usually possess one or more of the following components.

3.3.1.1 Cationic Surface

Due to anionic phosphate groups in the phospholipid bilayer of the cell membrane, it is expected that particles with positive ζ-potentials (usually in the +10 to +40 mV range) show better association with cell membranes, and therefore, a higher level of cell internalization (even though positive charges are also usually associated with more cytotoxicity). However, effective delivery has been reported for carriers with negative ζ-potentials as well. It should be clarified that rather than the charge of the carriers, it is the ζ-potential of the assembled siRNA/carrier particle that influences the membrane interactions. The ζ-potential obviously depends on the chemical nature of the carrier and carrier:siRNA ratio; however, other factors that impact the carrier and siRNA interaction could also affect the ζ-potential, as was shown in our experiments on lipid-modified PEIs, where the ζ-potential of complexes was enhanced with lipid-modified PEIs compared to naïve PEI.[25] This might be due to a robust affinity among the assembled lipid components under aqueous conditions,[11] or repulsion of negative ions (that usually surround the cationic particles and partially neutralize their charge) by the hydrophobic substituents. Bedi *et al.* reported a liposomal delivery system with a phage fusion coat protein as the targeting moiety for delivering siRNA against PRDM14 protein in MCF7 breast cancer cells.[129] This delivery system was able to silence the target protein significantly, despite a ζ-potential of approximately −40 mV. However, the efficient silencing could be partly due to the targeting moiety, since the liposomal formulation without the coat protein was not nearly as effective (ζ-potential of liposomes without the coat protein was not reported). Another study on a non-targeting liposomal formulation (with a peptide to facilitate cell penetration) with a ζ-potential of −10 mV has reported efficient *in vitro* and *in vivo* silencing of ACTB.[130] These results indicated that siRNA silencing with carriers with negative ζ-potentials could be possible with the help of cell penetrating moieties. However, in a recent study on the impact of lipoplex (lipid complexes of siRNA) surface charge on siRNA delivery, Lavigne *et al.* reported effective siRNA silencing of CXCR4 protein (a cell surface receptor on T-cells that is essential for viral entry, and a suitable target in HIV1 infection) using lipid carriers with a ζ-potential of −54 mV and no cell penetrating protein.[131] Lipoplexes with the same lipid composition were prepared using different siRNA loads which resulted in a range of ζ-potentials. While lipoplexes with positive surface charge were clearly more effective in siRNA internalization and silencing, carriers with negative charge were still able to create significant silencing.

3.3.1.2 Hydrophobicity

Due to the hydrophobic nature of phospholipid bilayers of the cell membrane, hydrophobic moieties in siRNA particles seem to stand a better chance of interacting with cell surfaces. Different lipid-based carriers, as well as liposomal formulations have been successful for siRNA delivery. Addition of hydrophobic moieties to a hydrophilic carrier has shown considerable success and our group has pursued hydrophobic modification of different polymers, including a low molecular weight (LMW) PEI with little efficiency in the naïve form.[18,25,36,132–138] We demonstrated a significant increase in siRNA delivery and silencing efficiency by substituting different fatty acids (including linoleic acid, caprylic acid and palmitic acid) on the amine groups of the polymer backbone with different substitution levels. Conversely, cholesterol, being a major component of the cell membrane and involved in membrane fusion, macropinocytosis, caveolin-mediated and lipid raft-mediated endocytosis,[139] could provide advantage to carriers for enhanced interactions with cell membrane. Cholesterol could also stabilize liposomal structures[82] and act as a targeting moiety for liver cells.[140] Substitution of cholesterol on LMW PEI has been reported.[141,142]

3.3.1.3 Cell Penetrating Capability

"Cell penetrating peptides" (CPPs) are usually small peptides (6–30 amino acids) that have the capability for membrane translocation.[143] CPPs have been used both for siRNA delivery on their own (by direct conjugation to siRNA or stable complex formation), and as a membrane-penetrating component in a delivery system. The highly basic guanidinim group on the arginine seems to play an important role in internalization of arginine-rich CPPs due to formation of hydrogen bonds with certain cell surface molecules (including phosphates and sulfates).[144,145] Perhaps that partially explains why arginine-rich CPPs, such as HIV1 TAT[146] and oligoarginines[147] are among the most frequently employed CPPs. Interaction of CPPs with proteoglycans and/or phospholipids on the cell membrane is summarized in Konate *et al.* and is not reviewed here.[148] Different mechanisms, including macropinocytosis, signal-activated endocytosis and direct translocation routes ("inverted micelle" model) have been proposed for the efficacy of CPPs.[149] The use of CPPs as the sole carrier or as a component of a delivery system will be further discussed later.

3.3.1.4 Cell Binding Ligands

Incorporating ligands into delivery systems that bind to cell surface components is an effective strategy due to two separate mechanisms: active targeting property to specifically bind to target cells, and general association with cell membrane that is a necessity for internalization.

Ligand-mediated cell surface binding is especially effective for delivery systems that have incorporated PEG for stealth properties, which was shown to diminish cell uptake and silencing efficiency of the delivered siRNA.[150] The beneficiary effect of cell binding ligands have been also reported for cationic liposomes modified with the K16GACYGLPHKFCG peptide,[123] folate-modified polymeric carriers,[151] PEI-hydrazone-doxorubicin-PEI-PEG-folate nanoparticles[152] and a "mannosylated" new polymer designed to interact with mannose receptors (CD206; primarily expressed on macrophages and dendritic cells and up-regulated in tumor-associated macrophages).[153]

3.3.2 Entering the Cell

The cellular entry of siRNA may occur by direct transfer through cellular membranes or by energy-dependent membrane buddings generally known as endocytosis. Endocytosis could occur through different pathways, including clathrin-mediated and clathrin-independent pathways such as caveolae-mediated, clathrin- and caveolae-independent pathways and macropinocytosis, which are all involved in siRNA uptake depending on the nature of the carrier.[139,154,155] The internalization pathway could be different in different cells, and might even vary in the same cells depending on the experimental (environmental) conditions,[156,157] such as the siRNA dose[148] and physical characteristics of the carrier, including size, charge, polydispersity and the presence of a binding ligand.[158–160] For instance, it has been shown that adding cardiac glycosides (such as ouabain and strophanthidin) on lipid carriers could stimulate endocytosis upon binding to plasma membrane Na^+/K^+ ATPase found in all mammalian cells.[161]

Zhou *et al.* have reported a comparative study on intracellular fate of cationic liposomes and SPANosomes (SP; a liposomal formulation with a non-ionic traditional surfactant, Span 80), where SP carriers showed enhanced delivery of siRNA to cytoplasm compared to cationic liposome formulation.[162] The authors attributed this difference to different entry pathways. Microscopy indicated that cationic liposome-mediated delivery mostly relied on clathrin-mediated endocytosis,[163] while SP-mediated siRNA delivery depended on a combination of pathways, including clathrin-mediated endocytosis, macropinocytosis and caveolae-mediated endocytosis.[164] Unlike clathrin-coated vesicles, contents of macropinosomes and caveosomes may not merge with late endosomes or lysosomes, and therefore the authors concluded that this longer time window for the release of siRNA into the cytoplasm might be partially responsible for the enhanced delivery by SP carriers.[162] For more details on internalization pathways and effect of carrier on the fate of siRNA during internalization, refer to our review paper on siRNA delivery.[11]

3.3.3 Permeating the Lipid Bilayer of Endosomes

Internalization of siRNA does not ensure siRNA delivery to its destination, as it can easily be degraded after being internalized. Therefore, in case of endocytosis, the siRNA needs to cross the lipid bilayer that forms the entrapping vesicle and enter the cytoplasm. Nucleic acids are not capable of spontaneously translocating across endosomal membranes, and a carrier that disrupts these lamellar structures may be required.[165] This crossing might be accomplished by (i) non-contact mechanisms (that could involve endosomal swelling and eruption) or (ii) direct interaction between the carrier and lipid bilayer. A common non-contact mechanism is the "proton sponge effect", where protonation of the carrier causes elevated H^+ concentration. Influx of Cl^- ions to neutralize the elevated H^+ concentration increases the osmotic pressure and eventually bursts the endosome.[166]

Membrane disruption, destabilization or fusion could be the result of direct contact of carrier with the entrapping vesicle. Lipid-based carriers can interact with the vesicle wall *via* a mechanism called mesomorphic phase behavior, where the cationic lipids form charge-neutral pairs with anionic lipids of cellular membrane through ionic interaction, and disrupt the usual lamellar structure to a hexagonal phase and allow the siRNA to pass through. The extent of this type of interaction depends on the pH of the vesicle environment; while it might be minimal for an amino lipid at the physiologic pH, it increases at acidic pH of a lysosome.[167] The degree of hydrophobicity of the siRNA/carrier particle, through this direct interaction with cellular membrane, may affect timely escape of siRNA from the late endosome. CPPs and fusogenic peptides can facilitate siRNA escape *via* peptide-mediated pore formation or rearrangement of the lipid bilayer.[168] Hydrophobic amino acids such as arginine[169] could promote siRNA escape by fusion with endosomal membranes. PEGylation could negatively impact escape from entrapping vesicles due to steric interference and hydration of siRNA particles.[165]

Total siRNA associated with the cell is not always an accurate predictor of silencing efficiency. A possible explanation for this discrepancy is the extent of entrapment in endocytic vesicles, and unavailability of siRNA at the site of action (cytoplasm).[122,170] A recent study showed that the RNAi efficiency of siRNA delivered by mesoporous silica nanoparticles was strongly dependent on the endosomal escape capability.[171] In this study, the endosomal entrapment of siRNA was evaluated using a fluorescein amidite (FAM)-labeled siRNA. Authors confirmed that adding chloroquine (known for its endosomal eruption effect) to cells after 6 h of siRNA incubation releases the siRNA molecules from endosomes; however, adding chloroquine after 36 h showed almost no effect, indicating that siRNA was degraded in endolysosomes at the latter time point. Therefore, efficient and timely endosomal escape seems to be an important factor in siRNA silencing. A significant increase in siRNA silencing efficiency after application of low-voltage pulses (with no electroporation effect)[172] or photosensitizing chemicals and blue light

(5.1 $mW\,cm^{-2}$; known to erupt endosomes)[173] were also reported, confirming the importance of this step in siRNA-based RNAi.

Recently, Zhou *et al*. reported a new method to quantify the amount of siRNA in cytoplasm using a molecular beacon (MB).[162] In this method, cells are exposed to FAM-labeled siRNA and Cy3-labeled MB delivered with a carrier, and the mean fluorescence of both labels were measured by flow cytometry at different time point. FAM-siRNA and Cy3-MB *vs*. time profiles were used to calculate pharmacokinetic parameters for cellular and cytoplasmic exposure, respectively. Pharmacokinetics profiles of siRNA delivered by cationic liposome and SP indicated a significant difference in the terminal half-lives of these formulations.[162] The half-life of cationic liposome carriers ($\sim$0.8 h) was consistent with the published data for intracellular half-life of naked siRNA,[174] while the half-life of SP the formulation was $\sim$2 h. The significantly higher cytoplasmic AUC for the SP formulation confirmed that more efficient release into cytoplasm was responsible for a higher siRNA silencing efficiency with SP carriers.[162]

Polymeric carriers have been designed to take advantage of the acidic pH of endolysosomes to facilitate cytoplasmic release. Lim *et al*. have reported efficient *in vitro* and *in vivo* delivery of siRNA using a new acid-degradable poly(ketal amidoamine) (PKAA).[175] The ketal linkages in the backbone of the biodegradable poly(amidoamine) are stable in physiologic conditions; however, the carrier exhibited buffering capacity and endosomolytic activity due to the secondary amine groups in the backbone, and rapid degradation in acidic endosomes, which released the siRNA into the cytoplasm.

3.3.4 Intra-cytoplasmic Trafficking

Once in the cytoplasm, siRNA is ready to be incorporated into the RISC, which is generally assumed to be readily available in the cytoplasm. However, while associated with the carrier, siRNA is not "available" for this step (with the exception of conjugates with minimal interference with the siRNA silencing effect[176]). While the strength of siRNA/carrier association is an advantage for intracellular delivery, it could work against RISC incorporation if the complex is not dissociated in a timely manner. In fact, the disassembly of the carrier/siRNA package could become a major barrier for efficient silencing. In the case of interionic interaction of siRNA and carrier, it is hypothesized that the competitive binding of anionic molecules in cytoplasm (*e.g.* cytoplasmic RNAs such as mRNAs and tRNAs, and glycosaminoglycans) to the cationic carrier is the main mechanism of dissociation.[123,177] Weakened siRNA/carrier linkages due to the interaction of the carrier with endosomal membranes[178] and/or swelling or charge alteration of siRNA/carrier complex in the endosome could also contribute to dissociation.[123] Incorporating biodegradable bonds in the carrier is a strategy that could facilitate timely dissociation of siRNA. Cross-linking LMW PEI using dithiobissuccinimidylpropionate (DSP) and dimethyl-3,3′-dithiobispropionimidate (DTBP)[179] or 1,3-butanediol (or 1,6-hexanediol) diacrylates[180] are examples of this strategy to create labile disulfide (–S–S–) and –C(=O)–N(H)-linkages. The disulfide linkers (and to lesser extent, thioether linkers) have been also used for conjugating CPP to siRNA, which could negatively impact the silencing efficiency if it is not degraded in the cytoplasm in time.[181] Adding a disulfide linker between the sense strand of siRNA and a biotin group in an aptamer-streptavidin-siRNA conjugate has been explored to enhance siRNA release from the carrier.[107] This type of linker has been also used to form siRNA polyplexes; increasing the molecular weight of siRNA, and therefore the total anionic charges was intended to enhance the multivalent electrostatic interaction with a cationic carrier. Lee *et al.* have reported on "polymerized" thiolated siRNA *via* reducible disulfide linkage for cleavage in the cytoplasm environment.[182]

It has been proposed that intra-cytoplasmic targeting could improve efficiency, since mRNA[183] and possibly RISC are asymmetrically distributed in the cytoplasm.[11] Using commercially available Lipofectamine™ 2000, it was shown that siRNA was localized close to p-bodies before being incorporated into the RISC.[184] In addition to targeting RISC components themselves, targeting p-bodies might be another possibility. Several studies have indicated that when activated, RISC is localized in perinuclear region,[108,123,169] which makes targeting the microtubules another possible strategy to increase silencing efficiency.[11]

Recently, Alabi *et al.* reported a method to track cytoplasmic siRNA dissociation, which is based on two different siRNA-based fluorescent probes whose fluorescence emission changes in response to the assembly state of the siRNA.[185] In this method, first probe is a redox-sensitive fluorescence-quenched probe (mimicking the activity of MB[186]) that only fluoresces after disassembly in a reductive environment. The second probe is a fluorescence resonance energy transfer-labeled siRNA pair that only fluoresces when the carrier/siRNA is intact. Their experiments with a lipid-based nanoparticle showed a two-phase kinetic decay and an almost identical release rate based on the data collected for two different probes. Their study also showed a 20-fold faster release rate for the first phase (known as the burst release) compared to second phase.[185] Using novel methods such as this approach might better reveal the dissociation kinetics of siRNA/carrier complexes, an understudied theme, and soon lead us to more effective strategies for timely release of siRNA inside the cytoplasm.

3.4   siRNA Delivery Systems

While naked chemically-modified siRNA has shown efficacy in certain physiological settings, many therapeutic strategies require a well-designed carrier to facilitate intracellular delivery.[39] Many reviews could be found in literature that present siRNA delivery systems in detail, including cationic polymers,[11,187–190] lipids,[165,191] peptides,[143] endosomes,[192] and carbon nanotubes.[192] Here, we briefly introduce the major categories of these carriers and focus mainly on the most recent achievements in this field.

3.4.1 Lipid-based Delivery Systems

Cationic lipids could interact with siRNA through ionic moieties to form a lipoplex.[193] The cationic nature of lipids not only results in spontaneous interaction with the siRNA, but also facilitates association with cell membrane, which is a crucial for cellular internalization.[194] Higher instability (compared to polymeric systems), a more pronounced toxicity compared to neutral particles, risk of immunogenicity (due to uptake by macrophages), and shorter serum half-life (partly due to uptake by MPS) are some of the disadvantages associated with lipoplexes,[195] which could be resolved to some extent through PEGylation of the lipid, or further modification of the lipid formulation. For example, a lipid-based nanoparticle coated with the glycosaminoglycan hyaluronan has been recently reported for P-gp silencing *via* siRNA in A549 and NAR cells (Table 3.1).[196] Cardiolipin, a cationic analog of phospholipids found in the cardiac muscle has also been used for siRNA-mediated C-raf silencing in animal models.[197,198] Malamas *et al.* have reported a new amphiphilic cationic lipid carrier with selective pH-sensitive endosomal membrane disruptive capabilities based on three domains (cationic head, hydrophobic tail, and amino acid-based linker).[199] In this study, *in vitro* silencing of the reporter genes GFP and luciferase is reported in Chinese hamster ovary (CHO) cells and HT29 cancer cells, respectively (Table 3.1).

Liposomes constitute another type of lipid-based delivery system, which comprises highly ordered aggregates with a lipid bilayer envelope that entraps an aqueous internal phase. They have been shown to be capable of delivering both hydrophilic and hydrophobic drugs (with more significant success with hydrophilic drugs encapsulated in the aqueous phase). The lipid bilayer could be PEGylated (by using PEG-conjugated lipid components) for stealth properties, or carry targeting moieties (including monoclonal antibodies, in which case they are known as immunoliposomes) on the outer surface.[200–202] Stealth liposomes have been shown to increase the circulation times and systemic AUC for the encapsulated cargo, due to a decrease in clearance and/or volume of distribution.[90] The effectiveness of liposomes for siRNA delivery depends on their size, charge, lipid composition, and additional moieties.[203] 1,2-dioleoyl-3-trimethylammonium-propane (DOTAP) and 1,2-dioleoyl-*sn*-glycero-3-phosphoethanolamine (DOPE) arecommon liposome-forming components, and a DOTAP–DOPE cationic liposome system in combination with CPPs has been recently reported for effective *in vitro* silencing of GFP and anti-apoptotic protein, survivin, in HT1080 and A549 cells, respectively.[204] Effective silencing by liposomal delivery of siRNAs has also been reported *in vivo*. Systemic delivery of intersectin-1 siRNA by cationic liposomes derived from dimethyl dioctadecyl ammonium bromide (DOAB) and cholesterol has demonstrated effective silencing in the lung vascular bed in mice after retro-orbital vein injection (Table 3.1).[205] Mainelis *et al.* have reported *in vivo* delivery of Bcl2 and multi-drug resistance protein-1 (MRP1)-targeting siRNA in combination with doxorubicin (DOX) using cationic liposomes composed of phosphatidylcholine, 1,2-dipalmitoyl-*sn*-glycero-3-phosphatidylcholine, cholesterol, and DOTAP.[206] In this study, siRNA/DOX delivery *via* inhalation of aerosols resulted in >90% tumor volume reduction compared to ~40% reduction by IV injection of DOX with the same dose (Table 3.1). Di Paolo *et al.* have also reported a PEGylated liposome containing 1,2-distearoyl-glycero-3-phosphoethanolamine (DSPE) with and without an antibody against GD_2 (a human neuroblastoma-selective marker).[200,201] Delivering ALK siRNA with theses liposomes induced apoptosis and inhibited angiogenesis in neuroblastoma cells *in vivo*.

Finally, a more complex lipid-based carrier for siRNA delivery is stable nucleic acid lipid particles (SNALP), which are typically composed of multiple lipids, including neutral, cationic and PEGylated lipids.[48] The multifunctionalized SNALP, due to the presence of a variety of components, enables tailoring for a variety of purposes; however, such a complicated structure could introduce new challenges in the developmental procedure.[11] SNALP formulations of siRNA have been studied for apolipoprotein B (ApoB) silencing in cynomolgus monkeys[207] and for PLK1 silencing in subcutaneous tumors in mice.[208]

3.4.2 Polymer-based Delivery Systems

Cationic polymers could form relatively stable siRNA complexes (known as polyplexes) based on ionic interactions. However, due to the repetitive nature of a polymer structure, the degree of interaction with the siRNA, as well as physical and chemical characteristics of the polymer (including size, ionic charge, and hydrophobicity) could be easily adjusted to engineer an efficient and non-toxic carrier. This flexibility is reflected in the overwhelming volume of studies on siRNA delivery using polymeric carriers. As shown in Figure 3.3c, more than half of the siRNA delivery studies were performed using polymers.

PEI is often considered the "gold standard" in non-viral nucleic acid delivery, and is one of the most studied polymers for siRNA delivery. The high charge density of PEI and its strong "proton sponge effect" provide effective siRNA protection and internalization; however, the efficacy and toxicity of PEI is dependent upon molecular weight. While non-toxic LMW PEI has demonstrated poor efficacy in siRNA delivery, the severe toxicity of high molecular weight PEI has marred its impressive efficacy.[11,25,209] The efficacy (and toxicity) of PEI also depends on the degree of branching in the polymer backbone. Branched PEI contains primary, secondary and tertiary amines, whereas the linear polymer is composed of all secondary amines except for the primary amines at the terminals.[210] Generally speaking, branched PEI is more effective in nucleic acid delivery compared to linear structure,[211] although some studies also report superior delivery with linear PEI. Many studies have undertaken different strategies to "functionalize" LMW PEI. We investigated the effect of hydrophobic moieties on 2 kDa branched PEI by creating a library of polymers using different fatty acids as substituents and substitution levels. Some modified polymers have shown superior efficacy for siRNA delivery to a variety of cell lines, with linoleic acid (LA), caprylic acid (CA), and palmitic acid (PA) as the most successful substituents, depending on the cell type and the target chosen.[18,25,27–30,138] Navarro *et al.* have reported modification of PEI with phospholipids, including the phosphoethanolamines DOPE and dipamitoyl-*sn*-glycero-3-phosphoethanolamine (DPPE), and phosphocholine (PC), which significantly increased the silencing efficacy of the polymer (*e.g.* 60%, 30% and 5% decrease in GFP expression, respectively).[212] LMW (1.8 kDa) PEI modification with deoxycholic acid (DA) has also been reported for delivering Src homology region 2 domain-containing tyrosine phosphatase-1 (SHP-1) siRNA to rat heart myoblasts (H9C2 cells) *in vitro* and *in vivo*.[213] *In vivo* siRNA delivery induced a significant reduction in myocardial apoptosis and infarct size in a rat myocardial infarction model (Table 3.1). Other recent attempt at modification strategies include (i) adding poly(PEG-histidine-PEG-glutamic acid) copolymer for pH sensitization, and improved stability and biocompatibility;[214] (ii) conjugation with Tris-(hydroxymethyl)acrylamidomethane (THA) to improve serum tolerance and reduce toxicity due to added hydroxyl groups;[215] (iii) employing a mixture of transferrin-conjugated PEI (for target specificity) and hexyl acrylate-conjugated oligoethyleneimine (OEI);[23] and (iv) utilizing a mixture with mono-olein/oleic acid to form a liquid crystalline system for *in vivo* siRNA delivery to skin[216] (Table 3.1).

Chitosan is a naturally-occurring polysaccharide that is derived from deacetylation of chitin[217] and is extensively studied for siRNA delivery, partially due to its biodegradability[218] and low toxicity[219] profile. Matokanovic *et al.* have reported ionic gelation of chitosan with tripolyphosphate (TPP) for delivering siRNA against heat shock protein 70 (Hsp70) in human leukemia Jurkat cells and MCF7 human breast cell lines.[220] In a recent study, de-*N*-acetylated chitosan was used for silencing of P-gp in immortalized rat endothelial cell line RBE4 as a blood–brain barrier model, which resulted in increased cellular delivery and efficacy of the P-gp substrate doxorubicin.[221] Khurana *et al.* have compared transfection efficiency and therapeutic activity of chitosan to a cationic lipid-based carrier in HEK 293 cells.[222] In this study, the chitosan carrier was slightly inferior to the lipoplexes in silencing GFP (~57% and ~70% suppression). However, *in vivo* studies involving this polymer are scarce, which is possibly due to its limited efficacy.[11]

Dendrimers are another popular category of polymeric carriers for siRNA delivery. Poly(amidoamine) (PAMAM) is a dendrimer that was developed in the 1980s[223] with sufficient charge density to form stable complexes with siRNA at appropriate concentrations.[224] It is commercially available as Polyfect™ and Superfect™ for siRNA transfection.[225] Posocco *et al.* have recently reported a fifth-generation triethanolamine (TEA)-core PAMAM dendrimer designed for delivery of "sticky" siRNA (ssiRNA) molecules.[226] While traditional siRNAs usually have two thymine nucleotide hangovers at 3′ end of each strand (to protect from degradation and help with the silencing effect[227]), ssiRNAs have longer (usually more than five) thymine and adenine hangovers that have shown better silencing in some studies.[228] Posocco *et al.* showed a significant silencing efficiency for their proposed system against Hsp27 (a molecular chaperone involved in resistance against anticancer drugs[229]) and translationally controlled tumor protein (TCTP; a highly conserved protein involved in cell survival regulation[230]) in a variety of cancer cell lines.[226] A fourth-generation cystamine core PAMAM conjugated to a targeting peptide (oligo-arginine R9) *via* a PEG linker has also been reported for angiotensin II type 1 receptor (AT1R) silencing *in vitro* and *in vivo*.[231] AT1R is the mediator for angiotensin II adverse effects and worsened cardiac function after ischemic injury;[232] *in vivo* silencing of this protein significantly improved the functional recovery compared to control groups (saline and unloaded polymer) after ischemia-reperfusion injury in a rat model.[231] A PEI-related polymer PEG-conjugated dendrimer called poly-propylene imine (PPI) has shown effective Bcl2 silencing *in vitro*, and an improved tumor accumulation *in vivo* using a luteinizing hormone-releasing hormone targeting moiety.[233]

3.4.3 Peptides

The repetitive molecular structure of peptides and the versatile character-istics of its building blocks (amino acids) offer flexibility similar to that of polymeric siRNA carriers. Peptides have been used as cellular uptake en-hancers (CPPs) and targeting moieties. Peptides have also been used to form complexes with siRNA based on electrostatic interaction of cationic amino acids (*e.g.* arginine and lysine) with siRNA.[11] However, enhanced MPS up-take has been reported for highly charged peptides, and incorporation of cysteine in a lysine-rich peptide was shown to lower the degree of opsoni-zation.[234] A recent study on a library of peptides showed that stearylation of the peptide (instead of the acetyl group), which induces amphipathic char-acteristics, could significantly increase the silencing efficiency against the luciferase.[235] This study also showed a lower efficacy for peptides with ex-tremely high or low overall charge, indicating an optimal range of charge density for effective siRNA delivery.

Tagalakis *et al.* have reported a comparative study on siRNA formulations containing linear and branched oligolysine or oligoarginine peptides.[236] Different peptides were combined with cationic liposomes for siRNA delivery against luciferase in Neuro-2A cells *in vitro*. While formulations containing linear peptides were more condensed and stable, the branched peptide-containing complexes showed higher silencing activity, indicating that the greater stability of the formulations with linear peptides might limit siRNA release within the cell.[236] Peptides have been also used successfully *in vivo*. An animal study in athymic nude mice with a CPP known as MPG has shown efficient cyclin B1 silencing and tumor growth suppression.[237] A D-arginine-coupled RVG protein was recently used for *in vivo* siRNA delivery to silence calpain2 in a mouse model.[238] This study was designed to investigate the role of calpain2 in hippocampal synaptic plasticity, and calpain2 gene silencing eliminated long-term potentiation and impaired learning and memory.

3.4.4 Recent Accomplishments in siRNA Delivery

3.4.4.1 PolysiRNA

Binding siRNA molecules together is a recent approach that was speculated to increase silencing efficiency. Jo *et al.* recently reported a "nunchucks" siRNA structure by creating a dimeric siRNA conjugate.[19] By hypothesizing that high stiffness and low spatial charge density of siRNA limit the effectiveness of electrostatic condensation by cationic carriers, the authors introduced a reductively cleavable alkyl chain to link two siRNA molecules together. Single-stranded dimeric siRNA conjugates were synthesized by joining the sense strands of a GFP siRNA with different spacers (C3, C6, and C12) modified with a thiol group at the 3′ end. The preparation was completed by annealing single-stranded sense dimers with monomeric single-stranded antisense strands (Table 3.1). The longest alkyl spacer (C24, formed by two C12 spacers on the sense strands) delivered by a linear PEI showed the highest GFP silencing in a GFP-expressing MDA-MB-435 breast carcinoma cell line.[19] Polymerizing siRNA has also been performed by adding thiol groups at the 5′-ends of sense and anti-sense strands of different siRNA sequences to form self-polymerized siRNA polymers.[21,239] In one study, the reporter gene RFP was targeted *via* delivering polysiRNA by thiolated gelatin nanoparticles *in vitro* and *in vivo* in a RFP-expressing murine melanoma cell line (B16F10) and squamous cell carcinoma (SCC-7) tumor-bearing athymic nude mice, respectively.[239] In another study, a thiolated human serum albumin delivery system was used for polysiRNA delivery and effective tumor accumulation and silencing was demonstrated by tumor growth suppression *via* VEGF silencing in PC-3 tumor-bearing Balb/c nude mice.[21]

3.4.4.2 Nanotubes

Carbon nanotubes (CNT) have attracted considerable attention as effective nanocarriers, partially due to high surface area (which could increase the loading efficiency on the surface or within the interior core[240]), the possibility of conjugation with drugs or delivery system components (such as folic acid as a targeting moiety[241] and antibodies[242]), and low toxicity.[243] The high aspect ratio of CNT was reported to facilitate transmembrane penetration and internalization of the cargo via the "nanoneedle" mechanism regardless of the cell type, and without any additional functionalization of the carrier.[244] They were also shown to be able to take advantage of the EPR effect, resulting in tumor accumulation.[245] Battigelli *et al.* have reported a series of multi-walled CNTs conjugates with positively charged dendrons (ammonium or guanidinium groups; Figure 3.5) for silencing PLK1 in A549 lung cancer cells.[246] The best silencing efficiency was achieved by an

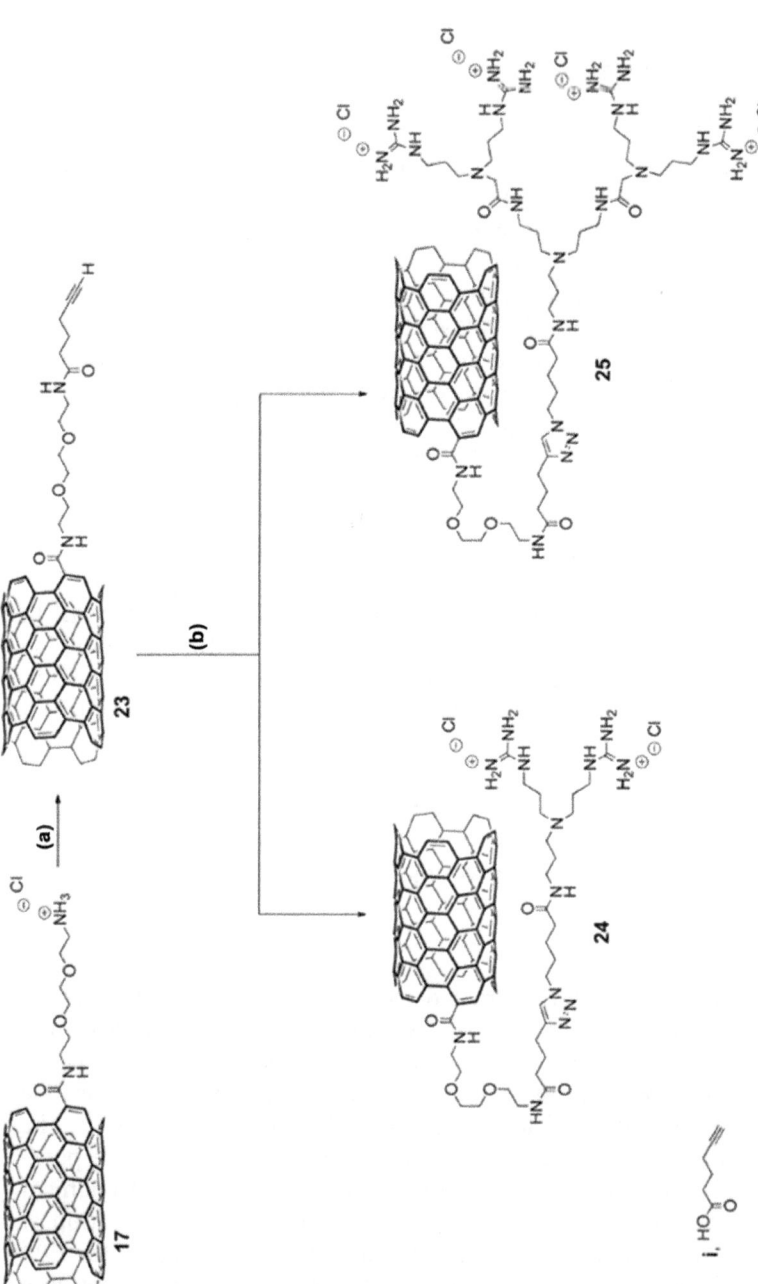

Figure 3.5 Multi-wall carbon nanotubes (MWCNTs). The synthetic scheme of functionalization of nanotubes with 1st and 2nd generation dendrons *via* click reaction. (a) i, EDC × HCl, HOBt,DIEA, DMF, 72 h; (b) 7 or 15, 2,6 lutidine, Cu(MeCN)₄PF₆, THPTA, NMP, 72 h; 2. HCl in dioxane, overnight. These nanotubes were used for silencing PLK1 in A549 lung cancer cells with significant efficiency.

Reprinted with permission from ref. 246. Copyright © 2013 WILEY-VCH Verlag GmbH & Co. KGaA, Weinheim.

ammonium derivative and the silencing effect seemed to be more stable compared to liposomal formulations (even though the PLK1 silencing was more significant at an early time point with liposomes). A single-walled CNT formulation was also studied for silencing human telomerase reverse transcriptase (hTERT) in PC-3 prostate cancer cells and BALB/c nude mice bearing PC-3 cell tumors.[247] Telomerase activity is reactivated in >90% of all human malignant tumors, making hTERT a natural target for RNAi strategies.[248] In this study, CNTs were functionalized with PEI and the tumor targeting NGR peptides. *In vivo* experiments showed enhanced tumor accumulation of single-walled CNT formulation, and combination of siRNA delivery and near-infrared photothermal therapy significantly enhanced the therapeutic efficacy.[247]

3.4.4.3 Exosomes

Evidence suggests that membrane vesicles, and in particular exosomes, play a crucial role in the horizontal transfer of RNA between cells as a part of communication and information exchange among cells.[249,250] Exosomes are more homogenous and generally smaller than microvesicles, with diameters ranging from 40 to 120 nm, and are derived from the internal endo-lysosomal compartments of cells. Exosomes are highly enriched in proteins found within microvesicular bodies, such as CD63, CD9, Alix and Lamp1.[192] In a recent study, exosomes were prepared from HEK293 cells by differential centrifugation of cell supernatants.[251] Exosomes carrying siRNA against luciferase significantly silenced the target protein in HCC70 cells *in vitro*. In another study, exosomes purified immature dendritic cells transfected with plasmids encoding RVG (peptide derived from the rabies virus glycoprotein that specifically binds to acetylcholine receptors expressed on neuronal cells) were used to deliver siRNAs against GAPDH and BACE-1 (enzyme β-secretase, a potential target in Alzheimer's disease therapy) to mouse brain cortex by IV injection.[252]

3.4.4.4 Silica

Different forms of silica have been investigated as siRNA carriers in the form of a porous nanoparticle. In this approach, the pores in the particle are usually functionalized with a cationic carrier (for ionic interaction with siRNA) and/or targeting moieties. The size of the pores could be in the 2–10 nm range (usually called mesoporous particles) or 20–60 nm range (known as nanoporous particles). Inorganic silica (SiO_2)-based nanoparticles, for example, were recently synthesized by a general method that includes hydrolysis and further condensation of silicate source in ethanol under basic conditions. The nanoporous particles were then covalently decorated with PEI.[253] Effective luciferase silencing was achieved in osteosarcoma U2OS

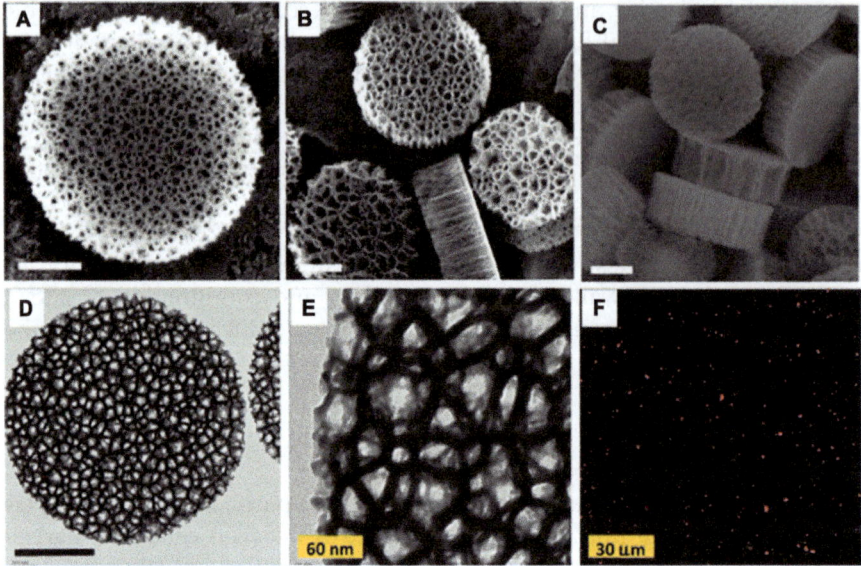

Figure 3.6 Nanoporous silicon particles. Morphological observation of discoidal silicon particles by scanning electron microscopy (SEM; A to C), transmission electron microscopy (TEM; D and E), and confocal laser scanning microscopy (CLSM; F). The SEM images represent unmodified particles (A), particles decorated with PEI (B), and the siRNA-loaded PEI-decorated particles (C). TEM images of siRNA-associated particles were taken at $20\,000 \times$ (D) and $80\,000 \times$ (E) magnifications, respectively. The confocal image illustrates Alexa Fluor 555-labeled siRNA particles (D). Scale barsrepresent 300 nm unless indicated.
Reprinted from *Biomaterials*, M. Zhang, R. Xu, X. Xia, Y. Yang, J. Gu, G. Qin, X. Liu, M. Ferrari and H. Shen, Polycation-functionalized nanoporous silicon particles for genesilencing on breast cancer cells, vol. 35, pp. 423–431, Copyright (2014), with permission from Elsevier.[254]

cells using this delivery system (Table 3.1). In another study, silica particles were prepared by electrochemical etching of silicon wafers, which were then oxidized to form the functional -OH groups for PEI binding[254] (Figure 3.6). Ataxia telangiectasia mutated (ATM) silencing in MDA-MB-435 cells was shown to be effective *in vitro* with these particles. Mesoporous silica particles were prepared by co-condensation reaction of tetraethylorthosilicate and 3-mercaptopyl-trimethoxysilane, which resulted in formation of surface thiol groups that was then transformed into disulphide bonds using S-(2-aminoethylthio)-2-thiopyridine. The bonds were then functionalized by adamantane, which forms stable complexes with ethylenediamine-modified β-cyclodextrin. The amino groups on modified cyclodextrin can in turn complex with siRNA *via* electrostatic interaction.[255] Bcl-2 and GFP silencing was demonstrated in HeLa cells and Zebrafish larvae *in vitro* and *in vivo*, respectively.

3.4.4.5 Other Approaches

Addition of hyaluronic acid (HA) into carrier formulations has been explored to enhance the biocompatibility and efficacy of siRNA. This approach could be beneficial due to the fact that: (i) HA can reduce non-specific interactions of the carrier with serum proteins;[256] (ii) HA can improve siRNA internalization for cells that express CD44, RHAMM, or HARLEC receptors;[257] (iii) after internalization, HA is sorted to non-lysosomal vesicles and accumulates in the perinuclear region, possibly co-localizing with RISC;[258] and (iv) HA could weaken the association of siRNA and the carrier, which could release siRNA more efficiently.[256] A recent study formulated chitosan with HA, which resulted in ~85% silencing of luciferase in A549 cells, without the usual toxicity associated with the chitosan.[259]

D-α-tocopheryl polyethylene glycol succinate (vitamin E TPGS or TPGS) is a water-soluble derivative of vitamin E with emulsifying characteristics. It has been shown to form micelles that are capable of delaying the release of the cargo and interacting with cell membrane.[260] Conjugation of TPGS to siRNA *via* disulfide bonds creates a labile conjugate after internalization due to elevated glutathione concentration in cytoplasm.[261] A conjugate of TPGS with herceptin (as targeting moiety) and PLK1 siRNA (on different TPGS molecules) *via* disulfide bonds was recently prepared for co-delivery of the siRNA and docetaxel. The IC_{50} of docetaxel dropped significantly compared

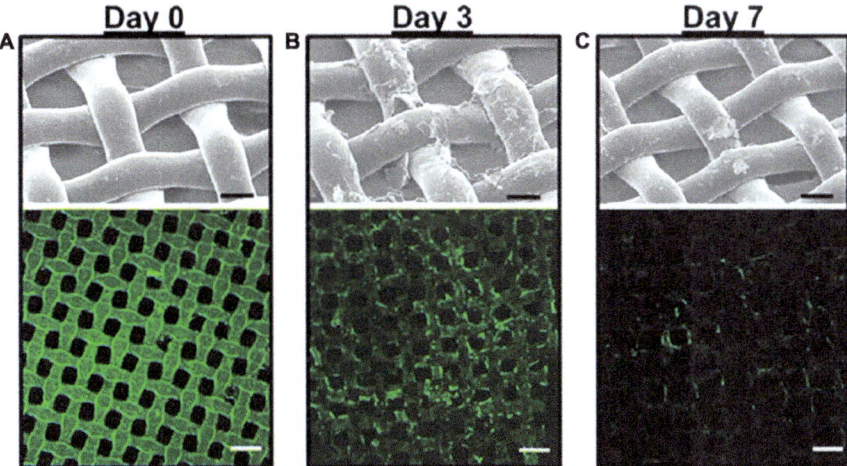

Figure 3.7 Layer-by-layer dressing for localized silencing. Side-by-side comparison of scanning electron microscopy (SEM) and confocal imaging showing the breakdown of the electrostatically assembled coating on a commercially available woven nylon dressing on day 0 (A), day 3 (B), and day 7 (C), in cell-conditioned media. SEM scale bar = 50 μm; confocal scalebar = 100 μm.
Reprinted with permission from S. Castleberry, M. Wang and P. T. Hammond, *ACS Nano*, 2013, 7, 5251–5261. Copyright(2013) American Chemical Society.[263]

to the commercially available formulation and docetaxel was delivered with TPGS micelles and simultaneous PLK1 siRNA silencing.[262]

Finally, localized delivery of siRNA is desirable for certain clinical applications, such as wound healing, and could help avoid the obstacles facing systemic delivery of siRNA. In a recent study, an ultrathin electrostatically assembled coating on a commercially available woven nylon dressing (Figure 3.7) was used for siRNA delivery *in vitro*.[263] Coating was performed layer-by-layer, starting with alternating base layers of PEI and a polyanion (laponite silicate and dextrose sulfate were examined), and continuing with layers of protamine sulfate, calcium phosphate and siRNA nanoparticles, protamine sulfate, and polyanion. Slow release of siRNA (over a 120 h period) and effective GFP silencing were reported;[263] theoretically, it would be possible to release siRNA locally at sites of tissue regeneration using such a system.

References

1. C. Napoli, C. Lemieux and R. Jorgensen, *Plant Cell*, 1990, **2**, 279–289.
2. A. Fire, S. Xu, M. K. Montgomery, S. A. Kostas, S. E. Driver and C. C. Mello, *Nature*, 1998, **391**, 806–811.
3. K. Raemdonck, R. E. Vandenbroucke, J. Demeester, N. N. Sanders and S. C. De Smedt, *Drug Discovery Today*, 2008, **13**, 917–931.
4. H. Guo, N. T. Ingolia, J. S. Weissman and D. P. Bartel, *Nature*, 2010, **466**, 835–840.
5. C. Z. Chen, *N. Engl. J. Med.*, 2005, **353**, 1768–1771.
6. Y. Zhang, Z. Wang and R. A. Gemeinhart, *J. Controlled Release*, 2013, **172**, 962–974.
7. S. Vasudevan, Y. Tong and J. A. Steitz, *Science*, 2007, **318**, 1931–1934.
8. D. Castanotto and J. J. Rossi, *Nature*, 2009, **457**, 426–433.
9. W. Hwang do, S. Son, J. Jang, H. Youn, S. Lee, D. Lee, Y. S. Lee, J. M. Jeong, W. J. Kim and D. S. Lee, *Biomaterials*, 2011, **32**, 4968–4975.
10. K. Sliva and B. S. Schnierle, *Virol. J.*, 2010, 7, 248.
11. H. M. Aliabadi, B. Landry, C. Sun, T. Tang and H. Uludag, *Biomaterials*, 2012, **33**, 2546–2569.
12. H. Bo, S. Zhang, L. Gao, Y. Chen, J. Zhang, X. Chang and M. Zhu, *BMC Cancer*, 2013, **13**, 496.
13. W. Zhao, J. P. Kruse, Y. Tang, S. Y. Jung, J. Qin and W. Gu, *Nature*, 2008, **451**, 587–590.
14. F. M. Gu, Q. L. Li, Q. Gao, J. H. Jiang, K. Zhu, X. Y. Huang, J. F. Pan, J. Yan, J. H. Hu, Z. Wang, Z. Dai, J. Fan and J. Zhou, *Mol. Cancer*, 2011, **10**, 150.
15. D. Harada, N. Takigawa, N. Ochi, T. Ninomiya, M. Yasugi, T. Kubo, H. Takeda, E. Ichihara, K. Ohashi, S. Takata, M. Tanimoto and K. Kiura, *Cancer Sci.*, 2012, **103**, 1795–1802.
16. J. Izrailit, H. K. Berman, A. Datti, J. L. Wrana and M. Reedijk, *Proc. Natl. Acad. Sci. U. S. A.*, 2013, **110**, 1714–1719.

17. S. Sevimli, S. Sagnella, M. Kavallaris, V. Bulmus and T. P. Davis, *Biomacromolecules*, 2013.
18. B. Landry, H. M. Aliabadi, A. Samuel, H. Gul-Uludag, X. Jiang, O. Kutsch and H. Uludag, *PLoS One*, 2012, 7, e44197.
19. S. D. Jo, J. S. Kim, C. O. Joe, H. Mok and Y. S. Nam, *Macromol. Biosci.*, 2013, **14**(2), 195–201.
20. H. Y. Yoon, H. R. Kim, G. Saravanakumar, R. Heo, S. Y. Chae, W. Um, K. Kim, I. C. Kwon, J. Y. Lee, D. S. Lee, J. C. Park and J. H. Park, *J Controlled Release*, 2013, **172**(3), 653–61.
21. S. Son, S. Song, S. J. Lee, S. Min, S. A. Kim, J. Y. Yhee, M. S. Huh, I. Chan Kwon, S. Y. Jeong, Y. Byun, S. H. Kim and K. Kim, *Biomaterials*, 2013, **34**, 9475–9485.
22. Y. Kapilov Buchman, E. Lellouche, S. Zigdon, M. Bechor, S. Michaeli and J. P. Lellouche, *Bioconjugate Chem*, 2013, **24**(12), 2076–2087.
23. M. Ganjalikhani Hakemi and M. Hashemi, *Iran. J. Basic Med. Sci.*, 2013, **16**, 973–978.
24. S. T. Chou, K. Hom, D. Zhang, Q. Leng, L. J. Tricoli, J. M. Hustedt, A. Lee, M. J. Shapiro, J. Seog, J. D. Kahn and A. J. Mixson, *Biomaterials*, 2014, **35**, 846–855.
25. H. M. Aliabadi, B. Landry, R. K. Bahadur, A. Neamnark, O. Suwantong and H. Uludag, *Macromol. Biosci.*, 2011, **11**, 662–672.
26. C. Gehin, J. Montenegro, E. K. Bang, A. Cajaraville, S. Takayama, H. Hirose, S. Futaki, S. Matile and H. Riezman, *J. Am. Chem. Soc.*, 2013, **135**, 9295–9298.
27. H. M. Aliabadi, B. Landry, P. Mahdipoor, C. Y. Hsu and H. Uludag, *Eur. J. Pharm. Biopharm.*, 2012, **81**, 33–42.
28. H. M. Aliabadi, P. Mahdipoor and H. Uludag, *Cancer Gene Ther.*, 2013, **20**, 169–177.
29. H. M. Aliabadi, R. Maranchuk, C. Kucharski, P. Mahdipoor, J. Hugh and H. Uludag, *J. Controlled Release*, 2013, **172**, 219–228.
30. H. Montazeri Aliabadi, B. Landry, P. Mahdipoor and H. Uludag, *Mol. Pharm.*, 2011, **8**, 1821–1830.
31. L. Miao, Y. Wang, H. Xia, C. Yao, H. Cai and Y. Song, *Biochem. Biophys. Res. Commun.*, 2013, **440**, 792–797.
32. J. Li, Y. Guo, X. Feng, Z. Wang, Y. Wang, P. Deng, D. Zhang, R. Wang, L. Xie, X. Xu, Y. Zhou, N. Ji, J. Hu, M. Zhou, G. Liao, N. Geng, L. Jiang and Q. Chen, *J. Cancer Res. Clin. Oncol.*, 2012, **138**, 563–571.
33. X. Wang, X. Chen, J. Fang and C. Yang, *Int. J. Clin. Exp. Pathol.*, 2013, **6**, 586–597.
34. A. L. Mulkeen, T. Silva, P. S. Yoo, J. C. Schmitz, E. Uchio, E. Chu and C. Cha, *Arch. Surg.*, 2006, **141**, 367–374, discussion 374.
35. A. Alshamsan, S. Hamdy, A. Haddadi, J. Samuel, A. O. El-Kadi, H. Uludag and A. Lavasanifar, *Transl. Oncol.*, 2011, **4**, 178–188.
36. M. Abbasi, A. Lavasanifar, L. G. Berthiaume, M. Weinfeld and H. Uludag, *Cancer*, 2010, **116**, 5544–5554.

37. J. Haupenthal, C. Baehr, S. Kiermayer, S. Zeuzem and A. Piiper, *Biochem. Pharmacol.*, 2006, **71**, 702–710.
38. D. M. Dykxhoorn, D. Palliser and J. Lieberman, *Gene Ther.*, 2006, **13**, 541–552.
39. K. A. Whitehead, R. Langer and D. G. Anderson, *Nat. Rev. Drug Discovery*, 2009, **8**, 129–138.
40. M. J. Tolentino, A. J. Brucker, J. Fosnot, G. S. Ying, I. H. Wu, G. Malik, S. Wan and S. J. Reich, *Retina*, 2004, **24**, 660.
41. J. Shen, R. Samul, R. L. Silva, H. Akiyama, H. Liu, Y. Saishin, S. F. Hackett, S. Zinnen, K. Kossen, K. Fosnaugh, C. Vargeese, A. Gomez, K. Bouhana, R. Aitchison, P. Pavco and P. A. Campochiaro, *Gene Ther.*, 2006, **13**, 225–234.
42. L. Singerman, *Retina*, 2009, **29**, S49–50.
43. J. Guo, K. A. Fisher, R. Darcy, J. F. Cryan and C. O'Driscoll, *Mol. Biosyst.*, 2010, **6**, 1143–1161.
44. T. Okuda, D. Kito, A. Oiwa, M. Fukushima, D. Hira and H. Okamoto, *Biol. Pharm. Bull.*, 2013, **36**, 1183–1191.
45. D. Palliser, D. Chowdhury, Q. Y. Wang, S. J. Lee, R. T. Bronson, D. M. Knipe and J. Lieberman, *Nature*, 2006, **439**, 89–94.
46. Y. Minakuchi, F. Takeshita, N. Kosaka, H. Sasaki, Y. Yamamoto, M. Kouno, K. Honma, S. Nagahara, K. Hanai, A. Sano, T. Kato, M. Terada and T. Ochiya, *Nucleic Acids Res.*, 2004, **32**, e109.
47. J. Yano, K. Hirabayashi, S. Nakagawa, T. Yamaguchi, M. Nogawa, I. Kashimori, H. Naito, H. Kitagawa, K. Ishiyama, T. Ohgi and T. Irimura, *Clin. Cancer Res.*, 2004, **10**, 7721–7726.
48. A. de Fougerolles, H. P. Vornlocher, J. Maraganore and J. Lieberman, *Nat. Rev. Drug Discovery*, 2007, **6**, 443–453.
49. G. Dorn, S. Patel, G. Wotherspoon, M. Hemmings-Mieszczak, J. Barclay, F. J. Natt, P. Martin, S. Bevan, A. Fox, P. Ganju, W. Wishart and J. Hall, *Nucleic Acids Res.*, 2004, **32**, e49.
50. M. C. Luo, D. Q. Zhang, S. W. Ma, Y. Y. Huang, S. J. Shuster, F. Porreca and J. Lai, *Mol. Pain*, 2005, **1**, 29.
51. C. Kriegel, H. Attarwala and M. Amiji, *Adv. Drug Delivery Rev.*, 2013, **65**, 891–901.
52. F. Alexis, E. Pridgen, L. K. Molnar and O. C. Farokhzad, *Mol. Pharm.*, 2008, **5**, 505–515.
53. C. R. Dass and P. F. Choong, *J. Drug Target*, 2008, **16**, 257–261.
54. B. Ballarin-Gonzalez, F. Dagnaes-Hansen, R. A. Fenton, S. Gao, S. Hein, M. Dong, J. Kjems and K. A. Howard, *Mol. Ther. Nucleic Acids*, 2013, **2**, e76.
55. Y. Tang, Y. B. Li, B. Wang, R. Y. Lin, M. van Dongen, D. M. Zurcher, X. Y. Gu, M. M. Banaszak Holl, G. Liu and R. Qi, *Mol. Pharm.*, 2012, **9**, 1812–1821.
56. C. N. Landen, W. M. Merritt, L. S. Mangala, A. M. Sanguino, C. Bucana, C. Lu, Y. G. Lin, L. Y. Han, A. A. Kamat, R. Schmandt, R. L. Coleman,

D. M. Gershenson, G. Lopez-Berestein and A. K. Sood, *Cancer Biol. Ther.*, 2006, **5**, 1708–1713.

57. B. Urban-Klein, S. Werth, S. Abuharbeid, F. Czubayko and A. Aigner, *Gene Ther.*, 2005, **12**, 461–466.

58. J. M. Layzer, A. P. McCaffrey, A. K. Tanner, Z. Huang, M. A. Kay and B. A. Sullenger, *RNA*, 2004, **10**, 766–771.

59. A. L. Jackson and P. S. Linsley, *Nat. Rev. Drug Discovery*, 2010, **9**, 57–67.

60. L. Aagaard and J. J. Rossi, *Adv. Drug Delivery Rev.*, 2007, **59**, 75–86.

61. M. A. Behlke, *Oligonucleotides*, 2008, **18**, 305–319.

62. P. S. Eder, R. J. DeVine, J. M. Dagle and J. A. Walder, *Antisense Res. Dev.*, 1991, **1**, 141–151.

63. S. Kennedy, D. Wang and G. Ruvkun, *Nature*, 2004, **427**, 645–649.

64. Y. Zou, P. Tiller, I. W. Chen, M. Beverly and J. Hochman, *Rapid Commun. Mass Spectrom.*, 2008, **22**, 1871–1881.

65. J. J. Turner, S. W. Jones, S. A. Moschos, M. A. Lindsay and M. J. Gait, *Mol. Biosyst.*, 2007, **3**, 43–50.

66. J. Haupenthal, C. Baehr, S. Zeuzem and A. Piiper, *Int. J. Cancer*, 2007, **121**, 206–210.

67. T. Kubo, Z. Zhelev, H. Ohba and R. Bakalova, *Oligonucleotides*, 2007, **17**, 445–464.

68. T. Kubo, Z. Zhelev, H. Ohba and R. Bakalova, *Biochem. Biophys. Res. Commun.*, 2008, **365**, 54–61.

69. J. Soutschek, A. Akinc, B. Bramlage, K. Charisse, R. Constien, M. Donoghue, S. Elbashir, A. Geick, P. Hadwiger, J. Harborth, M. John, V. Kesavan, G. Lavine, R. K. Pandey, T. Racie, K. G. Rajeev, I. Rohl, I. Toudjarska, G. Wang, S. Wuschko, D. Bumcrot, V. Koteliansky, S. Limmer, M. Manoharan and H. P. Vornlocher, *Nature*, 2004, **432**, 173–178.

70. J. Harborth, S. M. Elbashir, K. Vandenburgh, H. Manninga, S. A. Scaringe, K. Weber and T. Tuschl, *Antisense Nucleic Acid Drug Dev.*, 2003, **13**, 83–105.

71. B. A. Kraynack and B. F. Baker, *RNA*, 2006, **12**, 163–176.

72. D. V. Morrissey, J. A. Lockridge, L. Shaw, K. Blanchard, K. Jensen, W. Breen, K. Hartsough, L. Machemer, S. Radka, V. Jadhav, N. Vaish, S. Zinnen, C. Vargeese, K. Bowman, C. S. Shaffer, L. B. Jeffs, A. Judge, I. MacLachlan and B. Polisky, *Nat. Biotechnol.*, 2005, **23**, 1002–1007.

73. S. H. Kim, J. H. Jeong, S. H. Lee, S. W. Kim and T. G. Park, *J. Controlled Release*, 2008, **129**, 107–116.

74. J. P. Dassie, X. Y. Liu, G. S. Thomas, R. M. Whitaker, K. W. Thiel, K. R. Stockdale, D. K. Meyerholz, A. P. McCaffrey, J. O. McNamara 2nd and P. H. Giangrande, *Nat. Biotechnol.*, 2009, **27**, 839–849.

75. V. V. Ambardekar, H. Y. Han, M. L. Varney, S. V. Vinogradov, R. K. Singh and J. A. Vetro, *Biomaterials*, 2011, **32**, 1404–1411.

76. S. Gao, F. Dagnaes-Hansen, E. J. Nielsen, J. Wengel, F. Besenbacher, K. A. Howard and J. Kjems, *Mol. Ther.*, 2009, **17**, 1225–1233.

77. D. A. Hume, *Curr. Opin Immunol.*, 2006, **18**, 49–53.
78. C. Halma, M. R. Daha and L. A. van Es, *Clin. Exp. Immunol.*, 1992, **89**, 1–7.
79. Z. Wang, G. Liu, H. Zheng and X. Chen, *Biotechnol. Adv.*, 2013, **32**(4), 831–843.
80. A. D. Judge, G. Bola, A. C. Lee and I. MacLachlan, *Mol. Ther.*, 2006, **13**, 494–505.
81. D. Guo, B. Wang, F. Han and T. Lei, *Expert Opin Biol. Ther.*, 2010, **10**, 927–936.
82. M. A. Tran, R. J. Watts and G. P. Robertson, *Pigm. Cell Melanoma Res.*, 2009, **22**, 388–399.
83. W. Low, A. Mortlock, L. Petrovska, T. Dottorini, G. Dougan and A. Crisanti, *J. Biotechnol.*, 2007, **129**, 555–564.
84. L. Xu and T. Anchordoquy, *J. Pharm. Sci.*, 2011, **100**, 38–52.
85. A. Schroeder, C. G. Levins, C. Cortez, R. Langer and D. G. Anderson, *J. Intern. Med.*, 2010, **267**, 9–21.
86. T. Ishida and H. Kiwada, *Int. J. Pharm.*, 2008, **354**, 56–62.
87. K. Hatanaka, T. Asai, H. Koide, E. Kenjo, T. Tsuzuku, N. Harada, H. Tsukada and N. Oku, *Bioconjugate Chem.*, 2010, **21**, 756–763.
88. R. Kanasty, J. R. Dorkin, A. Vegas and D. Anderson, *Nat. Mater.*, 2013, **12**, 967–977.
89. H. Lee, A. K. Lytton-Jean, Y. Chen, K. T. Love, A. I. Park, E. D. Karagiannis, A. Sehgal, W. Querbes, C. S. Zurenko, M. Jayaraman, C. G. Peng, K. Charisse, A. Borodovsky, M. Manoharan, J. S. Donahoe, J. Truelove, M. Nahrendorf, R. Langer and D. G. Anderson, *Nat. Nanotechnol.*, 2012, 7, 389–393.
90. H. M. Aliabadi, M. Shahin, D. R. Brocks and A. Lavasanifar, *Clin. Pharmacokinet.*, 2008, **47**, 619–634.
91. K. Raemdonck, T. F. Martens, K. Braeckmans, J. Demeester and S. C. De Smedt, *Adv. Drug Delivery Rev.*, 2013, **65**, 1123–1147.
92. L. E. Prevette, M. L. Lynch, K. Kizjakina and T. M. Reineke, *Langmuir*, 2008, **24**, 8090–8101.
93. C. Lemarchand, R. Gref and P. Couvreur, *Eur. J. Pharm. Biopharm.*, 2004, **58**, 327–341.
94. J. Z. Du, C. Q. Mao, Y. Y. Yuan, X. Z. Yang and J. Wang, *Biotechnol. Adv.*, 2013, **32**(4), 789–803.
95. R. K. Jain, *Cancer Metastasis Rev.*, 1987, **6**, 559–593.
96. B. S. Sandanaraj, H. U. Gremlich, R. Kneuer, J. Dawson and S. Wacha, *Bioconjugate Chem.*, 2010, **21**, 93–101.
97. E. A. Azzopardi, E. L. Ferguson and D. W. Thomas, *J. Antimicrob. Chemother.*, 2013, **68**, 257–274.
98. X. Li, Q. Zhao and L. Qiu, *J. Controlled Release*, 2013, **171**, 152–162.
99. C. Tuerk and L. Gold, *Science*, 1990, **249**, 505–510.
100. A. D. Ellington and J. W. Szostak, *Nature*, 1990, **346**, 818–822.
101. L. Chaiet and F. J. Wolf, *Arch. Biochem. Biophys.*, 1964, **106**, 1–5.
102. E. O. Stapley, J. M. Mata, I. M. Miller, T. C. Demny and H. B. Woodruff, *Antimicrob Agents Chemother*, 1963, **161**, 20–27.

103. J. O. McNamara 2nd, E. R. Andrechek, Y. Wang, K. D. Viles, R. E. Rempel, E. Gilboa, B. A. Sullenger and P. H. Giangrande, *Nat. Biotechnol.*, 2006, **24**, 1005–1015.

104. N. Zhao, H. G. Bagaria, M. S. Wong and Y. Zu, *J. Nanobiotechnol.*, 2011, **9**, 2.

105. M. R. Alam, X. Ming, M. Fisher, J. G. Lackey, K. G. Rajeev, M. Manoharan and R. L. Juliano, *Bioconjugate Chem.*, 2011, **22**, 1673–1681.

106. P. Kumar, H. Wu, J. L. McBride, K. E. Jung, M. H. Kim, B. L. Davidson, S. K. Lee, P. Shankar and N. Manjunath, *Nature*, 2007, **448**, 39–43.

107. T. C. Chu, K. Y. Twu, A. D. Ellington and M. Levy, *Nucleic Acids Res.*, 2006, **34**, e73.

108. Y. L. Chiu, A. Ali, C. Y. Chu, H. Cao and T. M. Rana, *Chem. Biol.*, 2004, **11**, 1165–1175.

109. C. F. Xia, Y. Zhang, R. J. Boado and W. M. Pardridge, *Pharm. Res.*, 2007, **24**, 2309–2316.

110. K. F. Pirollo, A. Rait, Q. Zhou, S. H. Hwang, J. A. Dagata, G. Zon, R. I. Hogrefe, G. Palchik and E. H. Chang, *Cancer Res.*, 2007, **67**, 2938–2943.

111. J. F. Ross, P. K. Chaudhuri and M. Ratnam, *Cancer*, 1994, **73**, 2432–2443.

112. T. Yoshizawa, Y. Hattori, M. Hakoshima, K. Koga and Y. Maitani, *Eur. J. Pharm. Biopharm.*, 2008, **70**, 718–725.

113. S. Kawakami, F. Yamashita, K. Nishida, J. Nakamura and M. Hashida, *Crit. Rev. Ther. Drug Carrier Syst.*, 2002, **19**, 171–190.

114. D. W. Bartlett, H. Su, I. J. Hildebrandt, W. A. Weber and M. E. Davis, *Proc. Natl. Acad. Sci. U. S. A.*, 2007, **104**, 15549–15554.

115. A. Salvati, A. S. Pitek, M. P. Monopoli, K. Prapainop, F. B. Bombelli, D. R. Hristov, P. M. Kelly, C. Aberg, E. Mahon and K. A. Dawson, *Nat. Nanotechnol.*, 2013, **8**, 137–143.

116. C. Frantz, K. M. Stewart and V. M. Weaver, *J. Cell Sci.*, 2010, **123**, 4195–4200.

117. W. Li and F. C. Szoka Jr., *Pharm. Res.*, 2007, **24**, 438–449.

118. L. Schaefer and R. M. Schaefer, *Cell Tissue Res.*, 2010, **339**, 237–246.

119. R. S. Burke and S. H. Pun, *Bioconjugate Chem.*, 2008, **19**, 693–704.

120. D. W. Bartlett and M. E. Davis, *Nucleic Acids Res.*, 2006, **34**, 322–333.

121. R. Juliano, M. R. Alam, V. Dixit and H. Kang, *Nucleic Acids Res.*, 2008, **36**, 4158–4171.

122. Y. Zhang, H. Li, J. Sun, J. Gao, W. Liu, B. Li, Y. Guo and J. Chen, *Int. J. Pharm.*, 2010, **390**, 198–207.

123. A. D. Tagalakis, L. He, L. Saraiva, K. T. Gustafsson and S. L. Hart, *Biomaterials*, 2011, **32**, 6302–6315.

124. F. Labat-Moleur, A. M. Steffan, C. Brisson, H. Perron, O. Feugeas, P. Furstenberger, F. Oberling, E. Brambilla and J. P. Behr, *Gene Ther.*, 1996, **3**, 1010–1017.

125. H. Arima, Y. Aramaki and S. Tsuchiya, *J. Pharm. Sci.*, 1997, **86**, 786–790.

126. I. S. Zuhorn, D. Kalicharan, G. T. Robillard and D. Hoekstra, *Mol. Ther.*, 2007, **15**, 946–953.
127. M. S. Maginnis, J. C. Forrest, S. A. Kopecky-Bromberg, S. K. Dickeson, S. A. Santoro, M. M. Zutter, G. R. Nemerow, J. M. Bergelson and T. S. Dermody, *J. Virol.*, 2006, **80**, 2760–2770.
128. E. V. Vassilieva, K. Gerner-Smidt, A. I. Ivanov and A. Nusrat, *Am. J. Physiol.: Gastrointest. Liver Physiol.*, 2008, **295**, G965–976.
129. D. Bedi, T. Musacchio, O. A. Fagbohun, J. W. Gillespie, P. Deinnocentes, R. C. Bird, L. Bookbinder, V. P. Torchilin and V. A. Petrenko, *Nanomedicine*, 2011, **7**, 315–323.
130. Y. Sakurai, H. Hatakeyama, Y. Sato, H. Akita, K. Takayama, S. Kobayashi, S. Futaki and H. Harashima, *Biomaterials*, 2011, **32**, 5733–5742.
131. C. Lavigne, K. Slater, N. Gajanayaka, C. Duguay, E. Arnau Peyrotte, G. Fortier, M. Simard, A. J. Kell, M. L. Barnes and A. R. Thierry, *Expert Opin. Biol. Ther.*, 2013, **13**, 973–985.
132. M. Abbasi, H. Uludag, V. Incani, C. Y. Hsu and A. Jeffery, *Biomacromolecules*, 2008, **9**, 1618–1630.
133. M. Abbasi, H. Uludag, V. Incani, C. Olson, X. Lin, B. A. Clements, D. Rutkowski, A. Ghahary and M. Weinfeld, *Biomacromolecules*, 2007, **8**, 1059–1063.
134. A. Alshamsan, A. Haddadi, V. Incani, J. Samuel, A. Lavasanifar and H. Uludag, *Mol. Pharm.*, 2009, **6**, 121–133.
135. V. Incani, X. Lin, A. Lavasanifar and H. Uludag, *ACS Appl. Mater. Interfaces*, 2009, **1**, 841–848.
136. V. Incani, E. Tunis, B. A. Clements, C. Olson, C. Kucharski, A. Lavasanifar and H. Uludag, *J. Biomed. Mater. Res., Part A*, 2007, **81**, 493–504.
137. K. C. Bahadur, B. Landry, H. M. Aliabadi, A. Lavasanifar and H. Uludag, *Acta Biomater.*, 2011, **7**, 2209–2217.
138. J. Valencia-Serna, H. Gul-Uludag, P. Mahdipoor, X. Jiang and H. Uludag, *J. Controlled Release*, 2013, **172**, 495–503.
139. J. J. Lu, R. Langer and J. Chen, *Mol. Pharm.*, 2009, **6**, 763–771.
140. C. Lorenz, P. Hadwiger, M. John, H. P. Vornlocher and C. Unverzagt, *Bioorg. Med. Chem. Lett.*, 2004, **14**, 4975–4977.
141. W. J. Kim, C. W. Chang, M. Lee and S. W. Kim, *J. Controlled Release*, 2007, **118**, 357–363.
142. J. W. Yockman, A. Kastenmeier, H. M. Erickson, J. G. Brumbach, M. G. Whitten, A. Albanil, D. Y. Li, S. W. Kim and D. A. Bull, *J. Controlled Release*, 2008, **132**, 260–266.
143. I. Nakase, G. Tanaka and S. Futaki, *Mol. Biosyst.*, 2013, **9**, 855–861.
144. J. B. Rothbard, T. C. Jessop, R. S. Lewis, B. A. Murray and P. A. Wender, *J. Am. Chem. Soc.*, 2004, **126**, 9506–9507.
145. N. Sakai, T. Takeuchi, S. Futaki and S. Matile, *ChemBioChem*, 2005, **6**, 114–122.
146. H. Brooks, B. Lebleu and E. Vives, *Adv. Drug Delivery Rev.*, 2005, **57**, 559–577.

147. J. B. Rothbard, T. C. Jessop and P. A. Wender, *Adv. Drug Delivery Rev.*, 2005, **57**, 495–504.
148. K. Konate, L. Crombez, S. Deshayes, M. Decaffmeyer, A. Thomas, R. Brasseur, G. Aldrian, F. Heitz and G. Divita, *Biochemistry*, 2010, **49**, 3393–3402.
149. P. Jarver, I. Mager and U. Langel, *Trends Pharmacol. Sci.*, 2010, **31**, 528–535.
150. X. L. Wang, R. Xu and Z. R. Lu, *J. Controlled Release*, 2009, **134**, 207–213.
151. C. J. Cheng and W. M. Saltzman, *Biomaterials*, 2011, **32**, 6194–6203.
152. D. W. Dong, B. Xiang, W. Gao, Z. Z. Yang, J. Q. Li and X. R. Qi, *Biomaterials*, 2013, **34**, 4849–4859.
153. S. S. Yu, C. M. Lau, W. J. Barham, H. M. Onishko, C. E. Nelson, H. Li, C. A. Smith, F. E. Yull, C. L. Duvall and T. D. Giorgio, *Mol. Pharm.*, 2013, **10**, 975–987.
154. J. M. Barichello, S. Kizuki, T. Tagami, T. Asai, T. Ishida, H. Kikuchi, N. Oku and H. Kiwada, *Int. J. Pharm.*, 2011, **410**, 153–160.
155. Y. Shen, B. Wang, Y. Lu, A. Ouahab, Q. Li and J. Tu, *Int. J. Pharm.*, 2011, **414**, 233–243.
156. G. Sahay, D. Y. Alakhova and A. V. Kabanov, *J. Controlled Release*, 2010, **145**, 182–195.
157. Z. ur Rehman, D. Hoekstra and I. S. Zuhorn, *J. Controlled Release*, 2011, **156**, 76–84.
158. N. P. Gabrielson and D. W. Pack, *J. Controlled Release*, 2009, **136**, 54–61.
159. S. Hobel, A. Loos, D. Appelhans, S. Schwarz, J. Seidel, B. Voit and A. Aigner, *J. Controlled Release*, 2011, **149**, 146–158.
160. J. Rejman, V. Oberle, I. S. Zuhorn and D. Hoekstra, *Biochem. J.*, 2004, **377**, 159–169.
161. Y. Y. Tam, S. Chen, J. Zaifman, Y. K. Tam, P. J. Lin, S. Ansell, M. Roberge, M. A. Ciufolini and P. R. Cullis, *Nanomedicine*, 2013, **9**, 665–674.
162. C. Zhou, Y. Zhang, B. Yu, M. A. Phelps, L. J. Lee and R. J. Lee, *Nanomedicine*, 2013, **9**, 504–513.
163. J. Rejman, A. Bragonzi and M. Conese, *Mol. Ther.*, 2005, **12**, 468–474.
164. C. Zhou, Y. Mao, Y. Sugimoto, Y. Zhang, N. Kanthamneni, B. Yu, R. W. Brueggemeier, L. J. Lee and R. J. Lee, *Mol. Pharm.*, 2012, **9**, 201–210.
165. Z. Rehman, I. S. Zuhorn and D. Hoekstra, *J. Controlled Release*, 2013, **166**, 46–56.
166. O. Boussif, F. Lezoualc'h, M. A. Zanta, M. D. Mergny, D. Scherman, B. Demeneix and J. P. Behr, *Proc. Natl. Acad. Sci. U. S. A.*, 1995, **92**, 7297–7301.
167. J. Zhang, H. Fan, D. A. Levorse and L. S. Crocker, *Langmuir*, 2011, **27**, 1907–1914.
168. C. Plank, W. Zauner and E. Wagner, *Adv. Drug Delivery Rev.*, 1998, **34**, 21–35.

169. S. W. Kim, N. Y. Kim, Y. B. Choi, S. H. Park, J. M. Yang and S. Shin, *J. Controlled Release*, 2010, **143**, 335–343.
170. J. Heyes, L. Palmer, K. Bremner and I. MacLachlan, *J. Controlled Release*, 2005, **107**, 276–287.
171. M. Wang, X. Li, Y. Ma and H. Gu, *Int. J. Pharm.*, 2013, **448**, 51–57.
172. A. Sridharan, C. Patel and J. Muthuswamy, *Mol. Ther. Nucleic Acids*, 2013, **2**, e82.
173. J. A. Jorgensen, A. S. Longva, E. Hovig and S. L. Boe, *Nucleic Acid Ther.*, 2013, **23**, 131–139.
174. D. W. Bartlett and M. E. Davis, *Biotechnol. Bioeng.*, 2007, **97**, 909–921.
175. H. Lim, J. Noh, Y. Kim, H. Kim, J. Kim, G. Khang and D. Lee, *Biomacromolecules*, 2013, **14**, 240–247.
176. T. Endoh and T. Ohtsuki, *Adv. Drug Delivery Rev.*, 2009, **61**, 704–709.
177. S. Huth, F. Hoffmann, K. von Gersdorff, A. Laner, D. Reinhardt, J. Rosenecker and C. Rudolph, *J. Gene Med.*, 2006, **8**, 1416–1424.
178. J. Zhang, H. Fan, D. A. Levorse and L. S. Crocker, *Langmuir*, 2011, **27**, 9473–9483.
179. M. A. Gosselin, W. Guo and R. J. Lee, *Bioconjugate Chem.*, 2001, **12**, 989–994.
180. M. L. Forrest, J. T. Koerber and D. W. Pack, *Bioconjugate Chem.*, 2003, **14**, 934–940.
181. L. Crombez, A. Charnet, M. C. Morris, G. Aldrian-Herrada, F. Heitz and G. Divita, *Biochem. Soc. Trans.*, 2007, **35**, 44–46.
182. S. J. Lee, M. S. Huh, S. Y. Lee, S. Min, S. Lee, H. Koo, J. U. Chu, K. E. Lee, H. Jeon, Y. Choi, K. Choi, Y. Byun, S. Y. Jeong, K. Park, K. Kim and I. C. Kwon, *Angew Chem., Int. Ed.*, 2012, **51**, 7203–7207.
183. C. E. Holt and S. L. Bullock, *Science*, 2009, **326**, 1212–1216.
184. A. Jagannath and M. J. Wood, *Mol. Biol. Cell*, 2009, **20**, 521–529.
185. C. A. Alabi, G. Sahay, R. Langer and D. G. Anderson, *Integr. Biol.*, 2013, **5**, 224–230.
186. S. Tyagi and F. R. Kramer, *Nat. Biotechnol.*, 1996, **14**, 303–308.
187. J. Wu, W. Huang and Z. He, *Sci. World J.*, 2013, **2013**, 630654.
188. T. Musacchio and V. P. Torchilin, *Front. Biosci., Landmark Ed.*, 2013, **18**, 58–79.
189. K. Singha, R. Namgung and W. J. Kim, *Nucleic Acid Ther.*, 2011, **21**, 133–147.
190. N. Saranya, A. Moorthi, S. Saravanan, M. P. Devi and N. Selvamurugan, *Int. J. Biol. Macromol.*, 2011, **48**, 234–238.
191. T. K. Dash and V. B. Konkimalla, *Crit. Rev. Ther. Drug Carrier Syst.*, 2013, **30**, 469–493.
192. S. El Andaloussi, S. Lakhal, I. Mager and M. J. Wood, *Adv. Drug Delivery Rev.*, 2013, **65**, 391–397.
193. P. Guo, O. Coban, N. M. Snead, J. Trebley, S. Hoeprich, S. Guo and Y. Shu, *Adv. Drug Delivery Rev.*, 2010, **62**, 650–666.
194. R. C. Kane, A. T. Farrell, H. Saber, S. Tang, G. Williams, J. M. Jee, C. Liang, B. Booth, N. Chidambaram, D. Morse, R. Sridhara,

P. Garvey, R. Justice and R. Pazdur, *Clin. Cancer Res.*, 2006, **12**, 7271–7278.

195. M. Hashida, S. Kawakami and F. Yamashita, *Chem. Pharm. Bull.*, 2005, **53**, 871–880.

196. D. Landesman-Milo, M. Goldsmith, S. Leviatan Ben-Arye, B. Witenberg, E. Brown, S. Leibovitch, S. Azriel, S. Tabak, V. Morad and D. Peer, *Cancer Lett.*, 2013, **334**, 221–227.

197. P. Y. Chien, J. Wang, D. Carbonaro, S. Lei, B. Miller, S. Sheikh, S. M. Ali, M. U. Ahmad and I. Ahmad, *Cancer Gene Ther.*, 2005, **12**, 321–328.

198. A. Pal, A. Ahmad, S. Khan, I. Sakabe, C. Zhang, U. N. Kasid and I. Ahmad, *Int. J. Oncol.*, 2005, **26**, 1087–1091.

199. A. S. Malamas, M. Gujrati, C. M. Kummitha, R. Xu and Z. R. Lu, *J. Controlled Release*, 2013, **171**, 296–307.

200. D. Di Paolo, C. Ambrogio, F. Pastorino, C. Brignole, C. Martinengo, R. Carosio, M. Loi, G. Pagnan, L. Emionite, M. Cilli, D. Ribatti, T. M. Allen, R. Chiarle, M. Ponzoni and P. Perri, *Mol. Ther.*, 2011, **19**, 2201–2212.

201. D. Di Paolo, C. Brignole, F. Pastorino, R. Carosio, A. Zorzoli, M. Rossi, M. Loi, G. Pagnan, L. Emionite, M. Cilli, S. Bruno, R. Chiarle, T. M. Allen, M. Ponzoni and P. Perri, *Mol. Ther.*, 2011, **19**, 1131–1140.

202. S. Cressman, I. Dobson, J. B. Lee, Y. Y. Tam and P. R. Cullis, *Bioconjugate Chem.*, 2009, **20**, 1404–1411.

203. Y. Barenholz, C. Bombelli, M. G. Bonicelli, P. di Profio, L. Giansanti, G. Mancini and F. Pascale, *J. Colloid Interface Sci.*, 2011, **356**, 46–53.

204. A. M. Cardoso, S. Trabulo, A. L. Cardoso, S. Maia, P. Gomes, A. S. Jurado and M. C. Pedroso de Lima, *Mol. Pharm.*, 2013, **10**(7), 2653–2666.

205. C. Bardita, D. Predescu and S. Predescu, *J. Visualized Exp.*, 2013, **76**, 50316.

206. G. Mainelis, S. Seshadri, O. B. Garbuzenko, T. Han, Z. Wang and T. Minko, *J. Aerosol Med. Pulm. Drug Delivery*, 2013, **26**, 345–354.

207. T. S. Zimmermann, A. C. Lee, A. Akinc, B. Bramlage, D. Bumcrot, M. N. Fedoruk, J. Harborth, J. A. Heyes, L. B. Jeffs, M. John, A. D. Judge, K. Lam, K. McClintock, L. V. Nechev, L. R. Palmer, T. Racie, I. Rohl, S. Seiffert, S. Shanmugam, V. Sood, J. Soutschek, I. Toudjarska, A. J. Wheat, E. Yaworski, W. Zedalis, V. Koteliansky, M. Manoharan, H. P. Vornlocher and I. MacLachlan, *Nature*, 2006, **441**, 111–114.

208. A. D. Judge, M. Robbins, I. Tavakoli, J. Levi, L. Hu, A. Fronda, E. Ambegia, K. McClintock and I. MacLachlan, *J. Clin. Invest.*, 2009, **119**, 661–673.

209. L. Wightman, R. Kircheis, V. Rossler, S. Carotta, R. Ruzicka, M. Kursa and E. Wagner, *J. Gene Med.*, 2001, **3**, 362–372.

210. R. B. Shmueli, D. G. Anderson and J. J. Green, *Expert Opin. Drug Delivery*, 2010, 7, 535–550.

211. D. Fischer, Y. Li, B. Ahlemeyer, J. Krieglstein and T. Kissel, *Biomaterials*, 2003, **24**, 1121–1131.

212. G. Navarro, S. Essex, R. R. Sawant, S. Biswas, D. Nagesha, S. Sridhar, C. T. de Ilarduya and V. P. Torchilin, *Nanomedicine*, 2013, **10**(2), 411–419.
213. D. Kim, J. Hong, H. H. Moon, H. Y. Nam, H. Mok, J. H. Jeong, S. W. Kim, D. Choi and S. H. Kim, *J Controlled Release*, 2013, **168**, 125–134.
214. S. J. Tseng, Y. F. Zeng, Y. F. Deng, P. C. Yang, J. R. Liu and I. M. Kempson, *Chem. Commun.*, 2013, **49**, 2670–2672.
215. X. Dong, L. Lin, J. Chen, Z. Guo, H. Tian, Y. Li, Y. Wei and X. Chen, *Macromol. Biosci.*, 2013, **13**, 512–522.
216. F. T. Vicentini, L. V. Depieri, A. C. Polizello, J. O. Del Ciampo, A. C. Spadaro, M. C. Fantini, Vitoria Lopes and M. Badra Bentley, *Eur. J. Pharm. Biopharm.*, 2013, **83**, 16–24.
217. W. E. Rudzinski and T. M. Aminabhavi, *Int. J. Pharm.*, 2010, **399**, 1–11.
218. G. M. Escott and D. J. Adams, *Infect. Immun.*, 1995, **63**, 4770–4773.
219. T. Chandy and C. P. Sharma, *Biomater., Artif, Cells, Artif. Organs*, 1990, **18**, 1–24.
220. M. Matokanovic, K. Barisic, J. Filipovic-Grcic and D. Maysinger, *Eur. J. Pharm. Sci.*, 2013, **50**, 149–158.
221. J. Malmo, A. Sandvig, K. M. Varum and S. P. Strand, *PLoS One*, 2013, **8**, e54182.
222. B. Khurana, A. K. Goyal, A. Budhiraja, D. Aora and S. P. Vyas, *Drug Delivery*, 2013, **20**, 57–64.
223. Y. Wang, Z. Li, Y. Han, L. H. Liang and A. Ji, *Curr. Drug Metab.*, 2010, **11**, 182–196.
224. X. C. Shen, J. Zhou, X. Liu, J. Wu, F. Qu, Z. L. Zhang, D. W. Pang, G. Quelever, C. C. Zhang and L. Peng, *Org. Biomol. Chem.*, 2007, **5**, 3674–3681.
225. T. Minko, M. L. Patil, M. Zhang, J. J. Khandare, M. Saad, P. Chandna and O. Taratula, *Methods Mol. Biol.*, 2010, **624**, 281–294.
226. P. Posocco, X. Liu, E. Laurini, D. Marson, C. Chen, C. Liu, M. Fermeglia, P. Rocchi, S. Pricl and L. Peng, *Mol. Pharm.*, 2013, **10**, 3262–3273.
227. S. M. Elbashir, J. Martinez, A. Patkaniowska, W. Lendeckel and T. Tuschl, *EMBO J.*, 2001, **20**, 6877–6888.
228. A. L. Bolcato-Bellemin, M. E. Bonnet, G. Creusat, P. Erbacher and J. P. Behr, *Proc. Natl. Acad. Sci. U. S. A.*, 2007, **104**, 16050–16055.
229. A. Zoubeidi and M. Gleave, *Int. J. Biochem. Cell Biol.*, 2012, **44**, 1646–1656.
230. V. Baylot, M. Katsogiannou, C. Andrieu, D. Taieb, J. Acunzo, S. Giusiano, L. Fazli, M. Gleave, C. Garrido and P. Rocchi, *Mol. Ther.*, 2012, **20**, 2244–2256.
231. J. Liu, C. Gu, E. B. Cabigas, K. D. Pendergrass, M. E. Brown, Y. Luo and M. E. Davis, *Biomaterials*, 2013, **34**, 3729–3736.
232. C. M. Ferrario and W. B. Strawn, *Am. J. Cardiol.*, 2006, **98**, 121–128.
233. O. Taratula, O. B. Garbuzenko, P. Kirkpatrick, I. Pandya, R. Savla, V. P. Pozharov, H. He and T. Minko, *J. Controlled Release*, 2009, **140**, 284–293.
234. R. C. Adami and K. G. Rice, *J. Pharm. Sci.*, 1999, **88**, 739–746.

235. A. H. van Asbeck, A. Beyerle, H. McNeill, P. H. Bovee-Geurts, S. Lindberg, W. P. Verdurmen, M. Hallbrink, U. Langel, O. Heidenreich and R. Brock, *ACS Nano*, 2013, 7, 3797–3807.
236. A. D. Tagalakis, L. Saraiva, D. McCarthy, K. T. Gustafsson and S. L. Hart, *Biomacromolecules*, 2013, 14, 761–770.
237. L. Crombez, M. C. Morris, S. Dufort, G. Aldrian-Herrada, Q. Nguyen, G. Mc Master, J. L. Coll, F. Heitz and G. Divita, *Nucleic Acids Res.*, 2009, 37, 4559–4569.
238. S. Zadran, G. Akopian, H. Zadran, J. Walsh and M. Baudry, *Neuromol. Med.*, 2013, 15, 74–81.
239. S. J. Lee, J. Y. Yhee, S. H. Kim, I. C. Kwon and K. Kim, *J. Controlled Release*, 2013, 172, 358–366.
240. S. Boncel, P. Zajac and K. K. Koziol, *J. Controlled Release*, 2013, 169, 126–140.
241. X. Zhang, L. Meng, Q. Lu, Z. Fei and P. J. Dyson, *Biomaterials*, 2009, 30, 6041–6047.
242. E. Heister, V. Neves, C. Tilmaciu, K. Lipert, V. S. Beltran, H. M. Coley, S. R. P. Silva and J. McFadden, *Carbon*, 2009, 47, 2152–2160.
243. B. S. Wong, S. L. Yoong, A. Jagusiak, T. Panczyk, H. K. Ho, W. H. Ang and G. Pastorin, *Adv. Drug Delivery Rev.*, 2013, 65, 1964–2015.
244. K. Kostarelos, L. Lacerda, G. Pastorin, W. Wu, S. Wieckowski, J. Luangsivilay, S. Godefroy, D. Pantarotto, J. P. Briand, S. Muller, M. Prato and A. Bianco, *Nat. Nanotechnol.*, 2007, 2, 108–113.
245. A. K. Iyer, G. Khaled, J. Fang and H. Maeda, *Drug Discovery Today*, 2006, 11, 812–818.
246. A. Battigelli, J. T. Wang, J. Russier, T. Da Ros, K. Kostarelos, K. T. Al-Jamal, M. Prato and A. Bianco, *Small*, 2013, 9(21), 3610–3619.
247. L. Wang, J. Shi, H. Zhang, H. Li, Y. Gao, Z. Wang, H. Wang, L. Li, C. Zhang, C. Chen, Z. Zhang and Y. Zhang, *Biomaterials*, 2013, 34, 262–274.
248. W. Zhang and L. Xing, *Int. J. Oncol.*, 2013, 43, 1228–1234.
249. J. M. Aliotta, M. Pereira, K. W. Johnson, N. de Paz, M. S. Dooner, N. Puente, C. Ayala, K. Brilliant, D. Berz D. Lee, B. Ramratnam, P. N. McMillan, D. C. Hixson, D. Josic and P. J. Quesenberry *Exp. Hematol.*, 2010, 38, 233-245.
250. F. Collino, M. C. Deregibus, S. Bruno, L. Sterpone, G. Aghemo, L. Viltono, C. Tetta and G. Camussi, *PLoS One*, 2010, 5, e11803.
251. S. Ohno, M. Takanashi, K. Sudo, S. Ueda, A. Ishikawa, N. Matsuyama, K. Fujita, T. Mizutani, T. Ohgi, T. Ochiya, N. Gotoh and M. Kuroda, *Mol. Ther.*, 2013, 21, 185–191.
252. L. Alvarez-Erviti, Y. Seow, H. Yin, C. Betts, S. Lakhal and M. J. Wood, *Nat. Biotechnol.*, 2011, 29, 341–345.
253. Y. K. Buchman, E. Lellouche, S. Zigdon, M. Bechor, S. Michaeli and J. P. Lellouche, *Bioconjugate Chem.*, 2013, 24, 2076–2087.

254. M. Zhang, R. Xu, X. Xia, Y. Yang, J. Gu, G. Qin, X. Liu, M. Ferrari and H. Shen, *Biomaterials*, 2014, **35**, 423–431.

255. X. Ma, C. Teh, Q. Zhang, P. Borah, C. Choong, V. Korzh and Y. Zhao, *Antioxid. Redox Signaling*, 2013, **21**(5), 707–702.

256. T. Ito, N. Iida-Tanaka, T. Niidome, T. Kawano, K. Kubo, K. Yoshikawa, T. Sato, Z. Yang and Y. Koyama, *J. Controlled Release*, 2006, **112**, 382–388.

257. H. Mok, J. W. Park and T. G. Park, *Bioconjugate Chem.*, 2007, **18**, 1483–1489.

258. S. P. Evanko and T. N. Wight, *J. Histochem. Cytochem.*, 1999, **47**, 1331–1342.

259. S. Al-Qadi, M. Alatorre-Meda, E. M. Zaghloul, P. Taboada and C. Remunan-Lopez, *Colloids Surf. B Biointerfaces*, 2013, **103**, 615–623.

260. Y. Mi, Y. Liu and S. S. Feng, *Biomaterials*, 2011, **32**, 4058–4066.

261. P. S. Ghosh, C. K. Kim, G. Han, N. S. Forbes and V. M. Rotello, *ACS Nano*, 2008, **2**, 2213–2218.

262. J. Zhao, Y. Mi and S. S. Feng, *Biomaterials*, 2013, **34**, 3411–3421.

263. S. Castleberry, M. Wang and P. T. Hammond, *ACS Nano*, 2013, 7, 5251–5261.

264. H. L. Johns, C. Gonzalez-Lopez, C. L. Sayers, M. Hollinshead and G. Elliott, *Traffic*, 2013, **15**(2), 157–178.

265. M. Omedes Pujol, D. J. Coleman, C. D. Allen, O. Heidenreich and D. A. Fulton, *J. Controlled Release*, 2013, **172**, 939–945.

266. I. Vorackova, P. Ulbrich, W. E. Diehl and T. Ruml, *Arch. Virol.*, 2013, **159**(4), 677–688.

267. J. Han, Q. Wang, Z. Zhang, T. Gong and X. Sun, *Small*, 2013, **10**(3), 524–535.

268. Y. Wu, J. Ji, R. Yang, X. Zhang, Y. Li, Y. Pu and X. Li, *Drug Delivery*, 2013, **20**, 296–305.

269. R. Jiang, X. Lu, M. Yang, W. Deng, Q. Fan and W. Huang, *Biomacromolecules*, 2013, **14**, 3643–3652.

270. J. S. Kim, Y. K. Lee, H. Y. Jeong, S. J. Kang, M. W. Kim, S. H. Ryu, H. S. Kim, K. S. Kim, D. E. Kim and Y. S. Park, *Yonsei Med. J.*, 2013, **54**, 1149–1157.

271. J. Bae, M. Mie and E. Kobatake, *Biotechnol. Lett.*, 2013, **35**, 2081–2089.

272. A. Mitsueda, Y. Shimatani, M. Ito, T. Ohgita, A. Yamada, S. Hama, A. Graslund, S. Lindberg, U. Langel, H. Harashima, I. Nakase, S. Futaki and K. Kogure, *Biopolymers*, 2013, **100**, 698–704.

273. S. Sanna-Cherchi, R. V. Sampogna, N. Papeta, K. E. Burgess, S. N. Nees, B. J. Perry, M. Choi, M. Bodria, Y. Liu, P. L. Weng, V. J. Lozanovski, M. Verbitsky, F. Lugani, R. Sterken, N. Paragas, G. Caridi, A. Carrea, M. Dagnino, A. Materna-Kiryluk, G. Santamaria, C. Murtas, N. Ristoska-Bojkovska, C. Izzi, N. Kacak, B. Bianco, S. Giberti, M. Gigante, G. Piaggio, L. Gesualdo, D. Kosuljandic Vukic, K. Vukojevic, M. Saraga-Babic, M. Saraga, Z. Gucev, L. Allegri, A. Latos-Bielenska, D. Casu, M. State, F. Scolari, R. Ravazzolo, K. Kiryluk, Q. Al-Awqati, V. D. D'Agati,

I. A. Drummond, V. Tasic, R. P. Lifton, G. M. Ghiggeri and A. G. Gharavi, *N. Engl. J. Med.*, 2013, **369**, 621–629.

274. B. Xiao, H. Laroui, S. Ayyadurai, E. Viennois, M. A. Charania, Y. Zhang and D. Merlin, *Biomaterials*, 2013, **34**, 7471–7482.

275. A. M. O'Mahony, J. Ogier, R. Darcy, J. F. Cryan and C. M. O'Driscoll, *PLoS One*, 2013, **8**, e66413.

276. K. Guk, H. Lim, B. Kim, M. Hong, G. Khang and D. Lee, *Int. J. Pharm.*, 2013, **453**, 541–550.

277. J. Barman-Aksozen, C. Beguin, A. M. Dogar, X. Schneider-Yin and E. I. Minder, *Blood Cells, Mol., Dis.*, 2013, **51**, 151–161.

278. J. M. Li, Y. Y. Wang, W. Zhang, H. Su, L. N. Ji and Z. W. Mao, *Int. J. Nanomed.*, 2013, **8**, 2101–2117.

279. L. Qin, Y. Sun, P. Liu, Q. Wang, B. Han and Y. Duan, *J. Biomater. Sci. Polym. Ed.*, 2013, **24**, 1757–1766.

280. Y. L. Lin, Y. Yuksel Durmaz, J. E. Nor and M. E. Elsayed, *Mol. Pharm.*, 2013, **10**(7), 2730–2738.

281. A. Falamarzian, H. M. Aliabadi, O. Molavi, J. M. Seubert, R. Lai, H. Uludag and A. Lavasanifar, *J. Biomed. Mater. Res., Part A*, 2013, **102**(9), 3216–3228.

282. T. Wagner, M. Dieckmann, S. Jaeger, S. Weggen and C. U. Pietrzik, *Exp. Cell Res.*, 2013, **319**, 1956–1972.

283. N. Yoneda, A. Yasue, T. Watanabe and E. Tanaka, *J. Dent. Res.*, 2013, **92**, 716–720.

284. O. Taratula, A. Kuzmov, M. Shah, O. B. Garbuzenko and T. Minko, *J. Controlled Release*, 2013, **171**, 349–357.

285. D. D. Bruyn Ouboter, T. Schuster, V. Shanker, M. Heim and W. Meier, *J. Biomed. Mater. Res., Part A*, 2013, **102**(4), 1155–1163.

286. D. Lin, Q. Jiang, Q. Cheng, Y. Huang, P. Huang, S. Han, S. Guo, Z. Liang and A. Dong, *Acta Biomater.*, 2013, **9**, 7746–7757.

287. Y. Zhao, Y. Qin, Y. Liang, H. Zou, X. Peng, H. Huang, M. Lu and M. Feng, *Biomacromolecules*, 2013, **14**, 1777–1786.

288. L. Yin, Z. Song, Q. Qu, K. H. Kim, N. Zheng, C. Yao, I. Chaudhury, H. Tang, N. P. Gabrielson, F. M. Uckun and J. Cheng, *Angew Chem., Int. Ed.*, 2013, **52**, 5757–5761.

289. H. Wei, L. R. Volpatti, D. L. Sellers, D. O. Maris, I. W. Andrews, A. S. Hemphill, L. W. Chan, D. S. Chu, P. J. Horner and S. H. Pun, *Angew Chem., Int. Ed.*, 2013, **52**, 5377–5381.

290. H. Y. Cho, S. E. Averick, E. Paredes, K. Wegner, A. Averick, S. Jurga, S. R. Das and K. Matyjaszewski, *Biomacromolecules*, 2013, **14**, 1262–1267.

291. Y. Shen, H. Fang, K. Zhang, R. Shrestha, K. L. Wooley and J. S. Taylor, *Nucleic Acid Ther.*, 2013, **23**, 95–108.

292. S. Sciarretta, S. Marchitti, F. Bianchi, A. Moyes, E. Barbato, S. Di Castro, R. Stanzione, M. Cotugno, L. Castello, C. Calvieri, I. Eberini, J. Sadoshima, A. J. Hobbs, M. Volpe and S. Rubattu, *Circ. Res.*, 2013, **112**, 1355–1364.

293. J. Guo, A. M. O'Mahony, W. P. Cheng and C. M. O'Driscoll, *Int. J. Pharm.*, 2013, **447**, 150–157.
294. W. Wei, P. P. Lv, X. M. Chen, Z. G. Yue, Q. Fu, S. Y. Liu, H. Yue and G. H. Ma, *Biomaterials*, 2013, **34**, 3912–3923.
295. L. Rose, H. M. Aliabadi and H. Uludag, *Acta Biomater.*, 2013, **9**, 7429–7438.
296. M. Micaroni, A. C. Stanley, T. Khromykh, J. Venturato, C. X. Wong, J. P. Lim, B. J. Marsh, B. Storrie, P. A. Gleeson and J. L. Stow, *PLoS One*, 2013, **8**, e57034.
297. G. Lin, R. Hu, W. C. Law, C. K. Chen, Y. Wang, H. Li Chin, Q. T. Nguyen, C. K. Lai, H. S. Yoon, X. Wang, G. Xu, L. Ye, C. Cheng and K. T. Yong, *Small*, 2013, **9**, 2757–2763.
298. M. Mahato, P. Kumar and A. K. Sharma, *Mol. Biosyst.*, 2013, **9**, 780–791.
299. X. Dong, L. Lin, J. Chen, H. Tian, C. Xiao, Z. Guo, Y. Li, Y. Wei and X. Chen, *Acta Biomater.*, 2013, **9**, 6943–6952.
300. S. R. Jung, N. J. Song, D. K. Yang, Y. J. Cho, B. J. Kim, J. W. Hong, U. J. Yun, D. G. Jo, Y. M. Lee, S. Y. Choi and K. W. Park, *Nutr. Res.*, 2013, **33**, 162–170.
301. C. Zheng, M. Zheng, P. Gong, J. Deng, H. Yi, P. Zhang, Y. Zhang, P. Liu, Y. Ma and L. Cai, *Biomaterials*, 2013, **34**, 3431–3438.
302. A. M. Wilson, B. Morquette, M. Abdouh, N. Unsain, P. A. Barker, E. Feinstein, G. Bernier and A. Di Polo, *J. Neurosci.*, 2013, **33**, 2205–2216.
303. U. R. Michaelis, E. Chavakis, C. Kruse, B. Jungblut, D. Kaluza, K. Wandzioch, Y. Manavski, H. Heide, M. J. Santoni, M. Potente, J. A. Eble, J. P. Borg and R. P. Brandes, *Circ. Res.*, 2013, **112**, 924–934.
304. B. B. Lundy, A. Convertine, M. Miteva and P. S. Stayton, *Bioconjugate Chem.*, 2013, **24**, 398–407.
305. K. Lee, M. H. Oh, M. S. Lee, Y. S. Nam, T. G. Park and J. H. Jeong, *Int. J. Pharm.*, 2013, **445**, 196–202.
306. J. Qian and X. Gao, *ACS Appl. Mater. Interfaces*, 2013, **5**, 2845–2852.
307. J. Zhang, C. He, C. Tang and C. Yin, *Pharm. Res.*, 2013, **30**, 1228–1239.
308. P. Pierrat, G. Laverny, G. Creusat, P. Wehrung, J. M. Strub, A. VanDorsselaer, F. Pons, G. Zuber and L. Lebeau, *Chemistry*, 2013, **19**, 2344–2355.
309. H. Meng, W. X. Mai, H. Zhang, M. Xue, T. Xia, S. Lin, X. Wang, Y. Zhao, Z. Ji, J. I. Zink and A. E. Nel, *ACS Nano*, 2013, 7, 994–1005.
310. Y. Hattori, T. Nakamura, H. Ohno, N. Fujii and Y. Maitani, *Int. J. Pharm.*, 2013, **443**, 221–229.
311. K. M. Choi, G. L. Park, K. Y. Hwang, J. W. Lee and H. J. Ahn, *Mol. Pharm.*, 2013, **10**, 763–773.
312. Z. C. Feng, M. Riopel, J. Li, L. Donnelly and R. Wang, *Am. J. Physiol. Endocrinol. Metab.*, 2013, **304**, E557–565.
313. J. R. Doyle, J. M. Lane, M. Beinborn and A. S. Kopin, *J. Lipid Res.*, 2013, **54**, 823–830.

314. H. X. Wang, M. H. Xiong, Y. C. Wang, J. Zhu and J. Wang, *J. Controlled Release*, 2013, **166**, 106–114.
315. C. Kienzle, S. A. Eisler, J. Villeneuve, T. Brummer, M. A. Olayioye and A. Hausser, *Mol. Biol. Cell*, 2013, **24**, 222–233.
316. A. Fujimoto, M. Kurban, M. Nakamura, M. Farooq, H. Fujikawa, A. G. Kibbi, M. Ito, M. Dahdah, M. Matta, H. Diab and Y. Shimomura, *J. Dermatol. Sci.*, 2013, **69**, 159–166.
317. S. Vartanian, C. Bentley, M. J. Brauer, L. Li, S. Shirasawa, T. Sasazuki, J. S. Kim, P. Haverty, E. Stawiski, Z. Modrusan, T. Waldman and D. Stokoe, *J. Biol. Chem.*, 2013, **288**, 2403–2413.
318. S. Y. Tzeng and J. J. Green, *Adv. Healthcare Mater.*, 2013, **2**, 468–480.
319. K. H. Bae, J. Y. Lee, S. H. Lee, T. G. Park and Y. S. Nam, *Adv. Healthcare Mater.*, 2013, **2**, 576–584.
320. Y. H. Li, Q. S. Shi, J. Du, L. F. Jin, L. F. Du, P. F. Liu and Y. R. Duan, *Int. J. Mol. Med.*, 2013, **31**, 163–171.
321. B. M. Godinho, J. R. Ogier, R. Darcy, C. M. O'Driscoll and J. F. Cryan, *Mol. Pharm.*, 2013, **10**, 640–649.
322. F. A. Galaway and P. G. Stockley, *Mol. Pharm.*, 2013, **10**, 59–68.
323. T. Fujita, K. Yanagihara, F. Takeshita, K. Aoyagi, T. Nishimura, M. Takigahira, F. Chiwaki, T. Fukagawa, H. Katai, T. Ochiya, H. Sakamoto, H. Konno, T. Yoshida and H. Sasaki, *Cancer Sci.*, 2013, **104**, 214–222.
324. Y. Liu, Z. Liu, Y. Wang, Y. R. Liang, X. Wen, J. Hu, X. Yang, J. Liu, S. Xiao and D. Cheng, *Brain Res.*, 2013, **1490**, 43–51.
325. M. Chen, H. M. Cooper, J. Z. Zhou, P. F. Bartlett and Z. P. Xu, *J. Colloid Interface Sci.*, 2013, **390**, 275–281.
326. A. O. Tezgel, G. Gonzalez-Perez, J. C. Telfer, B. A. Osborne, L. M. Minter and G. N. Tew, *Mol. Ther.*, 2013, **21**, 201–209.
327. T. Kudlyk, R. Willett, I. D. Pokrovskaya and V. Lupashin, *Traffic*, 2013, **14**, 194–204.
328. J. M. Pelet and D. Putnam, *Pharm. Res.*, 2013, **30**, 362–376.
329. E. Kobayashi, A. K. Iyer, F. J. Hornicek, M. M. Amiji and Z. Duan, *Clin. Orthop. Relat. Res.*, 2013, **471**, 915–925.
330. B. Fang, L. Jiang, M. Zhang and F. Z. Ren, *Biochimie*, 2013, **95**, 251–257.
331. N. Niimi, K. Kohyama and Y. Matsumoto, *J. Neuroimmunol.*, 2013, **254**, 39–45.
332. K. M. Choi, K. Kim, I. C. Kwon, I. S. Kim and H. J. Ahn, *Mol. Pharm.*, 2013, **10**, 18–25.

CHAPTER 4

Magnetic Targeting as a Vehicle for the Delivery of Nanomedicines

MITSUO OCHI* AND GOKI KAMEI

Department of Orthopaedic Surgery, Graduate School of Biomedical Science, Hiroshima University, 1-2-3, Kasumi, Minami-ku, Hiroshima, Japan
*Email: ochim@hiroshima-u.ac.jp

4.1 Introduction

Regenerative medicine has progressed recently thanks to the use of various cells, scaffolds and growth factors. Most regenerative medicine research focuses on tissue engineering, but this usually requires technically demanding procedures with some special equipment or facilities and proper scaffolds or a growth factor. Therefore, intravenous or intra-articular cell transplantation without scaffolds is a potentially more attractive option for tissue repair. In terms of cell sources, our studies have confirmed that bone marrow derived mesenchymal stem cells (MSCs) are obtained particularly easily from the iliac crest or proximal tibia without any immunogenic re-action to autologous bone marrow. In addition, it has been reported that MSCs can differentiate into several lineages, including osteogenic, chon-drogenic, or adipogenic lineages and that they can be applied effectively to tissue repair. Therefore, we have employed bone marrow derived MSCs in our regenerative medicine research.[1,2] The results have shown the efficacy of bone marrow derived MSCs, with indications that a large number of MSCs is

RSC Drug Discovery Series No. 51
Nanomedicines: Design, Delivery and Detection
Edited by Martin Braddock
© The Royal Society of Chemistry 2016
Published by the Royal Society of Chemistry, www.rsc.org

needed for improved regeneration. However, scar tissue and loose bodies occur in the knee joint if we use a large number of cells to treat cartilage injury in the rat cartilage defect model.[2] Therefore, for effective cartilage repair we concluded that it is essential to accumulate the appropriate number of MSCs in the cartilage defect, meaning that avoiding an excess of cells reduces the rate of complications such as scar tissue formation. There have been several reports on the local adhesion technique using gravity, whereby an upward-facing cartilage defect was filled with synovial MSCs, and the position was held for 10 minutes. This technique comprises a less invasive procedure, with good results. However, the number of cells attached to the cartilage defect was about only 60% of the injected cells in an *ex vivo* trial.[3] Ochi was motivated by the belief that a more effective accumulation of MSCs would be possible if the MSCs were attached to the cartilage defect in the same way that iron is attracted by a magnet. In 2008, Ochi and colleagues achieved a world first by successfully accumulating MSCs labeled with supra-paramagnetic iron oxide (m-MSCs) in the cartilage defect under an external magnetic force in a rabbit and fresh frozen porcine model (Figure 4.1).[4] Furthermore, he proved that m-MSCs could be accumulated and engrafted at the cartilage defect site of an *in vivo* porcine model[5] (Figure 4.2).

4.2   External Magnetic Device

We made use of a bulk superconducting magnet system (Hitachi Ltd, Ibaraki, Japan), which is a device of 10 cm diameter, characterized as a permanent magnet system because the magnetic force is directed to the center of the disk surface, and its magnitude decreases away from the surface. This allows us to effectively accumulate a relatively small number of m-MSCs in a desired area (Figure 4.3). In our study, the magnetic force at the center of the magnet surface was 5 Tesla (T), and at 30 mm away from the surface it was approximately 0.5 T.[6]

After working on further downsizing, the new device fits into the palm of the hand, with a maximum magnetic field of 3 T and a diameter of 2 cm (Figure 4.3).

4.3   Labeling Mesenchymal Stem Cells with Supraparamagnetic Iron Oxide

It has been reported that MSCs can differentiate into several lineages, including osteogenic, chondrogenic, or adipogenic cells, and there are many reports that refer to the cell proliferation and differentiation into several lineages of magnetically labeled MSCs.[7,8] MSCs are labeled with super-paramagnetic iron oxide (SPIO) nanoparticles only or with SPIO complexed to transfection agents (TA), for example, ferumoxide, ferucarbotran, ferum-oxide-proramine, poly-L-lysine(PLL)-ferumoxide, *etc.* There have been various reports about cell toxicity and the blocking of cellular differentiation of SPIO and TA.[9–11] However, there has not been a clear standard regarding which

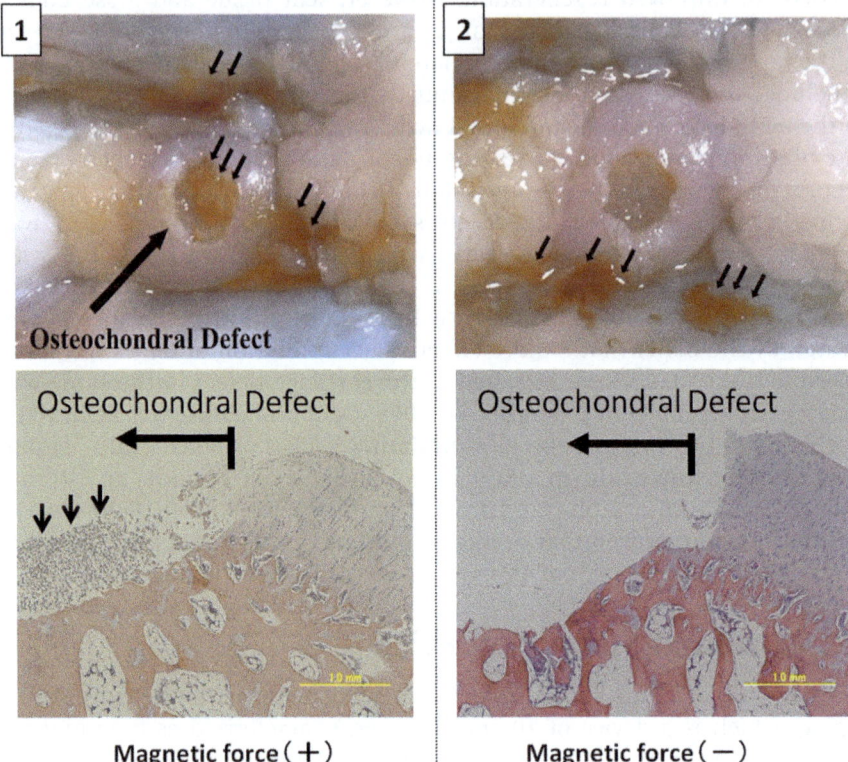

Figure 4.1   Rabbit model. 1: A magnetic force causes the m-MSCs to accumulate in
the cartilage defect. 2: Lack of a magnetic force meant that m-MSCs did
not accumulate in the cartilage defect.[4]
(Reprinted from *Arthroscopy: The Journal of Arthroscopic & Related Surgery*, vol. 24, T. Kobayashi, M. Ochi, S. Yanada, M. Ishikawa, N. Adachi,
M. Deie and K. Arihiro, A Novel Cell Delivery System Using Magnetically
Labeled Mesenchymal Stem Cells and an External Magnetic Device for
Clinical Cartilage Repair, pp. 69–76, Copyright (2008), with permission
from Elsevier.)

SPIO and TA medicines should be used. Recent studies have tended to show
that ferucarbotran is less toxic than Feridex and that protamine is less toxic
than PLL.[12,13] In our research, we always emphasise the need for safety, so in
this study we used ferucarbotoran and protamine, since they have been
approved by the US Food and Drug Administration, the former as a magnetic
resonance contrast agent and the latter as an antidote to heparin anti-
coagulation. However, no previous study has confirmed whether magnetically
labeled MSCs (exposed to external magnetic fields) are able to proliferate and
differentiate into several lineages. In addition, there have been few studies on
the effects of static magnetic fields at the cellular level, and the full effects of
strength and exposure time of the magnetic fields on magnetically labeled
MSCs remains unclear.[14,15] Nakamae and Ochi and colleagues proved that
an external magnetic force did not affect the cell proliferation of m-MSCs.[16]

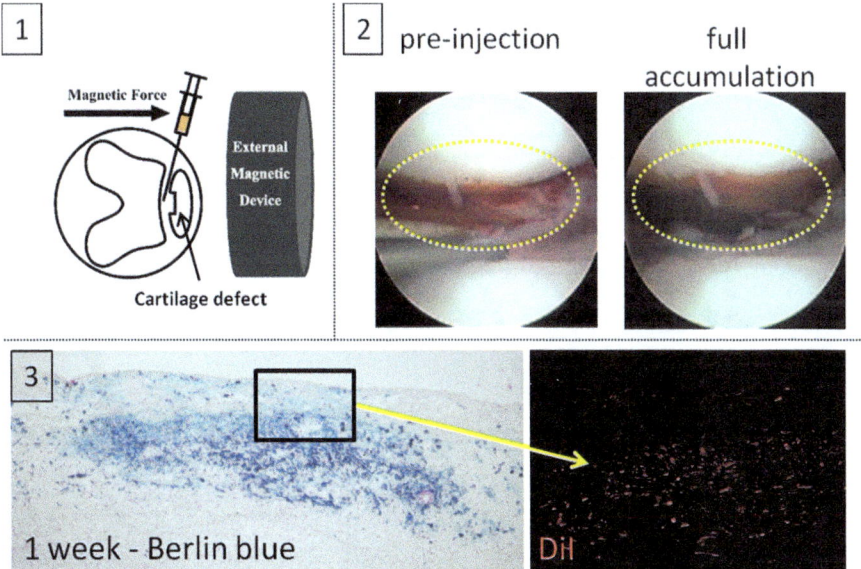

Figure 4.2 1: The cell injection method. 2: Arthroscopic view (left: pre-injection; right: m-MSCs accumulated at the cartilage defect). 3: Histological findings at 1 week after injection shows that m-MSCs were engrafted at the cartilage defect site (left: Berlin blue staining; right: DiI staining). (Reprinted with permission from *Am. J. Sports Med.*)

In this chapter, we also show that the proliferation and chondrogenic, osteogenic, and adipogenic differentiation of m-MSCs were not affected by magnetic labeling and exposure to a magnetic force.

4.3.1 Cell Proliferation

Cell proliferation was assayed using the 2-(2-methoxy-4-nitrophenyl)-3-(4-nitrophenyl)-5-(2,4-disulfophenyl)-2H-tetrazolium, monosodium salt (WST-8) procedure, as described previously.[17] The assay was performed using the Cell Counting Kit-8 (Dojindo Laboratories, Kumamoto, Japan). Approximately 1×10^4 of magnetically labeled MSCs either with or without exposure to an external magnetic force were seeded into each well of a 24-well plate. The same concentration of unlabeled MSCs (non-MSCs) was seeded into each well of a 24-well plate. At 0 and 7 days of culture, a working solution containing WST-8 and 1-methoxy PMS (5 and 0.2 mM, respectively, as the final concentration) was added into each well as a 1 : 10 volume of culture medium. After incubation for 4 h, the absorbance of each well was measured at 450 nm, using a SpectraMax 190 microplate spectrophotometer, (Molecular Devices, Sunnyvale, CA, USA). We compared the absorbance of day 0 with that of day 7 in order to evaluate the cell proliferation of non-MSCs, m-MSCs, and m-MSCs that were exposed to a magnetic force under the following conditions. Magnetic field: 0.6 T, 1.5 T, and 3.0 T; and

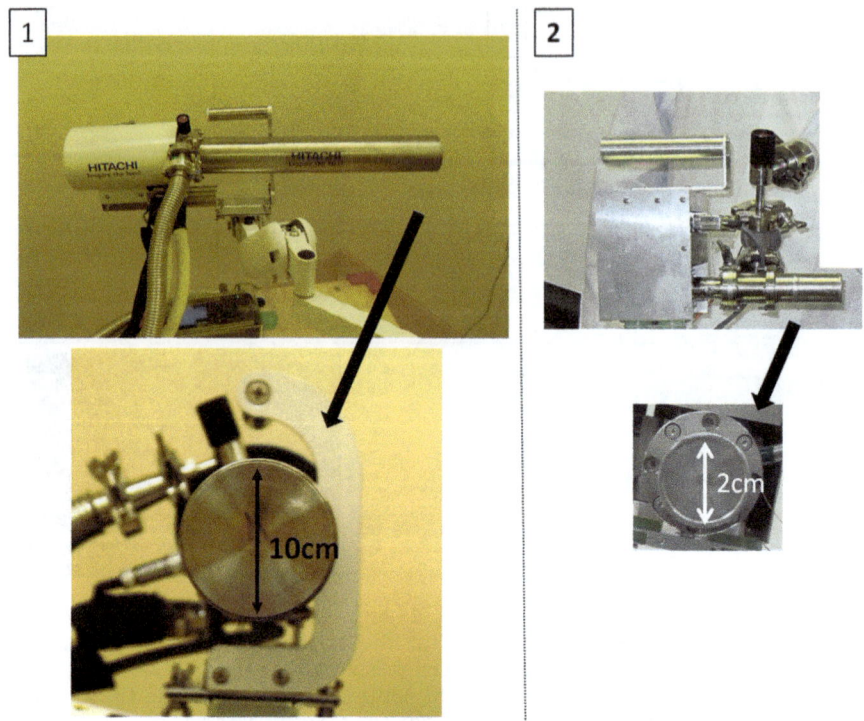

Figure 4.3   A bulk superconducting magnet system. 1: The diameter is 10 cm and
the maximum field is ~ 5.0 T. 2: The diameter is 2 cm and the maximum
field is ~ 3.0 T.
(Part 1 reprinted with permission from 'Senshin Iryo Navigator' in
Japanese.)

exposure time: 10, 30, and 60 min. After 7 days in culture, the proliferation of
m-MSCs was not significantly different from that of non-MSCs ($n = 8$).
However, the proliferation of m-MSCs exposed to the magnetic force was
significantly higher than that of m-MSCs or non-MSCs without magnetic
force. The proliferation of m-MSCs that had been exposed to a magnetic
field for 10 min tended to be greater than those exposed for 30 and 60 min in
all magnetic field conditions (Figure 4.4). As for 10 min exposure, the
proliferation at 3.0 T was greater than at 0.6 and 1.5 T (Figure 4.4).

These results showed that labeling of MSCs has no effect and that exposure
to magnetic fields has a beneficial effect on cell proliferation, especially when
the magnetic field is relatively strong and exposure relatively short.

4.3.2   Multipotential Differentiation Capacity

4.3.2.1   *Chondrogenic Differentiation*

The pellet culture system was used to assess the chondrogenic differentiation. The MSCs, m-MSCs and m-MSCs that were exposed to a magnetic

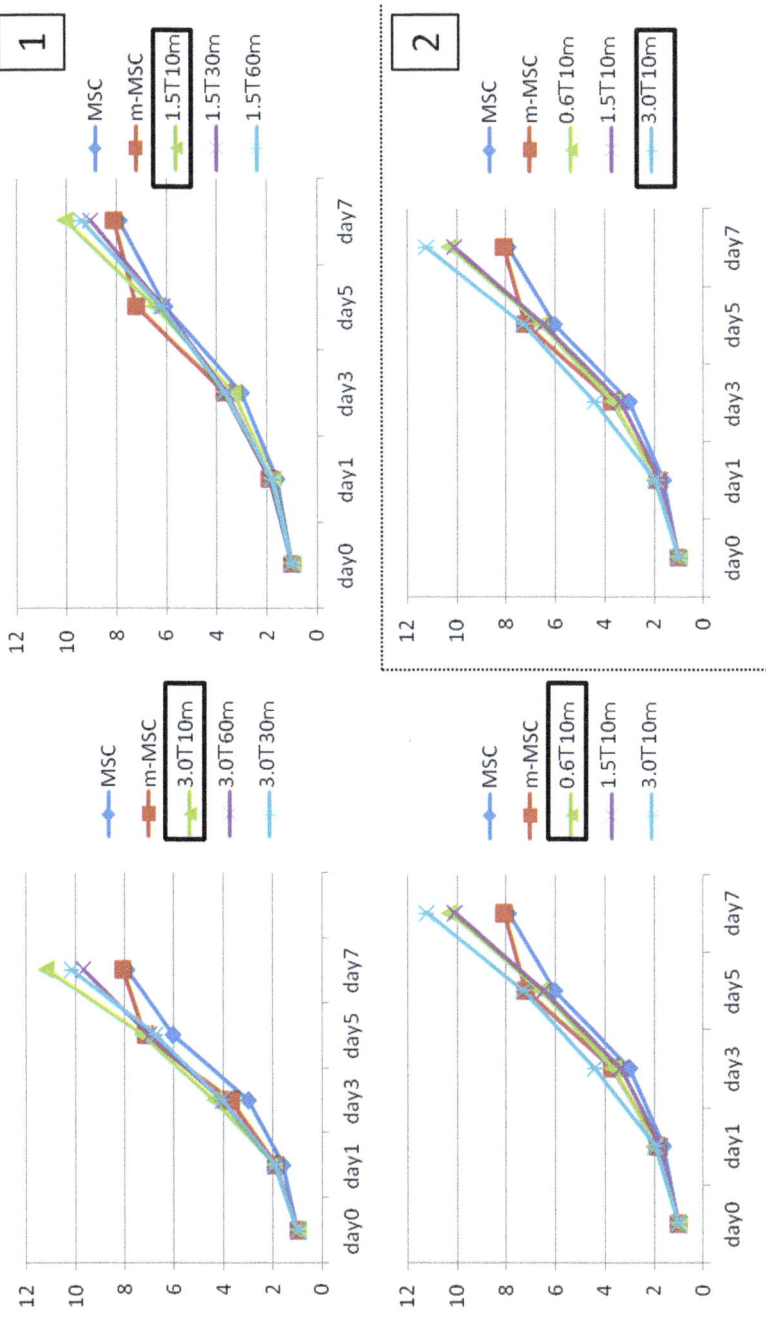

Figure 4.4 Cell proliferation. 1: The proliferation of m-MSCs exposed to the magnetic force was significantly greater than that of m-MSCs or non-MSCs without a magnetic force. The proliferation of m-MSCs exposed to a magnetic field for 10 min tended to be higher than that for 30 and 60 min in all magnetic field conditions. 2: As for 10 min exposure, the proliferation of m-MSCs exposed to 3.0 T was higher than those exposed to 0.6 and 1.5 T.

force (1.5 T, 10 min) were placed at a density of 5×10^5 cells cm^{-3} in a 15 mL polypropylene tube and cultured for 21 days in a chondrogenic medium. The chondrogenic differentiation medium was changed every 3 days. After 21 days, the chondrogenic cultured pellets were fixed in 4% paraformaldehyde and stained with safranin O solution. The level of differentiation was determined by evaluating the expression of cartilage-specific genes, type II collagen, aggrecan and sox-9 by real-time PCR.

4.3.2.2   Osteogenic Differentiation

The MSCs, m-MSCs and m-MSCs that were exposed to each magnetic force were then plated at a density of 5×10^3 cells cm^{-3} in 12-well plates and cultured in osteogenic medium for 21 days. Osteogenic culture was fixed in 4% paraformaldehyde and stained with alizarin red solution for 30 min. The expression of type I collagen and osteocalcin (evidence of osteogenic differentiation), was evaluated by real-time PCR.

There was no major difference in chondrogenic and osteogenic differentiation potential of MSCs after labeling by SPIO and exposure to magnetic fields. The results are shown in Figure 4.5.

4.4   Adhesion of m-MSCs to the Tissue Injured Site

4.4.1   Cell Adhesion Rate in *Ex vivo* Studies

Fresh frozen porcine patella was used to evaluate the adhesion rate of m-MSCs with the external magnetic force to the tissue injured site in this novel technique. A chondral defect was created in the center of the patella. To determine the time required for cell attachment to the defect, the patella was placed onto the dish, parallel to the surface of the external magnetic force device in the following conditions. Magnetic force: 0 T, 0.6 T, and 1.5 T; exposure time: 10 and 60 min. After this, the patella was turned with the defect side down for 10 min, and non-adhered cells fell into the culture medium. The cell number was counted and the attached cell number was extrapolated. The adhesion rate of the 0.6 T and 1.5 T groups was statistically significantly higher than that of the 0 T group, and there was no statistically significant difference between the four groups (0.6 T and 1.5 T, 10 and 60 min). Therefore, we determined that a 1.5 T and 10 min exposure to an external magnetic force should be used for this technique because the stronger magnetic force was more effective to accumulate m-MSCs and the shorter time was better to avoid the risk of infection.

4.4.2   Cell Distribution (Bioluminescence Imaging)

The *in vivo* bioluminescence imaging was performed using a NightOwl LB 981 system (Berthold Technologies, Oak Ridge, TN, USA), consisting of a highly sensitive Peltier air-cooled slow-scan capacitive couple device camera

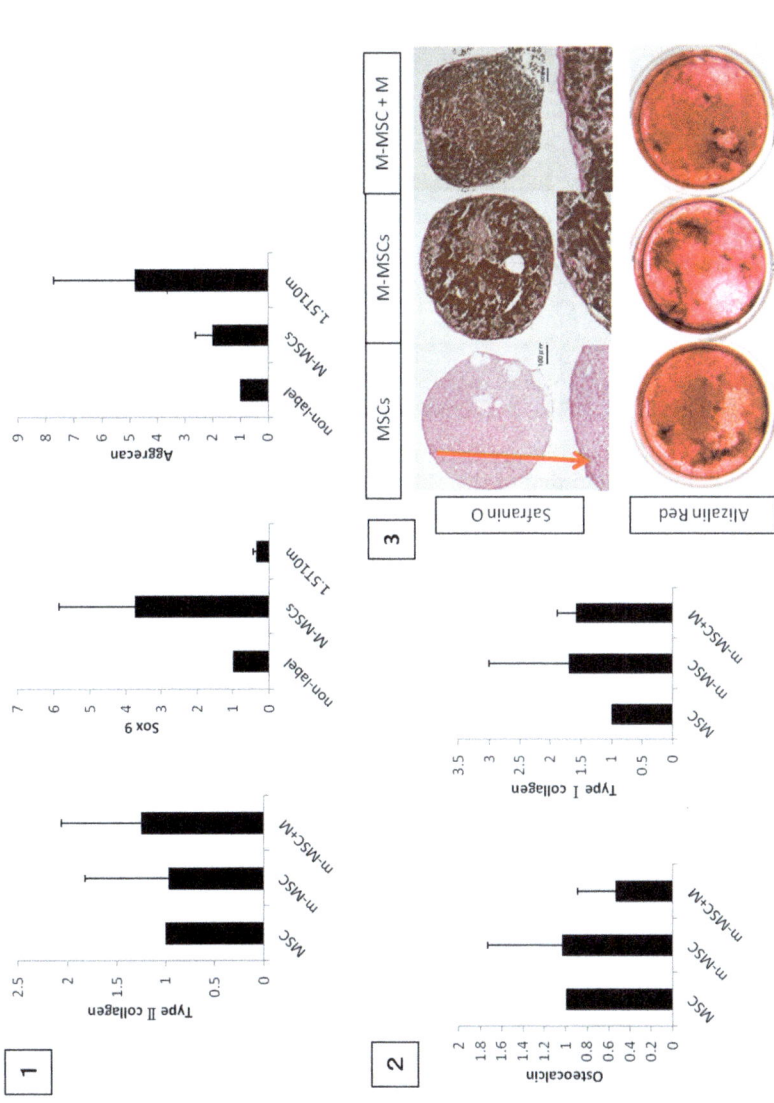

Figure 4.5 1: Chondrogenic differentiation by RT-PCR. 2: Osteogenic differentiation by RT-PCR. 3: Safrain O (for chondrogenic) and alizarin red (for osteogenic) staining. (Reprinted with permission from *Am. J. Sports Med.*)

Day0   Day1   Day3   Day7   Day14   Day28

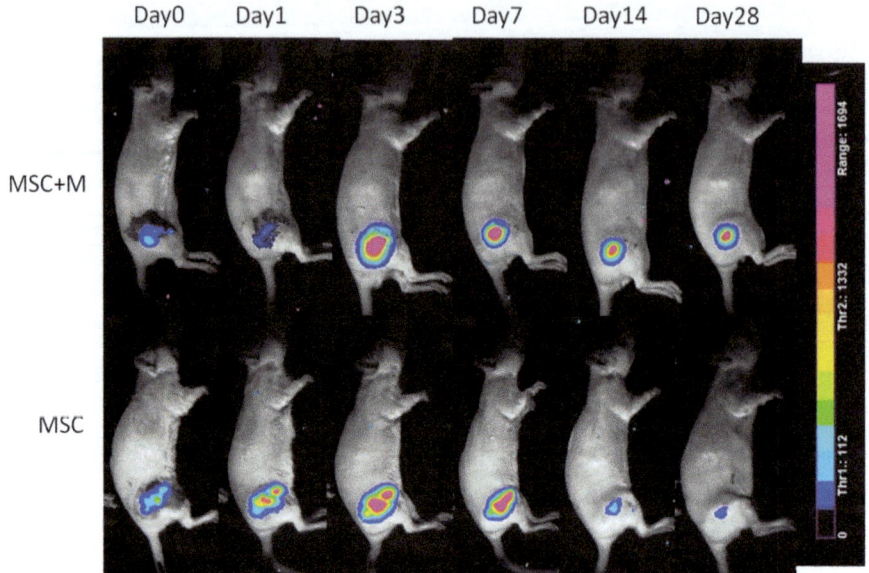

Figure 4.6   *In vivo* bioluminescence imaging. The intensity of the MSC + M group
was higher than that of the MSC group at every time point.
(Images courtesy of Kodama A., Hiroshima University, Japan.)

(2184×1472 pixels) in order to evaluate the kinetics of the transplanted
m-MSCs in a rat fracture model. Images were taken of the rats and the
luminescence intensity was evaluated 0, 1, 3, 7, 14 and 28 days after transplantation. The luminescence intensity was compared between the two
groups, one being the MSC + M group where m-MSCs were exposed to the
magnetic field (1.5 T, 10 min) and accumulated in the fracture site, and the
other being the MSC group where MSCs were injected into the fracture site
without the magnetic field. The intensity of the MSC + M group at 3 and
28 days was significantly higher than that of the MSC group, and the
intensity of the MSC + M group increased more steeply from day 1 to day 3
than that of the MSC group (Figure 4.6). The result indicated that a greater
number of the MSC + M group's injected cells accumulated and adhered to
the fracture site than those of the MSC group, thus showing that our magnetic delivery system promotes cell proliferation.[18]

4.5   Animal Studies

4.5.1   Cartilage Regeneration

A full-thickness cartilage defect with a diameter of 6 mm was created in the
center of the patella in mini-pigs aged 6–7 months. The cell transplantation
was performed at 4 weeks after the creation of the cartilage defect. In the
MSC + M group, magnetically labeled MSCs (5×10^6 cells) were injected and
accumulated in the cartilage defect under a 1.5 T external magnetic force for

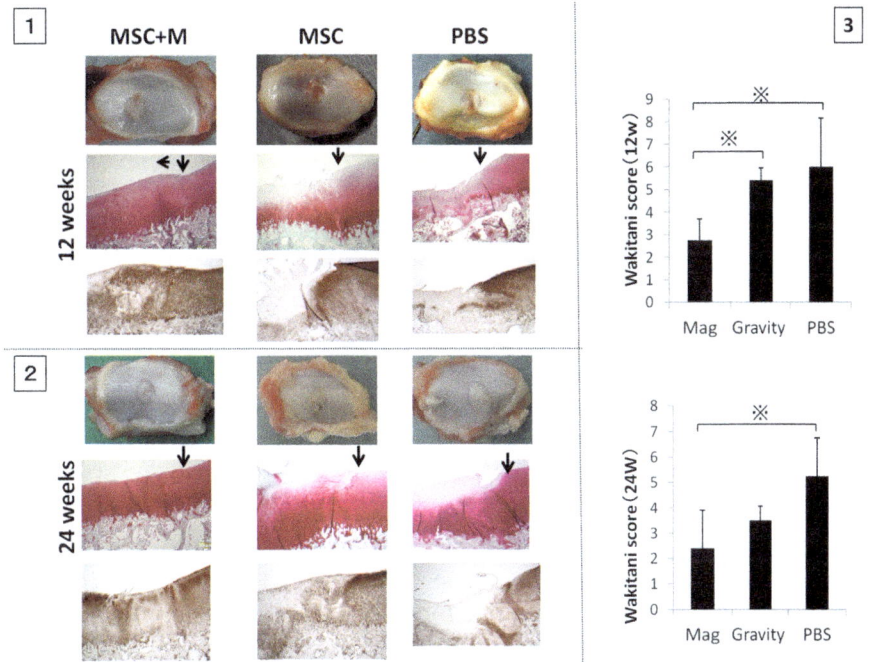

Figure 4.7 1: At 12 weeks after cell transplantation. 2: At 24 weeks after cell transplantation. Upper: macroscopic findings; middle: safranin O staining; bottom: type II collagen staining. 3: Wakitani score: a better histological score was obtained in the MSC + M group at 12 and 24 weeks. (Reprinted with permission from *Am. J. Sports Med.*)

10 min. In the MSC group, MSCs were accumulated under gravity and the position was held for 10 min. In the control group, only PBS was injected. The regenerated tissue was evaluated 12 and 24 weeks after cell transplantation, using macroscopic findings and the histological Wakitani grading. Macroscopic findings of the MSC + M group at 12 and 24 weeks revealed better integration of the border zone, and the surface of the regenerated tissue was smoother than in other groups. The stiffness of the regenerated tissue was nearly normal, but a little harder than other groups. At 12 and 24 weeks after cell transplantation, the histology of the MSC + M group demonstrated that the cartilage defect was repaired with hyaline-like cartilage (Figure 4.7).[5]

4.5.2 Bone Regeneration

Bone regeneration was evaluated using female Lewis rats with a mean weight of 185 g. The mid-shaft of the femur was cut at a depth of 3 mm after a 1.2 mm k-wire was inserted. Then, the periosteum of the femur was cauterised circumferentially and non-union was induced. The experimental model was divided into three groups. In the MSC + M group, m-MSCs $(1.0 \times 10^6$ cells) were injected percutaneously and a magnetic force was

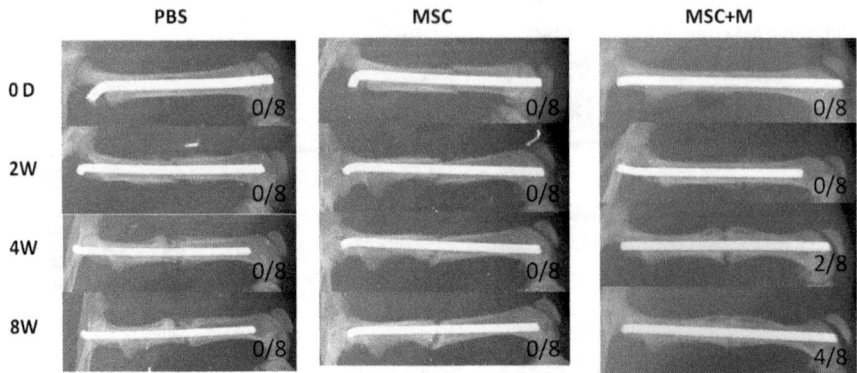

Figure 4.8   X-ray images showing that bone union was obtained in four of eight femur cases in the MSC + M group at 8 weeks after transplantation and in two cases at 4 weeks. However, bone union was not observed in any cases in the MSC and PBS groups, although a slight callus was observed. (Images courtesy of Kodama A, Hiroshima University, Japan.)

applied for 10 min. In the MSC group, MSCs $(1.0 \times 10^6$ cells) were injected percutaneously without a magnetic force. In the PBS group, only PBS was injected with the application of a magnetic force $(n = 8$ in each group). The fracture healing was evaluated by a histological score based on Allen's classification at 4 and 8 weeks and an X-ray at 2, 4 and 8 weeks after transplantation. The mean Allen score in the MSC + M group was significantly higher than that in other groups. In X-rays of the MSC + M group, bone union was obtained in four cases at 8 weeks and two cases at 4 weeks after cell transplantation and in the MSC and PBS groups, bone union was not observed in any cases, although a slight callus was observed (Figure 4.8).[6]

4.5.3   Muscle Regeneration

A muscle injury was created at the tibialis anterior muscle of 9 week old Lewis rats. The muscle belly was cut transversely in the mid-portion. The length of the injury was 6 mm, the width was 4 mm and the depth was 5 mm. The experimental model was divided into three groups. In the MSC + M group, m-MSCs $(1.0 \times 10^6$ cells) were injected percutaneously into the muscle defect site and a magnetic force was applied for 10 min. In the MSC group, m-MSCs $(1.0 \times 10^6$ cells) were injected percutaneously without a magnetic force. In the PBS group, only PBS was injected and a magnetic force applied. Histological and electromechanical evaluations were performed for the assessment of muscle regeneration at 1 and 4 weeks after cell transplantation. For the histological evaluation, Masson trichrome staining was performed for fibrosis and muscle healing. Desmin and isolectin B4 staining were performed for the assessment of muscle regeneration and vascularization. For the electromechanical evaluation, the fast-twitch and tetanus strength ratios were measured. Muscle healing and

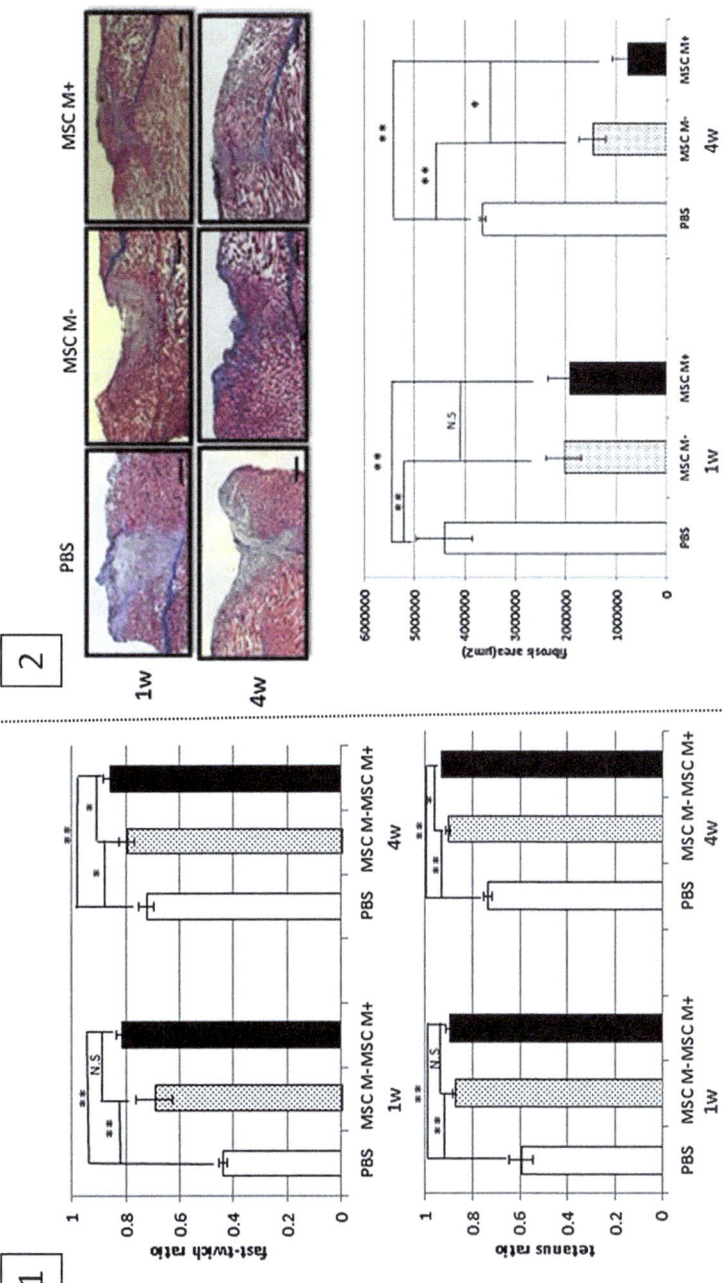

Figure 4.9 1: Electromechanical muscle function was significantly better in the MSC + M group than in other groups at 4 weeks. 2: Muscle healing evaluated by Masson trichrome staining was significantly more enhanced in the MSC + M group than in other groups at 1 and 4 weeks.[19]

(Reprinted with permission from A. Nakabayashi, N. Kamei, T. Sunagawa, O. Suzuki, S. Ohkawa, A. Kodama, G. Kamei and M. Ochi, *J. Orthop. Res.*, 2013, **31**, 754, John Wiley and Sons, Copyright © 2012 Orthopaedic Research Society.)

neovascularization were significantly more enhanced histologically in the MSC + M group than in the other groups at 1 and 4 weeks. Electromechanical muscle function improvement was significantly greater in the MSC + M group than in the other groups at 4 weeks (Figure 4.9).[19]

4.6   Conclusion

Our new cell delivery system with m-MSCs and an external magnetic force has the potential to become a novel therapeutic option for tissue regeneration in the foreseeable future.[20–24]

Acknowledgements

Since 2012, this research has been supported by The Project for Realization of Regenerative Medicine, awarded to Professor Mitsuo Ochi from the Ministry of Education, Culture, Sports, Science and Technology, Japan.

References

1. M. Nishimori, M. Deie, A. Kanaya, H. Exham, N. Adachi and M. Ochi, *J. Bone Jt. Surg., Br. Vol.*, 2006, **88**, 1236.
2. M. Agung, M. Ochi, S. Yanada, N. Adachi, Y. Izuta, T. Yamasaki and K. Toda, *Knee Surg. Sports Traumatol. Arthrosc.*, 2006, **14**, 1307.
3. H. Koga, M. Shimaya, T. Muneta, A. Nimura, T. Morito, M. Hayashi, S. Suzuki, Y. J. Ju, T. Mochizuki and I. Sekiya, *Arthritis Res. Ther.*, 2008, **10**, R84.
4. T. Kobayashi, T. M. Ochi, S. Yanada, M. Ishikawa, N. Adachi, M. Deie and K. Arihiro, *Arthroscopy*, 2008, **24**, 69.
5. G. Kamei, T. Kobayashi, S. Ohkawa, W. Kongcharoensombat, N. Adachi, K. Takazawa, H. Shibuya, M. Deie, K. Hattori, J. L. Goldberg and M. Ochi, *Am. J. Sports Med.*, 2013, **41**(6), 1255.
6. N. Saho, *et al.*, Development of portable superconducting bulk magnet system, *Teion Kogaku*, 2011, **46**, 102–110.
7. C. De Bari, F. Dell'Accio, P. Tylzanowski and F. P. Luyten, *Arthritis Rheum.*, 2001, **44**(8), 1928.
8. K. Pelttari, E. Steck and W. Richter, *Injury*, 2008, **39S1**, 58.
9. L. Kostura, D. L. Kraitchman, A. M. Mackay, M. F. Pittenger and J. W. Bulte, *NMR Biomed.*, 2004, **17**, 513.
10. A. S. Arbab, G. T. Yocum, A. M. Rad, A. Y. Khakoo, V. Fellowes, E. J. Read and J. A. Frank, *NMR Biomed.*, 2005, **18**, 553.
11. V. Mailander, M. R. Lorenz, V. Holzapfel, A. Musyanovych, K. Fuchs, M. Wiesneth, P. Walther, K. Landfester and H. Schrezenmeier, *Mol. Imaging Biol.*, 2008, **10**(3), 138.
12. F. L. Sorgi, S. Bhattacharya and L. Huang, *Gene Ther.*, 1997, **4**, 961.
13. J. W. Bulte, D. L. Kraitchman, A. M. Mackay and M. F. Pittenger, *Blood*, 2004, **104**, 3410.

14. D. Luciana and A. Luigi, *Micron*, 2005, **36**, 195.
15. L. Potenza, L. Ubaldi, R. De Sanctis, L. Cucchiarini and M. Dachà, *Mutat. Res.*, 2004, **561**, 53.
16. T. Nakamae, N. Adachi, T. Kobayashi, Y. Nagata, N. Tanaka and M. Ochi, *Sports Med. Arthrosc. Rehabil. Ther. Technol.*, 2010, **2**(1), 5.
17. L. Zou, X. Zou, L. Chen, H. Li, T. Mygind, M. Kassem and C. Bünger, *J. Orthop. Res.*, 2008, **26**, 56.
18. A. Kodama, N. Kamei, G. Kamei, W. Kongcharoensombat, S. Ohkawa, A. Nakabayashi and M. Ochi, *J. Bone Jt. Surg., Br. Vol.*, 2012, **94**, 998.
19. A. Nakabayashi, N. Kamei, T. Sunagawa, O. Suzuki, S. Ohkawa, A. Kodama, G. Kamei and M. Ochi, *J. Orthop. Res.*, 2013, **31**(5), 754.
20. T. Kobayashi, M. Ochi, S. Yanada, M. Ishikawa, N. Adachi, M. Deie and K. Arihiro, *Arthroscopy*, 2009, **25**, 1435.
21. J. Hori, M. Deie, T. Kobayashi, Y. Yasunaga, S. Kawamata and M. Ochi, *J. Orthop. Res.*, 2011, **29**, 531.
22. S. Oshima, M. Ishikawa, Y. Mochizuki, T. Kobayashi, Y. Yasunaga and M. Ochi, *J. Bone Jt. Surg., Br. Vol.*, 2010, **92**, 1606.
23. S. Ohkawa, N. Kamei, G. Kamei, M. Shi, N. Adachi, M. Deie and M. Ochi, *Tissue Eng., Part C*, 2013, **19**, 631.
24. Y. Fujioka, N. Tanaka, N. K. Nakanishi, N. Kamei, T. Nakamae, B. Izumi, R. Ohta and M. Ochi, *Spine*, 2012, **37**(13), E768.

CHAPTER 5

The Development of Theranostics – Imaging Considerations and Targeted Drug Delivery

WA'EL AL RAWASHDEH,[a,†] SIEM WOUTERS,[b,†] AND
FABIAN KIEßLING*[a]

[a] Department of Experimental Molecular Imaging, Rheinisch-Westfälische Technische Hochschule Aachen, Aachen, Germany; [b] Department of Biomedical NMR, Eindhoven University of Technology, Eindhoven, The Netherlands
*Email: fkiessling@ukaachen.de

5.1 Introduction

The field of nanomedicine has been in development for several decades, leading to clinically relevant nanoparticle formulations that have shown to improve drug efficacy and biodistribution.[1,2] In more recent years, a closely related field has come into fruition that aims to combine therapy and diagnostic imaging agents within the same material. The emerging field was coined by the term theranostics and has seen a tremendous rise in research activity since the early years of the 21st century.[3] This chapter aims to provide a detailed overview of the practical use of theranostics in the

---

[†] Contributed equally to this work.

RSC Drug Discovery Series No. 51
Nanomedicines: Design, Delivery and Detection
Edited by Martin Braddock
© The Royal Society of Chemistry 2016
Published by the Royal Society of Chemistry, www.rsc.org

development and application of nanomedicine. Furthermore, the carrier materials and imaging modalities that are used in the field of nano-theranostics at various temporospatial scales are outlined and discussed.

The amount of research effort invested into the topic of theranostics has exponentially increased due to its usefulness for development of nano-medicine, owing to its multi-functionality in the development process. Theranostics can simultaneously (quantitatively) image and monitor dis-eased tissue, drug delivery kinetics and accumulation, and drug efficacy on both the short- and long-term. This approach differs from image-guided therapy in that the latter uses imaging and therapy separately to guide the application of therapy rather than combining the two in one platform. It should be stressed that applying different imaging modalities in theranostics allows for application both *in vitro* and *in vivo*, enabling drug development and evaluation at multiple temporospatial scales. For example, high-sensitivity imaging modalities such as two-photon laser microscopy can image at a cellular level, whereas magnetic resonance imaging (MRI) is more suited for application at a tissue or corporal level.[4]

These functionalities may eventually assist in moving the field of nano-medicine towards personalized medicine in which therapy and drug dose can be tuned to individual requirements, thereby greatly increasing the efficacy of treatments.[5] The concept of personalized medicine is to first non-invasively image nanomedicine biodistribution and target site accumulation, then, based on the magnitude of target-site accumulation, divide the patients into two groups. The first group, with no to low target site accumulation includes the likely nonresponders, who will continue with conventional treatment, such as chemotherapy in the case of cancer patients. The second group, with moderate to high target-site accumulation includes likely responders to nanomedicine, and thus would receive nanomedicine treatment. Non-invasive imaging follows to determine whether personalized medicine has succeeded, or if patients must be moved to the first group.[6]

Nanoparticles have generated great interest as theranostic prospects due to many attractive characteristics that allow the delivery of a large drug payload. Nanoparticles can localize at the site of disease and deliver their payload passively or specifically, therefore increasing the drug efficacy, increasing contrast-to-noise ratio in imaging and limiting toxicity effects to the disease site. Nanoparticle size has been found to play a major role in the determination of the fate of nanoparticles. It has been shown that upon intravascular injection serum-proteins adhere to nanoparticles surfaces, creating a corona (opsonin) that allows recognition by phagocytes and eventual elimination from circulation. Opsonization was found to be size dependent: it is easier for proteins to adhere to large surface areas than to small ones. Hence, circulation time is size-dependent and was found to be increased for nanoparticles <200 nm. In contrast, ultra small nanoparticles can be excreted rapidly through glomerular filtering and eventual renal excretion, for which the general consensuses on size limit is ~5.5 nm.[7] However, circulation time and blood retention can be improved by coating the nanoparticle surface with stealth molecules, such as poly (ethylene glycol) (PEG), which reduce opsonization.[4,5,8,9] Thus, prolonged retention may improve theranostic delivery to the target site.

The process of disease targeting can be accomplished passively, for example in oncology by exploiting the irregularities of tumor vasculature (see Figure 5.1). The rapid angiogenic process initiated by growing tumors creates flaws in tumor vasculature in the form of flawed shape, structure and high permeability.[10] Nanoparticles can passively extravasate through the hyper-permeable vessels into the tumor tissue and become retained there due to lack of adequate lymphatic drainage systems. This process of passive targeting and accumulation is described as the enhanced permeability and retention (EPR) effect.[11]

Specific targeting can be achieved by the use of active targeting, a concept first described as Ehrlich's magic bullet, of an over-expressed marker within the diseased tissue.[12] By functionalizing the nanoparticle with targeting ligands, targeting site accumulation can be further improved and the application of theranostics expands to non-angiogenic diseases. These ligands are typically organic molecules, monoclonal antibodies, proteins, peptides and substrates which are conjugated to the nanoparticle surface in order to maximize targeting efficiency. Receptor-mediated endocytosis can be achieved by targeting receptors that regulate uptake by the cell of certain compounds (*e.g.* transferrin).[3,13] Active targeting has succeeded *in vitro*, but *in vivo* results are rather critical. In fact, Kunjachan *et al.* showed that because of the target being generally expressed elsewhere, active targeting leads to reduced overall tumor accumulation eventually.[14] Hence, it is of great importance to critically analyze and research active targeting *in vivo*.

Another advantage of nanoparticle carriers for theranostics that elegantly combines targeting and prolonged retention is increased payload and co-loading of therapeutic and diagnostic agents. This local increase in concentration enhances accumulation, and in the case of diagnostics, the imaging efficacy of the agent. This is might be useful for low-sensitivity modalities such as MRI.[3]

As previously mentioned, the field of theranostics is still at an early stage of development. There are a number of unresolved issues before a realistic application of theranostic nanoparticles in clinical settings can be performed. Ideally, the properties of the therapeutic and imaging agents are similar in a way that co-loading and simultaneous application are in-dependent of one another. In reality, the precise interaction between these two agents requires extensive characterization before and after combining them in a nanoparticle platform.[9]

The required concentration of therapeutic and diagnostic agents can differ significantly. For example, the positron emission tomography (PET) imaging agent 2-[^{18}F]-2-deoxyglucose (FDG) and the chemotherapeutic anticancer agent doxorubicin are applied in concentrations that differ by several orders of magnitude. The sensitivity of the imaging modality itself also plays a significant role; nuclear and optical imaging modalities have a

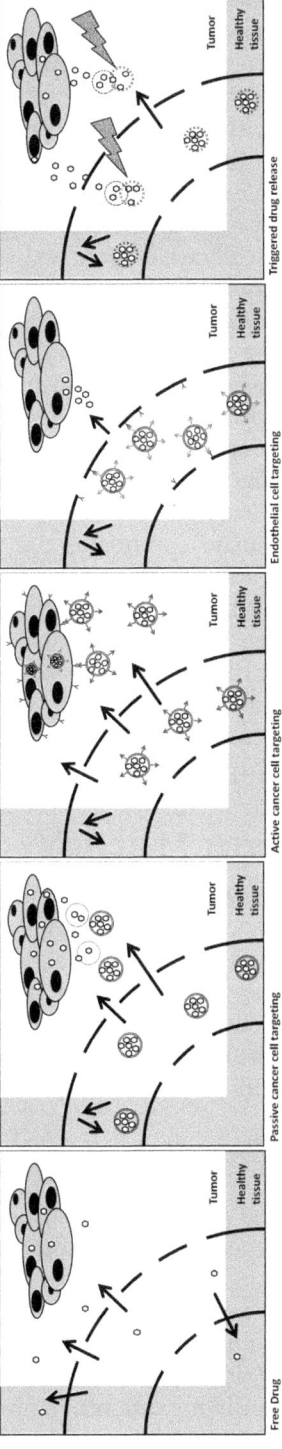

Figure 5.1   Schematic representation of various drug delivery strategies deployed in (nano)medicine.

considerably higher sensitivity than MRI, although this problem can be solved partially by the increased loading capacity of nanoparticles.[3] Similarly, the chemical properties, such as hydrophilicity/hydrophobicity, molecular charge and pH can differ between the therapeutic and diagnostic agent, creating problems within various aspects of nanotheranostics, including the stability of the theranostic nanoparticle and agent loading, release and efficacy. Undesirable toxicity issues may exist for both the imaging and therapeutic agent and dosage may therefore need to be adapted.[4]

As is the case with any clinically used agent, extensive toxicology analyses need to be performed for every theranostic nanoparticle before application, while the increased specificity of nanoparticle targeting may decrease the required overall dose. The nanoparticle material may induce toxicity effects, which can be aggravated by agent loading and prolonged blood retention. Another size-related issue with the use of nanoparticles as theranostic carriers is extravasation through blood vessels. Impaired vasculature in diseases is not always sufficient for nanoparticle extravasation.[4]

In this chapter, two main sections are presented. Firstly, various drug carrier materials and theranostic candidates are presented to exemplify synthesis methods, structure, advantages and disadvantages. Secondly, basic imaging techniques will be outlined for different modalities along with their advantages and shortcomings, and preclinical and clinical applicability, including prominent theranostic application examples for each modality. The aim of this chapter is to address the development of theranostics from all possible aspects and discuss the criteria that must be taken in consideration for realistic *in vivo* applications.

5.2   Theranostic Carrier Materials

In the past decades a wide array of carrier materials have been increasingly investigated as potential therapeutic and/or diagnostic systems with various degrees of success.[15,16] Moreover, new materials and constructs constantly arise offering new possibilities. The main carriers intensely investigated include, but are not limited to, polymeric nanoparticles, liposomes, (radiolabeled) antibodies, microbubbles and carbon nanoparticles. These six types are discussed in depth in this section (see Figure 5.2).

5.2.1   Polymeric Nanoparticles

Nanoparticles are colloidal-sized particles with a diameter range of 1–1000 nm, allowing various drugs to be encapsulated, adsorbed, dispersed or covalently bound to them.

Polymeric nanoparticles are promising theranostic candidates due to their versatility, biocompatibility and functionality. They offer a variety of biomedical applications where their properties can be tailored with relative ease to specific requirements by modifying different parameters during synthesis, such as nature of the polymer, concentration, reaction temperature and

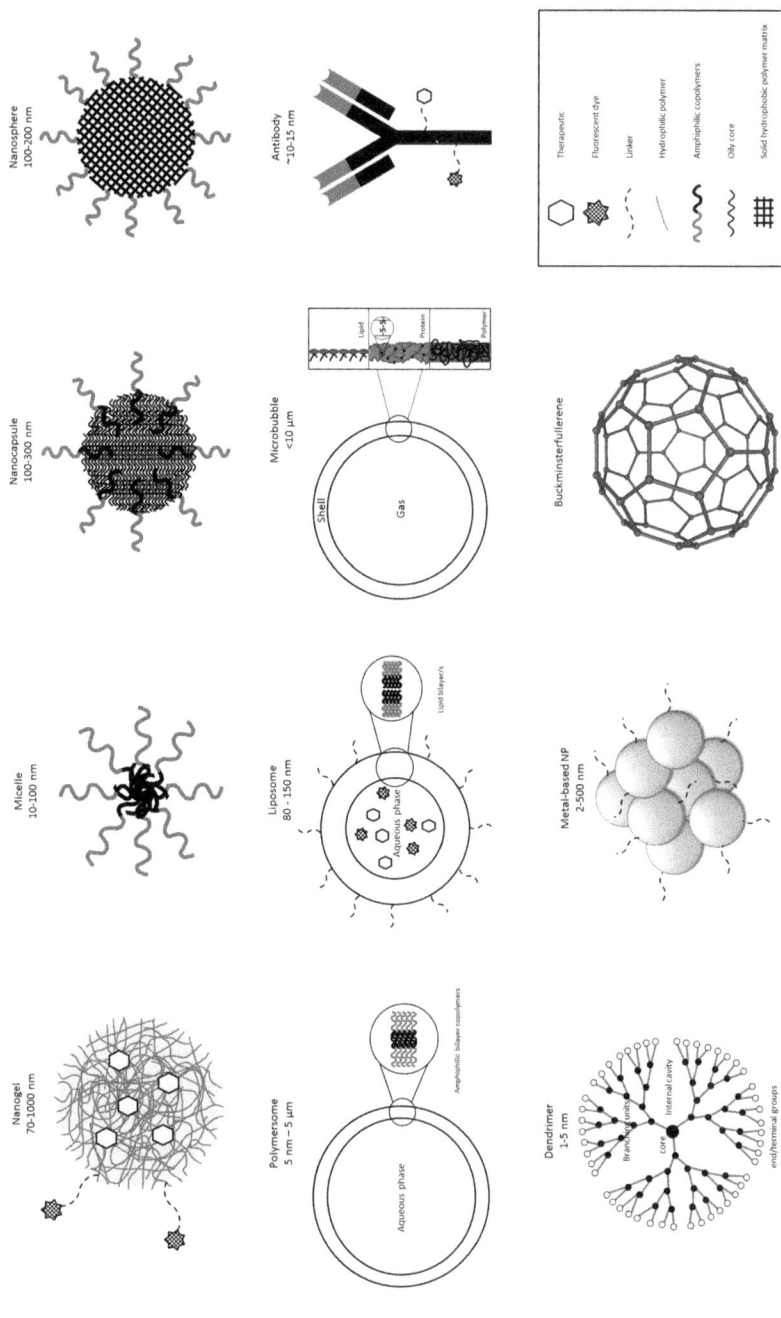

Figure 5.2 Schematic depiction of carriers routinely used for theranostic applications.

additives.[17] Furthermore, their biocompatible (in some cases biodegradable) nature suggests that polymeric nanoparticles might be more suitable for long-term *in vivo* applications than their heavy-metal based counterparts. One of the main polymeric nanoparticles emerging as a hot candidate for theranostics is the nanogel.

5.2.1.1   Nanogels

Nanogels are nanoscaled hydrogels formed by physical or chemical highly crosslinked hydrophilic polymer networks. Once dispersed in aqueous media, nanogels swell up to a soft network that can entrap a considerable volume of water. They are characterized by their smooth and elastic outer layers that do not cause friction with the endothelium of blood vessels; thus, they do not affect the biological functionalities of circulation cells.[18] Because of their large size range (70–1000 nm), they allow a high payload of several drugs to be entrapped and/or covalently conjugated within their core. Moreover, nanogels have been extensively investigated for controlled drug-release mechanisms, and various stimuli-responsive nanogel systems have been developed, including pH-responsive, thermo-sensitive, photochemical-sensitive, magnetic field-responsive and multi-responsive nanogels.[18,19] Nanogels can be prepared using various methods such as precipitation polymerization, chemical modification of polymers and emulsion photo-polymerization processes, among other methods.[17,19]

Polymeric particles are generally made of either biodegradable or bio-compatible materials, or a combination of both. Biodegradable polymeric particles allow controlled release of drug load by diffusion or polymer degradation. Accordingly, they might prolong the exposure of a drug to the target site after accumulating at the desired tissue of choice, generally tumors or sites of inflammation. Biocompatible polymeric particles provide a unique characteristic: they allow higher aqueous solubility for hydrophobic drugs, due to their hydrophobic core.

Combining a biodegradable hydrophilic monomer with a biodegradable hydrophobic monomer, such as PEG covalently bounded to polyester, creates an amphiphilic block copolymer. These compounds possess the advantages of both material types resulting in a particle with regions that have (opposite) affinities for an aqueous solvent. Amphiphilic block copolymers can be synthesized using different block polymer combinations or different block lengths, offering a variety of custom synthesized com-pounds, granting flexibility of synthesis according to the desired/intended use. In the following subsections, four amphiphilic block copolymers are discussed: micelles, nanospheres, nanocapsules and polymersomes.

5.2.1.2   Micelles

In aqueous solution amphiphilies try to minimize the interaction of the hydrophobic block with the solution. Hence, once a specific and narrow concentration of amphiphilies is reached, the critical micelle concentration, they self-assemble into colloidal-sized particles, with a size range of 10–100 nm, simply known as micelles. The hydrophobic block is oriented inwards, establishing the core and is effectively removed from the solution, while the hydrophilic block is oriented outwards, creating a corona. This configuration facilitates loading of hydrophobic drugs into the core of the micelle. Moreover, the hydrophilic corona becomes highly bound with water, which in turn suppresses opsonization by blood proteins, decreases phagocytosis and elimination by the reticuloendothelial system (RES), thus prolonging circulation time. Micelles are not considered solid particles. Micelles can be synthesized in several ways depending on the solubility of the copolymer used. In case of relatively water-soluble copolymers, either direct dissolution methods or film casting methods are used. In the case of relatively water-insoluble copolymers, a dialysis or oil-in-water emulsion procedure can be applied.[15] Note that micellular systems are also often composed of (phospho)lipids.

5.2.1.3 Nanospheres

Nanospheres are identified as polymeric solid matrix-type particles made of hydrophobic material, where drugs can be encapsulated, dissolved, covalently bound or adsorbed to the matrix. They tend to be larger than micelles in size (100–200 nm) and are more polydispersed. Although the size range prolongs circulation, elimination from circulation is relatively fast due to the hydrophobic surfaces that are highly susceptible to opsonization and consequent capture by the RES. Accordingly, there have been several attempts to make the surface more 'water-like' by adsorbing various surfactants, creating an amphiphilic particle. Success in prolonging circulation was achieved in the short-term, but in the long-term no difference was found between surface-modified and unmodified nanospheres, mostly due to surfactant desorption.

Nanospheres are prepared depending on the type of the polymers used. In the case of polymers that need polymerization, two methods can be used: emulsion polymerization or interfacial polymerization. For preformed polymers several preparation methods are used: emulsification/solvent evaporation, emulsification/solvent diffusion, salting out techniques or solvent displacement, which is the most popular method for nanosphere preparation.[15]

5.2.1.4 Nanocapsules

Nanocapsules are polymeric colloidal vesicular systems where drugs can be restricted to a basin or within a cavity surrounded by a polymer membrane, with a size range of 100–300 nm. Generally, it is an oily-liquid core with a hydrophobic polymer monolayer. Hence, nanocapsules are optimal systems for hydrophobic drug encapsulation. However, as explained earlier, such systems are eliminated from the circulation with a relative ease due to opsonization. Adding a hydrophilic copolymer, *e.g.* PEG to the polymer monolayer creates an amphiphilic vesicle with a hydrophilic exterior, similar to nanospheres, prolonging circulation time. Nanocapsules are normally prepared by interfacial deposition of performed polymers, but are less commonly prepared with a modification to the solvent displacement method.[15]

5.2.1.5   *Polymersomes*

Polymersomes are very similar to nanocapsules with the exception that the core consists of an aqueous phase surrounded by a polymer bilayer. Thus, polymersomes are drug delivery vesicles, optimized for encapsulation and delivery of water-soluble drugs which are entrapped in the aqueous phase core. Also, they have the broadest size range of all the amphiphilic particles discussed here (5 nm–5 μm). Polymersomes are analogous to liposomes; however, their external bilayer consists of amphiphilic bilayer copolymers. Moreover, polymersomes in general have higher PEG surface density in comparison to liposomes. Hence, they circulate longer than PEGylated liposomes. Similar to nanogels, various stimuli-responsive polymersomes have been developed.[20–22] Polymersomes are prepared similarly to liposomes, using a film rehydration technique.

5.2.1.6   *Dendrimers*

Dendrimers are hyper-branched, 'tree-like', 3D, monodispersed macromolecules with micelle-like behavior and nanoscale container properties.[23–25] They can be constructed with a defined architectural and chemical structure, and once decorated with bioactive ligands they turn into nanomaterials with attractive binding properties. Their flexible structure offers a high degree of surface functionality and branching, tunable uniform size, water solubility, well-defined molecular weight and available internal cavities, making dendrimers ideal carriers for creating functional nanomaterials, and thus promising candidates as drug delivery systems.[26,27]

Dendrimers consist of three distinct units: a central core, interior branches that are radically attached to the core and terminal functional groups attached to periphery series of braches. The interior branches, or 'generations', play an important role in determining physical chemical properties, while the terminal functional groups determine the properties (functionality) of the dendrimers.[25] Dendrimers are typically synthesized using three methods: the divergent method, the convergent method and the double exponential and mixed method.

Drugs, heavy metals, DNA or other molecules of interest can be either attached to the periphery or encapsulated in the internal cavity. Hence, dendrimers have been used in the biomedical field as MRI contrast agents, gene transfer reagents, in anti-tumor therapy and in drug delivery with broad applications such as transdermal-, oral-, ocular-, pulmonary- and targeted-drug delivery.[27]

Although dendrimers are generally considered biocompatible, the safe use of dendrimers is case-specific and depends largely on the specific structure and composition of the materials used, as well as possible unknown bioactivity. Hence, toxicity remains the main obstacle facing the *in vivo* applications of dendrimers. Therefore, more in-depth studies should be performed to address fundamental and practical issues and enhance the understanding of material–cell interactions and drug-delivery problems before dendrimers can be considered for clinical applications.[26]

5.2.2 Liposomes

Liposomes are self-assembling colloid particles that consist of one or more lipid bilayers (known as lamella) that surround an aqueous compartment. Of all nanoparticle platforms, liposomes are the most clinically established, since they offer numerous advantages over other nanoparticles. Lipid-based systems like these are biocompatible, biodegradable and are easily tunable in size. The aqueous lumen and lipid bilayer are very suitable for hydrophilic and hydrophobic compounds, respectively, making liposomes a very versatile platform for theranostic applications.[28] The large diameter, which is on average between 50 and 400 nm, makes it possible for a considerably large amount of drug to be encapsulated, which protects the drug from degradation and/or modification. Encapsulation of the compounds therefore improves the plasma stability, blood circulation and pharmacological fate, as opposed to free compounds. The large surface area can be functionalized with numerous targeting and stealth moieties, further increasing these characteristics and making active targeting possible. Liposome preparation is usually prepared using the hydration of (biocompatible) lipid film, followed by extrusion to homogenize the size distribution.[29] Adding theranostic compounds during the hydration process encapsulates the drugs, and subsequent filtration removes free drug from the solution. The biodistribution and compound release kinetics of the resulting liposomes can be adapted by varying the lipid composition and charge, bilayer stabilization using compounds such as cholesterol, and regulating the drug-to-lipid ratio.[13,28]

5.2.3 Antibodies

Antibodies play a multifunctional role in both therapy and imaging, and, by extension, in theranostics due to their ability to target almost any substance as (over-)expressed by diseases. Treatment is normally achieved using antibodies that activate components of the immune system, interfere with disease progression mechanisms and enhance the activity of other biological compounds.[30] Imaging is made possible by linking one or more imaging labels to the antibody, which can be conjugated directly to the antibody, and is suitable for highly sensitive imaging modalities such as nuclear and optical imaging.[31] In this case, a therapeutic antibody itself becomes the theranostic material. Additionally, radiolabeled non-therapeutic antibodies can become a theranostic agent due to the cytotoxic effects of radiation.[32,33] For less sensitive imaging modalities, antibodies are coupled to one of the previously described carrier materials that can carry large amounts of imaging labels and/or drugs. The use of antibodies as theranostic platforms is not necessarily without problems. Target expression specific to the diseased regions is a prerequisite for the use in theranostics and is in most cases the limiting factor in treatment and imaging effectiveness. Furthermore, their long blood circulation half-lives can complicate their use in imaging.[30,34]

5.2.4   Metal Nanoparticles

Metal nanoparticles have been the focus of biomedical research since the 1960s because of the breadth of their potential applications: biotechnology, biosensors, magnetic separation, diagnostic imaging and targeted drug delivery.[35] They have the largest nanoparticle size range, starting at just below 2 nm to 500 nm and, rarely, exceeding 700 nm. In order to fully utilize metal nanoparticles in diagnostic imaging and drug delivery, the particles should be stable, biocompatible and selectively targeted to desired sites.[35] Similar to other materials discussed, targeting can be achieved by conjugating the nanoparticle with an appropriate ligand that has specific binding to target cells. Moreover, multiple copies of therapeutic substances can be attached on the particles, creating a drug-delivery platform that facilitates increased concentration of therapeutic and diagnostic substances at the pathological site. Three metals are mainly used to synthesize nanoparticles: iron, gold and silver.

5.2.4.1   *Iron Oxide Nanoparticles*

Three oxides of iron mainly occur in nature: FeO, Fe_2O_3 and Fe_3O_4, of which the latter two are paramagnetic and superparamagnetic, respectively. They have emerged as promising MRI contrast agents and theranostic candidates. Fe_3O_4 make excellent imaging probes for high-resolution MRI images, where they induce a decrease in T_1 and T_2 relaxation times of surrounding water protons and hence manipulate the signal intensity of the imaged tissue. The most common synthesis procedure of the magnetite nanoparticle is the chemical coprecipitation of iron salts.[35]

5.2.4.2   *Gold Nanoparticles*

Colloidal gold is a suspension of nanosized particles of gold with properties that differ from bulk gold. They are used therapeutically as anti-inflammatory agents to treat arthritis, experimentally as cytotoxic agents, and in photodynamic therapy. Diagnostically, due to their strong absorption of X-rays, they are potent X-ray contrast agents. Furthermore, their optical properties make them attractive for Raman spectroscopy and photoacoustic imaging.

The gold nanoparticle suspension solution can have an intense red color for particles <100 nm or a dirty yellowish color for larger particles.[35] These different colors and optical characteristics are due to the interaction of the particles with light, where free electrons on the metal surface oscillate in the presence of an electromagnetic field of light in a process termed surface plasmon resonance. Plasmon resonance decays either radiatively or non-radiatively, producing light and heat, respectively. Furthermore, the shape of gold nanoparticles plays a major role in their properties and thus their possible applications.

The most common synthesis procedure for gold nanospheres (10–20 nm) is chemical reduction of gold salts such as hydrogen tetrachloroaurate using citrate as the reducing agent. For larger particles (30–100 nm), seeding of Au^{+3} by hydroxylamine is used. Other modifications of the synthesis procedure yields various particle shapes, such as rods, triangles and polygonal rods.

5.2.4.3 Silver Nanoparticles

Silver nanoparticles are typically spherical particles with a size range of 1–100 nm. Often they are actually composed of a large percentage of silver oxide.[35] Silver nanoparticles, like gold nanoparticles, have been used for hundreds of years in tinting glass. However, lately there have been solid efforts to use silver nanoparticles in the medical field due to their attractive physiochemical properties, especially their anti-bacterial characteristics. Hence, silver nanoparticles have been incorporated into various medical devices, surgical instruments, and surgical masks and used in the treatment of wounds.[35] Furthermore, due to their plasmon resonance and large effective scattering cross-section silver nanoparticles are now being investigated for molecular labeling.[36]

The synthesis procedure of silver nanoparticles is similar to that of gold nanoparticles, where a reduction agent such as sodium borohydride is used to reduce silver salt in the presence of a colloidal stabilizer such as polyvinyl alcohol, citrate or cellulose. Similarly to gold nanoparticles, size plays a major role in the potential application of silver nanoparticles, where particles only <10 nm have been shown to interact with the HIV-1 virus.[37]

5.2.5 Nanocarbons

Relatively new types of nanomaterial used in the field of theranostics are the nanocarbons, well-defined structures (in a size range of a few nanometers) that have attracted considerable interest in the field due to their distinctive physical and chemical properties. Among these, the optical properties (*e.g.* fluorescence) inherent to their structure make them useful for contrast generation in optical imaging. Additionally, nanocarbons such as fullerenes can act as photosensitizers in photodynamic therapy. Furthermore, the large availability of surface area makes nanocarbons useful for high-payload drug and contrast agent loading; particularly, hollow nanocarbons such as carbon nanotubes and fullerenes constitute interesting theranostic platforms and have been used in various research studies. Carbon dots (C dots), amorphous nanoparticles, have been shown to have high fluorescence levels, making them a possible alternative to heavy-metal based quantum dots. There are some issues with the use of nanocarbons. The synthesis of these materials often produces a low yield in comparison to other nanomaterials (although progress has already been made), and without modification they are insoluble in aqueous solution. The biodistribution and excretion pathways are heavily dependent on the particle size and modification, as well as the administration route, leading to distinct toxicity profiles for individual nanocarbons. Internalization by various cell types has been observed, making a broad cytotoxicity assessment an important part of nanocarbon development.[38,39]

5.2.6   Microbubbles

Microbubbles are spherical bubbles filled with free gas or vapor, typically air or perfluorocarbon, within a biocompatible shell made of lipids, proteins or polymers in an unbounded volume of liquid, with a size range <10 μm, typically 1–4 μm.[40,41] Microbubbles are the only particles discussed in this section that are in the micro-range, which makes them suitable only for intravascular targets.

Microbubbles are usually prepared using three methods; mechanical agitation and sonication are the most common methods, in addition to the use of microfluidic devices.[40] They have been established for more than 20 years now as the most effective intravascular contrast agent particle for ultrasound. Microbubbles have been widely applied and are established in clinical diagnosis for functional imaging such as assessing microcirculation in tumors, characterizing liver lesions and evaluating myocardial function and perfusion.[40,41] Moreover, microbubbles can be actively targeted to proliferating and inflamed endothelial cells in tumors, cardiovascular and other disorders.[40]

Unlike tissues, because of their compressibility, microbubbles undergo volumetric oscillation (resonance) in response to an ultrasound pulse. This oscillation creates a non-linear response to ultrasound waves, which serves as a means of distinguishing between microbubble and tissue ultrasound responses.[41,42] Applying an ultrasound pulse with a sufficiently high acoustic pressure amplitude leads to microbubble destruction and the generation of acoustic energy that can temporary increase blood vessel permeability, thereby enhancing local drug delivery.[40]

Microbubbles also hold potential as theranostic systems. Upon ultrasound-mediated destruction, they can enhance local drug delivery across biological barriers. Furthermore, the possibility of loading therapeutics into or onto their shells permits them to serve as vehicles for localized and triggered drug delivery.[39] Although microbubbles have been intensely investigated as drug delivery systems, more in-depth research must be undertaken to fully understand the underlying mechanisms governing microbubble-mediated permeabilization of biological barriers, to be able to harness their full potential.[40,41]

5.3 Theranostics and Imaging

The number of publications on theranostics has increased exponentially over the past 10 years, covering a wide array of imaging and therapeutic agents, in and on various materials. Current research into theranostics focusses mainly on the treatment and imaging of cancer, although other applications have been investigated (*e.g.* infections). For optimal design of a theranostic system, it is of importance to know the capabilities and limitations of the carrier and the therapeutic agent, as well as the imaging agent and modality. This section highlights the characteristics of various imaging modalities used in theranostics and examples in which they have been used in (pre)clinical research, using the nanomaterials described in Section 5.2. Herein, we primarily focus on the actual theranostic research, not on (nano-)applications of therapy or imaging that show promise for theranostics.

5.3.1 Nuclear Imaging

Nuclear imaging is mainly performed using single photon emission computed tomography (SPECT) and PET. Both techniques rely on the detection of highly energetic γ-photons (keV–MeV range) as emitted by administered radiopharmaceuticals. Nuclear imaging modalities are advantageous over other modalities due to being quantitative and highly sensitive techniques; the relationship between signal intensity and tracer concentration is relatively straightforward and there is little to no background signal present. Furthermore, the sensitivity of both techniques is exceptional, making single radioactive decay events detectable. This is particularly true for PET, of which the sensitivity is approximately 10 times higher than SPECT and lies in the picomolar range. In contrast, SPECT can be used for simultaneous imaging of multiple radiopharmaceuticals (with different γ-photon energies), which is impossible with PET due to the nature of positron annihilation photons. Disadvantages of both modalities are poor soft-tissue contrast and spatial resolution (in the range of (sub-)millimeter), which can be overcome partially by using hybrid imaging techniques with high resolution, such as computed tomography (CT) or MRI.[3]

A classic example of a theranostic radionuclide is ^{131}I, which is used in both treatment and SPECT imaging of thyroid carcinomas. However, most radionuclides have little use in theranostic nuclear imaging without some form of conjugation to a ligand. Binding the radionuclide to a targeting ligand (usually therapeutic antibodies) allows the compound to be used to diagnose and treat disease specifically.[34] The choice of radionuclide, and subsequently modality, is predominantly governed by the type of emission and the decay half-life. The type of emission is directly related to the choice in modality, *i.e.* direct γ-photon emission (SPECT) or indirect through annihilation reactions (PET). The decay half-life is usually shorter for PET radionuclides, compared to those for SPECT. It should be noted that the possibilities for conjugation to the ligand are also determinants of the choice of radionuclide; some can be conjugated directly (*e.g.* ^{18}F-FDG) while others require chelation (*e.g.* ^{113m}In).[3] For the latter, popular examples used in tumor imaging and treatment include DOTA chelates bound to soma-tostatin analogues: DOTATOC and DOTATATE.[43]

Nuclear imaging as used in the clinic provides a relatively straightforward means to quantify radioactive tracer concentrations, even at levels un-detectable by most other imaging modalities. In a phase I trial by Seymour *et al.*,[44] the SPECT tracer ^{123}I was coupled to *N*-(2-hydroxypropyl)-methacrylamide (HPMA) copolymers bearing doxorubicin as a chemother-apeutic agent and galactosamine as a moiety targeting the hepatic asialoglycoprotein receptor (ASGPR). The complex (dubbed PK2) was administered intravenously in patients with primary and metastatic liver cancer to assess toxicity issues, hepatic targeting efficiency and anti-neoplastic efficacy. Gamma scintigraphy was used to determine clearance kinetics, while SPECT and gamma camera imaging were used to determine biodistribution. Results showed that the PK2 complex greatly improved hepatic targeting through both active (ASGPR targeting) and passive (EPR) mechanisms, making a higher systemic dose possible, compared to non-polymer-bound doxorubicin.[44] Another example in which nuclear imaging was used to assess targeting efficacy was described by Bartlett *et al.*, who compared the biodistribution in tumor-bearing mice of transferrin-targeted *versus* non-targeted nanoparticles consisting of cyclodextrin-containing polycations and siRNA by conjugating a DOTA chelate to the siRNA for ^{64}Cu labelling for micro-PET/CT imaging. Furthermore, the siRNA targeted the luciferase mRNA as expressed by the tumor, which was used as an indicator of cell uptake using bioluminescence imaging (BLI). Interestingly, whilst the results of micro-PET/CT indicated a similar biodistribution for targeted and non-targeted complexes, BLI results pointed to a greater siRNA efficacy for transferrin-targeted complexes. Additionally, EPR-mediated passive target-ing was observed for both complexes.[45]

5.3.2 Computed Tomography

Computed tomography (CT) is a medical imaging technique based on the variable absorption of computer-processed X-rays in different tissues. An X-ray source with detectors placed opposite will rotate around a stationary object to capture a large series of 2D X-ray images around a single axis of rotation. 3D images are generated from that large series of 2D images a using digital geometry processing.[46] CT is routinely used for non-invasive visualization of morphological structures because it has excellent spatial resolution, high-throughput analysis, is user-friendly and user-independent.[47] Moreover, CT eliminates the superimposition of images of structure outside the areas of interest and allows for fast scanning. However, CT is considered a moderate to high-radiation imaging technique and children who received multiple CT scans were found to be three times more likely to develop leukemia and brain cancer, since radiation used in CT scans can damage cell and DNA molecules.[48] Hence, increased radiation dosage in the case of repeated scans presents a major disadvantage. Moreover, CT contrast is based on X-ray absorption, so tissues with similar absorption coefficients, like soft tissues, have similar contrast and thus are difficult to distinguish.[49]

Contrast-enhanced CT allows for morphological and functional imaging of vasculature at very high resolution, but CT suffers from low sensitivity and requires high concentrations of contrast agent, which can cause some undesired reactions such as vomiting or nausea.[49-51]

CT clinical applications have been predominantly diagnostic, using CT independently or in hybrid imaging devices such as PET/CT. Moreover, CT is used as a guide in interventional procedures. Preclinical applications have followed suit, where numerous CT contrast agents have been investigated with various degrees of success and CT has been fused with other modalities, such as fluorescence-meditated molecular tomography (FMT)/CT.[14,52] CT theranostic applications have been hindered by the low sensitivity of CT to contrast agents, but developments in carrier loading capacity are expected to have considerable impact on CT-based theranostic applications.[53]

The use of metal nanoparticles as carriers is one such development. In an example by von Maltzahn *et al.*, a theranostic complex was devised in which gold nanorods were coated with PEG chains to create long-circulating (~ 17 hours) near-infrared (NIR)-activated nanoantennas for CT-guided photothermal tumor therapy. The strongly X-ray attenuating gold particles were used to assess with CT the biodistribution of the nanorods in a mouse model with subcutaneously implanted human xenograft tumors. The nanorods achieved a passive intratumoral accumulation of $\sim 7\%$ injected dose (ID) g^{-1} (after 72 hours), which amplified the tumor NIR absorption approximately seven-fold compared to controls. Subsequently, mice treated with gold PEGylated nanorods showed complete remission of tumor tissue after irradiation, compared to the control groups.[54] Nevertheless, a lot of theranostic systems for CT still rely on the use of iodinated compounds, of which the effectiveness can be improved by loading them into nanoparticles. For example, Miyata *et al.*[55] used PEGylated, transferrin-conjugated liposomes loaded with iomeprol (iodine-based contrast agent) and sodium borocaptate (for boron neutron capture therapy) for glioma treatment and followed the boron delivery using real-time CT. The transferrin conjugation to the liposomes increased boron concentrations in glioma cells approximately four-fold and showed prolonged contrast as opposed to untargeted liposomes.[55]

CT may play a very important role to assess the fate of (normally radiolucent) microcapsules in embolization therapy, in which supplying blood vessels are blocked using micro-sized particles. Lu *et al.*[56] encapsulated clinically available Lipiodol, an iodine-containing CT contrast agent, into polyvinyl alcohol capsules. It was shown that these capsules possess multiple characteristics that make them suitable for embolization therapy, as well for contrast enhanced CT imaging. *In vitro* and *in vivo* experiments showed the compound to be cytocompatible and (cyto)toxicity or significant lipiodol release (1.73% after 72 hours), and the observed radiopacity was suitable to observe embolization.[56] Bartling *et al.* created a proof-of-principle microparticle consisting of a polymer core with triiodinated, aromatic side chains, which was then coated with ultrasmall iron oxide particles. The resulting complex was shown to generate sufficient contrast for not only CT (and X-ray fluoroscopy), but T_2-weighted MRI as well due to the presence of the iron oxide. This multimodal system could be used for real-time, high-sensitivity tracking of renal embolization in rabbits using CT and MRI. It was hypothesized that CT could be used in low proton-density regions (*e.g.* in the imaging of pulmonary reflux of the embolization material) or to avoid susceptibility artefacts commonly associated with the use of iron oxides in MRI, whereas MRI could be useful for repeated measurements or in regions that are more sensitive to radiation exposure.[57]

5.3.3   Magnetic Resonance Imaging

The principle of MRI is based on the precession of magnetic moments of atomic nuclei with a non-zero value of the quantum mechanical property of spin, within a strong external magnetic field (^1H being the most well-known and oft-used example). The application of radiofrequency pulses can move the net magnetic moment from its original state (aligned with the external magnetic field), after which it returns through relaxation processes that can be exploited to produce an image (contrast). The relaxation process is primarily governed by the relaxation times T_1 and T_2, which describe relaxation in the longitudinal and transversal plane.[4]

MRI contrast agents provide an indirect means of contrast enhancement by locally shortening the relaxation times of the hydrogen nuclei. These contrast agents can be divided into two categories based on their magnetic properties: paramagnetic and superparamagnetic compounds. Paramagnetic compounds, such as gadolinium and manganese ions (Gd^{3+} and Mn^{2+}, respectively), have a relatively larger T_1-shortening effect compared to their T_2-shortening effect and are therefore used to provide positive contrast on T_1-weighted images. These compounds are usually conjugated to a carrier material using a chelate. Conversely, superparamagnetic compounds such as iron oxide particles are used to obtain negative contrast on T_2-weighted images, due to their relatively large T_2-shortening effect. These materials often simultaneously act as the carrier material.

It is of great relevance that MRI contrast agents provide indirect contrast: rather than being imaged, they reduce the relaxation times of ^1H atoms to provide contrast. This is expressed as the relaxivity, the quantitative term that describes the ability of an MRI contrast agent to shorten T_1 and T_2 relaxation times. The relaxivity (in $mM^{-1}\,s^{-1}$) exists for both the longitudinal (r_1) and transversal (r_2) relaxation processes and shortens relaxation times through eqn (5.1):

$$\frac{1}{T_{i,\text{obs}}} = \frac{1}{T_{i,\text{bas}}} + r_i \cdot [\text{CA}], \quad \text{with } i \in \{1, 2\} \tag{5.1}$$

In this equation, $T_{i,\text{obs}}$ (in seconds) represents observed value of the relaxation time, $T_{i,\text{bas}}$ (in seconds) the baseline value of the relaxation time, r_i the relaxivity and [CA] the concentration of the (para)magnetic compound in mM.[58] The actual value of relaxivity is dependent on many factors, including the magnetic field strength/precession frequency, temperature and pH. The Solomon–Bloembergen–Morgan equations further describe the deciding factors for relaxivity, such as the water–ion exchange rates, (nano)particle flexibility and the rotational correlation time. It is particularly the latter that makes nanoparticles ideal carriers for MRI contrast agents, by slowing down rotation and thus increasing the relaxivity. Subsequently, following eqn (5.1), a smaller dose is needed to induce the same change in the relaxation times.[59] The water–ion exchange rate can be exploited by (temporarily) blocking water access to the agent, which is often used for triggered theranostics.[60]

The advantage of using MRI for theranostics manifests itself in the form of superior resolution and soft-tissue contrast. However, its sensitivity is inherently low, making relatively high doses of contrast agent a requisite for detectable signal changes (in the range of micromolar concentrations). This problem can be overcome by incorporating large amounts of contrast agent into the carrier and by applying active or passive targeting.[3] However, the use of nanoparticle carriers complicate the use of particularly gadolinium-based complexes, which require complete renal excretion due to the high toxicity of free Gd^{3+} ions. Slow and/or incomplete excretion increases the chance of transmetallation, causing accumulation in the hepatobiliary system.[61] The accumulation of Gd^{3+} ions has been linked to the nephrogenic systemic fibrosis and has thus hindered the development of nanosized MRI contrast agents for clinical use.[62] Iron oxide based agents are less toxic, but have an r_2/r_1 ratio that may complicate T_1-weighted imaging methods.[63]

Another approach to produce MRI contrast of is to use nanoparticles labeled with ^{19}F, since this compound does not occur naturally in the body and therefore does not suffer from background signal (*i.e.* optimal contrast). However, sufficient ^{19}F atoms have to be present in a voxel for detection of signal.[64]

Quantitative imaging of agent concentration using MRI contrast agents is possible, provided that the baseline T_i and the agent relaxivity are known. This is also the basis for dynamic contrast-enhanced MRI methods. However, this is less than straightforward, since the measured MRI signal in itself is dependent on many factors, such as repetition times, flip angles and scan sequence; it is not solely a product of T_1 or T_2.[3,4]

MRI is arguably the most versatile imaging modality for the use in theranostics since, unlike most imaging modalities, the contrast can be activated by both internal and external stimuli and is mainly limited by MRI sensitivity. An interesting example employing paramagnetic compounds is given by de Smet *et al.*, in which [Gd(HPDO3A)(H$_2$O)] and doxorubicin were incorporated in temperature-sensitive liposomes for image-guided local drug delivery using high-intensity focused ultrasound hyperthermia. The *in vitro* temperature-dependent release of drug was ascertained by fluorimetric measurements, on the basis of the difference of fluorescence emission for encapsulated and free doxorubicin. The release of imaging agent was determined using T_1 relaxation time measurements, since the relaxivity of encapsulated agent is lower than that of free agent (due to limited water access). These measurements showed that rapid co-release of drug and agent occurred $>40\,^\circ$C, whereas no leakage was observed at $37\,^\circ$C. A proof-of-principle *in vivo* study further showed a linear correlation between T_1, and Gd and doxorubicin concentrations, making these systems an interesting tool for monitoring and controlling local drug delivery.[65]

In the application of photodynamic therapy (PDT), it is of key importance that the used photosensitizer accumulates primarily in the tumor post-administration and is activated by light to form cytotoxic reactive oxygen species (ROS). To achieve the latter, Liu *et al.* employed a C$_{60}$ fullerene nanocarbon modified with four PEG-chains and three Gd-DTPA groups as a photosensitizer which is water-soluble, large enough to profit from passive targeting *via* the EPR effect and gives significant contrast in MRI images. Compared with a standard contrast agent (Magnevist®), this complex showed prolonged blood circulation and increased tumor accumulation which could be accurately tracked using MRI, and the latter was confirmed by replacing the Gd^{3+} ion with ^{59}Fe and performing gamma counting of tumor and muscle. This is favorable for PDT, as the highest therapeutic effect is obtained when the intra-tumor concentration is at its maximum due to the short half-life of ROS. Indeed, there was a good accordance between the time profiles of the MRI contrast and the therapeutic effect of PDT in *in vivo* experimental setups of Meth AR1 fibrosarcoma tumor-bearing mice. Furthermore, the Gd-DTPA moiety did not show signs of gadolinium dissociation and did not alter the PDT efficacy. These findings indicate that the C$_{60}$–PEG–GdDTPA complex as a theranostic complex has several advantages over standard PDT photosensitizers.[66]

A multifunctional MRI theranostic platform reported by Medarova *et al.* employed a superparamagnetic nanoparticle with a dextran coating, to which thiolated siRNA (targeting the green fluorescent protein (GFP)) was coupled *via* a linker, as well as the fluorescent dye Cy5.5 and a membrane translocation peptide. MRI was performed to visualize the (comparable) EPR-mediated particle accumulation in a mouse model bearing bilateral 9L-GFP and 9L-RFP tumors (the latter expressing red fluorescent protein). NIR imaging showed a decrease in imaging signal only in the GFP-expressing tumor. Subsequent experiments in which the siRNA was switched to target the gene encoding the antiapoptotic protein survivin, showed a drop in survivin transcription of 97% and increased tumor apoptosis and necrosis.[67]

5.3.4 Ultrasound

Medical ultrasonography, commonly referred to as ultrasound (ultrasound), is a medical imaging technique utilizing oscillating sound pressure waves with frequency in the MHz range. Images are produced by sending a pulse of ultrasound waves, produced by a transducer, into tissue and recording the reflected echoes from tissue borders with different acoustic impedances. ultrasound possesses numerous advantages in that it is non-invasive, easy to use, low-cost and non-ionizing, making ultrasound the second most clinically used medical imaging technique. In clinical practice, ultrasound is used in many fields, including but not limited to cardiology, oncology, urology, neurology, gynecology and gastroenterology.[68] Recent miniaturizing efforts have succeeded in producing portable ultrasound scanners that permit use in limited space or in-field, adding a unique advantage to ultrasound.[69]

In addition to the diagnostic use of ultrasound, a therapeutic use has emerged in recent years. High-intensity focused ultrasound focuses high intensity acoustic energy, as the name suggests, on a particular area that is moved along the pathological tissue. Accordingly, the acoustic energy is absorbed by the tissue, resulting in temperature increase (>60 °C), thus locally heating and destroying pathological cells, a process referred to as thermal ablation.[68]

Furthermore, ultrasound diagnostic and therapeutic potential is enhanced when using contrast agents. Microbubbles, gas filled shells of lipids, polymers or proteins (discussed in Part 5.2.6) have significantly broadened the diagnostic potential of ultrasound. Microbubbles allow characterizing pathologies based on the functional and molecular characteristics of blood vessels. Microbubbles can attach, encapsulate or entrap therapeutic entities, transforming contrast-enhanced ultrasound from a solely diagnostic application into a theranostic application.[68] Once injected, microbubbles can be tracked by ultrasound and upon reaching the target site a destructive ultrasound pulse is applied releasing the microbubbles' therapeutic contents. Additionally, the interaction between ultrasound and microbubbles can potentially enhance drug delivery *via* temporally increasing vascular and cellular membrane permeability through stable or inertial microbubble cavitation.[68]

Drug-loaded microbubbles have the advantage of reducing systematic drug exposure, therefore reducing healthy tissue damage. Also, microbubbles slow nucleic acid degradation upon system administration in the case of nucleic acids delivery. Nevertheless, image-guided drug delivery with microbubbles has some limitations. In particular, the absolute amount of drugs delivered to the pathological site is relatively low. This is attributed to the very short half-life of circulating microbubbles (typically a few minutes), inability to extravasate (microbubbles are limited to intravascular targeting) and limited amount of therapeutic entities loaded within the microbubble shell.[68] It is important to mention that image-guided drug delivery using microbubbles and ultrasound can be performed by co-injection of drugs with microbubbles; however, this process does not constitute theranostic application.

Theranostic applications of image-guided drug delivery using microbubbles and ultrasound remain preclinical, but comprehensive evaluation of these theranostic microbubbles is being undertaken. A prominent example is vascular endothelial growth factor receptor-2-targeted polymeric microbubbles with model drugs (rhodamine-B and coumarin-6) encapsulated within the shell, which could be deposited in colon cancer xenografts upon ultrasound.[70]

Rapoport *et al.*[71] describe a theranostic system consisting of a mixture of doxorubicin-loaded micelles consisting of PEGylated L-lactide- or caprolactone-based block copolymers, and perfluoropentane (PFP) nanodroplets stabilized by the same block copolymers. Upon heating to physiological temperatures, the PFP nanodroplets vaporize inside the copolymer shell, creating nano- and/or microbubbles with doxorubicin partitioning between the bubbles and the micelles. *In vitro* experiments using samples of various cancers showed that under ultrasound sonication of the microbubble/micelle combination, intracellular doxorubicin uptake improved significantly compared to delivery without microbubbles or ultrasound sonication. *In vivo* experiments using the sonication of the mixture in murine tumors led to substantial tumor growth reduction and/or regression; this was not observed in the control group. Additionally, the nanobubbles were shown to passively extravasate into the tumor *via* the EPR effect and coalesce intro highly echogenic microbubbles, thereby producing a strong contrast within the tumor.[71]

Ultrasound also shows great promise for use in targeted gene delivery. Christiansen *et al.* showed that a luciferase reporter plasmid could be charge-coupled to a microbubble consisting of a perfluorobutane (PFB) core stabilized by cationic lipids. Rats injected with the bubbles and treated at the hind limb muscle with high-power intermittent ultrasound displayed perivascular luciferase transfection due to microbubble sonoporation, as opposed to rats not treated with ultrasound.[72] Lentacker *et al.*, employing a similar system which used a microbubble with a PFB core and albumin shell coated with poly(allylamine hydrochloride), demonstrated that DNA bound this way is protected from nuclease activity.[73] Hauff *et al.* tested the feasibility of therapeutic gene delivery in two different animal models using gas-filled poly(D,L-lactide-*co*-glycolide) (PLGA) microbubbles in which the plasmid was encapsulated into the polymer shell. In the first model, these microbubbles were loaded with the plasmid encoding β-galactosidase and delivered to CC531 liver tumors in rats using ultrasound-mediated sonoporation. Only rats treated with ultrasound and microbubbles showed pDNA expression. The second animal model employed Capan-1 tumors in nude mice, which were injected with micro-bubbles carrying a plasmid encoding for the tumor suppressor gene p16 (normally not present in this type of tumor). Again, mice treated with this microbubble system and ultrasound were found to have an increased p16 expression in the tumor, and consequently a strong inhibition of tumor growth.[74]

5.3.5 Optical Imaging

Optical imaging consists of various techniques that use visible, ultraviolet and infrared light imaging. Optical imaging is a no-ninvasive, non-ionizing and low-cost technique, but it is limited to few centimeters (millimeters in case of fluorescence-based optical imaging) in penetration depth, which restricts its potential clinical application. Preclinically and clinically relevant optical imaging techniques, generally referred to as diffuse optical imaging or near-infrared optical tomography, employ NIR light that is detected after transmission across a biological tissue to produce images and resolve spectroscopic information about the composition of the tissue.[75–77]

Light in the visible range and photons in the infrared region are partially absorbed by natural fluorochromes and water, respectively. However, the NIR region (600–900 nm) offers a window of opportunity where absorbance of NIR light is low enough to allow penetration up to few centimeters within biological tissues.[78,79]

Many different imaging approaches have been developed. Here we discuss two approaches. Planar or reflectance fluorescent imaging, the technically simpler approach, has the light source and the detector placed on top of the specimen. The light source is at a single wavelength and is illuminated on top of the specimen exciting fluorophores that emit fluorescent photons at a different wavelength, which are recorded by the detector. The emitted light reaches the surface of the tissue where it is measured using a filter that filters away the excitation light.

Reflectance imaging systems are easy to use, acquire and generate data rapidly, require low computational power, provide easy and straightforward data analysis and can detect subnanomolar amounts of fluorochromes making optical imaging attractive in terms of sensitivity.[78] However, reflectance imaging is not fundamentally quantitative and has very limited penetration depth. Recent improvement efforts have used point-illumination instead of line-illumination and have achieved deeper tissue penetration (10 mm) and more quantitative data analysis.[80,81] Moreover, multiple fluorescent channels allow better quantification of fluorescence concentration.[80] Reflectance imaging in mice can be considered analogous to fluorescence endoscopy in humans, where near-surface epithelial cancers can be imaged with ease. Moreover, the technology can be applied as intra-operative tumor margin visualization assistance in surgery using goggles.[78]

The more technically complex approach is tomographic (trans-illumination or epi-illumination) where the light source and the detector are placed at opposite sides of the specimen.[76] FMT is a commercially available preclinical transillumination based system.[82] FMT directs excitation NIR light at multiple locations (incident illumination array) into an object to transilluminate a particular portion of that object, then detects fluorescence emitted from the fluorescent probe within that object.[82] FMT uses inversion techniques that take into account propagation of photons into tissue and produce 3D images *in vivo*.[6]

Fluorescent probes can be of two types: unquenched probes which will produce signal under all circumstances, and quenched probes that will only produce signal after activation (*e.g.* enzyme activation).[52] It is important to note that fluorescent probes can emit signal more than once, in contrast to near-infrared probes where emission only occurs once. This characteristic explains the high sensitivity of optical imaging using fluorescent probes despite the strong absorption.

FMT systems can employ different excitation channels; a commercially available system employs up to four channels (635, 680, 750 and 790 nm). It has been shown that at wavelengths around 750 nm, the background signal is minimal, therefore the signal detected is mostly from the fluorescent probe.[83] This is an advantage where the background signal does not need to be subtracted during quantification.

FMT is relatively quantitative, measures nanomolar amounts, images deep tissues and produces 3D images. However, fluorescence signal reconstruction is not yet fully optimized, although serious improvement efforts are underway.[84] Moreover, FMT requires high computational power and suffers from inaccurate signal-organ localization. Therefore, a novel hybrid imaging protocol fusing micro-computed tomography (μCT) imaging (providing anatomical data) with FMT has significantly improved signal to organ localization.[85] FMT has numerous preclinical applications including oncology, inflammation and cardiovascular diseases. For instance, accumulation of near-infrared fluorophore-labeled nanomedicines in tissues other than superficial tumors and signal organ biodistribution were non-invasively visualized and quantified.[85] Moreover, non-invasive staging of squamous cell carcinoma and effects of therapy were monitored and quantified *in vivo* in mice.[52]

Fluorescent quantum dots are mainly suitable for *in vitro* imaging of theranostic systems on a cellular scale, due to the fact that they have a high quantum yield in optical imaging, but the penetration depth of their excitation and emission wavelengths is in the range of only a few millimeters. In a system devised by Yuan *et al.*,[86] CdTe quantum dots with excitation and emission wavelengths at 400 nm and 570 nm, respectively, were synthesized as a sensing system for the interaction between the anthracene anti-cancer drug mitoxantrone (MTX) and DNA. By adsorption of MTX on the quantum dot surface, fluorescence of the quantum dots was quenched through the photoinduced electron-transfer process to 1% at micromolar concentrations

Table 5.1 An overview of the imaging modalities applicable to the field of theranostics, including their most relevant characteristics and (dis)advantages.[a]

Imaging technique	Type of electromagnetic radiation used	Spatial resolution	Penetration depth	Sensitivity	Type of molecular probes	(Dis)advantages
Nuclear imaging	γ-rays	0.5–2 mm	No limit	10^{-10}–10^{-12} M	Radio(labeled) isotopes	+ High sensitivity − Low resolution − Radiation exposure
Computed tomography	X-rays	2–200 μm	No limit	Not well defined	Iodine-containing nanoparticles Metal nanoparticles (*e.g.* gold)	+ High resolution + Good soft tissue contrast − Radiation exposure − High cost
Magnetic resonance imaging	Radiowaves	25–100 μm	No limit	10^{-3}–10^{-5} M	(Super)paramagnetic compounds Hyperpolarized probes Chemical exchange saturation transfer agents	+ High resolution + Non-invasive + Multi-functional imaging possible − Low sensitivity (except for hyperpolarized probes) − High cost
Ultrasound	High-frequency sound waves	50–500 μm	No limit	Single MB detection possible	Microbubbles	+ Widely available + Cheap and easy to use + Therapeutic applications possible + Non-invasive − User dependent
Optical	Visible light	1 μm–5 mm[b]	<4 cm	10^{-9}–10^{-17}[c]	Fluorescent compounds (*e.g.* quantum dots)	+ Very high sensitivity + Multi-functional imaging possible + Activatable probes − Limited penetration depth − Quantification difficult

[a]Adapted from ref. 4, with permission from Elsevier © 2011.
[b]Highest for optical microscopy techniques, lowest for optical tomography methods.
[c]Highest for bioluminescence, lowest for fluorescence.

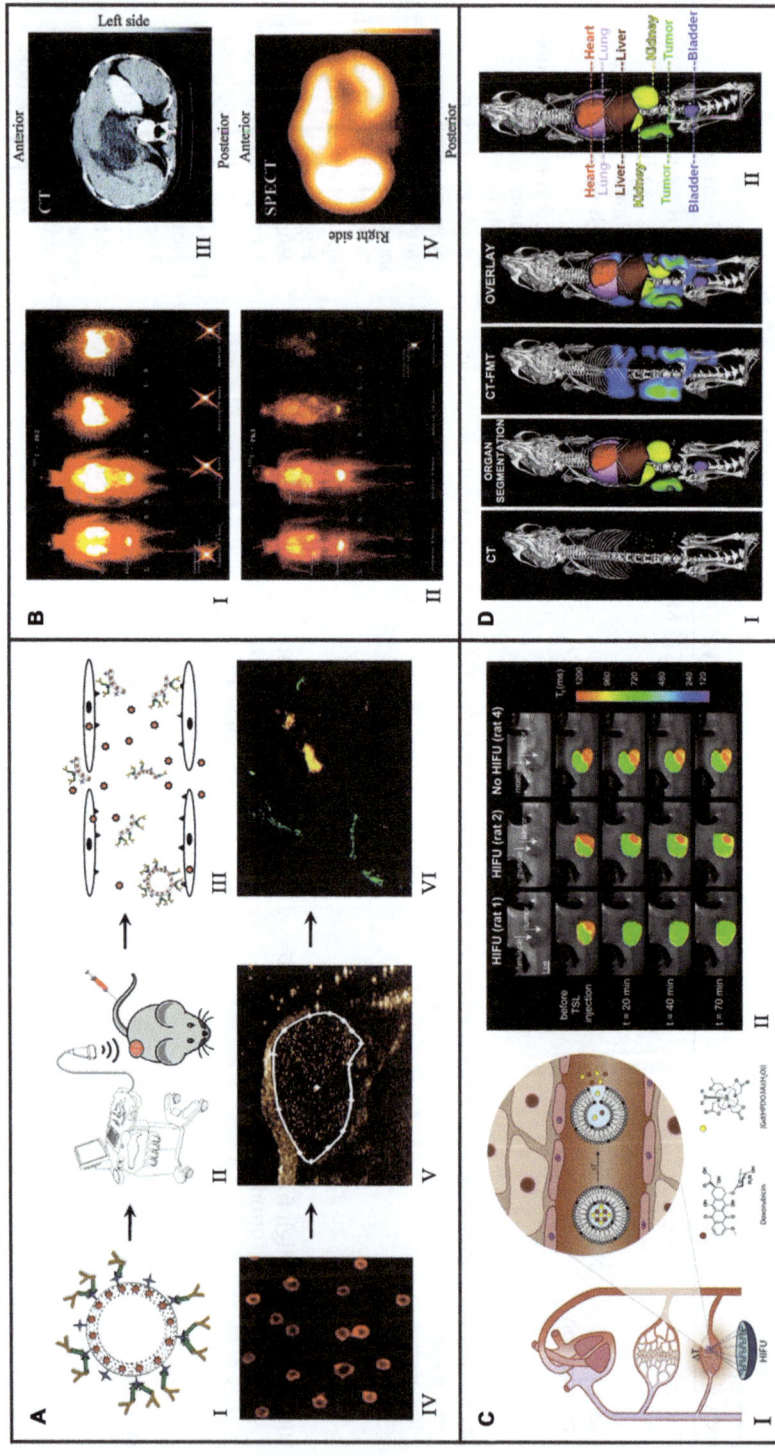

of MTX. Upon introduction of calf thymus DNA (ctDNA) into the mixture, MTX was able to bind to the DNA *via* various mechanisms, thereby leaving the quantum dot surface and undoing the quenching effect. The same theranostic system was subsequently used to determine the binding constant and binding site size of the interaction of MTX and ctDNA, on the basis of quantum dot fluorescence measurements. The sensitivity, relative incubation time independence and resistance to photobleaching might make this system a valuable tool in drug delivery imaging.[86]

The main disadvantage of metal-based quantum dots as nanotheranostics is their high toxicity, hindering their application in clinical settings.[7] Therefore, smaller organic fluorescent dyes may be more suitable for (eventual) *in vivo* applications, and already have found many

---

Figure 5.3   Examples of (nano)theranostic imaging. A: Schematic (I) and two-photon laser scanning microscopy images (IV) of vascular endothelial growth factor receptor (VEGFR)2-antibody-targeted PBCA microbubbles loaded with the model drug rhodamine-B were injected into a CT26 tumor-bearing mouse model, allowing for non-invasive ultrasound imaging (II) of microbubble accumulation within the tumor upon binding to VEGFR2 (V). Subsequent microbubble destruction using high mechanical index ultrasound pulses caused extravasation of the model drug into the tumor interstitium (III), as imaged by fluorescence microscopy. B: Clinical application of theranostics, in which non-invasive gamma camera imaging was performed of an ^{123}I-labeled, asialoglycoprotein receptor-targeted HPMA polymer carrying doxorubicin (I) and an untargeted control (II) on patients with primary and metastic liver cancer, providing information on biodistribution and tumor accumulation at various timepoints. Subsequent single photon emission computed tomography imaging (IV) registered to an anatomical computed tomography image (III) showed that compound accumulation occurred mainly in healthy liver tissue, but a substantial increase in the tumor (as compared to the control) was similarly found. C: High-intensity focused ultrasound (HIFU)-mediated local hyperthermia using theranostics: temperature-sensitive liposomes (TSLs) loaded with doxorubicin and [Gd(HPDO3A)(H$_2$O)]. Upon arrival in the heated tissue, the intraliposomal content is released and extravasation of drug and imaging agent occurs (I). Rats carrying subcutaneous 9L gliosarcoma were injected with TSLs and imaged using magnetic resonance imaging at 3 T. A T$_1$ map overlaid on the anatomical images showed that tumors treated with HIFU showed a decrease of T$_1$ values (either in the tumor as a whole or the rim), correspondent with imaging agent release, whereas no HIFU treatment did not lead to a decrease (II). D: High-resolution 3D μCT showing anatomical structures and presegmented organs of a CT26 tumor bearing mouse (I, first two panels and II), and fluorescence-mediated molecular tomography-based biodistribution data of an injected HPMA polymer labelled with the model drug Dy750, registered to the μCT data (I, third and fourth panel).

applications in preclinical research where they often fulfil a supporting role (see the example by Medarova *et al.*[67] in Section 5.3.3). Bardhan *et al.* employed a similar setup, in which they used a core–shell complex consisting of silica nanoparticles covered by a gold shell. These nanoshells greatly enhanced the quantum yield of indocyanine-green encapsulated within the shell, and MRI contrast was achieved by the incorporation of iron oxide nanoparticles into the silica core. The complex was PEGylated and targeted to human epithelial growth factor receptor 2 (HER2), as over-expressed by breast tumors, using antibodies. The complex proved suitable for NIR imaging to track the complex *in vivo* in tumor-bearing mouse models and showed significant accumulation in tumors over-expressing HER2 compared to those without HER2 over-expression. Similarly, MRI could be used to follow the particle on the basis of T_2-weighted contrast from the iron oxides. Although not used in this study, the gold nanoshell could be used for NIR-activated tumor therapy.[87] Kunjachan *et al.* used HPMA polymers as drug carriers carrying the fluorescent dye Dy750 as a model drug in a study to determine the efficacy of a hybrid protocol consisting of high-resolution μCT and FMT for non-invasive imaging and quantification of the compound in various deep-lying tissues and tumors, in a tumor-bearing mouse model. μCT images were used to pre-segment various organs relevant to nanoparticle accumulation and fate. FMT data were successfully registered to these images and segmentations, and shown to be very similar to the expected biodistribution, making the system suitable for *in vivo* optical quantitative assessment of nanoparticle distribution of both superficial and deep-lying tissues.[85] In a subsequent study, Kunjachan *et al.* compared tumor accumulation of untargeted and vascularly targeted nanocarriers using FMT. They found that adding peptidic ligands to the polymers resulted in enhanced liver uptake and reduced blood half-life to such a degree that the EPR-based accumulation dropped significantly. This lead to a reduced tumor uptake of the targeted compounds as compared to their non-targeted counterparts, illustrating that active targeting not always leads to higher accumulation efficacy of nanomedicines.[14]

Table 5.1 and Figure 5.3 summarize the characteristics of imaging techniques applicable to theranostics.

5.4 Conclusions

The use of nanomaterials has become established in both imaging and therapy. More recently, the rapidly developing field of theranostics attempts to combine the two within the same nanoparticle system. In this chapter, various aspects relating to theranostics have been addressed. It has been shown that theranostics can be used in various stages of therapy development, such as imaging and monitoring drug delivery and efficacy in various temporospatial domains. Various nanoparticle materials and imaging modalities that can be employed in theranostic research have been discussed, highlighting the versatility of the field. However, as illustrated by the examples, the theranostic formulation design heavily depends on the specific requirements of its application. The choice in carrier material also depends on the interaction between imaging agents, therapeutic agents and targeting moieties, as well as the influence of physiochemical properties and biocompatibility.

The majority of theranostic systems are preclinical and many obstacles have to be overcome before a clinical translation can be made. In particular, it is critical to consider where a theranostic agent is needed or where companion diagnostics may suffice. Making systems more complex (*e.g.* by adding an imaging label) will usually lead to worse pharmacokinetic and pharmacodynamic properties. This problem may be less evident for nanomedicines than small molecules due to their large size and the considerably lower impact of a small imaging marker. Nevertheless, it is expected that theranostics will play a crucial role in modern nanomedicine, such as acquiring accurate data on the biodistribution and clearance kinetics of nanomedicines. Theranostics can also be of great use for defining the optimal dosing and combination of drugs. Finally, since the accumulation of nanomedicines is highly variable between patients and even heterogeneous within the same tumor and its metastases, theranostics may support the implementation of personalized therapy regimens. Ultimately, only such theranostic concepts that improve the therapeutic outcome of patients over a non-imagable conduct will gain broad clinical acceptance.

References

1. T. Lammers, W. Hennink and G. Storm, *Br. J. Cancer*, 2008, **99**, 392.
2. A. S. Gupta, *Nanomed. Nanotechnol. Biol. Med*, 2011, 7, 763.
3. S. S. Kelkar and T. M. Reineke, *Bioconjug. Chem.*, 2011, **22**, 1879.
4. S. M. Janib, A. S. Moses and J. A. MacKay, *Adv. Drug Delivery Rev.*, 2010, **62**, 1052.
5. J. V. Jokerst and S. S. Gambhir, *Acc. Chem. Res.*, 2011, **44**, 1050.
6. T. Lammers, L. Y. Rizzo, G. Storm and F. Kiessling, *Clin. Cancer Res.*, 2012, **18**, 4889.
7. H. S. Choi, W. Liu, P. Misra, E. Tanaka, J. P. Zimmer, B. I. Ipe, M. G. Bawendi and J. V. Frangioni, *Nat. Biotechnol.*, 2007, **25**, 1165.
8. T. Lammers, S. Aime, W. E. Hennink, G. Storm and F. Kiessling, *Acc. Chem. Res.*, 2011, **44**, 1029.
9. J. Xie, S. Lee and X. Chen, *Adv. Drug Delivery Rev.*, 2010, **62**, 1064.
10. H. Maeda, J. Wu, T. Sawa, Y. Matsumura and K. Hori, *J. Controlled Release*, 2000, **65**, 271.
11. H. Maeda, G. Bharate and J. Daruwalla, *Eur. J. Pharm. Biopharm.*, 2009, **71**, 409.
12. Y. H. Bae and K. Park, *J. Controlled Release*, 2011, **153**, 198.
13. D. Peer, J. M. Karp, S. Hong, O. C. Farokhzad, R. Margalit and R. Langer, *Nat. Nanotechnol.*, 2007, **2**, 751.

14. S. Kunjachan, R. Pola, F. Gremse, B. Theek, J. Ehling, D. Moeckel, B. Hermanns, M. Pechar, K. Ulbrich and W. E. Hennink, *Nano Lett.*, 2014, **14**(2), 972–981.
15. K. Letchford and H. Burt, *Eur. J. Pharm. Biopharm.*, 2007, **65**, 259.
16. K. S. Soppimath, T. M. Aminabhavi, A. R. Kulkarni and W. E. Rudzinski, *J. Controlled Release*, 2001, **70**, 1.
17. S. Kray, A. Pich, S. Pargen, A. Balaceanu, M. Lenz, F. Spöler, F. Kiessling and W. Lederle, *J. Nanopart. Res.*, 2012, **14**, 1.
18. S. V. Vinogradov, *Nanomedicine*, 2010, **5**, 165.
19. D. Dorwal, *Int. J. Pharm. Pharm. Sci.*, 2012, **4**(3), 67–74.
20. D. E. Discher and F. Ahmed, *Annu. Rev. Biomed. Eng.*, 2006, **8**, 323.
21. M.-H. Li and P. Keller, *Soft Matter*, 2009, **5**, 927.
22. F. Meng, Z. Zhong and J. Feijen, *Biomacromolecules*, 2009, **10**, 197.
23. B. Klajnert and M. Bryszewska, *Acta Biochim. Pol.*, 2000, **48**, 199.
24. V. Gajbhiye, V. K. Palanirajan, R. K. Tekade and N. K. Jain, *J. Pharm. Pharmacol.*, 2009, **61**, 989.
25. A. Malik, S. Chaudhary, G. Garg and A. Tomar, *Adv. Biol. Res.*, 2012, **6**, 165.
26. J. Liu, W. D. Gray, M. E. Davis and Y. Luo, *Interface Focus*, 2012, **2**, 307.
27. H. Patel and P. Patel, *Int. J. Pharma Bio Sci.*, 2013, **4**(2), 454–463.
28. W. T. Al-Jamal and K. Kostarelos, *Acc. Chem. Res.*, 2011, **44**, 1094.
29. D. B. Fenske, A. Chonn and P. R. Cullis, *Toxicol. Pathol.*, 2008, **36**, 21.
30. T. A. Waldmann, *Science*, 1991, **252**, 1657.
31. A. Fernandez-Fernandez, R. Manchanda and A. J. McGoron, *Appl. Biochem. Biotechnol.*, 2011, **165**, 1628.
32. S. Del Vecchio, A. Zannetti, R. Fonti, L. Pace and M. Salvatore, *Q. J. Nucl. Med. Mol. Imaging*, 2007, **51**, 152.
33. Z. Liu and X.-J. Liang, *Theranostics*, 2012, **2**, 235.
34. T. Olafsen and A. M. Wu, *Semin. Nucl. Med.*, 2010, **40**, 167.
35. V. V. Mody, R. Siwale, A. Singh and H. R. Mody, *J. Pharm. Bioallied Sci.*, 2010, **2**, 282.
36. S. Schultz, D. R. Smith, J. J. Mock and D. A. Schultz, *Proc. Natl. Acad. Sci. U. S. A.*, 2000, **97**, 996.
37. J. L. Elechiguerra, J. L. Burt, J. R. Morones, A. Camacho-Bragado, X. Gao, H. H. Lara and M. J. Yacaman, *J. Nanobiotechnol.*, 2005, **3**, 1.
38. D. M. Goldenberg, *Am. J. Med.*, 1993, **94**, 297.
39. Z. Chen, L. Ma, Y. Liu and C. Chen, *Theranostics*, 2012, **2**, 238.
40. F. Kiessling, S. Fokong, P. Koczera, W. Lederle and T. Lammers, *J. Nucl. Med.*, 2012, **53**, 345.
41. E. Stride and N. Saffari, *Proc. Inst. Mech. Eng., Part H*, 2003, **217**, 429.
42. T. S. Kang and C.-k. Yeh, *Chang Gung Med. J.*, 2012, **35**(2), 125–139.
43. F. Forrer, H. Uusijärvi, C. Waldherr, M. Cremonesi, P. Bernhardt, J. Mueller-Brand and H. R. Maecke, *Eur. J. Nucl. Med. Mol. Imaging*, 2004, **31**, 1257.

44. L. W. Seymour, D. R. Ferry, D. Anderson, S. Hesslewood, P. J. Julyan, R. Poyner, J. Doran, A. M. Young, S. Burtles and D. J. Kerr, *J. Clin. Oncol.*, 2002, **20**, 1668.

45. D. W. Bartlett, H. Su, I. J. Hildebrandt, W. A. Weber and M. E. Davis, *Proc. Natl. Acad. Sci. U. S. A.*, 2007, **104**, 15549.

46. G. T. Herman, *Fundamentals of computerized tomography: image reconstruction from projections*, Springer, 2009.

47. J. Ehling, T. Lammers and F. Kiessling, *Thromb. Haemostasis*, 2013, **109**, 375.

48. D. J. Brenner and E. J. Hall, *N. Engl. J. Med.*, 2007, **357**, 2277.

49. F. Gremse, C. Grouls, M. Palmowski, T. Lammers, A. de Vries, H. Grüll, M. Das, G. Mühlenbruch, S. Akhtar and A. Schober, *Radiology*, 2011, **260**, 709.

50. J. J. Pasternak and E. E. Williamson, *Mayo Clin. Proc.*, 2012, **87**(4), 390–402.

51. J. Ehling, B. Theek, F. Gremse, S. Baetke, D. Möckel, J. Maynard, S.-A. Ricketts, H. Grüll, M. Neeman and R. Knuechel, *Am. J. Pathol.*, 2014, **184**, 431.

52. W. Al Rawashdeh, S. Arns, F. Gremse, J. Ehling, R. Knüchel-Clarke, S. Kray, F. Spöler, F. Kiessling and W. Lederle, *Neoplasia*, 2014, **16**, 235.

53. M. Dunne, J. Zheng, J. Rosenblat, D. A. Jaffray and C. Allen, *J. Controlled Release*, 2011, **154**, 298.

54. G. von Maltzahn, J.-H. Park, A. Agrawal, N. K. Bandaru, S. K. Das, M. J. Sailor and S. N. Bhatia, *Cancer Res.*, 2009, **69**, 3892.

55. S. Miyata, S. Kawabata, R. Hiramatsu, A. Doi, N. Ikeda, T. Yamashita, T. Kuroiwa, S. Kasaoka, K. Maruyama and S.-I. Miyatake, *Neurosurgery*, 2011, **68**, 1380.

56. X.-J. Lu, Y. Zhang, D.-C. Cui, W.-J. Meng, L.-R. Du, H.-T. Guan, Z.-Z. Zheng, N.-Q. Fu, T.-S. Lv and L. Song, *Int. J. Pharm.*, 2013, **452**, 211.

57. S. H. Bartling, J. Budjan, H. Aviv, S. Haneder, B. Kraenzlin, H. Michaely, S. Margel, S. Diehl, W. Semmler and N. Gretz, *Invest. Radiol.*, 2011, **46**, 178.

58. P. Caravan, J. J. Ellison, T. J. McMurry and R. B. Lauffer, *Chem. Rev.*, 1999, **99**, 2293.

59. P. Caravan, *Chem. Soc. Rev.*, 2006, **35**, 512.

60. H. Grüll and S. Langereis, *J. Controlled Release*, 2012, **161**, 317.

61. J. M. Idée, M. Port, I. Raynal, M. Schaefer, S. Le Greneur and C. Corot, *Fundam. Clin. Pharmacol.*, 2006, **20**, 563.

62. T. Grobner, *Nephrol. Dial. Transplant.*, 2006, **21**, 1104.

63. P. Reimer and T. Balzer, *Eur. Radiol.*, 2003, **13**, 1266.

64. J. Ruiz-Cabello, B. P. Barnett, P. A. Bottomley and J. W. Bulte, *NMR Biomed.*, 2011, **24**, 114.

65. M. de Smet, E. Heijman, S. Langereis, N. M. Hijnen and H. Grüll, *J. Controlled Release*, 2011, **150**, 102.

66. J. Liu, S.-i. Ohta, A. Sonoda, M. Yamada, M. Yamamoto, N. Nitta, K. Murata and Y. Tabata, *J. Controlled Release*, 2007, **117**, 104.
67. Z. Medarova, W. Pham, C. Farrar, V. Petkova and A. Moore, *Nat. Med.*, 2007, **13**, 372.
68. F. Kiessling, S. Fokong, J. Bzyl, W. Lederle, M. Palmowski and T. Lammers, *Adv. Drug Delivery Rev.*, 2014, **72**, 15–27.
69. B. P. Nelson and K. Chason, *Int. J. Emerg. Med.*, 2008, **1**, 253.
70. S. Fokong, B. Theek, Z. Wu, P. Koczera, L. Appold, S. Jorge, U. Resch-Genger, M. van Zandvoort, G. Storm and F. Kiessling, *J. Controlled Release*, 2012, **163**, 75.
71. N. Rapoport, Z. Gao and A. Kennedy, *J. Natl. Cancer Inst.*, 2007, **99**, 1095.
72. J. P. Christiansen, B. A. French, A. L. Klibanov, S. Kaul and J. R. Lindner, *Ultrasound Med. Biol.*, 2003, **29**, 1759.
73. I. Lentacker, B. G. De Geest, R. E. Vandenbroucke, L. Peeters, J. Demeester, S. C. De Smedt and N. N. Sanders, *Langmuir*, 2006, **22**, 7273.
74. P. Hauff, S. Seemann, R. Reszka, M. Schultze-Mosgau, M. Reinhardt, T. Buzasi, T. Plath, S. Rosewicz and M. Schirner, *Radiology*, 2005, **236**, 572.
75. R. Weissleder and M. J. Pittet, *Nature*, 2008, **452**, 580.
76. D. R. Leff, O. J. Warren, L. C. Enfield, A. Gibson, T. Athanasiou, D. K. Patten, J. Hebden, G. Z. Yang and A. Darzi, *Breast Cancer Res. Treat.*, 2008, **108**, 9.
77. V. Ntziachristos, J. Ripoll, L. V. Wang and R. Weissleder, *Nat. Biotechnol.*, 2005, **23**, 313.
78. U. Mahmood and R. Weissleder, *Mol. Cancer Ther.*, 2003, **2**, 489.
79. U. Mahmood, C.-H. Tung, A. Bogdanov Jr and R. Weissleder, *Radiology*, 1999, **213**, 866.
80. L. Abou-Elkacem, S. Björn, D. Doleschel, V. Ntziachristos, R. Schulz, R. M. Hoffman, F. Kiessling and W. Lederle, *Eur. Radiol.*, 2012, **22**, 1955.
81. S. Björn, V. Ntziachristos and R. Schulz, *Opt. Express*, 2010, **18**, 8422.
82. V. Ntziachristos, C.-H. Tung, C. Bremer and R. Weissleder, *Nat. Med.*, 2002, **8**, 757.
83. E. M. Sevick-Muraca and J. C. Rasmussen, *J. Biomed. Opt.*, 2008, **13**, 041303.
84. F. Gremse, B. Theek, S. Kunjachan, W. Lederle, A. Pardo, S. Barth, T. Lammers, U. Naumann and F. Kiessling, *Theranostics*, 2014, **4**(10), 960–971.
85. S. Kunjachan, F. Gremse, B. Theek, P. Koczera, R. Pola, M. Pechar, T. Etrych, K. Ulbrich, G. Storm and F. Kiessling, *ACS Nano*, 2012, **7**, 252.
86. J. Yuan, W. Guo, X. Yang and E. Wang, *Anal. Chem.*, 2008, **81**, 362.
87. R. Bardhan, W. Chen, C. Perez-Torres, M. Bartels, R. M. Huschka, L. L. Zhao, E. Morosan, R. G. Pautler, A. Joshi and N. J. Halas, *Adv. Funct. Mater.*, 2009, **19**, 3901.

CHAPTER 6

The Role of Imaging in Nanomedicine Development and Clinical Translation

JINZI ZHENG,*[a,b] RAQUEL DE SOUZA,[c] MANUELA VENTURA,[a] CHRISTINE ALLEN[a,c] AND DAVID JAFFRAY[a,b,d]

[a] TECHNA Institute for the Advancement of Technology for Health, University Health Network, Toronto, Ontario, Canada; [b] Institute of Biomaterials and Biomedical Engineering, University of Toronto, Toronto, Ontario, Canada; [c] Faculty of Pharmaceutical Sciences, University of Toronto, Toronto, Ontario, Canada; [d] Department of Radiation Oncology, University of Toronto, Toronto, Ontario, Canada
*Email: jinzi.zheng@rmp.uhn.on.ca

6.1 Introduction

There has been significant investment in the advancement of nanoparticles in the context of diagnosis and therapy. Several of these nanoplatforms have achieved success in both the research and clinical arenas.[1–8] Overall, there are three main rationales for employing nanoparticles instead of traditional small molecules as a new class of agents that have the potential to lead to improved diagnosis and treatment. First, the critical size of nanoparticles and their tunable surface characteristics result in pharmacokinetics profiles that enable applications requiring longitudinal imaging and enhanced drug delivery. Second, their extended circulation lifetime allows for increased

RSC Drug Discovery Series No. 51
Nanomedicines: Design, Delivery and Detection
Edited by Martin Braddock
© The Royal Society of Chemistry 2016
Published by the Royal Society of Chemistry, www.rsc.org

tumor tissue targeting resulting in greater target-to-background signal ratio in imaging applications and an enhanced therapeutic index when used as delivery vectors. Third, the high payload of imaging and/or therapeutic agents that nanoparticles carry can be exploited for amplification of imaging signal or therapeutic effect, especially when used in conjunction with lower sensitivity imaging systems or less cytotoxic drugs. In addition, various strategies have been developed for selectively directing the nanoparticles to target specific cell populations and sub-cellular compartments.[9,10] These include the engineering of nanoparticles responsive to different triggers that are either inherently present in the tumor microenvironment (*i.e.* pH[11] and matrix metalloproteinases[12]) or that can be induced externally (*i.e.* temperature[13,14] and light irradiation[15]). Molecularly targeted surface ligands can also be incorporated onto the outer layer of nanoparticles. These enable the retention of nanoparticles either on the surface of the cells of interest or to induce cellular uptake. Furthermore, appropriate surface modifications also lead to successful targeting to intracellular compartments such as the nucleus or the mitochondria.[16–18] All of these exciting advances support the clinical advancement of new nanosystems. This chapter illustrates how established, non-invasive and quantitative imaging methods can further accelerate the translational process for both existing and new nanomedicine platforms. While similar efforts are underway across different disease sites, it is important to note that the focus here is directed at the development and use of organic nanoparticles for oncology applications.

6.2 Imaging for *In vivo* Evaluation of the Spatio-temporal Distribution Characteristics of Nanomedicines

6.2.1 Rationale for Spatio-temporal Biodistribution Assessment

In drug development, pharmacokinetics and biodistribution profiles are often used as a surrogate to evaluate the potential effectiveness of a new therapeutic or diagnostic agent. For example, the blood concentration of a drug has often been correlated with its efficacy and toxicity.[19] As a result, the agent should be designed to reach the desired therapeutic or diagnostic effect at the lowest administration dose possible. The development process therefore aims to select the formulation that yields the highest agent concentration at the desired target site (*i.e.* tumor) and the lowest agent concentration elsewhere (*i.e.* healthy organs and background tissues).

The temporal component of a biodistribution assessment is also important. In the case of a diagnostic agent, the biodistribution kinetics define the optimal imaging window for obtaining information on a specific biological or physiological process. For example, in a routine functional computed tomography (CT) or dynamic contrast-enhanced magnetic resonance (MR) imaging session, it is important to characterize the distribution and clearance kinetics of the agent in order to accurately define the arterial and venous phases. If the imaging probe employed is involved in an active biological process, such as fluorodeoxyglucose (FDG) in cellular metabolism or ultra-small superparamagnetic iron oxide (USPIO) in macrophage phagocytosis, the timelines of these processes must be defined in order to set the time gap between probe administration and imaging (*i.e.* 1 hour for FDG-positron emission tomography (PET) and 24 hours for USPIO-MR). Similarly, in the case of a therapeutic agent, its pharmacokinetics and biodistribution influence its efficacy and toxicity. For example, studies conducted in mice to compare the efficacy of liposome-encapsulated doxorubicin *versus* free doxorubicin found a gain in the plasma area under the curve (AUC) of at least 60-fold[20–23] and an increase of 14-fold in the peak tumor drug concentration for the liposomal drug.[21] This resulted in a six-fold enhancement in treatment efficacy[24] and a significant decrease in toxicity.[20,25] Characterization of the temporal profile of the biodistribution and clearance of the therapeutic agent of interest will better enable the setting of appropriate dosing regimens.

6.2.2 Imaging as a Non-invasive Method for Nanoparticle Biodistribution Assessment

Traditional nanoparticle pharmacokinetics and biodistribution studies rely on plasma and tissue sampling. The invasive nature of these procedures can change the biological system under observation resulting in unreliable representation of the *in vivo* conditions. Furthermore, animals often need to be sacrificed at each sampling time point in order to either provide enough plasma or tissue for analysis or because their vital organs are removed during the biodistribution assessment process. These limitations associated with traditional pharmacokinetics and biodistribution studies can be overcome through the use of non-invasive imaging techniques in conjunction with appropriate labeling of the nanoparticle of interest. Image-based measurements, when successfully correlated with tissue agent concentrations, can be used to collect meaningful data on the same animal over multiple time points with minimal perturbations to biological and physiological processes.[26] Therefore, not only are animal-to-animal variations avoided, but also the total number of animals required for each study can be reduced.[27] Furthermore, the increasing availability of small-animal scanners allows the imaging assays employed in the preclinical environment to be more readily translated to the clinical setting.

To date, non-invasive nuclear imaging techniques such as PET and single photon emission computed tomography (SPECT) have been extensively explored for pharmacokinetics and biodistribution studies.[28,29] PET isotopes have shown an advantage over SPECT isotopes for radiolabeling of small molecules because atom replacement is possible with positron emitters such as ^{11}C, ^{15}O, ^{13}N and ^{18}F in a compound without modification of its pharmaceutical, biological or biochemical properties.[30] However, for long-circulating nanoparticles such as liposomes, PET labeling is unsuitable because the positron emitters often have much shorter physical half-lives (10 min for ^{13}N, 20.4 min for ^{11}C, 110 min for ^{18}F and 12.7 hours for ^{64}Cu) compared to the vascular half-life of the nanoparticles (~ 55 hours for liposomal doxorubicin, Doxil®). As such, positron emitter-labeled pharmaceutical carriers cannot be tracked over their full circulation lifetime *in vivo* and repeated imaging over multiple time points is greatly limited. Some SPECT radiotracers have physical half-lives that are suitable for longitudinal carrier therapeutics studies, such as ^{111}In with a half-life of 67.9 hours. The presence of a low amount of metal chelating agent such as diethylene triamine pentaacetic acid (DTPA) when labeling a macromolecular agent is also less of a concern compared to its use for the labeling of small molecules as it usually does not significantly affect the pharmacokinetics and distribution of the macromolecule.[31] In addition, advanced SPECT systems allow for collection of photons emitted at different energy windows. This is very valuable for applications requiring simultaneous imaging of multiple radiotracers (*i.e.* monitoring the biodistribution of multiple species of carriers labeled with radionuclides of different and resolvable gamma energies). However, as the radioactivity of the tracer decays, the imaging time needs to be increased in order to maintain sufficient count statistics in order to produce adequate image quality. Harrington *et al.*[32] conducted a clinical study administering ^{111}In-DTPA-labeled liposomes to 17 patients with locally advanced cancers. Serial whole-body gamma camera images were acquired up to 7 days post-injection showing liposome localization in the tumor lesions as well as in healthy tissues. The image quality of the data set acquired at day 7 was significantly deteriorated due to both physical decay of ^{111}In and biological clearance of liposomes. This report clearly demonstrates the challenge of changes in performance of SPECT over the course of a longitudinal biodistribution assessment. In addition, just like PET, the use of SPECT imaging to map the colloidal carrier tissue distribution is greatly limited by its inability to provide structural and anatomical information. The recent widespread adoption of integrated imaging systems such as PET/CT and SPECT/CT has enabled acquisition and fusion of anatomy with the nuclear imaging data. Furthermore, the development of PET/MR and SPECT/MR systems is likely to provide an even more fertile ground for innovations in the area of nuclear medicine-based monitoring of biodistribution of long-circulating nanoparticles.

MR has also been explored as a potential tool to image the tissue distribution of nanocarriers.[33–35] Viglianti *et al.*[36,37] reported a particularly interesting set of studies where MR was successfully used to visualize and quantify drug release from temperature-sensitive liposomes labeled with Mn^{2+}. Advantages of MR over other imaging modalities (*i.e.* CT, PET and

SPECT) include the absence of ionizing radiation and high soft-tissue contrast. However, its sensitivity for measuring T_1 and T_2-shortening contrast agent concentrations (*i.e.* 10^{-5} M for Mn^{2+}[37]) is approximately 10^5 to 10^7 times lower than SPECT (10^{-10} M[38]) and PET (10^{-11} to 10^{-12} M[38]), respectively. In addition, although a number of fast mapping techniques have been described for quantification of T_1[39–41] and T_2,[42] data collection times are still generally lengthier and image resolution lower than that of conventional qualitative MR acquisitions.[43,44]

More recently, quantitative image-based assessment using CT methods have also emerged to be feasible for longitudinal monitoring of nanoparticle distribution *in vivo*.[45–47] Zheng *et al.*[47] reported the successful use of liposomes containing iohexol and gadoteridol for quantification of whole-body distribution over a 14-day period in a rabbit tumor model. Image-based determination of iodine concentration in the blood, kidneys, liver, spleen and tumor, together with elemental analysis of iodine content in the urine and feces, accounted for ~100% and ~80% of the total injected dose of iodine in liposomes at 30 min and 3 days post-injection, respectively. Furthermore, the high resolution of CT imaging allowed for visualization and quantification of the heterogeneous intratumoral distribution of liposomes over time (Figure 6.1). Specifically, the highest percentage of tumor volume occupied by liposomes ($72 \pm 5\%$) was determined to occur at 2 days post-injection. It is important to consider, when performing repeated CT imaging, the amount of radiation dose delivered to an animal, especially small-size rodents such as mice. Investigation on the effect of serial microCT radiation dose to animals has found that there was no significant effect on tumor growth, mouse survival or ability to form metastasis.[48,49] Figueroa *et al.*[50] reported in their thermo-luminescent dosimeter assessment of mouse dosimetry an absorbed dose of 76 ± 5 mGy for one full rotation ($360°$) of microCT scan (MicroCAT IITM, Siemens Preclinical Solutions, Malvern, PA, USA) at 80 kVp and 54 mA with 0.5 mm of aluminum filtration. This is equivalent to approximately 1/5 of the whole-body absorbed dose delivered to a 20 g mouse with the administration of one of the longer lived SPECT isotopes ^{111}In (360 mGy, assuming no biological washout[51]). These dose considerations make it feasible to design serial high-resolution microCT imaging studies.

The whole-body imaging techniques described above have the ability to quantify drug tissue distribution non-invasively. Their resolution limitations make them inadequate for providing information on the intracellular localization of the administered agents. Microscopes with fluorescence detectors are currently the most widely used tool for visualization of molecule distribution within a cell.[52] Because optical imaging techniques lack depth penetration and are heavily affected by scattering effects, quantitative *in vivo* imaging cannot be performed. Hence, if the administered drug and carrier were either inherently fluorescent (*i.e.* doxorubicin) or labeled with a fluorescent probe, then by collecting a small biopsy sample from the location of interest, the intracellular localization of both the drug and carrier

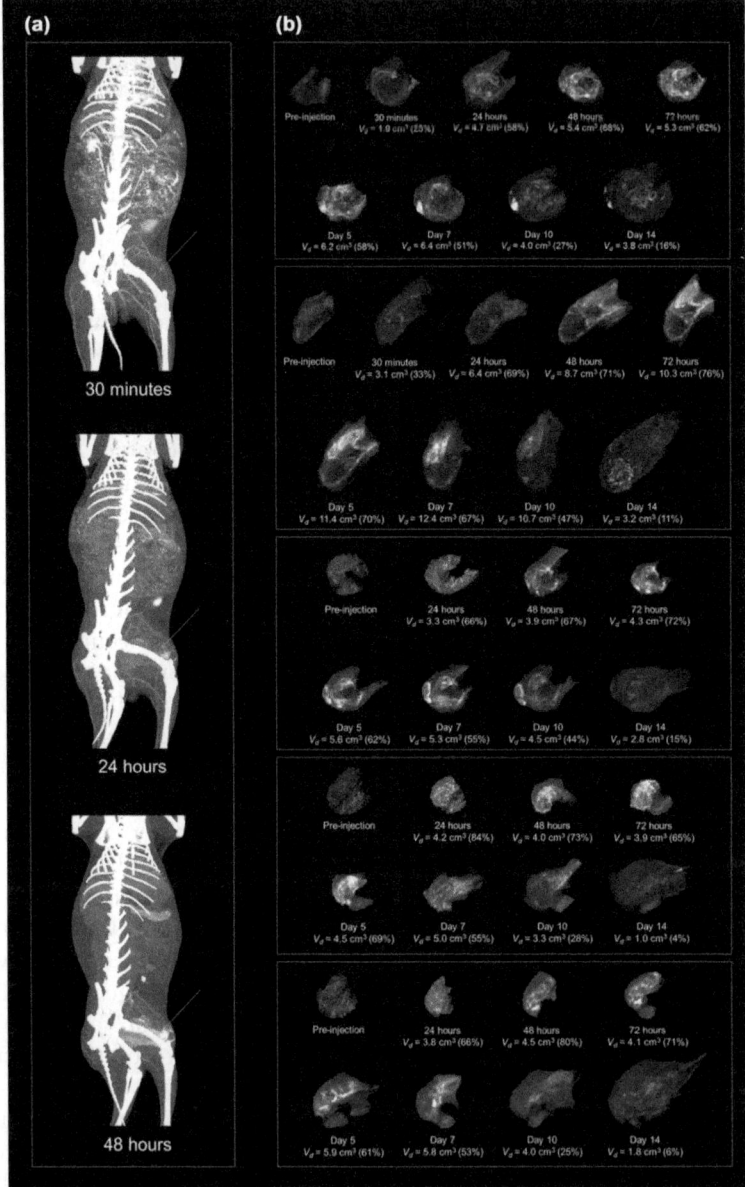

Figure 6.1 (a) Anterior views of 3D CT maximum intensity projections of a representative VX2 carcinoma-bearing male New Zealand White rabbit (3 kg) at 30 min, 24 hours and 48 hours post-liposome administration. The arrows indicate the site of the tumor, and the EPR effect is visualized through the gradual opacification of the tumor area resulting from the accumulation of the iohexol- and gadoteridol-containing liposomes. (b) The five right quadrants represent data acquired from five distinct animals, with each quadrant displaying 3D maximum intensity projections of the segmented tumor volumes pre- and up to 14 days post-liposome injection.
Figure and caption reproduced with permission from Zheng *et al.*[47]

can be visualized using a confocal microscope. However, accurate quantification of fluorescence is sometimes difficult even *ex vivo* due to changes in optical property. For example, doxorubicin fluorescence is partially quenched when the drug binds to DNA.[53]

6.3   Use of Imaging to Understand and Optimize Nanomedicine Performance

6.3.1   Investigation of Size-dependence and Lesion Targeting Ability

The employment of nanoparticles for drug delivery emerged in part from the critical nanometersize range that enables effective targeting to different disease sites of interest. For example, tumor targeting *via* the passive enhanced permeation and retention (EPR) effect, first described by Matsumura and Maeda in 1986,[54] requires particles to have a prolonged vascular residency time (*i.e.* maintain high plasma AUC for >6 hours in mice and rats[54–56]). The optimal pharmacokinetics of a nanoparticle necessary to exploit EPR can be achieved through alterations in particle size, charge, surface properties and shape.[19] Currently available evidence supports the development of nanoparticles that are PEGylated, of approximately 100 nm in diameter, and with a neutral surface charge for applications requiring prolonged blood circulation lifetime. It has been agreed that the degree of macromolecule accumulation in tumors is directly proportional to the blood AUC (or exposure) and inversely proportional to the rate of urinary clearance.[57–59] Once the prerequisite of high exposure has been achieved, the transport of macromolecules, such as nanoparticles, into tumor tissues is further affected by the tumor vascular pore size (up to 400 nm[60]). However, their subsequent intratumoral retention is a function of the particle diffusivity in the tumor interstitial space,[61] the speed of the tumor venous return (usually slower than normal tissue[62,63]), as well as the presence of a poor lymphatic drainage system.[62,63] Altogether, macromolecules and nanoparticles not only preferentially accumulate in tumors *via* the enhanced vascular permeability, but they are also preferentially retained there (for multiple hours to days). Conversely, low-molecular-weight agents are distributed systemically following administration, rapidly cleared from the circulating blood *via* renal clearance, and their tumor accumulation is only transient (in the order of minutes). Their small size allows them to readily return to blood circulation following extravasation into the tumor interstitial space.[56,64] The ability of the EPR effect to significantly increase tumor accumulation *versus* healthy tissue distribution results in increased target-to-background signal ratio for imaging and an enhanced therapeutic ratio for treatment. EPR is the hallmark of nanoparticle-based delivery of diagnostic and therapeutic agents to tumors.[59]

In addition to the prerequisite for a prolonged vascular residence time, tumor deposition of nanoparticles is also highly dependent on tumor microenvironment characteristics such as vascular density, blood flow, permeability and interstitial fluid pressure (IFP). These tumor micro-environment parameters cause heterogeneity in nanoparticle deposition across tumor types, lesions and intratumoral subregions.[46] Toy *et al.*[65] exploited a combination of microCT and multi-wavelength fluorescence molecular tomography (FMT) to investigate the dependence between the regional tumor blood flow rate and tumor deposition of liposomes of different sizes (30 nm, 65 nm and 100 nm). Each group of liposomes was labeled with a distinct fluorophore (Vivotag 750, 680 and 635) and the three different groups of liposomes of distinct sizes were administered as a single cocktail in order to enable monitoring of their distribution in the same tumor at the same time using the same imaging technique (although the tissue absorption, scattering and emission characteristics of the three fluorophores slightly differ and result in small differences in detection sensitivity and accuracy). The investigators concluded that higher regional tumor blood flow improves liposome deposition as it helps overcome high IFP and that blood flow has the most significant impact on the deposition of larger liposomes (*i.e.* 100 nm). For example, for 100 nm liposomes, a significantly higher deposition (*i.e.* 340-fold) was measured in tumor regions with fast blood flow (*i.e.* $\sim$ 350 mL min^{-1} per 100 mL of tissue) compared to those with slower blood flow (*i.e.* $\sim$ 30 mL min^{-1} per 100 mL of tissue). For the 65 nm liposomes, although the same trend was observed, the increase in deposition was roughly half (*i.e.* 180-fold) of that observed for the 100 nm liposomes for the same range of blood flow rates. However, for the 30 nm liposomes, no improvement in deposition was seen in regions of high blood flow. It is important to keep in mind that the amount of liposome deposition in regions of low blood flow is much greater ($\sim$ 10-fold) when the particle size is 30 nm compared to 65 nm and 100 nm (Figure 6.2). Based on these findings, it can be concluded that if the dominant tumor hemodynamic profile is known prior to liposome-based therapy, then the treatment could be personalized through the administration of the optimal liposome size to maximize tumor deposition.

Similar to the EPR effect present in tumors, inflammatory processes also exhibit an increase in blood perfusion and permeability.[66] Long-circulating nanoparticles have therefore also been explored for the treatment and imaging of inflammation and infection. Boerman *et al.*[67] investigated the effect of particle size on vascular circulation time and ultimately abscess-to-background signal ratio by employing ^{99m}Tc-labeled liposomes of 90, 120, 160 and 220 nm and SPECT imaging (Figure 6.3). The study concluded that liposomes with a diameter of 90 nm were the best candidate for effective imaging of infection as they provided good uptake in the lesion while having relatively low accumulation in the spleen. The ability of liposomes in the 90 nm size range to target inflammatory lesions is further confirmed by Zheng *et al.*[68] who demonstrated that iohexol-labeled liposomes allowed for more sensitive detection compared to FDG-PET. Furthermore, in this study, there were distinct differences in the pattern and kinetics of liposome

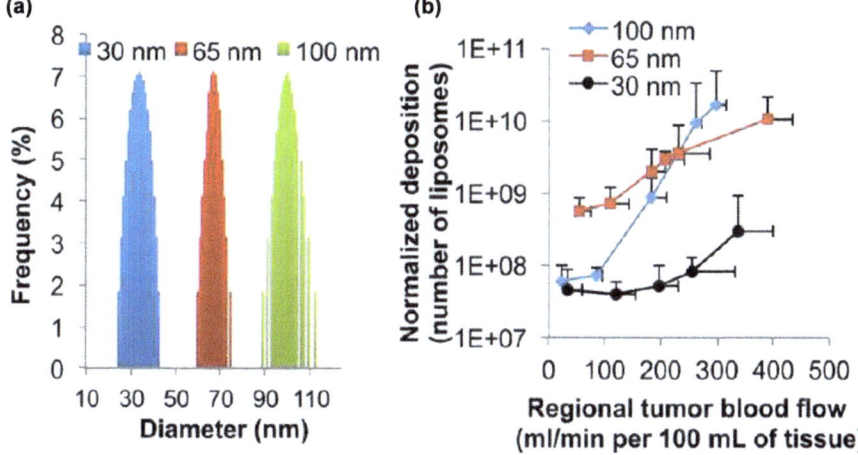

Figure 6.2 Dependence of liposome extravasation into tumors on the liposome size and blood flow on a region-by-region basis. Intratumoral deposition of the different liposome classes (*i.e.* different sizes, targeted or non-targeted) was measured in an orthotopic mouse mammary tumor (4T1) using fluorescence molecular tomography (FMT). To image all four liposome classes in the same tumors, distinct near infrared fluorophores were used to distinguish each class of liposome inside the tumor. (a) Size distribution of three different liposome classes (30, 65 and 100 nm) as measured by dynamic light scattering. (b) Intratumoral deposition of liposomes with the three different sizes is shown as a function of regional blood flow in tumors. Following tumor blood flow mapping using perfusion CT, the tumor deposition of the three different liposome classes (30, 65 and 100 nm) was measured noninvasively using FMT imaging at 24 hours after injection. The injection dose of each liposome class contained an equal number of particles. The deposition was normalized to the fractional blood volume of each region. The data of blood flow and liposome deposition is presented as mean $\pm$ SD in a given 3D region of interest ($n = 5$ animals; 5–6 tumor regions/flow zones per animal per liposome class). The scale of the y-axis is logarithmic.
Figure and caption reproduced with permission from Toy *et al.*[65]

deposition between the EPR-mediated tumor accumulation and uptake in the inflammatory lesion (Figure 6.4).

Nanoparticles have also been explored for transporting therapeutics to the brain. However, the blood–brain barrier (BBB) constitutes a challenging physiological barrier for effective drug delivery to the central nervous system. Liu *et al.*[69] investigated the transport of fluorescently labeled (with Rubpy dye) PEGylated silica nanoparticles of three different sizes (25, 50 and 100 nm) across the intact BBB in mice following injection into the carotid artery using optical imaging. The report concluded that the smallest size nanoparticles were most effectively internalized and transported across the BBB by brain capillary endothelial cells (Figure 6.5). The overall trend

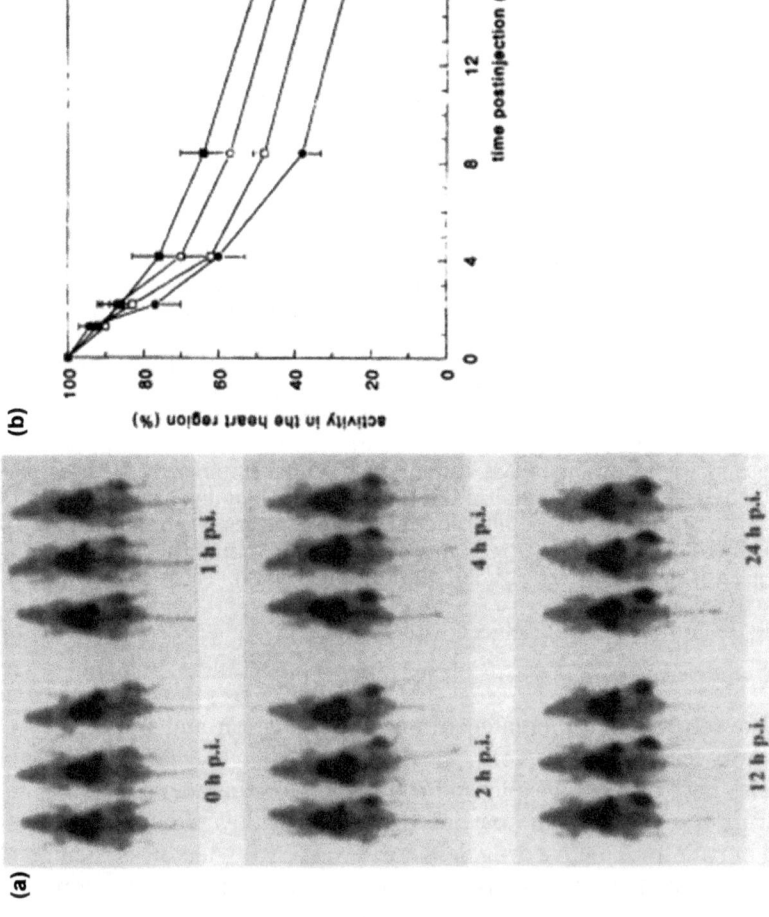

Figure 6.3 (a) Scintigrams of rats with unilateral *Staphylococcus aureus* infection in calf muscle imaged at 5 min and 1, 2, 4, 12 and 24 hours post-injection of ^{99m}Tc-labeled PEG liposomes with a mean diameter of 90 nm. (b) Quantitative analysis of the scintigraphic images of rats (three rats per group) injected with ^{99m}Tc-labeled PEG liposomes of different diameter. Blood pool activity measured 5 min post-injection was set at 100%. Error bars represent standard deviation. Figure and caption reproduced with permission from Boerman *et al.*[67]

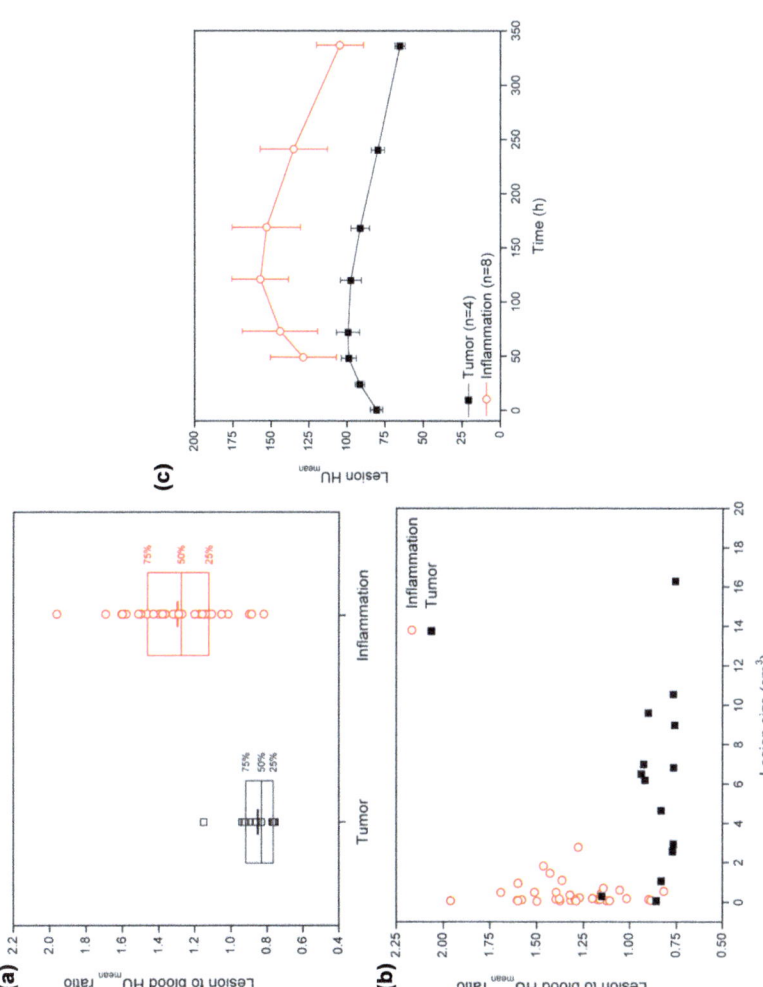

Figure 6.4   Imaging signal intensities of neoplastic and inflammatory lesions: (a) ratio of mean HU measured in an entire lesion over that measured in the blood from the liposome-CT dataset; (b) the difference in the normalized CT signals between the tumor and inflammatory sites is independent of lesion size; and (c) kinetic profiles of liposome contrast agent accumulation and clearance in tumor and inflammatory lesions. There are no data points before 48 hours post-liposome injection for the inflammatory lesions because there was insufficient contrast enhancement to identify the lesion from surrounding muscle. Note that five additional VX2-carcinoma bearing rabbits with eight inflammatory lesions from a separate study were included in the CT dataset (a).
Figure and caption reproduced with permission from Zheng *et al.*[68]

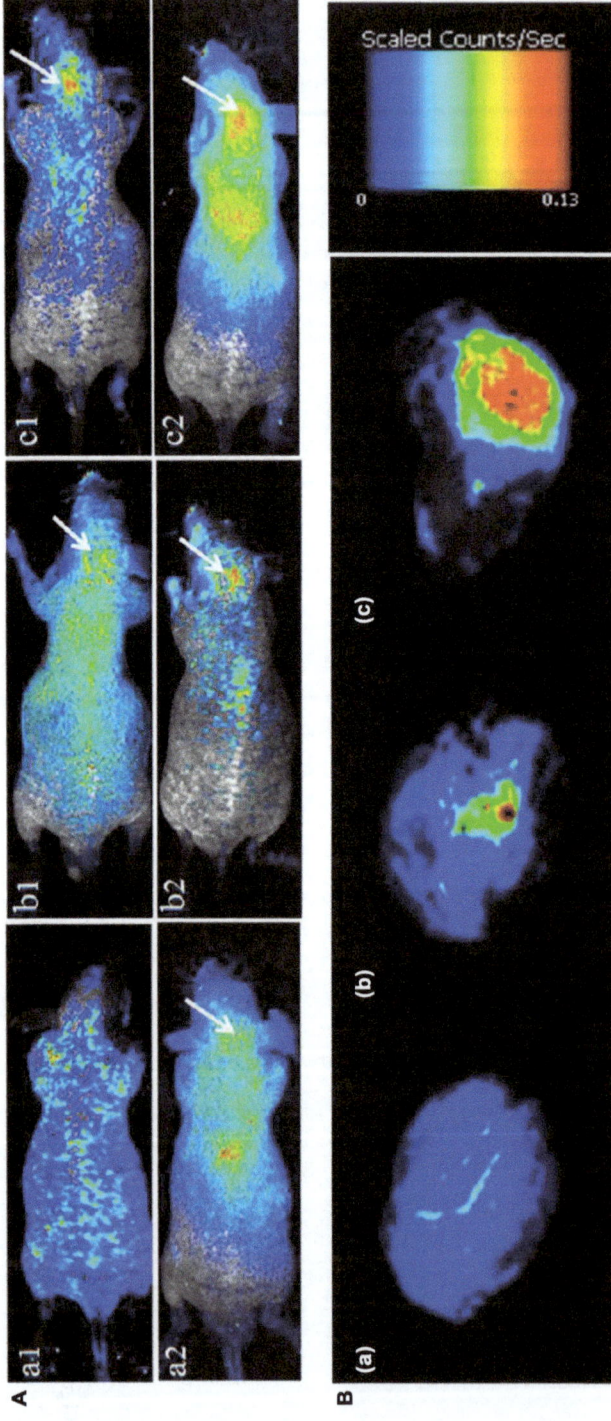

Figure 6.5 (A) *In vivo* imaging of brain deposition of PEGylated silica nanoparticles at different time points (1 representing 15 min; 2 representing 1 h) after injection *via* the carotid artery. Arrows mark the location of the brain. (B) *Ex vivo* optical imaging of resected brain. (a), (b) and (c) representing PEGylated silica nanoparticles of 100, 50 and 25 nm, respectively. Figure and caption reproduced with permission from Liu *et al.*[69]

observed was that of decreased BBB transport with increased silica nanoparticle size. These findings are important for the optimization of nanoparticle-based delivery as BBB crossing is desirable when targeting brain lesions, but it needs to be avoided in other applications, in order to minimize neurotoxicity.

In addition to vascular transport, nanocolloids can also be employed for lymphatic transport to lymph nodes following interstitial, subcutaneous and intradermal administration.[70] In comparison to small molecular weight dyes, due to their size, nanoparticles have the advantage of increased retention time and spread to lymph nodes that are downstream from the sentinel lymph node (SLN).[71] To date, many reports have agreed that the size range optimal for lymphatic uptake is between 10 and 100 nm, as smaller particles can extravasate into the bloodstream and larger particles are retained at the injection site.[72–75] Kjellman *et al.*[76] investigated the retention time and downstream lymph node localization of fluorescently-labeled USPIOs of three different sizes (15, 27 and 58 nm). The authors concluded that USPIOs of 15 nm resulted in more rapid and increased accumulation in the SLN (*i.e.* popliteal lymph node) compared to larger particles (Figure 6.6a). However, there was no significant difference in the amount of USPIOs detected in the downstream lymph nodes (*i.e.* inguinal and iliac, Figure 6.6b).

The ability to manipulate and tailor the size of nanoparticles in order to optimize their delivery to specific anatomical and disease sites using the desired injection route is one of the many rationales which support further development of nanomedicines. Advancements in imaging will continue to play an integral role in optimization and validation of nanoparticle performance both *in vitro* and *in vivo*.

6.3.2 Investigation of the Effectiveness of Active *Versus* Passive Targeting

The EPR effect is the theoretical basis behind nanoparticle deposition in tumors through passive targeting. Nanoparticles can escape the vasculature through irregularly angiogenic, fenestrated tumor blood vessels, and their clearance is delayed due to defective or absent tumor lymphatics. To further enhance the ability of nanoparticles to home to tumor sites and deliver their therapeutic load to tumor cells, different strategies to actively target specific cell populations have been explored. These mostly involve the conjugation of recognition moieties onto the surface of nanoparticles which result in preferential binding and internalization into cells of interest. *In vitro*, active targeting demonstrated effective and specific delivery of drugs to cells and subcellular compartments of interest.[77,78] However, preclinical reports have demonstrated that the addition of a surface moiety can change the physicochemical characteristics (*i.e.* surface charge) of the nanocarrier resulting in altered *in vivo* pharmacokinetics and biodistribution, which can negatively

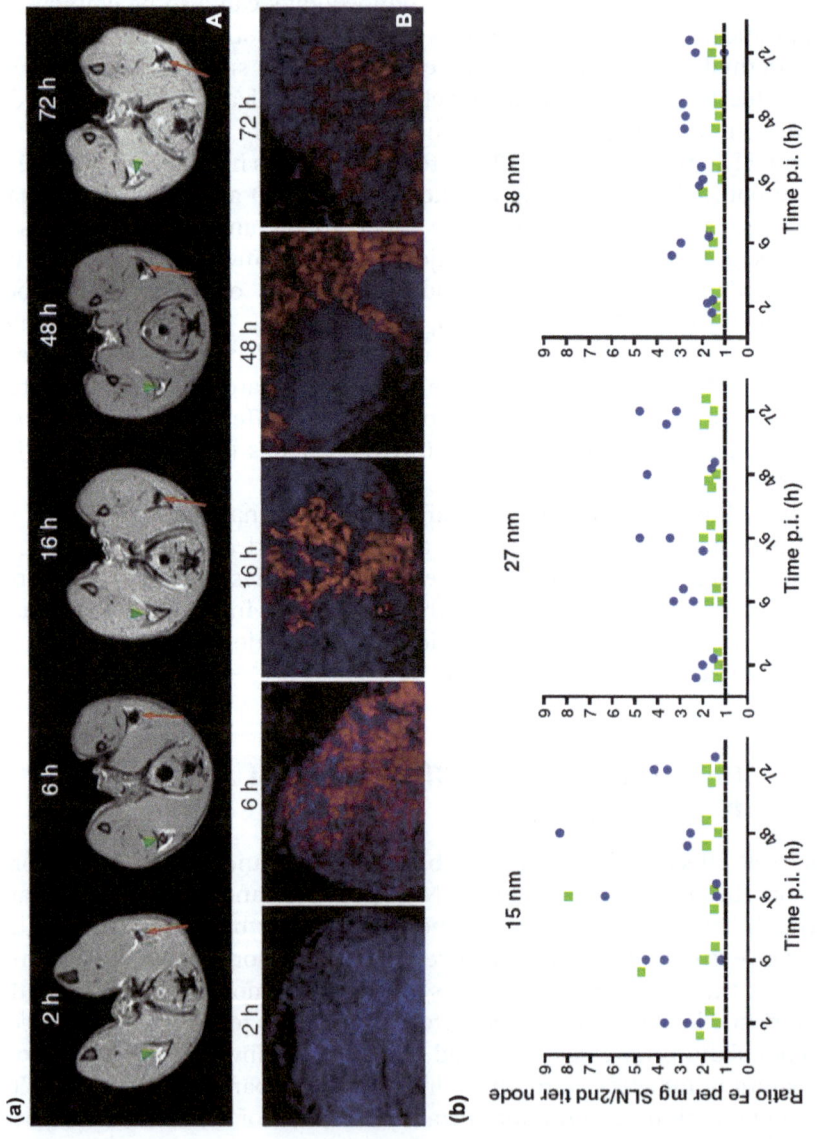

affect tumor deposition.[79] Furthermore, even in the absence of significant physico-chemical changes, active targeting of nanoparticles to tumor cells does not necessarily result in increased tumor accumulation, but can prolong intratumoral residence time[80] while also increasing cellular uptake and ultimately leading to an improved therapeutic effect.[81,82] An alternative active targeting approach that is less reliant on the tumor EPR effect involves the incorporation of moieties that are directed to tumor vascular endothelial cells. This is a promising strategy as endothelial target cells can be successfully reached before extravasation occurs and select peptide sequences have been identified to be highly specific to these tumor vascular cells.[83–85]

Imaging is being increasingly employed by drug delivery researchers for non-invasive investigation of the spatio-temporal distribution of actively targeted nanoparticles and to assess their therapeutic benefit when compared to their non-targeted counterparts. Numerous examples in the literature illustrate image-based assessment of nanoparticles targeted to highly expressed epitopes on tumor cells[9,86,87] and tumor endothelial cells.[45,88–90] In addition, the combined use of different imaging modalities as well as the use of multi-wavelength fluorescence imaging have enabled monitoring of the tumor targeting performance of actively and passively targeted nanoparticles in the same animal and tumor over time.

For example, Toy *et al.*[65] employed microCT and multi-wavelength FMT to elucidate the effect of nanoparticle size on their active targeting capability with respect to regional tumor blood flow. By using two different fluorophores, they were able to separately monitor EGFR-targeted and non-targeted liposomes co-administered to the same mouse (4T1 mammary tumor) in a single cocktail. This investigation confirmed that for liposomes of 100 nm in size, active targeting with epidermal growth factor (EGF) showed no benefit across a wide range of tumor blood flow rates (~ 30 mL min^{-1} to ~ 350 mL min^{-1}). This was consistent with previously reported results for human epidermal growth factor (HER)2-targeted liposomes.[82,91–95] However, when the liposome size was reduced to 30 nm, the authors saw a significant increase of ~ 12-fold in deposition in regions of slow tumor blood flow when

---

Figure 6.6 (a) Top panel A: axial sections of popliteal lymph nodes of animals injected with 15 nm USPIOs. MR images acquired at 2, 6, 16, 48 and 72 hours post-injection, respectively. Red arrows: lymph node on injection side. Green arrowheads: lymph node on control side. Maximal signal distortion is registered at 6 hours post-injection. Some chemical shift artifacts can be observed from the lymph nodes on the control side. Bottom panel B: corresponding fluorescence microscope images of 5 μm cryosections of the popliteal lymph nodes on the injection side. Red signal: DY-647 labeled USPIOs. Blue signal: DAPI stained cell nuclei. Magnification: 10×. (b) Amount of iron per mg tissue of the SLN divided by amount of iron per mg tissue of second-tier nodes, for all animals. Blue dots: popliteal/inguinal. Green squares: popliteal/iliac. Figure and caption reproduced with permission from Kjellman *et al.*[76]

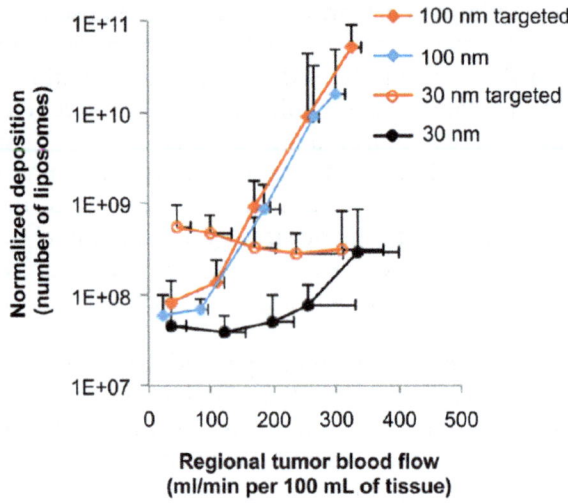

Figure 6.7 Dependence of liposome extravasation into tumors on the liposome size, active targeting toward the EGF receptor, and blood flow on a region-by-region basis. (a) Following tumor blood flow mapping using perfusion CT, the intratumoral deposition of four different liposome classes (30 and 100 nm with or without EGFR-targeting ligands) was quantitatively measured in the orthotopic mouse (4T1) mammary tumor using FMT imaging at 24 hours after injection. To image all four liposome classes in the same tumors, distinct near infrared fluorophores were used to distinguish each class of liposome inside the tumor. The intratumoral deposition of liposomes is shown as a function of regional blood flow in tumors.

Figure and caption reproduced with permission from Toy *et al.*[65]

active targeting was employed (Figure 6.7). They attributed this enhanced deposition to an increase in retention achieved by liposome binding and internalization into target cells that in turn prevented their washout back into the systemic circulation. These findings demonstrate that active targeting to tumor cells can enhance overall tumor deposition and facilitate the penetration of smaller sized drug carriers into core tumor regions with inherently slow blood flow. However, the clinical relevance of 30 nm liposomes may be somewhat limited due to their inability to carry a high therapeutic payload.

Dunne *et al.*[45] demonstrated that even ~80 nm liposomes could benefit from a surface conjugated active targeting moiety, in this case a tumor neovascular targeting peptide sequence, asparagine–glycine–arginine (NGR), in terms of overall tumor deposition and prolonged tumor retention. Conjugation of an optimal amount of NGR ligand (*i.e.* 0.64 mol%) resulted in a more than two-fold increase in %ID g^{-1} of liposome accumulation at the tumor site at 48 hours post-injection compared to the non-targeted formulation (Figure 6.8a). An enhanced tumor penetration by almost two-fold was also observed as CT-based assessment of iohexol-encapsulated liposomes

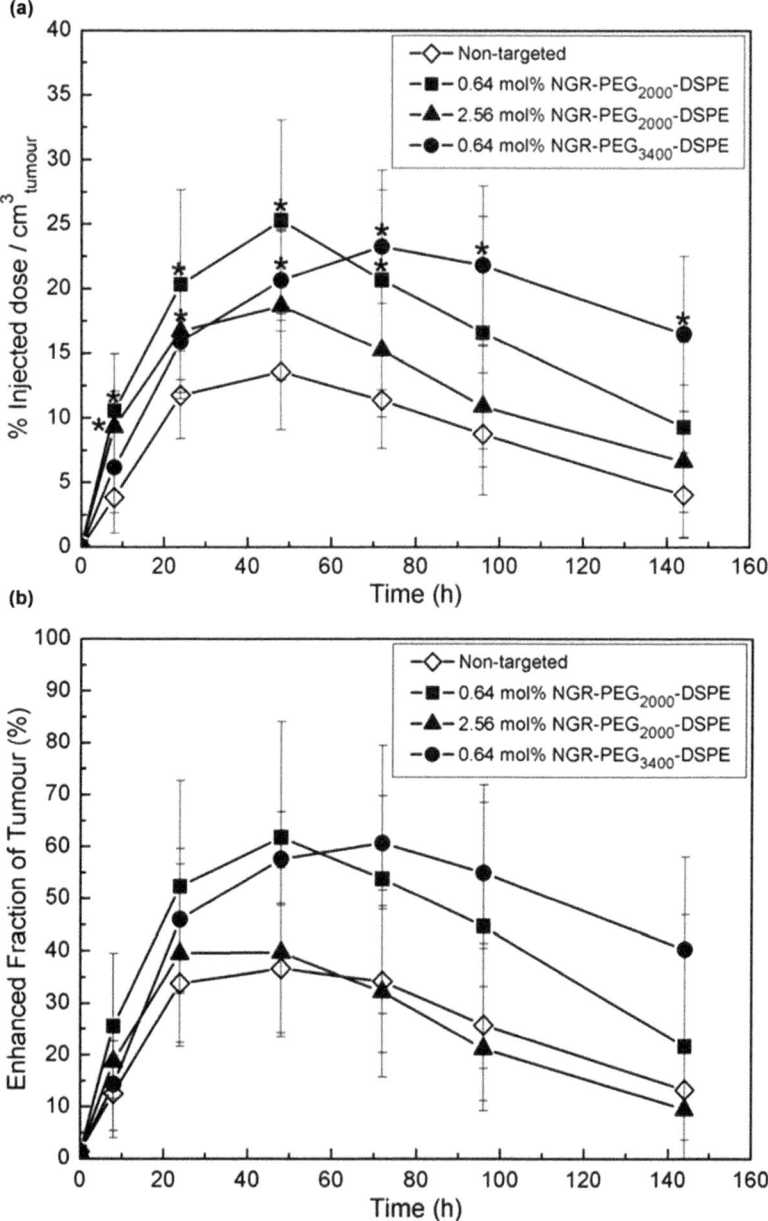

Figure 6.8   (a) Time-dependent tumor accumulation profiles for the non-targeted and NGR-conjugated liposome formulations containing iohexol in mice bearing H520 xenografts. (b) Evaluation of the tumor volume occupied by liposomes. Mean percentage of voxels within the tumor volume with signal enhancement (*i.e.* greater than two standard deviations over the mean of the pre-contrast tumor) is plotted *versus* time. Error bars represent standard deviation.
Figure and caption reproduced with permission from Dunne *et al.*[45]

showed that >60% of the tumor volume was reached by NGR-targeted liposomes, while the peak tumor volume occupancy for the non-targeted liposomes was only ~35% (Figure 6.8b). While highly promising, it remains unknown whether vascular targeting strategies can directly translate increased tumor deposition into improved therapeutic efficacy. Pastorino *et al.*[89] proposed and investigated a combined approach to active targeting by administering both tumor cell targeted (*via* the anti-GD$_2$ monoclonal antibody) and tumor vasculature targeted (*via* the NGR peptide) liposomal doxorubicin sequentially to mice bearing metastatic neuroblastoma. The authors reported that an additive effect in therapeutic efficacy was achieved compared with either therapy used alone (Figure 6.9). The authors postulated that the NGR-targeted liposomal doxorubicin caused the shutdown of the angiogenic vascular network that feeds the inner portions of the tumor, and indirectly lead to the death of tumor cells that rely on that blood supply. In addition, the anti-GD$_2$-targeted liposomes were able to achieve direct tumor cell kill in the tumor periphery. These findings support the complementary use of the two targeting mechanisms and demonstrate feasibility of performing combination therapies using the same drug but two distinct targeting mechanisms.

Active targeting has also been combined with strategies that increase cell penetration and circumvent multidrug resistance. Wang *et al.*[96] devised a multifunctional nanoparticle with the ability to target and penetrate tumor cells, as well as achieve intracellular drug release and circumvent the multidrug efflux transporter P-glycoprotein (Figure 6.10). The pH-sensitive PLGA-based nanoparticle carrying the chemotherapy drug vincristine was conjugated with R$_7$ cell penetrating peptides and folate for active targeting. With this approach, the nanoparticles can target tumor cells that over-express the folate receptor actively, while R$_7$ ensures the subsequent penetration of the nanoparticle into the tumor cells *via* endocytosis bypassing the P-glycoprotein drug efflux pumps. Once intracellular, the pH sensitive nanoparticle will release vincristine and provoke the desired cytotoxic effect. The authors used the fluorophore rhodamine-123 as a drug surrogate imaging agent to assess the effectiveness of this complex multi-step targeting, penetration, efflux-avoidance and release strategy. Fluorescence imaging allowed for whole-body biodistribution and bulk tumor targeting assessment through the use of an optical imager as well as intratumoral and intracellular distribution monitoring using a high-resolution confocal laser scanning microscope.

Comprehensive assessment of the performance of active targeting strategies has greatly benefited from the use of different non-invasive imaging techniques which enabled investigation of pharmacokinetics, biodistribution, tumor accumulation, cellular and vascular targeting, as well as intracellular localization. Monitoring of the multiple steps of the targeting process *in vivo* in the same animal allowed researchers to gain improved understanding of the active targeting process and to objectively assess its effectiveness and value for different applications.

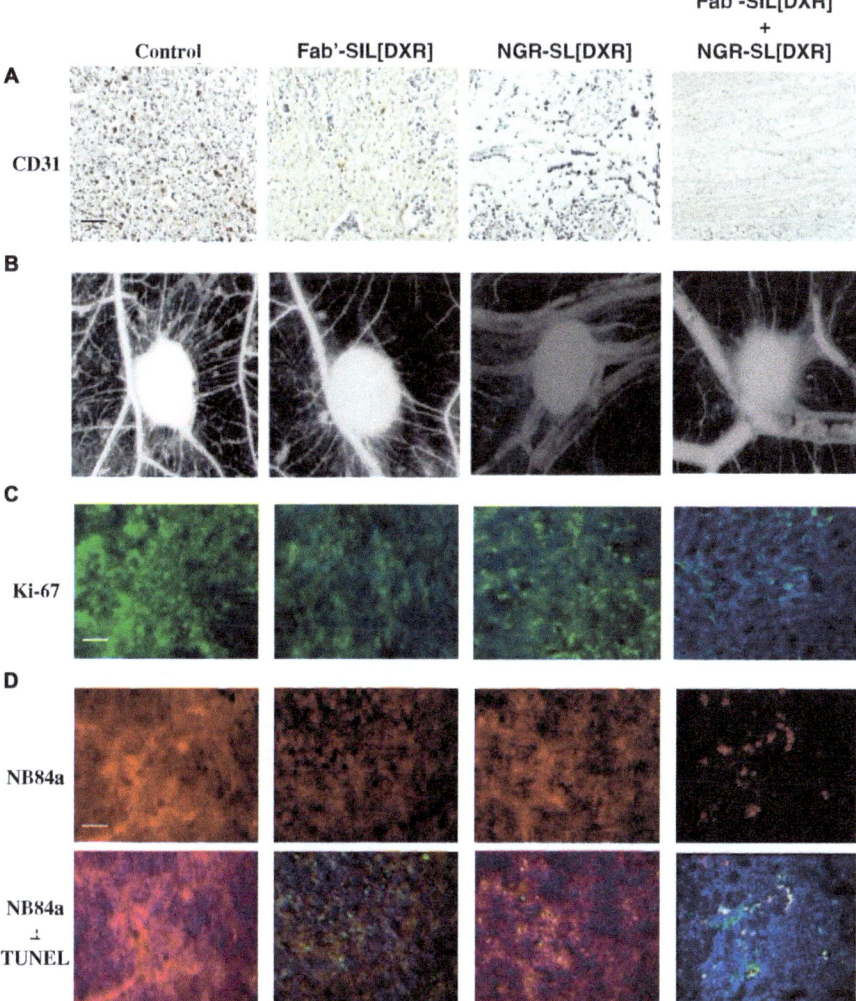

Figure 6.9 Effect of combination therapies on angiogenesis, tumor cell proliferation and apoptosis *in vivo*. A, C and D: immunohistochemical analysis of neuroblastoma metastases removed from untreated mice or mice treated with individual or combined liposomal formulations. Tumors were harvested on day 50 and tissue sections were immunostained for (A) CD31 to show endothelial cells, (C) Ki-67 to show tumor proliferating cells, (D) NB84a to show neuroblastoma cells, or else double-labeled for NB84a and TUNEL, to detect tumor apoptosis. Red: NB84a + neuroblastoma cells; yellow: merging of red NB84a signal and green TUNEL signal. Cell nuclei were stained with 4′,6-diamidino-2-phenylindole. Bars: 200 Am (A); 100 Am (C and D). B: angiogenesis inhibition *in vivo* in the chorioallantoic membrane model for combinations of Fab′-SIL(DXR) and NGR-SL(DXR). Original magnification: ×50.
Figure and caption reproduced with permission from Pastorino *et al.*[89]

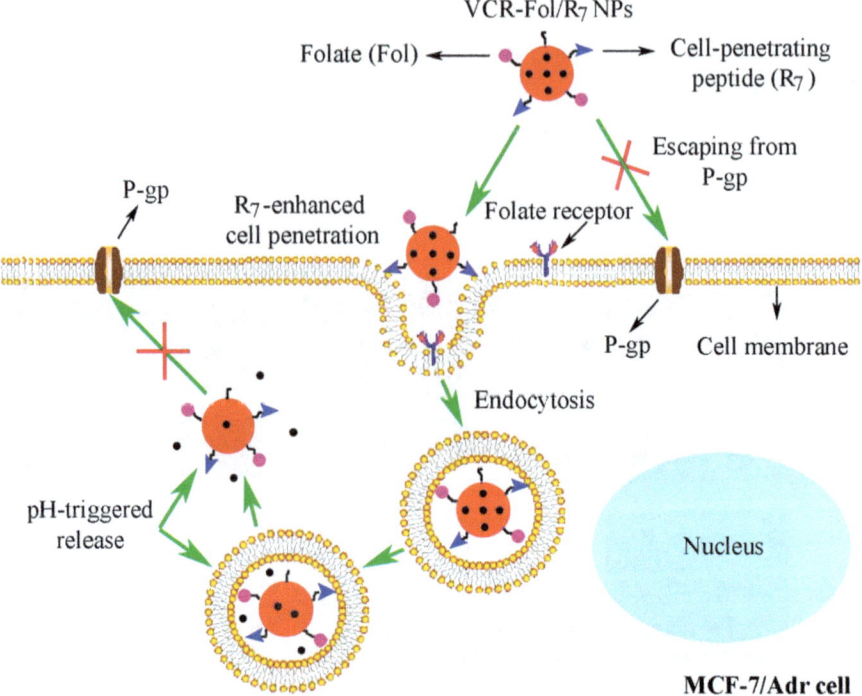

Figure 6.10   Schematic representation of the pH-sensitive PLGA-based nanoparticle carrying the chemotherapy drug vincristine targeted with R$_7$ cell penetrating peptides and folate for overcoming multidrug resistance of MCF-7/Adr cells.
Figure reproduced with permission from Wang *et al.*[96]

6.3.3   Stroma Modification to Enhance Nanomedicine Delivery and Efficacy

One of the most critical barriers for nanoparticle transport to target cells is found in the complex tumor microenvironment. Abnormal, highly fenestrated tumor blood vessels produce an environment with inconsistent blood flow, high IFP and hypoxia, creating a host of challenges for drug delivery and leading to poor prognosis.[97] The EPR effect may assist in nanoparticle deposition at the periphery of tumors, but penetration beyond the periphery is challenged by high IFP, which decreases the pressure gradient that drives convection-mediated nanoparticle transport in tumors.[98] Different techniques have been explored to modify the tumor stroma and decrease IFP in order to achieve enhanced nanoparticle delivery and penetration into tumors.

Geretti *et al.*[99] reported on the use of cyclophosphamide as an effective tumor stroma modifier. Pre-dosing with cyclophosphamide enhanced the tumor deposition of subsequent liposomal doxorubicin administrations in three preclinical mouse models of human breast cancer. Specifically, the increase in liposomal delivery was tumor-specific (*i.e.* no increase in heart or skin), liposome-specific (*i.e.* no increase in free doxorubicin deposition in tumors) and resulted in higher tumor growth inhibition when cyclophosphamide was given as a pre-dose compared to co-administration. Mechanistically, the authors theorized that the improved liposomal drug delivery achieved from cyclophosphamide pre-treatment resulted from cyclophosphamide-induced cell apoptosis, reduction in overall tumor cell density, decrease in IFP and an increase in vascular perfusion. PET-based assessment was also performed to confirm the enhanced tumor deposition and penetration of liposomes in cyclophosphamide pre-treated animals compared to those without pre-treatment over two doses of ^{64}Cu-labeled liposomal doxorubicin administered on a weekly basis.[100] This image-based evaluation was important to validate the observed efficacy priming effect of cyclophosphamide in the same tumors and in the same animals.

Another strategy explored to modulate tumor stroma is by targeting the vasculature through inhibition and decrease in the density of pericyte coverage. Pericytes cover as much as 70% of tumor blood vessel surface area, blocking the fenestrations through which nanoparticles can reach the tumor interstitium *via* EPR.[101] Pericytes are especially relevant in pancreatic ductal adenocarcinoma, a particularly stroma-rich tumor type. Meng *et al.*[101] sought to target pericytes by blocking tumor growth factor (TGF)-β, which supports pericytes and also contributes to tumor vasculature abnormalities. The group used mesoporous silica nanoparticles carrying a TGF-β inhibitor, followed by "hard" (120 nm PEI-PEG-coated mesoporous silica nano-particles) and "soft" (137 nm liposomal gemcitabine) nanoparticles. This two-wave administration approach (Figure 6.11) led to a higher and more prolonged tumor accumulation of both "hard" and "soft" nanoparticles, suggesting enhanced vascular access, as observed by fluorescence imaging. The liposome accumulation and clearance kinetics were faster than that observed for the silica nanoparticles. In addition, pre-treatment with TGF-β inhibitor loaded nanoparticles led to an improvement in intratumoral liposome deposition both in terms of overall accumulation and an increase in distribution homogeneity. Therefore, the enhanced delivery of the lipo-somal gemcitabine successfully translated into improved antitumor effects in a stroma dense pancreatic xenograft model. This report further confirms the viability and potential of employing stroma modifying agents as pre-treatment to improve the effectiveness of nanomedicines.

6.3.4 Assessment of the Performance of Activatable Nanomedicines

Imaging is a valuable tool to help understand and optimize the different steps involved in the development and implementation of drug delivery systems. One of the main factors which can influence the efficacy of a therapeutic agent is its ability to achieve sufficient delivery to the tumor in order to guarantee a therapeutically relevant local concentration. At the same time, non-specific accumulation of the drug-carrying nanoparticles in healthy organs needs to be minimized as it can lead to significant systemic

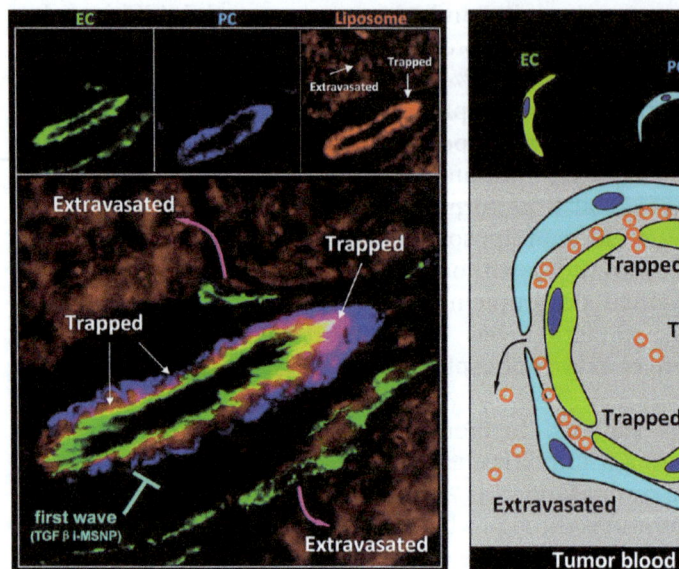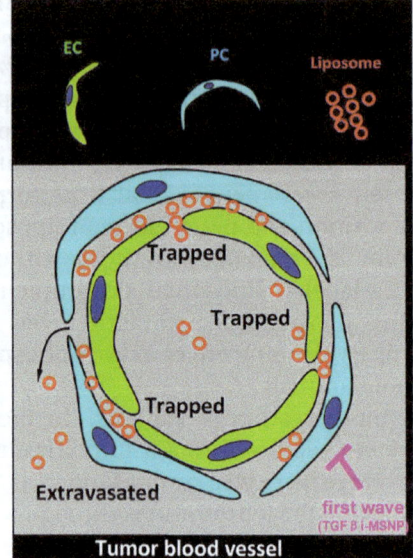

Figure 6.11   Changing the pericyte (PC) coverage of vascular fenestrations is import-ant for delivery of the liposomal gemcitabine carrier. This fluorescence-stained histological section shows that the pancreatic ductal adenocar-cinoma tumor uses a dense stromal barrier, which includes pericytes blocking the vascular access of intravenously injected red fluorescent liposomes. The fluorescence microscopy image obtained from the tumor site of an animal receiving prior TGF-β inhibitor (i)–mesoporous silica nanoparticle (MSNP) therapy shows a region of a blood vessel where pericytes are trapping some liposomes just beyond their point of egress from the vascular fenestrations. In contrast, other blood vessel regions not showing pericyte coverage allow liposome extravasation to the tumor interstitium and cancer cells.
Figure and caption reproduced with permission from Meng *et al.*[101]

toxicities. To date, chemotherapy still struggles with the challenge of per-forming within the optimal narrow range of concentrations needed to both achieve therapeutic efficacy and minimize healthy tissue toxicity. The em-ployment of nanoparticle carriers has, in part, helped overcome some of these delivery issues through improvement of the drug pharmacodynamics. However, nanoparticle penetration into tumors is hindered owing to limi-tations in convection and diffusion-based transport that occur as a result of the elevated IFP and nature of the tumor stroma.[102] Clinically available liposome formulations such as Doxil® and Lipoplatin™ show limited drug release at the tumor site due to their high *in vivo* stability.[103,104] As a result, although liposomal chemotherapeutics have demonstrated significant benefit in terms of reduced systemic toxicity, their moderate therapeutic efficacy has prevented them from displacing conventional chemotherapy.[104] To overcome the barriers of poor tumor penetration and bioavailability, a number of different strategies have been explored to increase the

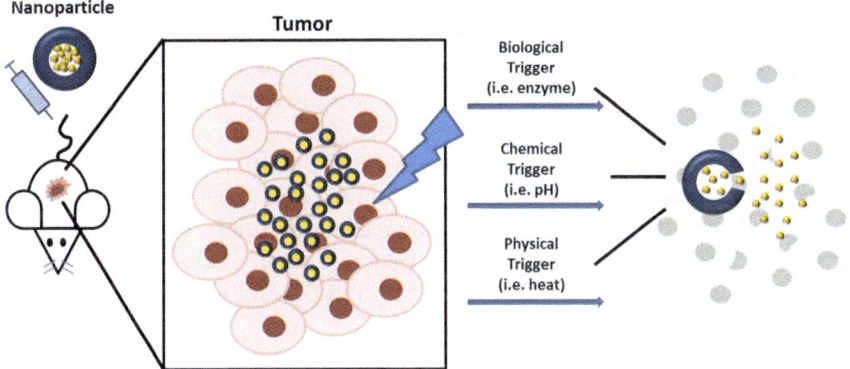

Figure 6.12   Schematic illustration (not to scale) of the drug release mechanism of activatable nanocarriers. Upon their arrival to the tumor site, the release of their drug load can be triggered *via* biological, chemical or physical cues.

concentration of nano-carriers and the bioavailability of the active drug at tumor sites, including active targeting (Section 6.3.2), as well as the application of biological, chemical and physical triggers to induce drug release (*i.e.* enzymes, pH and heat; Figure 6.12).[105–107] The design, optimization and evaluation of these new complex drug delivery systems can benefit from tailored imaging techniques aimed to (i) better target the tumor or area of potential disease spread; (ii) avoid non-target tissues (*i.e.* application of external heat sources to activate thermo-responsive carriers at the tumor site to minimize release in normal tissue); and (iii) quantitatively evaluate the drug delivery performance (*i.e.* chemodosimetry). Although there is still a lack of integrated image-based platforms specifically designed for real-time quantification of the drug or drug carrier concentration at the target site immediately before triggered release and immediately after activation with the ability to report on the dose of drug delivered, significant research is focused in this area. In fact, the concept of image-guided drug delivery (IGDD) has gained much momentum and there are targeted federal funding initiatives by the National Institutes of Health aimed at advancing and translating IGDD for clinical drug delivery applications.

Biological activation can be accomplished using two distinct strategies. First, a pro-drug can be either linked or encapsulated into a nanoparticle, and once they reach the tumor site and encounter tumor specific enzymes, they can be successfully cleaved and converted into their therapeutically active counterparts *in situ*. Second, drug delivery systems can be engineered to experience effective destabilization and release of their drug load in the presence of specific hydrolases such as proteases, lipases and oxido-reductases.[105,108,109] Ansari *et al.*[110] used a peptide linker containing a cleavage site specific for matrix metalloproteinases (MMPs) to attach the anticancer drug azademethylcolchicine to iron oxide nanoparticles for delivery and MR imaging of drug accumulation and localization *in vivo*.

Azademethylcolchicine remained inactive until the linker peptide was cleaved at the tumor site. The proof-of-principle investigation was conducted in a model of mammary adenocarcinoma in mice. A similar approach was exploited by Lee *et al.*,[111] who chemically conjugated the MMP-responsive peptide and a ^{64}Cu-DOTA complex onto chitosan nanoparticles in order to develop an activatable PET/optical imaging probe to quantify the presence of MMPs *in vivo*. If these approaches are optimized and successfully translated into the clinic, then chemotherapy has the potential to become more personalized and more effective, and may ultimately achieve the goal of delivering the right drug to the right location in the right patient at the right time.[112]

Numerous examples have been reported in the literature describing nanosystems that are destabilized upon encountering a chemical cue such as a change in pH.[113–116] These pH-sensitive systems are based on the fact that tumors and intracellular endosomal/lysosomal compartments have a more acidic pH compared to healthy tissues with a normal physiological pH. Specifically, pH in healthy extracellular compartments is ~ 7.4, while the intracellular pH is ~ 7.2. The pH measured in tumor tissues is even lower (~ 6.8) due to the high rate of aerobic and anaerobic glycolysis in cancer cells.[117] Drug delivery systems that are pH-sensitive can be engineered, for example, by incorporation of pH-responsive polymers in the hydrophobic core of micelles. Lee *et al.*[118] prepared micelles based on a triblock co-polymer PLA-b-PEG-b-polyHis, which released the encapsulated doxorubicin when the pH was lowered from 7.4 to 6.0, due to ionization of the polyHis blocks. Mechanisms of pH triggered polymeric ionization and subsequent micelle dissociation are described in detail elsewhere.[119] The *in vivo* image-based assessment of pH-triggered nanosystems is often accomplished by the combined use of an MR imageable nanoparticle such as iron oxide and an inherently fluorescent drug such as doxorubicin. For example, Lim *et al.*[115] achieved 90% of localized drug release within the first 24 hours at the tumor site (pH 5.5) using *N*-naphthyl-*O*-dimethymaleoyl chitosan-based doxorubicin-loaded magnetic nanoparticles in NIH3T6.7 tumor-bearing mice. However, a number of obstacles need to be overcome before pH-triggered drug release can be advanced for clinical implementation. First, it is known that there is high variability in extracellular pH in different sub-regions within the tumor. As a result, this heterogeneity in pH distribution can lead to sub-optimal drug release in selected tumor subregions. Furthermore, as many chemotherapeutic drugs are weak bases (*i.e.* anthra-cyclines), a mechanism termed "ion trapping" can occur in regions of high acidity, where basic drugs are retained in the acidic tumor interstitial space and ultimately their intracellular uptake limited even following successful release from nanocarriers.[120]

Physical triggers utilized for destabilization of nanosystems and *in situ* drug release include light and heat. Light production from laser systems is attractive because it allows for good spatial and temporal control of light delivery to be achieved.[121,122] In addition, high resolution and real-time optical imaging enables mapping of the distribution and activation of photoabsorbers in tumors.[123] When choosing a light-based approach, wavelength selection is a key parameter. Visible and UV wavelengths are not useful since tissue penetration is limited due to light photon scattering and absorption. Light in the near-infrared (NIR) range penetrates deeper into the tissue and there is relatively low tissue autofluorescence in the NIR optical window.[124] NIR light can be also used as a stimulus for localized lipid micelle destabilization and rupture, thus increasing the specificity of drug delivery. Specifically, micelles can be rendered light-sensitive by the incorporation or conjugation of a chromophore into their hydrophobic core.[125] NIR irradiation then changes the hydrophilic/hydrophobic balance of the light-sensitive system, thereby leading to their dissociation and release of therapeutic cargo.[119]

To date, the most successful triggered drug release nanomedicine is the thermosensitive liposome formulation known as ThermoDox® which responds to heat-mediated drug release in the mild hyperthermia temperature range (<43 °C). ThermoDox® has demonstrated significant improvements in the concentration of doxorubicin and its bioavailability at tumor sites.[126] A number of clinical trials are currently underway to investigate the clinical benefit of activation by radiofrequency ablation, microwave hyperthermia and high intensity focused ultrasound.[127–132] Image guidance played an important role in the development, optimization and preclinical evaluation of ThermoDox® through the use of MR imaging of manganese sulfate that was co-loaded into the liposomes with doxorubicin.[36,133] Viglianti *et al.*[36,37] reported on the use of T_1 MR-based Mn^{2+} concentration measurement as a surrogate for evaluation of *in vivo* drug release from thermosensitive liposomal doxorubicin. This allowed for optimization of the timing and duration of local hyperthermia with respect to temperature-sensitive liposome administration. Their findings suggest that thermosensitive liposomes are best administered during steady-state local hyperthermia for optimal intratumoral spatial distribution of the released drug molecules (Figure 6.13). Not only does MR imaging allow for image-based demonstration of temperature-dependent drug release, but Dewhirst and colleagues also pioneered image-based chemodosimetry.[37] Although MR image-guidance with respect to temperature monitoring and quantification of drug delivery and release was not included in the initial ThermoDox® clinical trials, there is a high probability that they will incorporated in subsequent trials as a means of quality control (personal communication, Mark Dewhirst).

6.4 Clinical Experience and Future Considerations

Despite the extensive preclinical experience in assessing the *in vivo* performance of nanosystems using different imaging techniques, the amount of clinical data remains limited. The most comprehensive clinical assessment of pharmacokinetics and biodistribution of a nanoparticle remains that performed by Harrington *et al.*[32] This report used a whole-body gamma camera to image the blood, tissue and tumor distribution of ^{111}In-DTPA

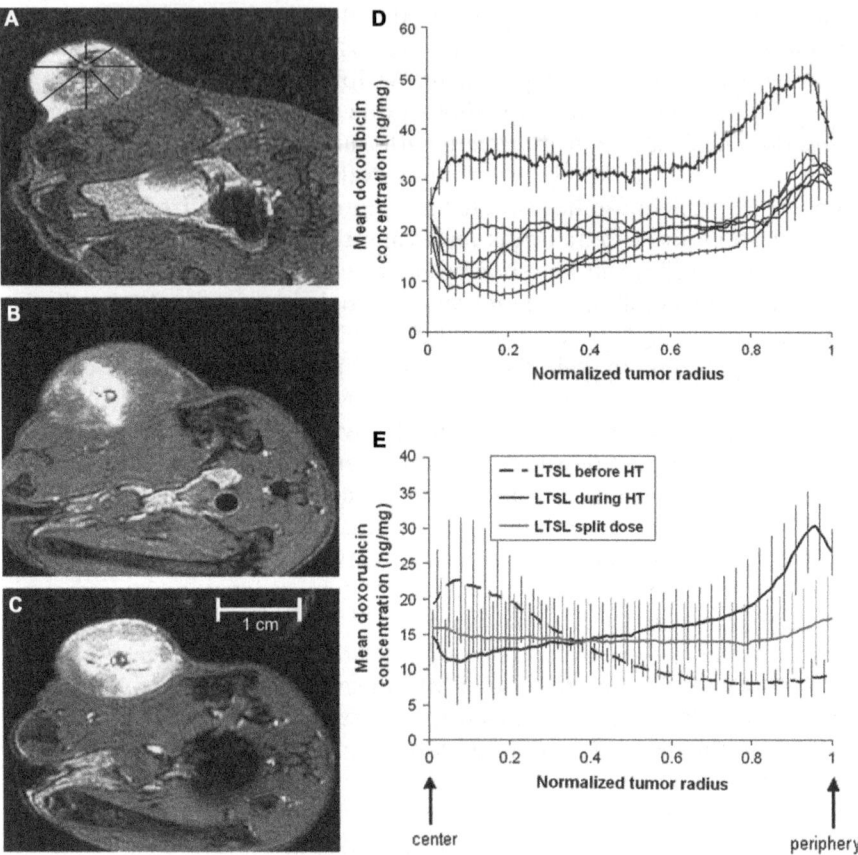

Figure 6.13   Tumor drug distribution after administration of doxorubicin- and manganese-containing lysolipid-based temperature-sensitive liposomes (LTSLs) and hyperthermia (HT) by three different schedules. (A–C) Axial pelvic magnetic resonance images show rats bearing flank fibrosarcomas (top left). Radial lines in (A) show the orientations of doxorubicin concentration profiles in (D) and (E). LTSLs administered during steady-state hyperthermia resulted in peripheral enhancement (liposome content release; white) at the edge of the tumor (A); LTSLs administered before hyperthermia resulted in central enhancement (B); and LTSLs administered in two equal doses, half before hyperthermia and the remainder after steady-state hyperthermia was reached, resulted in uniform enhancement (C). (D) T1-based mean tumor doxorubicin concentration (ng per mg of tissue) after treatment with LTSLs during hyperthermia, shown as a function of the normalized tumor radius for each rat. The bold profile is for the tumor shown in (A). Mean values are for 80 line profiles from each rat. (E) Mean doxorubicin concentration (ng per mg of tissue) profiles for each of the three therapeutic groups as a function of the normalized tumor radius ($n = 6$–7 rats per group). Vertical lines in (D) and (E) correspond to 95% confidence intervals.

Figure and caption reproduced with permission from Ponce *et al.*[133]

labeled liposomes in 17 patients with advanced solid cancers including breast, head and neck, lung, brain and cervix. The pharmacokinetics and healthy organ distribution of ^{111}In-DTPA labeled liposomes showed a high level of consistency among patients. However, the deposition of liposomes in solid tumors was found to be highly variable among patients both with different tumor types and within the same tumor type (Figure 6.14a, data

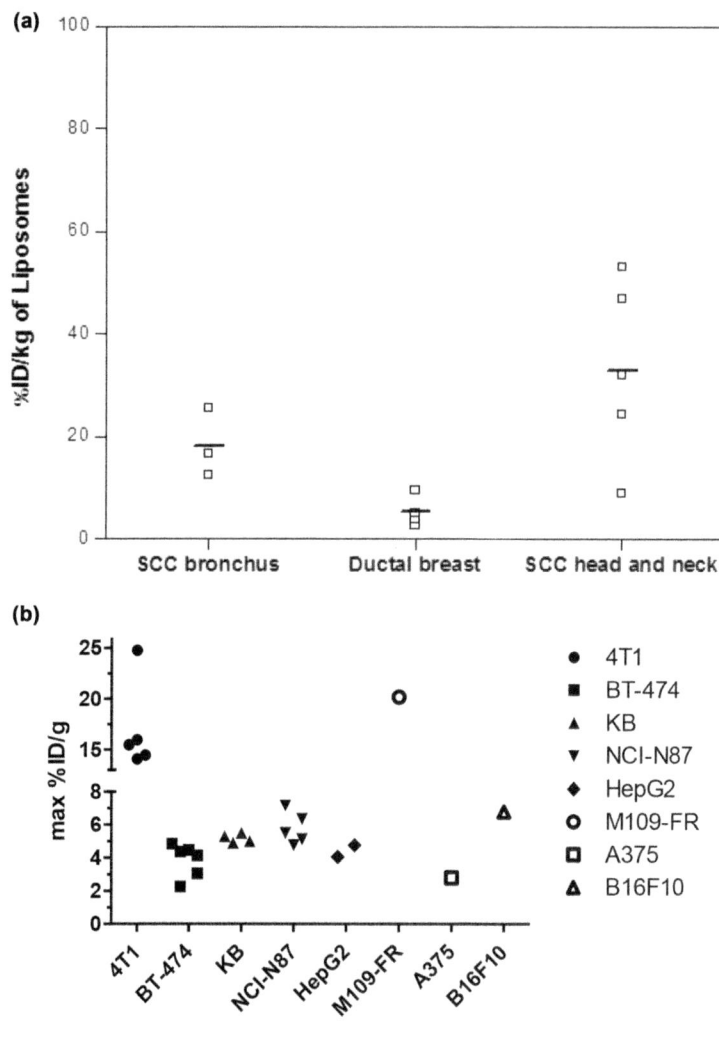

Figure 6.14   High variability in liposome deposition (a) among patients with different tumor types and within the same tumor type (plot generated from data from Harrington *et al.*[32]), and (b) in different mouse xenograft models and among tumors within the same model (reproduced with permission from Hendriks *et al.*[136]). SCC: squamous cell cancer.

re-plotted from Harrington *et al.*[32]). These results were not surprising, as high variability in liposome deposition in different mouse xenograft models and between different animals within the same tumor model have been observed[134] (Figure 6.14b).

These findings support the clinical development of an imaging assay to determine the EPR status of a tumor in order to attempt to predict the tumor deposition of a nanoparticle-based therapeutic. Lee *et al.*[135] developed a ^{64}Cu loading technique compatible with existing drug-loaded liposomes (Doxil®, HER2-targeted PEGylated liposomal doxorubicin and liposomal irinotecan) to allow for PET imaging of liposomal chemotherapeutic biodistribution and tumor deposition. Merrimack Pharmaceuticals (Cambridge, MA, USA) is currently evaluating the use of PET imaging of ^{64}Cu-labeled liposomes as a potential pre-selection strategy to quantify liposomal tumor deposition in patients.[136,137] If successful, the employment of an imageable nanoparticle together with its therapeutic counterpart, whether together as a theranostic product or separately as a diagnostic-therapeutic pair, has the potential to significantly impact personalized nano-therapeutic delivery, as it allows for pre-selection of patients with the greatest likelihood of benefiting from the nanomedicine treatment.

References

1. H. Green *et al.*, *Cancer Lett.*, 2011, **313**(2), 145–153.
2. M. Martin *et al.*, *Ann. Oncol.*, 2011, **22**(12), 2591–2596.
3. S. Stewart *et al.*, *J. Clin. Oncol.*, 1998, **16**(2), 683–691.
4. T. Boulikas, *Expert Opin. Invest. Drugs*, 2009, **18**(8), 1197–1218.
5. J. R. Infante *et al.*, *Cancer Chemother. Pharmacol.*, 2012, **70**(5), 699–705.
6. K. S. Lee *et al.*, *Breast Cancer Res. Treat.*, 2008, **108**(2), 241–250.
7. R. Plummer *et al.*, *Br. J. Cancer*, 2011, **104**(4), 593–598.
8. R. T. Poon and N. Borys, *Expert Opin. Pharmacother.*, 2009, **10**(2), 333–343.
9. J. D. Byrne, T. Betancourt and L. Brannon-Peppas, *Adv. Drug Delivery Rev.*, 2008, **60**(15), 1615–1626.
10. P. Debbage, *Curr. Pharm. Des.*, 2009, **15**(2), 153–172.
11. R. M. Sawant *et al.*, *Bioconjug. Chem.*, 2006, **17**(4), 943–949.
12. R. L. Scherer *et al.*, *Mol. Imaging*, 2008, 7(3), 118–131.
13. G. Kong and M. W. Dewhirst, *Int. J. Hyperthermia*, 1999, **15**(5), 345–370.
14. M. Bikram and J. L. West, *Expert Opin. Drug Delivery*, 2008, 5(10), 1077–1091.
15. L. Paasonen *et al.*, *J. Controlled Release*, 2007, **122**(1), 86–93.
16. W. Xie *et al.*, *Bioconjug. Chem.*, 2009, **20**(4), 768–773.
17. G. G. D'Souza *et al.*, *J. Drug Target*, 2008, **16**(7), 578–585.
18. M. Breunig, S. Bauer and A. Goepferich, *Eur. J. Pharm. Biopharm.*, 2008, **68**(1), 112–128.
19. S. D. Li and L. Huang, *Mol. Pharm.*, 2008, 5(4), 496–504.
20. P. K. Working and A. D. Dayan, *Hum. Exp. Toxicol.*, 1996, **15**(9), 751–785.

21. T. Siegal, A. Horowitz and A. Gabizon, *J. Neurosurg.*, 1995, **83**(6), 1029–1037.
22. A. A. Gabizon, Y. Barenholz and M. Bialer, *Pharm. Res.*, 1993, **10**(5), 703–708.
23. M. Amantea *et al.*, *Hum. Exp. Toxicol.*, 1999, **18**(1), 17–26.
24. G. T. Colbern *et al.*, *J. Inorg. Biochem.*, 1999, 77(1–2), 117–120.
25. P. K. Working *et al.*, *J. Pharmacol. Exp. Ther.*, 1999, **289**(2), 1128–1133.
26. J. K. Willmann *et al.*, *Nat. Rev. Drug Discovery*, 2008, 7(7), 591–607.
27. C. K. Abbey *et al.*, *Proc. Natl. Acad. Sci. U. S. A.*, 2004, **101**(31), 11438–11443.
28. A. Bhatnagar, R. Hustinx and A. Alavi, *Adv. Drug Delivery Rev.*, 2000, **41**(1), 41–54.
29. A. J. Fischman, N. M. Alpert and R. H. Rubin, *Clin. Pharmacokinet.*, 2002, **41**(8), 581–602.
30. S. Rottey, A. Signore and C. Van de Wiele, *Q J. Nucl. Med. Mol. Imaging*, 2007, **51**(2), 139–151.
31. M. Hamoudeh *et al.*, *Adv. Drug Delivery Rev.*, 2008, **60**(12), 1329–1346.
32. K. J. Harrington *et al.*, *Clin. Cancer Res.*, 2001, 7(2), 243–254.
33. P. C. Wu *et al.*, *Bioconjug. Chem.*, 2008, **19**(10), 1972–1979.
34. D. Zhu *et al.*, *Invest. Radiol.*, 2008, **43**(2), 129–140.
35. Y. B. Yu, *J. Drug Target*, 2006, **14**(10), 663–669.
36. B. L. Viglianti *et al.*, *Magn. Reson. Med.*, 2004, **51**(6), 1153–1162.
37. B. L. Viglianti *et al.*, *Magn. Reson. Med.*, 2006, **56**(5), 1011–1018.
38. P. J. Cassidy and G. K. Radda, *J. R. Soc. Interface*, 2005, **2**(3), 133–144.
39. N. J. Shah *et al.*, *Neuroimage*, 2001, **14**(5), 1175–1185.
40. S. Steinhoff *et al.*, *Magn. Reson. Med.*, 2001, **46**(1), 131–140.
41. N. J. Shah *et al.*, *Hepatology*, 2003, **38**(5), 1219–1226.
42. S. C. Deoni *et al.*, *Magn. Reson. Med.*, 2004, **52**(2), 435–439.
43. A. M. Blamire, *Br. J. Radiol.*, 2008, **81**(968), 601–617.
44. S. Clare and P. Jezzard, *Magn. Reson. Med.*, 2001, **45**(4), 630–634.
45. M. Dunne *et al.*, *J. Controlled Release*, 2011, **154**(3), 298–305.
46. S. Stapleton *et al.*, *PLoS One*, 2013, **8**(12), e81157.
47. J. Zheng, D. Jaffray and C. Allen, *Mol. Pharm.*, 2009, **6**(2), 571–580.
48. S. K. Carlson *et al.*, *Mol. Imaging Biol.*, 2007, **9**(2), 78–82.
49. A. Adams *et al.*, *Mol. Imaging Biol.*, 2006, **8**(2), 76–77.
50. S. D. Figueroa *et al.*, *Med. Phys.*, 2008, **35**(9), 3866–3874.
51. T. Funk, M. Sun and B. H. Hasegawa, *Med. Phys.*, 2004, **31**(9), 2680–2686.
52. J. M. Lanao and M. A. Fraile, *Curr. Pharm. Des.*, 2005, **11**(29), 3829–3845.
53. J. E. R. Riggs and R. Bachur, Clinical pharmacokinetics of anthracycline antibiotics, in *Pharmacokinetics of anticancer agents in humans*, ed. M. M. Ames, G. Powis and J. S. Kovach, 1983, Elsevier, Amsterdam, pp. 229–278.
54. Y. Matsumura and H. Maeda, *Cancer Res.*, 1986, **46**(12 Pt 1), 6387–6392.
55. Y. Noguchi *et al.*, *Jpn. J. Cancer Res.*, 1998, **89**(3), 307–314.
56. H. Maeda and Y. Matsumura, *Crit. Rev. Ther. Drug Carrier Syst.*, 1989, **6**(3), 193–210.

57. L. W. Seymour *et al.*, *Br. J. Cancer*, 1994, **70**(4), 636–641.
58. Y. Takakura and M. Hashida, *Pharm. Res.*, 1996, **13**(6), 820–831.
59. H. Maeda *et al.*, *J. Controlled Release*, 2000, **65**(1–2), 271–284.
60. F. Yuan *et al.*, *Cancer Res.*, 1995, **55**(17), 3752–3756.
61. F. Yuan *et al.*, *Cancer Res.*, 1994, **54**(13), 3352–3356.
62. A. K. Iyer *et al.*, *Drug Discovery Today*, 2006, **11**(17–18), 812–818.
63. S. A. Skinner, P. J. Tutton and P. E. O'Brien, *Cancer Res.*, 1990, **50**(8), 2411–2417.
64. L. W. Seymour *et al.*, *Eur. J. Cancer*, 1995, **31A**(5), 766–770.
65. R. Toy *et al.*, *ACS Nano*, 2013, 7(4), 3118–3129.
66. H. Maeda, *Cancer Sci.*, 2013, **104**(7), 779–789.
67. O. C. Boerman *et al.*, *J. Nucl. Med.*, 1997, **38**(3), 489–493.
68. J. Zheng *et al.*, *Contrast Media Mol. Imaging*, 2010, 5(3), 147–154.
69. D. Liu *et al.*, *ACS Appl. Mater. Interfaces*, 2014, **6**(3), 2131–2136.
70. G. Ravizzini *et al.*, *Wiley Interdiscip. Rev. Nanomed. Nanobiotechnol.*, 2009, **1**(6), 610–623.
71. S. T. Proulx *et al.*, *Cancer Res.*, 2010, **70**(18), 7053–7062.
72. L. Bergqvist, S. E. Strand and B. R. Persson, *Semin. Nucl. Med.*, 1983, **13**(1), 9–19.
73. Y. Mori *et al.*, *Magn. Reson. Med. Sci.*, 2012, **10**(4), 219–227.
74. C. Oussoren *et al.*, *Biochim. Biophys. Acta*, 1997, **1328**(2), 261–272.
75. M. A. Swartz, *Adv. Drug Delivery Rev.*, 2001, **50**(1–2), 3–20.
76. P. Kjellman *et al.*, *Nanomedicine*, 2014, **10**(5), 1089–1095.
77. S. Cressman *et al.*, *Bioconjug. Chem.*, 2009, **20**(7), 1404–1411.
78. V. A. Sethuraman and Y. H. Bae, *J. Controlled Release*, 2007, **118**(2), 216–224.
79. K. M. McNeeley, A. Annapragada and R. V. Bellamkonda, *Nanotechnology*, 2007, **18**(38), 385101.
80. M. K. Yu *et al.*, *Small*, 2011, 7(15), 2241–2249.
81. D. B. Kirpotin *et al.*, *Cancer Res.*, 2006, **66**(13), 6732–6740.
82. J. W. Park *et al.*, *Clin Cancer Res.*, 2002, **8**(4), 1172–1181.
83. W. Arap, R. Pasqualini and E. Ruoslahti, *Science*, 1998, **279**(5349), 377–380.
84. E. Koivunen, D. A. Gay and E. Ruoslahti, *J. Biol. Chem.*, 1993, **268**(27), 20205–20210.
85. R. Pasqualini *et al.*, *Cancer Res.*, 2000, **60**(3), 722–727.
86. N. T. Huynh *et al.*, *Nanomedicine*, 2010, 5(9), 1415–1433.
87. N. Karra and S. Benita, *Curr. Drug Metab.*, 2011, **13**(1), 22–41.
88. F. Bai *et al.*, *Biomaterials*, 2013, **34**(26), 6163–6174.
89. F. Pastorino *et al.*, *Cancer Res.*, 2006, **66**(20), 10073–10082.
90. K. Takara *et al.*, *J. Controlled Release*, 2012, **162**(1), 225–232.
91. A. Gabizon *et al.*, *Clin. Cancer Res.*, 2003, **9**(17), 6551–6559.
92. A. Gabizon *et al.*, *Adv. Drug Delivery Rev.*, 2004, **56**(8), 1177–1192.
93. D. Goren *et al.*, *Br. J. Cancer*, 1996, **74**(11), 1749–1756.
94. X. Huang *et al.*, *ACS Nano*, 2010, 4(10), 5887–5896.
95. J. W. Park *et al.*, *J. Controlled Release*, 2001, **74**(1–3), 95–113.

96. Y. Wang *et al.*, *Mol. Pharm.*, 2014, **11**(3), 885–894.
97. F. Kratz and A. Warnecke, *J. Controlled Release*, 2012, **164**(2), 221–235.
98. V. P. Chauhan *et al.*, *Nat. Nanotechnol.*, 2012, 7(6), 383–388.
99. E. Geretti *et al.*, *Cancer Res.*, 2013, **73**(8, Suppl. 1), 3271.
100. E. Geretti *et al.*, *Mol. Cancer Ther.*, 2015, **14**(9), 2060–2071.
101. H. Meng *et al.*, *ACS Nano*, 2013, 7(11), 10048–10065.
102. V. P. Chauhan *et al.*, *Annu. Rev. Chem. Biomol. Eng.*, 2011, **2**, 281–298.
103. K. M. Laginha *et al.*, *Clin. Cancer Res.*, 2005, **11**(19 Pt 1), 6944–6949.
104. S. C. White *et al.*, *Br. J. Cancer*, 2006, **95**(7), 822–828.
105. R. de la Rica, D. Aili and M. M. Stevens, *Adv. Drug Delivery Rev.*, 2012, **64**(11), 967–978.
106. H. Grull and S. Langereis, *J. Controlled Release*, 2012, **161**(2), 317–327.
107. P. Venugopalan *et al.*, *Pharmazie*, 2002, **57**(10), 659–671.
108. D. Aili *et al.*, *Nano Lett.*, 2010, **11**(4), 1401–1405.
109. M. J. Vicent *et al.*, *Angew Chem., Int. Ed.*, 2005, **44**(26), 4061–4066.
110. C. Ansari *et al.*, *Small*, 2013, **10**(3), 566–575, 417.
111. S. Lee *et al.*, *Bioconjug. Chem.*, 2014, **25**(3), 601–610.
112. L. Y. Rizzo *et al.*, *Curr. Opin. Biotechnol.*, 2013, **24**(6), 1159–1166.
113. G. H. Gao *et al.*, *Biomaterials*, 2012, **33**(35), 9157–9164.
114. E. K. Lim *et al.*, *Adv. Mater.*, 2011, **23**(21), 2436–2442.
115. E. K. Lim *et al.*, *Nanoscale Res. Lett.*, 2013, **8**(1), 467.
116. L. Ma, M. Liu and X. Shi, *J. Biomed. Mater. Res. B Appl. Biomater.*, 2012, **100**(2), 305–313.
117. E. S. Lee, Z. Gao and Y. H. Bae, *J. Controlled Release*, 2008, **132**(3), 164–170.
118. E. S. Lee *et al.*, *J. Controlled Release*, 2007, **123**(1), 19–26.
119. C. Oerlemans *et al.*, *Pharm. Res.*, 2010, **27**(12), 2569–2589.
120. N. Raghunand, B. P. Mahoney and R. J. Gillies, *Biochem. Pharmacol.*, 2003, **66**(7), 1219–1229.
121. R. Tong and D. S. Kohane, *Wiley Interdiscip. Rev. Nanomed. Nanobiotechnol.*, 2012, **4**(6), 638–662.
122. R. Xiong *et al.*, *Theranostics*, 2013, **3**(3), 141–151.
123. R. Bardhan *et al.*, *Adv. Funct. Mater.*, 2009, **19**, 3901e9.
124. A. G. Skirtach *et al.*, *Angew Chem., Int. Ed.*, 2006, **45**(28), 4612–4617.
125. Y. Zhao, *Chem. Rec.*, 2007, 7(5), 286–294.
126. J. P. May and S. D. Li, *Expert Opin. Drug Delivery*, 2013, **10**(4), 511–527.
127. *Clinical Trial NCT00346229*, 2006–2011.
128. *Clinical Trial NCT00441376*, 2007–2009.
129. *Clinical Trial NCT00617981*, 2008–2012.
130. *Clinical Trial NCT00826085*, 2009–2012.
131. *Clinical Trial NCT01464593*, 2011–2012.
132. *Clinical Trial NCT01640847*, 2012.
133. A. M. Ponce *et al.*, *J. Natl. Cancer Inst.*, 2007, **99**(1), 53–63.
134. B. S. Hendriks *et al.*, *CPT Pharmacometrics Syst. Pharmacol.*, 2012, **1**, e15.
135. H. Lee *et al.*, *Nanomedicine*, 2015, **11**(1), 155–165.
136. B. Hendriks *et al.*, *IMPAKT Breast Cancer Conference*, 2014, 55P.
137. P. Munster *et al.*, *San Antonio Breast Cancer Symposium*, 2013, P4-12-29.

CHAPTER 7

Anticancer Agent-Incorporating Polymeric Micelles: from Bench to Bedside

YASUHIRO MATSUMURA

Developmental Therapeutics, Exploratory Oncology Research & Clinical Trial Center, National Cancer Center, 6-5-1 Kashiwanoha, Kashiwa City 277-8577, Japan
Email: yhmatsum@east.ncc.go.jp

7.1 Introduction

Drugs categorized under the drug-delivery system (DDS) are made primarily by utilizing nanotechnology. In the field of oncology, DDS drugs have been produced and evaluated in preclinical or clinical trials, with some already approved for clinical use. More specifically, DDS can be used for active or passive targeting of tumour tissues. Active targeting refers to the development of monoclonal antibodies (mAbs) directed against tumour-related molecules, allowing targeting of a tumour using the specific binding of antibodies with their respective antigens. Recent promising advances in active targeting strategies include the development of antibody–drug conjugates (ADCs)[1] However, since no more than three anticancer agents-molecules can be bound at a time to mAbs without compromising the affinity of the antibodies, use of ADCs must be confined to only highly toxic anticancer agents and must not be extended to relatively less toxic

RSC Drug Discovery Series No. 51
Nanomedicines: Design, Delivery and Detection
Edited by Martin Braddock
© The Royal Society of Chemistry 2016
Published by the Royal Society of Chemistry, www.rsc.org

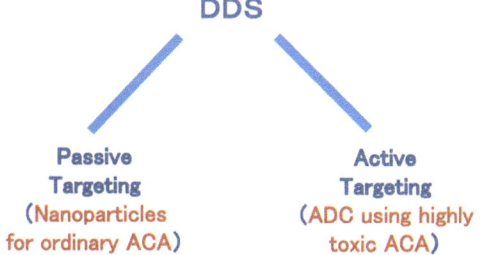

Figure 7.1 An antibody–drug conjugate (ADC) strategy should be confined to highly toxic anticancer agents (ACAs), but not to ordinary ACAs such as paclitaxel, doxorubicin and others, because only three or fewer ACA molecules should be conjugated to the monoclonal antibody (mAb), otherwise the affinity of the mAb is diminished if too many molecules of ACA are attached to the mAb. According to this principle, an unrealistic amount of ADC should have to be administered if the mAbs are conjugated with an ordinary ACA.

anticancer agents such as taxanes, doxorubicin (Adriamycin), *etc.* An unrealistic amount of ADCs would need to be administered if each of the anticancer agents were to be conjugated with a mAb. Therefore, for an ordinary anticancer agent, nanoparticles may serve as the best DDS. Furthermore, the application of DDS using monoclonal antibodies is, of course, restricted to tumours expressing high levels of related antigens (Figure 7.1).

Passive targeting can be achieved by utilizing the enhanced permeability and retention (EPR) effect.[2] This effect is based on the pathophysiological characteristics of solid tumour tissues, namely hypervascularity, incomplete vascular architecture, secretion of vascular permeability factors stimulating extravasation within cancer tissue and absence of effective lymphatic drainage from tumours that impedes the efficient clearance of macromolecules accumulated in solid tumour tissues (Figure 7.2).

Several techniques have been developed to maximally utilize the EPR effect, such as modification of drug structures and the development of drug carriers. Polymeric micelle-based anticancer drugs were originally developed by Kataoka *et al.* in the late 1980s and early 1990s.[3–5] Polymeric micelles were expected to increase the accumulation of drugs in tumour tissues by utilizing the EPR effect as well as to incorporate various kinds of drugs into their inner core with relatively high stability by chemical conjugation or physical entrapment. In addition, the size of micelles can be controlled within the diameter range of 20–100 nm to ensure that they do not penetrate normal vessel walls. With this development, it is expected that the incidence of drug-induced side effects may be decreased owing to the reduced drug distribution in normal tissues.

In this chapter, recent developments in polymeric micelle systems presently being evaluated in clinical trials are reviewed (Figure 7.3).

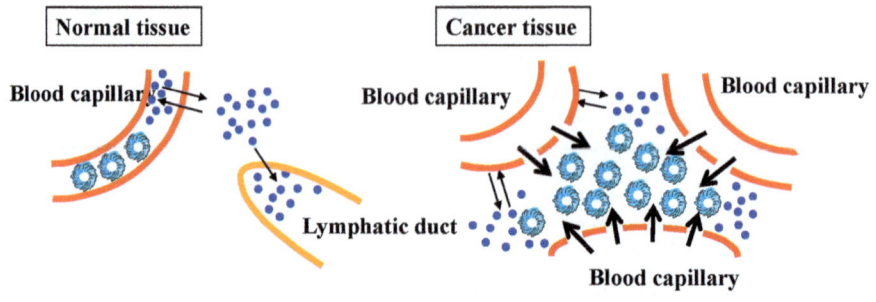

 : LMW anticancer agents   : DDS

Figure 7.2   Small molecules easily leak from normal vessels, which gives small molecules a short plasma half-life. In contrast, macromolecules have a long plasma half-life because they are too large to pass through the normal vessel walls, unless they are trapped by the reticuloendothelial system. Solid tumours generally possess several pathophysiological characteristics, namely hypervasculature, secretion of vascular permeability factors stimulating extravasation of macromolecules within the cancer, and absence of effective lymphatic drainage from tumours, which impedes the efficient clearance of macromolecules accumulated in solid tumour tissues. These characteristics of solid tumours are the basis of the EPR effect. LMW: low molecular weight; DDS: drug delivery system.

7.2   Anticancer Agents Incorporating Micelles under Clinical Evaluation

7.2.1   NK105, a Paclitaxel-incorporating Micelle

Paclitaxel (PTX) is one of the most useful anticancer agents against various cancers, including ovarian, breast, and lung cancers.[6,7] However, it produces serious adverse effects such as neutropenia and peripheral sensory neuropathy. In addition, anaphylaxis and other severe hypersensitive reactions have been reported in 2–4% of patients receiving the drug, even after premedication with anti-allergic agents; these adverse reactions have been attributed to the mixture of Cremophor EL and ethanol used for solubilizing PTX.[8] Of the adverse reactions, neutropenia can be prevented or managed effectively by administering a granulocyte colony-stimulating factor. There are no effective therapies to prevent or reduce nerve damage associated with PTX-induced peripheral neuropathy. Thus, neurotoxicity constitutes a significant dose-limiting toxicity of the drug.[9]

7.2.1.1   *Preparation and Characterization of NK105*

To construct NK105 micellar nanoparticles, block copolymers consisting of polyethylene glycol (PEG) and polyaspartate were used.[10] PTX was

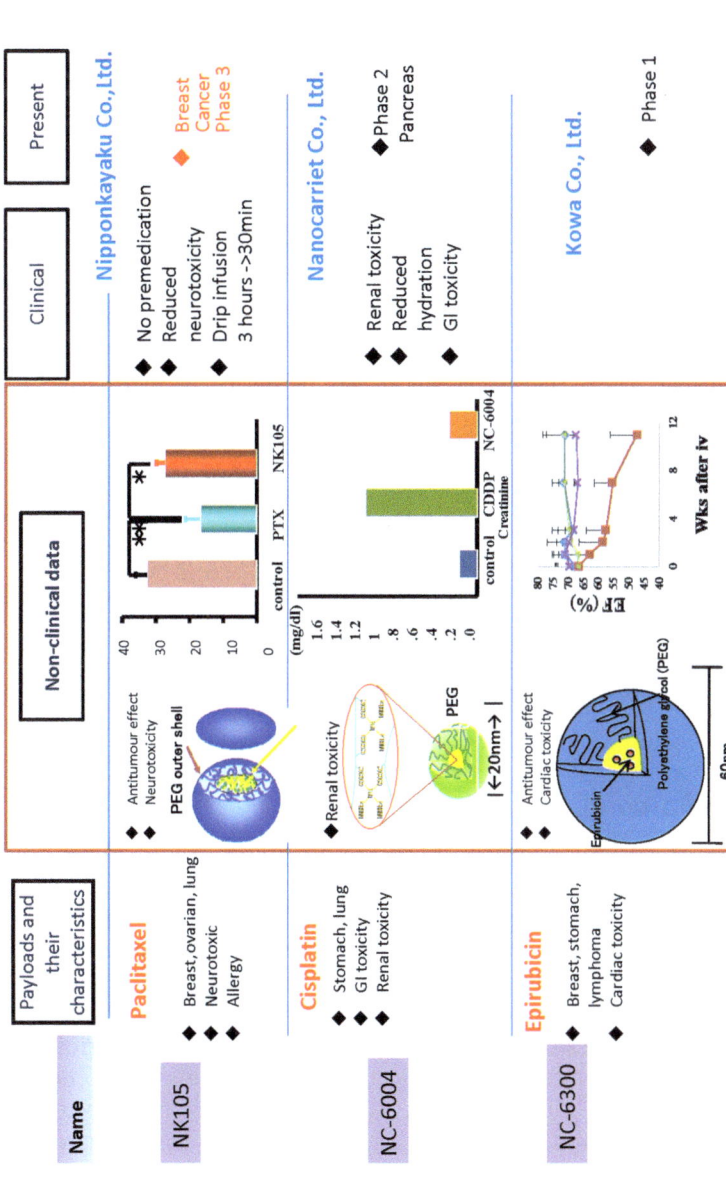

Figure 7.3 Anticancer agent-incorporated micelles under clinical evaluation. NK105 is a paclitaxel-incorporating micelle. Preclinical studies indicated higher antitumour activity and reduced neurotoxicity. Phase 1 and 2 trials showed that NK105 can be administered by 30 min drip infusion without any premedication and showed that NK105 reduced neurotoxicity. The phase 3 trial of NK105 is now underway in patients with metastatic breast cancer. NC-6004 is a cisplatin-incorporating micelle. Preclinical studies showed that NC-6004 had no renal toxicity. Phase 1 and 2 trials revealed that NC-6004 can be administered with minimum hydration and the gastrointestinal (GI) toxicity was reduced remarkably. NC-6300 is an epirubicin-incorporating micelle. A basic study of NC-6300 showed significant higher antitumour activity and significantly reduced cardiotoxicity of epirubicin. The phase 1 trial is now underway.

incorporated into polymeric micelles formed by physical entrapment utilizing hydrophobic interactions between PTX and the block copolymer polyaspartate chain. NK105 was prepared by facilitating the self-association of NK105 polymers and PTX. NK105 was obtained as a freeze-dried formulation and contained *ca.* 23% (w/w) of PTX. The weight-average diameter of the nanoparticles was approximately 85 nm with a narrow size distribution.[11]

7.2.1.2 Preclinical Studies

In an *in vivo* pharmacokinetics study using colon 26 tumour-bearing CDK mice, the plasma concentration at 5 min (C_{5min}) and the area under the curve (AUC) of NK105 were 11- to 20-fold and 50- to 86-fold higher, respectively, than those of PTX. The maximum concentration (C_{max}) and AUC of NK105 in colon 26 tumours were approximately three times and 25 times higher, respectively, than those of PTX. NK105 continued to accumulate in the tumours until 72 h post-injection. In BALB/c mice bearing subcutaneous HT-29 colon cancer tumours, NK105 exhibited superior antitumour activity compared with PTX ($P < 0.001$). Tumour suppression by NK105 increased in a dose-dependent manner.

PTX treatment has been shown to cause cumulative sensory-dominant peripheral neurotoxicity in humans, characterized clinically by numbness and/or paraesthesia of the extremities. Pathologically, axonal swelling, vesicular degeneration, and demyelination have also been observed. Both electrophysiological and morphological methods revealed that the peripheral neurotoxicity was significantly reduced in the NK105 administration group in rats.

7.2.1.3 Clinical Studies

A phase I study was designed to determine the maximum tolerated dose (MTD), dose-limiting toxicities (DLTs), and recommended dose (RD) of NK105 for phase II, as well as its pharmacokinetics.[12] NK105 was administered by intravenous infusion for 1 h every 3 weeks without antiallergic premedication. It should be noted that the clinically used PTX preparation is a mixture of Cremophore EL and ethanol because PTX has poor water solubility. However, the use of Cremophore EL is known to cause serious acute hypersensitivity.[8] Therefore, premedication with steroids and other agents is inevitable in the PTX injection. Since NK105 can be injected without the use of Cremophore EL or ethanol, NK105 was administered without any premedication. The starting dose was 10 mg PTX equivalent m^{-2}, and the dose escalated according to the accelerated titration method. DLTs were observed in two patients at 180 mg m^{-2} (grade 4 neutropenia lasting for more than 5 days), which was determined as the MTD. Allergic reactions were observed in only one patient. The RD was 150 mg m^{-2}.

A partial response was observed in one pancreatic cancer patient who received more than 12 courses of NK105. Colon and gastric cancer patients experienced stable disease lasting for 10 and seven courses of treatment, respectively. Despite long-term administration, only grade 1 or 2 neuropathy was observed when the dose or period of drug administration was modified. The C_{max} and AUC of NK105 showed dose-dependent characteristics. The plasma AUC of NK105 at 150 mg m^{-2} was approximately 30-fold higher than that of the commonly used PTX formulation. The DLT was grade 4 neutropenia. NK105 facilitates prolonged systemic PTX exposure in plasma. Tri-weekly 1 h infusion of NK105 was feasible and well tolerated, with antitumour activity.

Next, a phase 2 clinical trial was conducted in patients with previously treated advanced stomach cancer in order to see the anticancer activity and the safety profile.[13] The most common grade 4 hematological toxicity was neutropenia.

The grade 3/4 non-hematological toxicities were infrequent. Although NK105 was administered without any premedication, there were no patients who experienced grade 3/4 hypersensitive reaction. It is noteworthy that during the whole period of this study, only one patient experienced grade 3 neuropathy (1.8%). The incidences of grade 3 or 4 neuropathy of other PTX formulations, namely conventional PTX, Xyotax and Abraxane, were 10%, 15% and 11%, respectively, in each phase II setting.

The overall response rate was 25% with two complete responses and 12 partial responses.

Median overall survival as a second-line therapy in stomach cancer was reported to be 3–10 months. Compared to previously published data, overall survival of 14.4 months in stomach cancer as a second-line therapy in this phase 2 trial is remarkable.

In conclusion, the first phase 2 study with NK105 at 150 mg PTX equivalent m^{-2} provides positive proof of concept for high activity and tolerability of a new DDS formulation for PTX. A phase 3 study of NK105 compared with PTX is now underway in patients with previously treated advanced breast cancer.

7.2.2 NC-6004, Cisplatin-incorporating Micelle

Cisplatin [*cis*-dichlorodiammineplatinum (II): CDDP] is a key drug in the chemotherapy of various cancers, including lung, gastrointestinal and genitourinary cancers.[14,15] However, it is often necessary to discontinue cisplatin treatment because of its adverse reactions (*e.g.* nephrotoxicity and neurotoxicity) despite its persisting effects.[16] To date, platinum analogues (*e.g.* carboplatin and oxaliplatin) have been developed to overcome these cisplatin-related disadvantages.[17] Consequently, these analogues have become the standard drugs for ovarian and colon cancers.[18,19] However, these regimens including cisplatin constitute the standard treatment for lung, gastric, testicular and urothelial cancers.[20,21] Therefore, the development of DDS technology is anticipated, which would enable better selective cisplatin accumulation in solid tumours while lessening its distribution in normal tissue.

7.2.2.1 Preparation and Characterization of NC-6004

NC-6004 was prepared according to a slightly modified procedure reported by Nishiyama *et al.*[22] NC-6004 consists of PEG, a hydrophilic chain constituting the outer shell of micelles, and the coordinate complex of poly(glutamic acid) and CDDP, a polymer-metal complex-forming chain constituting the inner core of micelles. A narrowly distributed size of polymeric micelles (20 nm) was confirmed by dynamic light scattering measurement. Also, static light scattering measurement revealed that the CDDP-loaded micelles showed no dissociation upon dilution and the critical micellar concentration was $<5 \times 10^{-7}$ M, suggesting remarkable stability compared with typical micelles from amphiphilic block copolymers.

7.2.2.2 Preclinical Studies

NC-6004 showed a very long blood retention profile compared with CDDP. The AUC_{0-t} and C_{max} values were significantly higher in animals given NC-6004 than in animals given CDDP, namely, 65-fold and eight-fold, respectively $(P<0.001$ and $P<0.001$, respectively).[23] Regarding platinum accumulation in the tumour, platinum concentrations peaked at 10 min following CDDP administration and at 48 h following NC-6004 administration. The C_{max} in the tumour was 2.5-fold higher for NC-6004 than for CDDP $(P<0.001)$. Furthermore, the tumour AUC was 3.6-fold higher for NC-6004 than for CDDP $(81.2 \ \mu g \ mL^{-1} \ h^{-1}$ and $22.6 \ \mu g \ mL^{-1} \ h^{-1}$, respectively).[24]

In nude mice implanted with the human gastric cancer cell line MKN-45, NC-6004 administration groups $(5 \ mg \ kg^{-1}$ of CDDP) showed no significant difference in tumour growth rate compared with CDDP administration groups. Regarding time-course changes in body weight change rate, the CDDP $(5 \ mg \ kg^{-1})$ administration group showed a significant decrease $(P<0.001)$ in body weight compared with the control group. In contrast, the NC-6004 administration group showed no decrease in body weight compared with the control group.

Regarding renal function in rats, the CDDP $(10 \ mg \ kg^{-1})$ administration group showed significantly higher plasma concentrations of blood urea nitrogen and creatinine than the control group $(P<0.05$ and $P<0.001$, respectively) and the NC-6004 $(10 \ mg \ kg^{-1})$ administration group $(P<0.05$ and $P<0.001$, respectively).

Upon neurological examination, rats given NC-6004 showed no delay in sensory nerve conductive velocity (SNCV) compared with animals given 5% glucose. In contrast, rats given CDDP showed a significant delay $(P<0.05)$ in SNCV compared with animals given NC-6004. Sciatic nerve concentrations of platinum were significantly $(P<0.05)$ lower in rats given NC-6004. This finding is believed to be a factor in reduced neurotoxicity following NC-6004 administration compared with CDDP administration.

7.2.2.3 Clinical Studies

A phase I clinical trial of NC-6004 has recently been completed in the UK.[24] The starting dose of NC-6004 was 10 mg m^{-2}. NC-6004 was administered once every 3 weeks with only 1000 ml water loading on the day of administration. Administration of doses up to 120 mg m^{-2} was performed without inducing significant nephrotoxicity. Although nausea and vomiting are typical CDDP adverse effects, those caused by NC-6004 were generally mild. However, hypersensitivity reactions caused by NC-6004 occurred more frequently than those caused by CDDP.

Next, a phase 1/2 study was conducted in Asia. Although hypersensitivity reactions occurred in the UK phase 1 trial of NC-6004, no patient experienced the hypersensitivity reactions using oral steroid premedication. Compared to CDDP, gastrointestinal (GI) toxicity of NC-6004 was significantly reduced and no hydration is necessary for the NC6004 injection.[25] Since it is clear that no hydration or minimum hydration is necessary for the NC-6004 administration and the GI toxicity is significantly reduced compared to CDDP, it can be considered that a non-inferiority trial against CDDP is feasible in various types of cancer.

7.2.3 NC-6300, Epirubicin-incorporating Micelle

Anthracyclines such as epirubicin (EPI) and doxorubicin (DXR) are widely used and highly effective anticancer agents for the treatment of various human tumours including hepatocellular cancer, breast cancer, and gastric cancer.[26–30] However, anthracyclines induce several adverse effects, among which acute or chronic cardiotoxicity canbe irreversible.[31] Cardiotoxicity of EPI is ~66% of that of DXR, but it is still the most serious problem in oncological treatment; this side effect can sometimes require cessation of EPI treatment, resulting in an insufficient antitumour effect.[32] To address these issues, various anthracyclines have been developed to reduce cardiotoxicity without loss of the antitumour effect. However, to date, no such anthracyclines are available in a clinical context. Against this background, NC-6300 was synthesized.

7.2.3.1 Preparation and Characterization of NC-6300

NC-6300 comprises EPI covalently bound to PEG polyaspartate block copolymer through an acid-labile hydrazone bond. The conjugate spontaneously forms a micellar structure with a diameter of 40–80 nm in aqueous media, as reported previously.[33] *In vitro* findings indicated that it exhibited pH-dependent EPI release: the release of EPI from NC-6300 accelerated under increasingly acidic conditions.

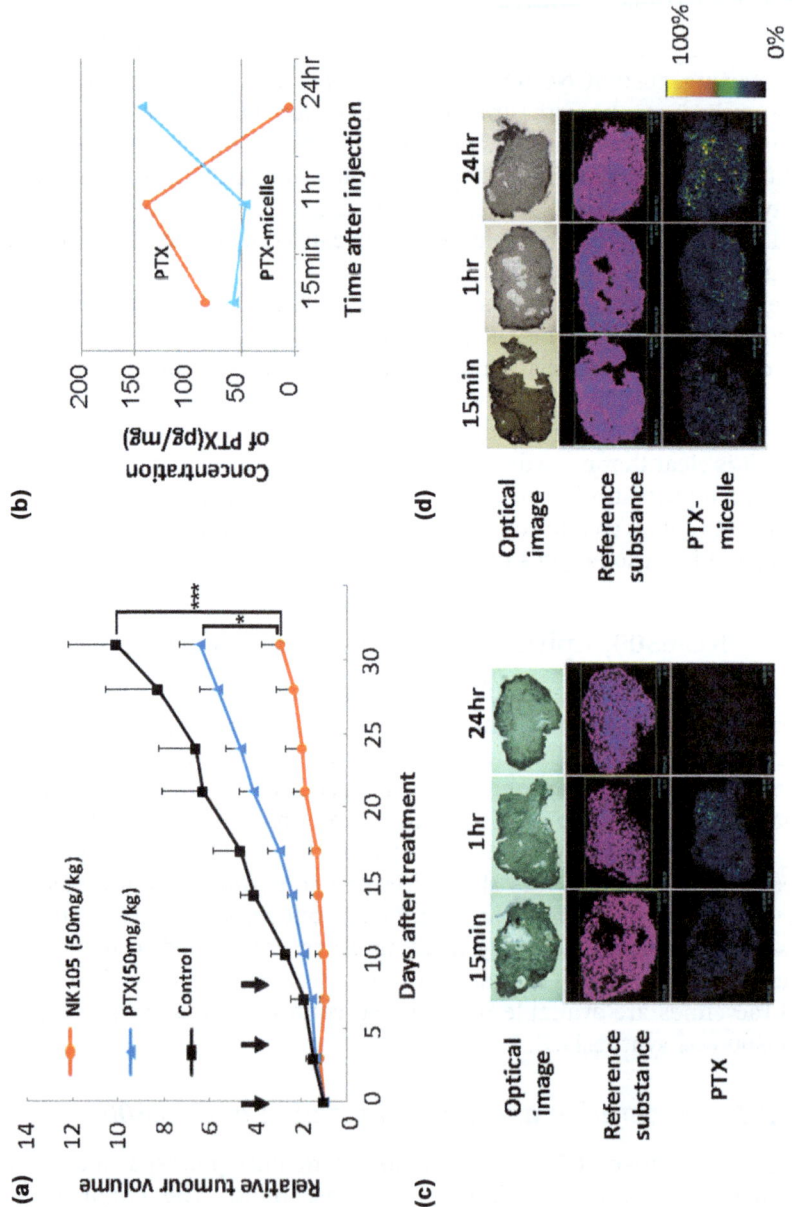

7.2.3.2   Preclinical Studies

Antitumour activity was observed in mice bearing Hep3B/Luc liver ortho-topic tumours. There was a significant difference between the control group and the groups treated with EPI and NC-6300.[34] Kaplan–Meier analysis showed that there was a significant improvement in the survival rate of the group administered NC-6300 at 10 mg kg^{-1} compared with that of the group given EPI at 10 mg kg^{-1} ($P = 0.002$).

Tissue distribution experiments were conducted to evaluate the toxicity and efficacy data obtained for both NC-6300 and EPI in terms of the plasma and tissue concentrations of each formulation. The concentration–time profiles in plasma and various tissues were obtained. After the injection of EPI, its concentration declined rapidly. NC-6300 exhibited significantly slower clearance. The clearance rate of NC-6300 in the orthotopic Hep3B tumour was significantly slower than that of other normal organs. In a similar manner to other drugs categorized in the DDS, NC-6300 showed higher accumulation in organs of the reticuloendothelial system. In the heart, a significantly higher concentration of EPI was obtained for several hours after the administration of conventional EPI compared with that of NC-6300. NC-6300 produced increases in EPI concentration, particularly in the plasma, liver, spleen and tumour, whereas NC-6300 decreased the free EPI concentration in the kidney, lung and heart.

Cardiotoxicity was evaluated by echocardiography in C57BL/6 mice during and following a total of nine administrations of NC-6300 (10 mg kg^{-1}) and conventional EPI (10 mg kg^{-1}) for 12 weeks. The ejection fraction of mice treated with conventional EPI (10 mg kg^{-1}) was significantly reduced compared with those of the control ($P = 0.0019$) and NC-6300 (10 mg kg^{-1}) treatment groups ($P = 0.0081$). Fractional shortening was also significantly lower in the conventional EPI treatment group.

7.2.3.3   Clinical Studies

Following the several preclinical evaluations, a phase 1 trial has begun in the National Cancer Center Hospital East, Kashiwa, Japan.

Figure 7.4   Antitumour activity and visualisation of paclitcasel (PTX) and NK105 distribution with mass spectroscopy analysis. (a) Antitumour activity was examined in an animal model with BXPC3 xenografts. NK105, PTX, or saline (as a control) was administered at a PTX equivalent dose of 50 mg kg^{-1} on days 0, 4, and 8. *: $P < 0.05$ (PTX *vs.* NK105), ***: $P < 0.001$ (saline *vs.* NK105). Bar = SD. (b) Liquid chromatography mass spectroscopy analysis of the PTX concentration in the tumours treated with PTX or NK105 . The samples are the same as those shown in (c) and (d). (c) (d) Imaging of PTX within the tumour was performed after (c) PTX or (d) NK105administration at a dose of 100 mg kg^{-1}. The upper, middle, and lower columns display the optical images, reference substance (an arbitrary signal of *m/z* 824.6), and PTX (specific signal of *m/z* 892.3 $[M + K]^{+}$), respectively.

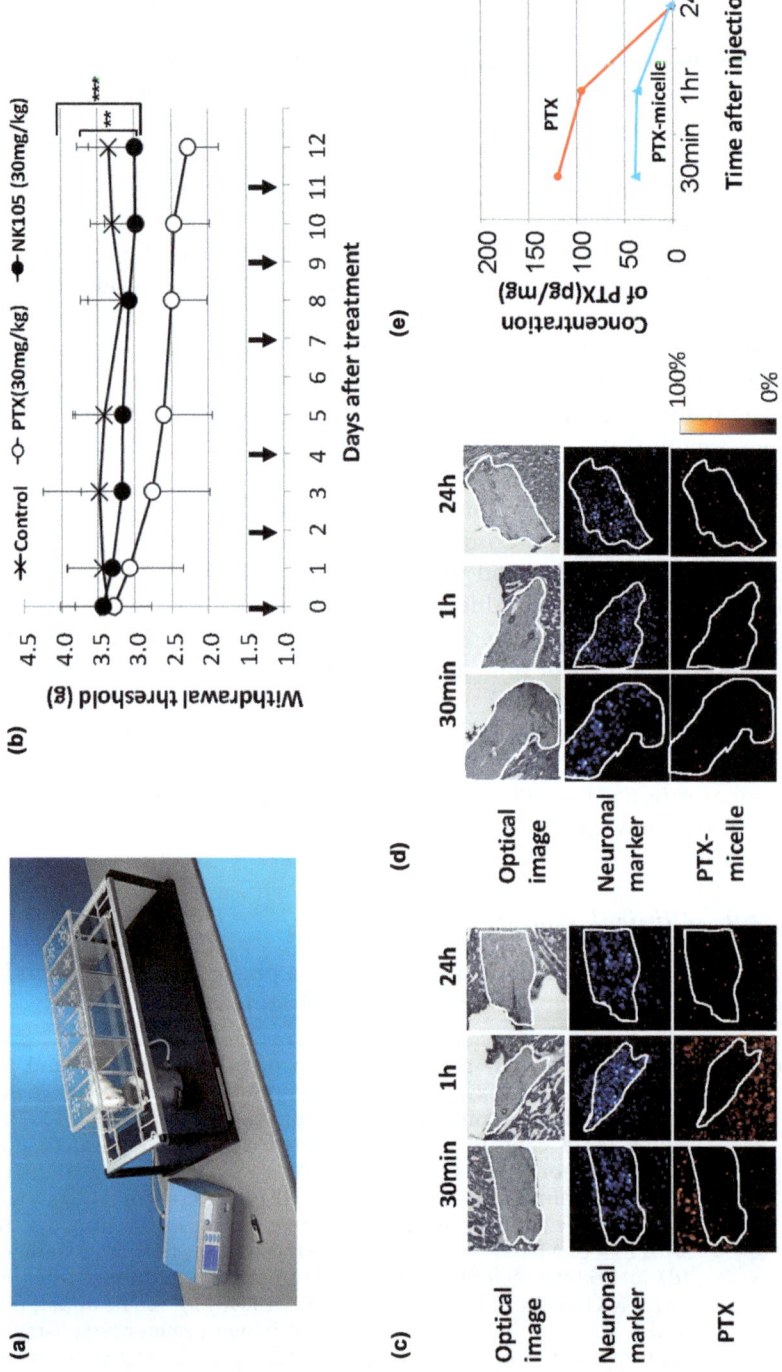

7.3 Verification of the EPR Effect using Imaging Mass Spectrometry

Although many studies have indicated that NK105 accumulates selectively in tumour tissue compared to PTX using high-performance liquid chromatography or liquid chromatography mass spectrometry analysis, it was not known whether NK105 could deliver PTX to cancer-cell clusters within the tumour tissue. Cancer tissue is heterogeneous and consists not only of cancer cells but also of abundant tumour stroma, the latter of which can act as a barrier against macromolecules, including NK105.[35,36] In a study using matrix-assisted laser desorption ionisation imaging mass spectrometry (MALDI-IMS), significant levels of PTX were observed following NK105 administration, even in the core of the tumour tissue, and the NK105 was retained for a long period of time (Figure 7.4).[37]

Low molecular weight anticancer agents, including molecular targeting agents, can easily extravasate from normal blood vessels and cause various adverse effects. DDS drugs such as NK105, which exhibit low short-term accumulation in normal tissues that lack the EPR effect, can minimise this drug toxicity. Our data gathered by IMS clearly demonstrate that the distribution of released PTX from NK105 in the peripheral nerve and surrounding tissues was quite low compared with PTX alone (Figure 7.5). These observations support the low incidence of peripheral neuropathy when PTX is administered as NK105.

This is the first report describing the precise distribution of a DDS drug by IMS, a new technique developed by our group and others. Notably, we successfully visualised and quantified the distribution of a non-radiolabeled and non-chemically modified drug in various frozen tissue slices microscopically. In addition to PTX, we have successfully visualised other anticancer agents, including SN-38, epirubicin, and monomethyl auristatin E. This success indicates that the MALDI-IMS technique can be applied to clinical biopsy specimens or surgically resected tissues after neo-adjuvant chemotherapy. In addition, the data obtained by MALDI-IMS can be utilised to facilitate drug design.

Figure 7.5 Peripheral neurotoxicity and visualisation of paclitaxel (PTX) and NK105 distribution by mass spectroscopy analysis. (a) Dynamic plantar aesthesiometer. (b) Mechanical sensory stress was assayed in an animal model of PTX-induced peripheral neuropathy. NK105, PTX, or saline was administered at $30 \, \mathrm{mg \, kg^{-1}}$ on days 0, 2, 4, 7, 9, and 11. **: $P < 0.01$ (PTX *vs.* NK105), ***: $P < 0.001$ (saline *vs.* PTX). Bar = SD. (c) (d) PTX within neuronal tissue was imaged after (c) PTX or (d) NK105 administration at a dose of $50 \, \mathrm{mg \, kg^{-1}}$. The upper, middle, and lower columns show the optical images, a neuronal marker (sphingomyelin-specific signal of 851.6 *m/z*), and PTX (specific signal of *m/z* 892.3 $[M + K]^{+}$), respectively. The neuronal area is delineated by a white line. (e) Analysis of the PTX concentration by liquid chromatography mass spectrometry. The samples are the same as those shown in (c) and (d).

7.4 Discussion and Conclusion

Due to the success of trastuzumab emtansine (TDM-1) in patients with metastatic breast cancer, active targeting, or ADC is again receiving attention in a robust fashion.[1] This ADC strategy should be confined to highly toxic anticancer agents but not to ordinary anticancer agents such as taxane, adriamycin and others, because only three or fewer anticancer agent molecules should be conjugated to the mAb, otherwise the affinity of the mAb is diminished. According to this principle, an unrealistic amount of ADC should be administered if the mAbs were conjugated with an ordinary anticancer agent. Therefore, for an ordinary anticancer agent, nanoparticles should be considered as a DDS tool.

As described earlier, several preclinical studies have demonstrated the advantages of using anticancer agent-incorporating polymeric micelles, and some are currently under clinical evaluation as single micelle agents alone[13,24,38-40] or single micelle agents in combination with non-micelle conventional anticancer agents.[41-43] However, there are scarcely any reports of combined use of micellar anticancer agents.[44]

Multidrug regimens have been developed to enhance the anticancer efficacy and suppress the emergence of drug resistance. However, patients receiving such multidrug regimens sometimes experience additional and stronger drug adverse effects that could necessitate discontinuation of the treatment. In this regard, combined use of anticancer agent-incorporating polymeric micelles may be a good alternative therapeutic modality, because micellar formulations are delivered selectively to tumour tissues *via* the EPR effect while their distribution to normal tissues is suppressed. Very recently we successfully demonstrated that combined use of NC-6300 and NC-4016, 1,2-diaminocyclohexane platinum (II)-incorporating micelles exerted a stronger antitumour effect and lower toxicity in a human gastric cancer model compared to epirubicin and oxaliplatin.[45] Several types of nanoparticles, in which multiple anticancer agents are loaded on to one particle, have been developed recently and been shown to exhibit potent anticancer activity.[46-48] These fixed drug ratio combinations are designed to maintain synergistic molar ratios to enhance the therapeutic benefit, and these multidrug-loaded nanoparticles may be useful in future clinical applications. However, there are some clinical concerns that need to be addressed in regard to this type of nanoparticles. Since the drugs are loaded in a fixed ratio prior to administration, flexible adjustment of the doses of each component drug encapsulated in the particles according to the profiles and severity of the emerging toxicities is difficult. Moreover, the types of cancer that are suitable for such treatments are probably limited. In contrast, in the case of combined use of single drug-loaded micelles, it is easier to adjust the dose ratio and modify the dose of each micellar preparation as necessary during treatment.

In addition to the development of combined use of micellar anticancer agents, we have succeeded in developing an antibody conjugated micellar particle containing epirubicin. We are now reaching the most interesting part of the development of micellar nanoparticles.

References

1. J. R. Junutula, H. Raab, S. Clark, S. Bhakta, D. D. Leipold, S. Weir, Y. Chen, M. Simpson, S. P. Tsai, M. S. Dennis, Y. Lu, Y. G. Meng, C. Ng, J. Yang, C. C. Lee, E. Duenas, J. Gorrell, V. Katta, A. Kim, K. McDorman, K. Flagella, R. Venook, S. Ross, S. D. Spencer, W. L. Wong, H. B. Lowman, R. Vandlen, M. X. Sliwkowski, R. H. Scheller, P. Polakis and W. Mallet, *Nature Biotech.*, 2008, **26**, 925–932.
2. Y. Matsumura and H. Maeda, *Cancer Res.*, 1986, **46**, 6387–6392.
3. K. Kataoka, G. S. Kwon, M. Yokoyama, T. Okano and Y. Sakurai, Block copolymer micelles as vehicles for drug delivery, *J. Controlled Release*, 1993, **24**, 119–132.
4. M. Yokoyama, M. Miyauchi, N. Yamada *et al.*, Polymer micelles as novel drug carrier: Adriamycin-conjugted poly(ethylene glycol)-poly(aspartic acid) block copolymer, *J. Controlled Release*, 1990, **11**, 269–278.
5. M. Yokoyama, T. Okano, Y. Sakurai, H. Ekimoto, C. Shibazaki and K. Kataoka, Toxicity and antitumor activity against solid tumors of micelle-forming polymeric anticancer drug and its extremely long circulation in blood, *Cancer Res.*, 1991, **51**, 3229–3236.
6. D. Khayat, E. C. Antoine and D. Coeffic, Taxol in the management of cancers of the breast and the ovary, *Cancer Invest.*, 2000, **18**, 242–260.
7. D. N. Carney, Chemotherapy in the management of patients with inoperable non-small cell lung cancer, *Semin. Oncol.*, 1996, **23**, 71–75.
8. R. B. Weiss, R. C. Donehower, P. H. Wiernik *et al.*, Hypersensitivity reactions from taxol, *J. Clin. Oncol.*, 1990, **8**, 1263–1268.
9. E. K. Rowinsky, V. Chaudhry, A. A. Forastiere *et al.*, Phase I and pharmacologic study of paclitaxel and cisplatin with granulocyte colony-stimulating factor: neuromuscular toxicity is dose-limiting, *J. Clin. Oncol.*, 1993, **11**, 2010–2020.
10. M. Yokoyama, T. Okano, Y. Sakurai, H. Ekimoto, C. Shibazaki and K. Kataoka, Toxicity and antitumor activity against solid tumors of micelle-forming polymeric anticancer drug and its extremely long circulation in blood, *Cancer Res.*, 1991, **51**, 3229–3236.
11. T. Hamaguchi, Y. Matsumura, M. Suzuki *et al.*, NK105, a paclitaxel-incorporating micellar nanoparticle formulation, can extend in vivo antitumour activity and reduce the neurotoxicity of paclitaxel, *Br. J. Cancer*, 2005, **92**, 1240–1246.
12. T. Hamaguchi, K. Kato, Y. Matsumura *et al.*, A Phase I and Pharma-cokinetic Study of NK105, a Paclitaxel-incorporating Micellar Nano-particle Formulation, *Br. J. Cancer.*, 2007, **97**, 170–176.
13. K. Kato, K. Chin, T. Yoshikawa, K. Yamaguchi, Y. Tsuji, T. Esaki, K. Sakai, M. Kimura, T. Hamaguchi, Y. Shimada, Y. Matsumura and R. Ikeda, Phase II study of NK105, a paclitaxel-incorporating micellar nanoparticle, for previously treated advanced or recurrent gastric cancer, *Invest. New Drugs*, 2012, **30**, 1621–1627.
14. A. Horwich, D. T. Sleijfer, S. D. Fossa, S. B. Kaye, R. T. Oliver, M. H. Cullen, G. M. Mead, R. de Wit, P. H. de Mulder, D. P. Dearnaley,

P. A. Cook, R. J. Sylvester and S. P. Stenning, Randomized trial of bleomycin, etoposide, and cisplatin compared with bleomycin, etoposide, and carboplatin in good-prognosis metastatic nonseminomatous germ cell cancer: a Multiinstitutional Medical Research Council/European Organization for Research and Treatment of Cancer Trial, *J. Clin. Oncol.*, 1997, **15**, 1844–1852.

15. B. J. Roth, Chemotherapy for advanced bladder cancer, *Semin. Oncol.*, 1996, **23**, 633–644.

16. V. Pinzani, F. Bressolle, I. J. Haug, M. Galtier, J. P. Blayac and P. Balmes, Cisplatin-induced renal toxicity and toxicity-modulating strategies: a review, *Cancer Chemother. Pharmacol.*, 1994, **35**, 1–9.

17. M. J. Cleare, P. C. Hydes, B. W. Malerbi and D. M. Watkins, Anti-tumor platinum complexes: relationships between chemical properties and activity, *Biochimie*, 1978, **60**, 835–850.

18. A. du Bois, H. J. Luck, W. Meier, H. P. Adams, V. Mobus, S. Costa, T. Bauknecht, B. Richter, M. Warm, W. Schroder, S. Olbricht, U. Nitz, C. Jackisch, G. Emons, U. Wagner, W. Kuhn and J. Pfisterer, A randomized clinical trial of cisplatin/paclitaxel versus carboplatin/paclitaxel as first-line treatment of ovarian cancer, *J. Natl. Cancer Inst.*, 2003, **95**, 1320–1329.

19. J. Cassidy, J. Tabernero, C. Twelves *et al.*, XELOX (capecitabine plus oxaliplatin): active first-line therapy for patients with metastatic colorectal cancer, *J. Clin. Oncol.*, 2004, **22**, 2084–2091.

20. A. Horwich, D. T. Sleijfer, S. D. Fossa *et al.*, Randomized trial of bleomycin, etoposide, and cisplatin compared with bleomycin, etoposide, and carboplatin in good-prognosis metastatic nonseminomatous germ cell cancer: a Multiinstitutional Medical Research Council/European Organization for Research and Treatment of Cancer Trial, *J. Clin. Oncol.*, 1997, **15**, 1844–1852.

21. J. Bellmunt, A. Ribas, N. Eres *et al.*, Carboplatin-based versus cisplatin-based chemotherapy in the treatment of surgically incurable advanced bladder carcinoma, *Cancer*, 1997, **80**, 1966–1972.

22. N. Nishiyama, S. Okazaki, Y. Matsumura *et al.*, Novel cisplatin-incorporated polymeric micelles can eradicate solid tumors in mice, *Cancer Res.*, 2003, **63**, 8977–8983.

23. H. Uchino, Y. Matsumura, T. Negishi *et al.*, Cisplatin-Incorporating Polymeric Micelles (NC-6004) Can Reduce Nephrotoxicity and Neurotoxicity of Cisplatin in Rats, *Br. J. Cancer*, 2005, **93**, 678–687.

24. R. Plummer, R. H. Wilson, H. Calvert *et al.*, A Phase I clinical study of cisplatin-incorporated polymeric micelles (NC-6004) in patients with solid tumours, *Br. J. Cancer*, 2011, **104**, 593–598.

25. Su Wu-Chou, L. Chen, P. L. Chung, J. Chen, Y. Lin, C. S. Pin and Y. Matsumura, Phase I/II study of NC-6004, a novel micellar formulation of cisplatin in combination with gemcitabin in patients with pancreatic cancer in Asia, *ESMO*, 2012, (Abs# 746P).

26. J. M. Llovet and J. Bruix, Systematic review of randomized trials for unresectable hepatocellular carcinoma: Chemoembolization improves survival, *Hepatology*, 2003, **37**, 429–442.

27. L. Biganzoli, Doxorubicin and Paclitaxel Versus Doxorubicin and Cyclophosphamide as First-Line Chemotherapy in Metastatic Breast Cancer: The European Organization for Research and Treatment of Cancer 10961 Multicenter Phase III Trial, *J. Clin. Oncol.*, 2002, **20**, 3114–3121.

28. M. Martin, A. Rodriguez-Lescure, A. Ruiz *et al.*, Randomized phase 3 trial of fluorouracil, epirubicin, and cyclophosphamide alone or followed by Paclitaxel for early breast cancer, *J. Natl. Cancer Inst.*, 2008, **100**, 805–814.

29. H. Roche, P. Fumoleau, M. Spielmann *et al.*, Sequential adjuvant epirubicin-based and docetaxel chemotherapy for node-positive breast cancer patients: the FNCLCC PACS 01 Trial, *J. Clin. Oncol.*, 2006, **24**, 5664–5671.

30. D. Cunningham, N. Starling, S. Rao *et al.*, Capecitabine and oxaliplatin for advanced esophagogastric cancer, *N. Engl. J. Med.*, 2008, **358**, 36–46.

31. S. E. Lipshultz, J. A. Alvarez and R. E. Scully, Anthracycline associated cardiotoxicity in survivors of childhood cancer, *Heart*, 2008, **94**, 525–533.

32. E. de Azambuja, M. Paesmans, M. Beauduin *et al.*, Long-term benefit of high-dose epirubicin in adjuvant chemotherapy for node-positive breast cancer: 15-year efficacy results of the Belgian multicentre study, *J. Clin. Oncol.*, 2009, **27**, 720–725.

33. M. Harada, I. Bobe, H. Saito *et al.*, Improved anti-tumor activity of stabilized anthracycline polymeric micelle formulation, NC-6300, *Cancer Sci.*, 2011, **102**, 192–199.

34. A. Takahashi, Y. Yamamoto, M. Yasunaga, Y. Koga, J. Kuroda, M. Takigahira, M. Harada, H. Saito, T. Hayashi, Y. Kato, T. Kinoshita, N. Ohkohchi *et al.*, NC-6300, an epirubicin-incorporating micelle, extends the antitumor effect and reduces the cardiotoxicity of epirubicin, *Cancer Sci.*, 2013, **104**, 920–925.

35. Y. Matsumura, Cancer stromal targeting (CAST) therapy, *Adv. Drug Delivery Rev.*, 2012, **64**, 710–719.

36. A. Dimou, K. N. Syrigos and M. W. Saif, Overcoming the stromal barrier: technologies to optimize drug delivery in pancreatic cancer, *Ther. Adv. Med. Oncol.*, 2012, **5**, 271–279.

37. M. Yasunaga, M. Furuta, K. Ogata, Y. Koga, Y. Yamamoto, M. Takigahira and Y. Matsumura, The significance of microscopic mass spectrometry with high resolution in the visualisation of drug distribution, *Sci. Rep.*, 2013, **3**, 3050.

38. Y. Matsumura and K. Kataoka, Preclinical and clinical studies of anticancer agent-incorporating polymer micelles, *Cancer Sci.*, 2009, **100**, 572–579.

39. Y. Matsumura, Preclinical and clinical studies of NK012, an SN-38-incorporating polymeric micelles, which is designed based on EPR effect, *Adv. Drug Deliv. Rev.*, 2011, **63**, 184–192.

40. T. Hamaguchi, T. Doi, T. Eguchi-Nakajima, K. Kato, Y. Yamada, Y. Shimada, N. Fuse, A. Ohtsu, S. Matsumoto, M. Takanashi and

Y. Matsumura, Phase I study of NK012, a novel SN-38-incorporating micellar nanoparticle, in adult patients with solid tumors, *Clin. Cancer Res.*, 2010, **16**, 5058–5066.

41. T. E. Nakajima, M. Yasunaga, Y. Kano, F. Koizumi, K. Kato, T. Hamaguchi, Y. Yamada, K. Shirao, Y. Shimada and Y. Matsumura, Synergistic antitumor activity of the novel SN-38-incorporating polymeric micelles, NK012, combined with 5-fluorouracil in a mouse model of colorectal cancer, as compared with that of irinotecan plus 5-fluorouracil, *Int. J. Cancer*, 2008, **122**, 2148–2153.

42. T. Nagano, M. Yasunaga, K. Goto, H. Kenmotsu, Y. Koga, J. Kuroda, Y. Nishimura, T. Sugino, Y. Nishiwaki and Y. Matsumura, Antitumor activity of NK012 combined with cisplatin against small cell lung cancer and intestinal mucosal changes in tumor-bearing mouse after treatment, *Clin. Cancer Res.*, 2009, **15**, 4348–4355.

43. T. Nagano, M. Yasunaga, K. Goto, H. Kenmotsu, Y. Koga, J. Kuroda, Y. Nishimura, T. Sugino, Y. Nishiwaki and Y. Matsumura, Synergistic antitumor activity of the SN-38-incorporating polymeric micelles NK012 with S-1 in a mouse model of non-small cell lung cancer, *Int. J. Cancer*, 2010, **127**, 2699–2706.

44. H. S. Na, Y. K. Lim, Y. I. Jeong, H. S. Lee, Y. J. Lim, M. S. Kang, C. S. Cho and H. C. Lee, Combination antitumor effects of micelle-loaded anticancer drugs in a CT-26 murine colorectal carcinoma model, *Int. J. Pharm.*, 2010, **383**, 192–200.

45. Y. Yamamoto, I. Hyodo, M. Takigahira, Y. Koga, M. Yasunaga, M. Harada, T. Hayashi, Y. Kato and Y. Matsumura, Effect of combined treatment with the epirubicin-incorporating micelles (NC-6300) and 1,2-diaminocyclohexane platinum (II)-incorporating micelles (NC-4016) on a human gastric cancer model, *Int. J. Cancer*, 2014, **135**, 214–223.

46. H. Xiao, W. Li, R. Qi, L. Yan, R. Wang, S. Liu, Y. Zheng, Z. Xie, Y. Huang and X. Jing, Co-delivery of daunomycin and oxaliplatin by biodegradable polymers for safer and more efficacious combination therapy, *J. Control Release*, 2012, **163**, 304–314.

47. E. J. Feldman, J. E. Lancet, J. E. Kolitz, E. K. Ritchie, G. J. Roboz, A. F. List, S. L. Allen, E. Asatiani, L. D. Mayer, C. Swenson and A. C. Louie, First-in-man study of CPX-351: a liposomal carrier containing cytarabine and daunorubicin in a fixed 5:1 molar ratio for the treatment of relapsed and refractory acute myeloid leukemia, *J. Clin. Oncol.*, 2011, **29**, 979–985.

48. P. Parhi, C. Mohanty and S. K. Sahoo, Nanotechnology-based combinational drug delivery: an emerging approach for cancer therapy, *Drug Discovery Today*, 2012, **17**, 1044–1052.

Polymeric Nanoparticles and Cancer: Lessons Learnt from CRLX101

ISMAEL GRITLI,[a] EDWARD GARMEY,[b] SCOTT ELIASOF,[b] ANDRES TELLEZ,[b] MARK E. DAVIS[c] AND YEN YUN*[d]

[a] College of Medical Science and Technology, Taipei Medical University, 250-Wu-xing St. Xinyi District, Taipei 11031, Taiwan; [b] Cerulean Pharma Inc., 840 Memorial Drive, Cambridge, MA 02139, USA; [c] Chemical Engineering, California Institute of Technology, Pasadena, CA 91125, USA; [d] City of Hope National Medical Center, 1500 Duarte Rd, Duarte, CA 91010, USA
*Email: YYen@coh.org

8.1 Introduction

CRLX101 (formerly IT-101) is a novel cyclodextrin-based polymer that self-assembles into nanoparticles. CRLX101 was designed to deliver sustained levels of the cytotoxic agent camptothecin (CPT) to tumour cells while minimizing delivery to normal tissue. Covalent conjugation of hydrophobic CPT to the cyclodextrin polymer increases its water solubility by three orders of magnitude and prevents inactivation through spontaneous lactone ring opening, a process that occurs rapidly at physiologic pH.[1]

CPT derivatives such as irinotecan (CPT-11, Camptosar®; Pfizer) and topotecan (TPT, Hycamtin®; GlaxoSmithKline) demonstrate clinical utility for the treatment of advanced solid tumours. The primary cellular target of

RSC Drug Discovery Series No. 51
Nanomedicines: Design, Delivery and Detection
Edited by Martin Braddock

CPT is the topoisomerase-1 (TOP1)–DNA cleavage complex which, when stabilised, prevents TOP1-mediated unwinding and subsequent DNA repair. Exposure of cancer cells to CPT leads to replication-mediated accumulation of double-stranded DNA breaks and apoptosis. However, the interaction of CPT with DNA is non-covalent and reverses within minutes of drug removal.[2] More recently, CPT and prolonged low-dose TPT have been shown to down-regulate hypoxia-inducible factor 1 (HIF-1α), which is associated with angiogenesis, metastasis, and resistance to vascular endothelial growth factor (VEGF) inhibition. Both mechanisms of action have important pharmacological implications that favour sustained exposure of tumours to active concentrations of these compounds.[3] However, CPT derivatives are associated with considerable toxicity, including diarrhoea and myelosuppression.[4,5]

In preclinical studies, CRLX101 nanoparticles demonstrated reduced rapid renal clearance compared to CPT, resulting in a prolonged plasma half-life compared to small-molecule CPT analogues. Furthermore, in tumour xenograft models, CRLX101 accumulated preferentially in tumour tissue because of the enhanced permeability of tumour neo-vasculature (Figure 8.1). Notably, prolonged release of active CPT from CRLX101 nanoparticles has been observed in tumour tissue resulting in enhanced TOP1 inhibition, potential HIF-1α effect, and increased antitumour activity against multiple human tumour xenografts (*e.g.* lymphoma, breast cancer, ovarian cancer, lung cancer, and colon cancer) compared to CPT-11 and TPT.[6,7]

In the first-in-human phase I clinical trial of CRLX101 that was designed to determine the toxicity, safety profile, and pharmacokinetics of CRLX101, the maximally tolerated dose (MTD) was determined and recommended to test in phase II studies with a weekly or bi-weekly dosing schedule.[8]

In this chapter, we try to position CRLX101 among relevant CPT polymeric nanoparticles and compare their mechanisms of action, chemical and physical properties and physiological outcomes. Thus, the first three sections cover a wide range of nanotherapeutic strategies to target TOP1, HIF-1α, and cancer stem cells. In the fourth section, we highlight the value of combining antiangiogenics with HIF-1α inhibitors. Finally, we cover in detail the preclinical and clinical evaluations of CRLX101 and discuss future directions.

8.2   Topoisomerase 1 Inhibitors

CPT was first semi-synthesised from the Chinese tree, *Camptotheca acuminata* in 1966.[9] CPT is a potent alkaloid cytotoxic agent that inhibits TOP1 enzymatic activity, blocks cell cycle during the S-phase and induces cellular apoptotic death. Under physiological conditions, TOP1 cuts one DNA strand, allowing the DNA to uncoil, and then re-ligates the DNA back into its supercoiled form. However, when CPT is present, it binds to the TOP1–DNA complex and prevents the re-ligation of the DNA, which eventually leads to apoptosis. CPT has low aqueous solubility in the lactone form and can be highly toxic in its carboxylate form.[10] In 1972, a phase II clinical study from

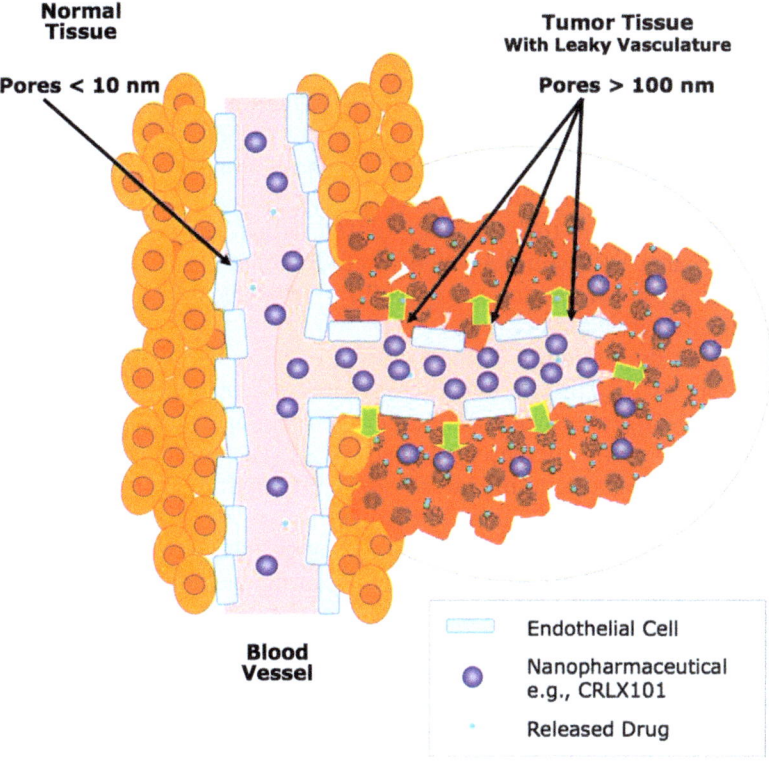

Figure 8.1 CRLX101-mediated dynamic tumour targeting. CRLX101 nanoparticles target tumours *via* leaky vasculature and releases their camptothecin (CPT) payload within tumour cells. This results in profound pharmacokinetic improvement, a sustained drug exposure and low systemic distribution.

the Mayo Clinic evaluated the sodium carboxylate form of CPT in 61 patients, and objective responses after 2 months were observed in only two (3%) patients with advanced gastrointestinal cancer. CPT was administered intravenously at a dose range of 90–180 mg m^{-2} every 21 days. The dose-limiting toxicity was reversible bone marrow suppression (leukopaenia and thrombocytopaenia). Other toxic effects included nausea, vomiting, diarrhoea, stomatitis, dermatitis, alopecia, and haemorrhagic cystitis. Based on the high degree of toxicity and the low response rate, further clinical evaluation of CPT was discontinued.[11]

Hertzberg *et al.*[12] synthesized several CPT derivatives in which the hydroxy lactone ring was modified and evaluated for inhibition of TOP1 and cytotoxicity to mammalian cells. The hydroxy lactone ring of CPT was critical for *in vivo* antitumour activity, consistent with earlier results, and correlated with the structure–activity relationships of TOP1 inhibition. Thus, in retrospect, it is not surprising that all prior clinical trials of CPT, in which the soluble, open ring sodium carboxylate form of CPT (*i.e.* lacking the hydroxy lactone ring) was administered, resulted in poor clinical outcome.[12] A biochemical study demonstrated in 1994 that human serum albumin preferentially binds the carboxylate of CPT with a 150-fold higher affinity than the lactone form. As a result of this interaction, the lactone ring opens rapidly and fully to the carboxylate form in human plasma under physiological conditions. Because earlier studies have shown that the carboxylate form of CPT is clinically inactive, even dosing the active form of CPT (which was not done in the earlier clinical trials) would not have resulted in clinical activity, because CPT would have been rapidly converted to the inactive form upon dosing. Thus, lactone ring stabilization is required to obtain meaningful clinical activity.[13]

Several studies attempted to modify CPT chemical structure in order to lower its toxicity profile and increase its water solubility. Most notably, CPT-11 and TPT are two examples of CPT chemically modified to keep lactone stability and increase water solubility. CPT-11 is a prodrug that is enzymatically converted into active 7-ethyl-10-hydroxy-CPT (SN38), which is 100–1000 times more potent but less water soluble. In Japan, the use of CPT-11 was approved for small cell lung cancer (SCLC) and non-small cell lung cancer (NSCLC), cervical cancer, and ovarian cancer. In Europe, the use CPT-11 was approved for second-line monotherapy and combination with fluoropyrimidines in patients with colon cancer. In the USA, TPT was approved for use in patients with second-line ovarian cancer and SCLC.[14] Water-soluble 7-modified CPT, gimatecan (ST1481) is another example that has been tested in a panel of different tumour types including platinum-resistant cancer cells. Compared to TPT, gimatecan showed superior anti-proliferative activity against many cancer types, except for glioma.[15] In 2008, Franco Zunino and his group at the IRCCS Foundation National Cancer Institute in Milan, Italy synthesised a novel hydrophilic CPT analogue with anti-tumour features, ST1968. To increase CPT solubility, ST1968 was synthesised by oxyiminomethylation on position 7 of CPT. Although ST1968 is less potent than other CPT derivatives such as active SN38, it showed remarkable activity against CPT-11-resistant tumour models and yeast cells transfected with mutant TOP1.[16] In 2008, another study from the same group identified a novel TOP1 inhibitor, topopyrone C, isolated from the fungi *Phoma* and *Penicillium* and successfully synthesized and modified it *in vitro*.[17,18] Cananzi *et al.* described in 2011 a novel analogue of lamellarin D that inhibited TOP1 and exhibited anti-tumour activity against NSCLC cells.[19] Recently, Cincinelli *et al.* synthesized a novel hybrid CPT-platinum anticancer agent. 7-oxyiminomethyl CPT was linked to diaminedichloroplatinum and tested for activity against a panel of cancer cells including TPT- and platinum-resistant types. *In vivo* studies on H460 human NSCLC xenografts showed 82% tumour volume inhibition. This hybrid agent was well tolerated at higher doses than those established for cisplatin but at lower doses than those established for CPT-11.[20]

Novel strategies to modify CPT water solubility and lower its toxicity also include the use of polymeric nanoparticles. In order to achieve a successful nanotherapeutic approach, three criteria have been suggested. (1) Nanoparticles must evade immunorecognition. For example, some approaches use polyethylene glycol (PEG) to mask nanoparticles from the immune system. (2) Nanoparticles must be eliminated after the biological effect has occurred. For instance, biodegradable materials can be used to control cellular uptake and the release rate of the free drug. (3) Nanoparticles must be delivered and accumulated preferentially in target tissue.[21] Besides active targeting that involves certain molecules with specific affinity to tumours, high molecular weight nanoparticles can passively accumulates in tumour tissue due to an enhanced permeation and retention (EPR) effect, and this accumulation can be sustained for a substantial amount of time since tumour tissues severely lack functional lymphatic vasculature.[22]

8.2.1   Carbohydrate-based Polymeric Nanoparticles

Mersana Therapeutics developed a carbohydrate-derived polymeric nanoparticle called XMT-1001, which uses a 40–70 kDa biodegradable hydrophilic polyacetal, poly(1-hydroxymethylethylene hydroxymethylformal) "Fleximer" conjugate to increase the solubility of CPT. XMT-1001 loaded with 7.5 wt% CPT showed improved antineoplastic efficacy against LS174T human colon tumour and A2780 human ovarian tumour xenografts in mice. Although accumulation in tumour tissue was detected, most carrier nanoparticles remained in circulation and free CPT levels were substantially high in liver tissue.[23] In a dose-escalation phase I study of 49 patients with advanced solid tumours (ClinicalTrials.gov identifier NCT00455052), intravenous administration of XMT-1001 at a dose range of 1–85 mg m^{-2} of CPT equivalent showed favourable pharmacokinetics with sustained free CPT plasma levels beyond 24 hours post-administration. Although no haemorrhagic cystitis or diarrhoea was detected, the highest dose caused grade 3 and grade 4 neutropaenia in two patients. 12 patients with refractory tumours had stable disease for more than 12 weeks, most of whom had pancreatic or NSCLC.[24]

Ochi and colleagues at Daiichi Sankyo Company (Japan) described a novel CPT-analogue nanoparticle DE-310 with potent antitumour activity. DE-310 is a carboxymethyldextran polyalcohol (CM-Dex-PA)—a class of polyhydroxylated carboxymethylpolysaccharides, which is linked to a hexacyclic CPT-derivative compound called DX-8951 monomethanesulfonate dehydrate (DX-8951f, also known as exatecan). In the nanoparticle, exatecan payload was linked *via* a Gly-Gly-Phe-Gly tetrapeptide spacer and achieved a loading capacity of 5–7 wt%. The DE-310 nanoparticle has a relatively large molecular size (360 kDa) and thus could passively target tumour tissue through the EPR effect and efficiently delay renal filtration. In a murine malignant fibrosarcoma ascites model, DE-310 showed prolonged survival up to 300% and free DX-8951 could be detected in both serum and ascites fluid. DE-310 seemed to be preferentially taken up by tumour cells than macrophages nearby. One possible explanation for this is the nanoparticle's high hydrophilicity, which could have caused decreased endocytic response of macrophages. The rationale for using the Gly-Gly-Phe-Gly spacer in DE-310 was to provide a sustained release of DX-8951 at a slow rate into the tumour tissue without prematurely releasing the drug into the peripheral circulation. Gly-Gly-Phe-Gly tetrapeptide is predominantly cleaved by cysteine proteases, which are reportedly overexpressed in a wide range of malignant tumours, including colon cancer (cathepsin H), metastatic bone marrow tumours (cathepsin L), and many other malignant tumour types (cathepsin B). Hence, once DE-310 is endocytosed by tumour cells, proteolytic lysosomal activity releases the free drug.[25,26]

T-0128 (MEN4901; Tanabe Seiyaku, Japan) is a 130 kDa polymeric nanoparticle conjugate of a novel CPT analogue (7-ethyl-10-aminopropyloxy-CPT or T-2513) to carboxymethyl dextran (CM-Dex). Unlike DE-310, this non-PA-polysaccharide nanoparticle's Gly-Gly-Gly linker allowed T-0128 to be preferentially taken up by macrophages or macrophage-like tumour cells. Tumour-associated cathepsins were shown to be responsible for the lysosomal drug release.[27] *In vitro* studies showed that T-0128 did not accumulate in malignant tumour cells including highly refractory Walker-256 rat carcinoma and B16 melanoma cells, while macrophage-like cells, such as J774.1, internalized T-0128 very efficiently. Fluorescein isothiocyanate dextran uptake suggested that T-0128 was possibly taken up by macrophages through fluid-phase pinocytosis.[28] Surprisingly, other studies showed that intravenous administration of T-0128 in several xenograft models including Walker-256 carcinoma, MX-1 mammary carcinoma, LX-1 lung carcinoma, St-4 gastric, and HT-29 colorectal tumours showed prolonged survival and tumour growth delay, many of which are highly refractory to CPT-11 and TPT.[29] Pharmacokinetic studies demonstrated that T-0128 could achieve relatively long plasma half-life and efficient tumour targeting. Nevertheless, substantial amounts of T-0128 nanoparticles were taken up by hepatic and splenic tissue and could potentially cause severe toxicities.[30]

8.2.2 Polyamine Polymeric Nanoparticles

Developed by Cell Therapeutics (Seattle, WA, USA), CT-2106 is a poly(L-glutamic acid) conjugate to 20(*S*)-camptothecin (PG-CPT). The 20(*S*)-hydroxyl group of CPT is linked to multiple γ-carboxylate sites on the PG polymer. Using a glycine linker, CPT loading on the polymer achieved up to 50 wt%. The high molecular weight of the PG anionic polymer not only increased CPT solubility, but also improved tumour penetration through the EPR effect and enhanced tumour growth inhibition compared to free CPT. Another feature of PG-CPT is its biodegradability which allows for the slow release of active CPT from nanoparticles once they are endocytosed by target cells.[31] Preclinically, CT-2106 showed potent antitumour activity against a drug-resistant human NSCLC cell line. Intravenous injection resulted in delayed tumour growth and prolonged median overall survival (mOS) up to four-fold in a nude mouse xenograft model. Additionally, CT-2106 showed favourable pharmacokinetics and a slow CPT release rate (1% per day) under pH-neutral conditions.[32] In 2003, phase I studies of CT-2106 in patients with advanced solid tumours were conducted (ClinicalTrials.gov identifier NCT00059917). Endpoints were to determine the MTD, toxicity, pharmacokinetics, and response rate of CT-2106. Patients with melanoma were the predominant population in the study. Fewer than 20% of patients had haematology-related toxicities (mostly anaemia). Although the study showed favourable pharmacokinetics and slower urinary excretion, no objective response (neither partial nor complete) was observed.[33] In 2006, CT-2106 entered phase II clinical trials for the second-line treatment of ovarian cancer. Primary endpoints were to measure response rate and secondary measurements were of toxicity, response duration, time to progression (TTP), and survival (ClinicalTrials.gov identifier NCT00291837).

Dallavalle *et al.* described in 2006 the synthesis of a series of novel CPT-conjugates with a polyamine chain linked to position 7 of CPT *via* an iminomethyl group. The compounds showed TOP1 inhibition and antitumour activity against H460 human NSCLC cells. Moreover, polyamine conjugation to CPT enhanced its solubility and tumour growth inhibitor effects.[34]

Dal Pozzo *et al.* designed novel tumour-targeted RGD (Arg-Gly-Asp) peptide–camptothecin conjugates. RGD cyclopeptides showed high affinity to α_V integrin receptors, which are preferentially overexpressed in tumour cells. By conjugating RGD cyclopeptides to CPT, selective tumour tissue targeting could be anticipated, and thus toxicity reduced. In fact, RGD peptide–CPT conjugates showed high affinity to integrin α_V and internalised into tumour cells overexpressing integrin α_V [A498 renal cell carcinoma (RCC) and A2780 ovarian carcinoma], but not tumour cells with low levels of integrins (PC3 prostate carcinoma). Moreover, tumour cells slowly internalised RGD peptide–CPT resulting in improved drug accumulation compared to free CPT treatment. In an A2780 human ovarian carcinoma xenograft model, RGD peptide–CPT conjugates were injected intraperitoneally or subcutaneously, resulting in 40% tumour growth inhibition and well-tolerated toxicities at a dose of 48 mg kg^{-1}. Free-drug release rate in blood circulation was $\sim 2.5\%$ and high levels of conjugate were detected inside tumours.[35]

8.2.3 HPMA Copolymeric Nanoparticles

HPMA copolymeric nanoparticles have been used to conjugate several cytotoxic drugs including paclitaxel (PNU166945), CPT (PNU166148 or MAG-CPT) and two platinates [AP5280 and AP5346 (ProLindac)].[36] All these nanoparticles share a common *N*-(2-hydroxypropyl)methacrylamide (HPMA) copolymer core. PNU166148 was developed by Pharmacia using HPMA and methacryloyl-glycine-ONp 95 : 5 or 90 : 10, thus the acronym MAG-CPT. CPT was modified at C-20 α-hydroxy group with glycine, which could be hydrolysed by esterase during endocytosis. In order to conjugate HPMA to CPT, a tetrapeptide (Gly-Phe-Leu-Gly) linker was attached, which provided both lactone ring stability in blood circulation and enhanced lysosome-mediated release of free CPT inside cells. Depending on its formulation, MAG-CPT had molecular weight of 20.9–27.8 kDa and CPT loading capacity of 5–10 wt%. When mixed with plasma, MAG-CPT did not aggregate and its interaction with serum albumin was weaker than that of free CPT.[37,38] Pharmacokinetic studies showed significant renal elimination of hydrolysed nanoparticles. The MAG-CPT linker was later modified to Gly-C6-Gly and was selected for dose-escalation phase I studies (ClinicalTrials.gov identifier NCT00004076). The study included 23 patients who received an intravenous infusion over 30 min every 4 weeks. The reported recommended MTD was 200 mg m^{-2}, and dose-limiting toxicities included grade 4 neutropaenia and thrombocytopaenia and grade 3 diarrhoea, and severe and unpredictable cystitis was also observed in few patients. The most pronounced adverse effect was bladder toxicity (dysuria and microscopic and macroscopic haematuria) and could only be resolved by withdrawal of treatment. One treatment-associated death was also reported. Moreover, no objective clinical responses were observed. Due to the severe bladder toxicities associated with MAG-CPT, new formulations attempted to decrease the polymer molecular weight in order to enhance urinary excretion. However, this new formula also meant that free CPT levels became lower in the tumour (12.2 ng g^{-1}) compared to the normal tissue (21.9 ng g^{-1}). Based on these results, clinical development of MAG-CPT was terminated.[39,40]

8.2.4   PEG Polymeric Nanoparticles

The use of PEG in drug delivery dates back to the 1970s. In theory, attaching PEG chains to the drug molecule of interest could achieve improved drug solubility, reduced dosage frequency, reduced toxicity, extended circulating life, increased drug stability, and enhanced protection from proteolytic degradation.[41,42]

Prothecan is a CPT-conjugated polymeric nanoparticle that was developed by Enzon Pharmaceuticals (Plantation, FL, USA) and entered phase II studies in patients with SCLC. In 1996, Greenwald *et al.* conceptualized the importance of increasing water solubility of CPT by conjugating it to PEG in a similar manner to that previously described for paclitaxel. Using an alanine linker, CPT could be conjugated to PEG giving a molecular weight of 40 kDa and dramatically improved solubility (2 mg mL^{-1} in water compared with 0.0025 mg mL^{-1} free CPT in water). Superior efficacy against P388 murine leukaemia and HT-29 human colorectal xenograft models and favourable pharmacokinetics ultimately led to choosing prothecan for phase I clinical studies in patients with solid tumours. Single doses of as much as 7000 mg m^{-2} could be well tolerated and major toxicities were restricted to neutropaenia and leukopaenia. Five out of 14 patients (35.7%) exhibited stable disease and one patient (7%) had a partial response.[43–45]

Enzon's second generation PEGylated CPT, pegamotecan, was developed few years later. Pegamotecan used an alanitate ester to link the hydrophobic CPT to hydrophilic PEG. Preclinical investigations showed superior anti-tumour activity in a human colon cancer xenograft model. Green fluorescent protein-labelled PEG polymers were used to study biodistribution. As expected, high molecular weight PEG passively accumulated in tumour tissue through the EPR effect and induced apoptosis in cancer cells. Importantly, liver and kidney toxicities were minimal compared to free CPT. Change in expression of BAX and Bcl-2, tumour-associated pro- and anti-apoptotic factors, indicated that caspase-dependent apoptotic pathways were responsible for pegamotecan cytotoxicity.[46] To determine the MTD and dose-limiting toxicity of pegamotecan, a phase I dose-escalation study was conducted in 27 patients with advanced solid tumours (ClinicalTrials.gov identifier NCT00080002). At the highest administrated dose $(4300 \, \text{mg m}^{-2})$, one patient had grade 4 gastrointestinal haemorrhage, grade 4 neutropaenia, and grade 3 haematuria and died 4 days later. Additional dose-limiting toxicities included grade 3 toxicities including anaemia, thrombocytopaenia, fatigue, prolonged partial thromboplastic time, haemorrhagic cystitis, dysuria, and urinary frequency in two out of four patients. Haematuria also occurred in eight out of 15 patients at a range of doses. Terminal elimination half-life was at 46 ± 12.8 hours and the MTD was determined to be $3240 \, \text{mg m}^{-2}$ every week for 3–4 weeks. No confirmed objective responses were detected. Due to the unanticipated high toxicity, a previously published dosing schedule with subjects receiving a higher dose in longer cycles $(7000 \, \text{mg m}^{-2}$ every 3 weeks) was advanced instead.[45] In 2009, phase II study results of second-line pegamotecan monotherapy in patients with advanced and metastatic gastric and gastro-oesophageal junction adenocarcinoma were published. Of the 34 patients who received the recommended dose, the partial response rate was 14.3% with a median TTP of 11.9 weeks and mOS at 38.1 weeks. Reported grade 3/4 toxicities included neutropaenia, thrombocytopaenia, fatigue, nausea, vomiting, and anorexia. Compared to CPT-11, pegamotecan had favourable pharmacokinetic and toxicity profiles, which made pegamotecan a good candidate for combination therapy. Although no drug-related mortality was reported in this study, genitourinary toxicities persisted in 35% of patients and microscopic haematuria was observed in 14.2% of patients.[47] In 2004, a safety and efficacy phase II study (ClinicalTrials.gov identifier NCT00079950) was conducted in patients with advanced or metastatic soft tissue.

Enzon also developed a PEG-prodrug that is loaded with SN38 (EZN-2208). Since only 3–4% of an injected dose of CPT-11 is converted to SN38 in the liver, and 55% of unconverted CPT-11 is excreted through renal clearance, it has been hypothesized that PEG conjugation would protect the active SN38 and offer antitumour features superior to CPT-11. Preclinical studies showed that EZN-2208 was active against several *in vivo* tumour models that are resistant to CPT-11 and showed prolonged circulation in the blood and thus longer tumour exposure compared to the free drug.[48,49] Moreover, EZN-2208 exhibited antitumour activity against TPT-resistant malignant tumours.[50] A dose-escalation phase I study was conducted to evaluate the safety, tolerability, pharmacokinetics, and activity of EZN-2208 in 39 patients with advanced malignancies (ClinicalTrials.gov identifiers NCT00520637 and NCT00520390). The nanoparticle was infused with or without granulocyte colony-stimulating factor (G-CSF) for 1 hour once every 21 days. MTD was determined at 16.5 mg m^{-2} with G-CSF infusion and 10 mg m^{-2} without. EZN-2208 was well tolerated and the dose-limiting toxicity was febrile neutropaenia. Adverse effects included fatigue (41%), alopecia (33%), diarrhoea (33%), nausea (33%), neutropaenia (23%), and vomiting (21%). The nanoparticle achieved a terminal half-life of 19.4 ± 3.4 hours. Sixteen patients (41%) achieved stable diesease, including 15% in whom disease was stable for more than 4 months.[51] Subsequently, a multiarm phase II clinical study has been conducted in second-line treatment of metastatic colon cancer (ClinicalTrials.gov identifier NCT00931840). The first arm was infused with EZN-2208 alone (9 mg m^{-2} on days 1, 8 and 15 every 28 days) in 93 patients with KRAS-mutation. The second and third arms included patients with KRAS wild-type tumour randomised to receive 400 mg m^{-2} cetuximab (Erbitux®; Bristol-Myers Squibb/Lilly/Merck) in combination with either 9 mg m^{-2} EZN-2208 (80 patients) or 125 mg m^{-2} CPT-11 (38 patients). In the KRAS-mutation arm, response rate was 0% and median progression-free survival (mPFS) was only 1.8 months. In the KRAS-wild-type arms, cetuximab plus CPT-11 or EZN-2208 showed no significant difference in response rate (10.7% vs. 14.3%), mPFS (4.9 months vs. 3.7 months) or mOS (9.1 months vs. 9.8 months). EZN-2208 was well tolerated in combination with cetuximab. Grade 3/4 adverse events included neutropaenia (30% in EZN-2208 alone, 35% in cetuximab plus EZN-2208, and 16% in cetuximab plus CPT-11); anaemia (10% in EZN-2208 alone, 1% in cetuximab plus EZN-2208, and 13% in cetuximab plus CPT-11); leukopaenia (8% in EZN-2208 alon, 9% in cetuximab plus EZN-2208, and 0% in cetuximab plus CPT-11); and other non-haematological adverse effects.[52] Another phase II clinical study has been conducted using EZN-2208 in patients with metastatic breast cancer (ClinicalTrials.gov identifier NCT01036113).

Developed by Nektar Therapeutics, NKTR-102 (etirinotecan pegol) is a polymeric nanoparticle with a proprietary PEG core conjugated to CPT-11 by a biodegradable linker that is slowly hydrolysed to form active SN38. This allows for a continuous exposure to SN38 and thus a prolonged TOP1 inhibition. NKTR-102 entered phase II clinical studies for colorectal cancer, breast cancer, ovarian cancer, and cervical cancer. Awada et al. at the Jules Bordet Institute, Université Libre de Bruxelles (Brussels, Belgium) recently reviewed results from a randomised phase II study of NKTR-102 in patients with second-line metastatic breast cancer (ClinicalTrials.gov identifier NCT00802945). The study included 70 patients randomly divided into two equal groups, where two dosage schedules (145 mg m^{-2} every 14 days or 21 days) were chosen for treatment. Although results indicated that 29% of patients achieved an objective response, grade 3 or worse adverse events were reported. On the 14-day dosing schedule, drug-related toxicities led to the discontinuation of 20% of subjects and caused two possible drug-related deaths (renal failure and septic shock). Therefore, the less toxic 21-day schedule was chosen instead for a subsequent phase III trial in 840 patients with metastatic breast cancer who previously received an anthracycline, a taxane and capecitabine (Xeloda®; Roche) (BEACON trial,

ClinicalTrials.gov identifier NCT01492101). This phase III has completed and the clinical data did not demonstrate an improvement in overall survival for etirinotecan pegol compared with treatment of physician's choice in patients with pre-treated breast cancer.[53–55,151] Vergote *et al.* recently published results from a phase II clinical trial of NKTR-102 in patients with recurrent platinum-resistant/refractory epithelial ovarian cancer (ClinicalTrials.gov identifier NCT00806156). Two dosage schedules (145 mg m^{-2} every 14 days or 21 days) were given to 71 eligible patients divided into two groups. The overall response rates (ORR) were 20% and 19%; mPFS was 4.1 months and 5.3 months, and mOS was 10.0 months and 11.7 months for the 14-day and 21-day schedules, respectively). Grade 3/4 toxicities included dehydration (24%) and diarrhoea (23%), whereas neutropaenia was less frequent (11.3%).[56] On a side note, the study suggested the use of the TOP1 gene as a biomarker prior to treatment in order to predict sensitively and reduce potential toxicities.[53] Other NKTR-102 clinical studies include an open-label multicenter extension study of NKTR-102 in subjects previously enrolled in NKTR-102 studies (ClinicalTrials.gov identifier NCT01457118); a phase IIa/IIb multicenter open-label study to evaluate NKTR-102 in combination with cetuximab *vs.* irinotecan in combination with cetuximab in second-line colorectal cancer patients (ClinicalTrials.gov identifier NCT00598975); a phase II study in cancer patients with hepatic impairment to evaluate the pharmacokinetics and safety of NKTR-102 (ClinicalTrials.gov identifier NCT01991678); a phase I study of NKTR-102 in bevacizumab-resistant high grade glioma (ClinicalTrials.gov identifier NCT01663012); a phase II study in patients with relapsed SCLC (ClinicalTrials.gov identifier NCT01876446); a new phase II study in patients with second-line, irinotecan-naïve, KRAS-mutant colorectal cancer (ClinicalTrials.gov identifier NCT00856375); and a phase II study in NSCLC (ClinicalTrials.gov identifier NCT01773109).

Other polymeric nanoparticle approaches include CPT-loaded poly-(DL-lactic acid) and poly(ethylene glycol)-*block*-poly(propylene glycol)-*block*-poly(ethylene glycol) copolymer (PEG-PPG-PEG), which has been shown to enhance the anti-tumour efficacy of CPT.[57]

8.2.5 Amphiphilic Polymeric Nanoparticles

In 2004, Zhang *et al.* at Nanjing University in China described a new family of biodegradable copolymers called poly(caprolactone-*co*-lactide)-*b*-PEG-*b*-poly(caprolactone-*co*-lactide) or PCLLA-PEG-PCLLA. This amphiphilic copolymer self-assembles into 70–180 nm nanoparticles that encapsulate a new CPT derivative 10-hydroxycamptothecin-10,20-diisobutyl dicarbonate or HCPT-1. Depending on copolymer composition and preparation conditions, HCPT-1 loading could achieve up to 7 wt% and water solubility increased 3000 times compared to the free drug. When dispersed in PBS, free HCPT-1 was rapidly released (up to 30% was released in the first hour). After 8 hours, HCPT-1 release steadily slowed down. The fast release could be due to the large surface area of the nanoparticle or the low T_g of the hydrophobic core of PCLLA-PEG-PCLLA nanoparticles, causing high mobilization of HCPT-1. However, biodistribution analysis showed that HCPT-1 blood concentration was relatively stable after 24 hours of intravenous injection. The highest levels of HCPT-1 accumulation were detected in mouse lung and spleen tissue and low levels were also detected in the kidney and liver. More preclinical studies are needed to verify the pharmacodynamics of this drug.[58,59]

Another interesting approach was taken by a 2010 Korean study at Kyung Hee University using pH-sensitive amphiphilic nanoparticles to encapsulate CPT. The pH-responsive polymeric micelle consisted of hydrophilic methyl ether PEG and pH-responsive poly(β-amino ester) that copolymerized, self-assembled and encapsulated CPT at neutral conditions (pH 7.4). The nano-sized (214 nm) particle depolymerizes in response to the tumoural acidic conditions (pH 6.4). Amphiphilic nanoparticles with an optimal drug loading of 10 wt% were shown to passively target breast tumour xenografts and release CPT load in response to a slightly acidic microenvironment. Importantly, antitumour activity was detected and >70% of CPT in the lactone ring form could be maintained for 24 hours. Using fluorescence imaging, the pH-PM biodistribution result clearly showed substantial tumour accumulation and an overall favourable pharmacokinetic profile.[60]

8.2.6   Bioconjugates

Minko *et al.* at Rutgers (USA) studied the cytotoxicity and antitumour efficacy of biotinylated PEG-conjugates to CPT. Biotin was included as a moiety to increase intestinal absorption *via* sodium-dependent multivitamin transporters. In this way CPT cytotoxicity increased up to 60 times in sensitive A2780 and 30 times in A2780/AD multidrug-resistant ovarian cancer cells. *In vitro* toxicity studies showed that PEG–biotin alone did not cause any cell toxicity and CPT–PEG–biotin induced caspase-dependent apoptosis.[61]

Another novel bioconjugate was described in 2004 used a PEG backbone conjugated to CPT *via* a glycine linker, and to folic acid *via* either of its two available carboxylate groups on the distal end. This double conjugation served two purposes. First, CPT water solubility was enhanced without destabilizing its lactone structure. Second, the folic acid could hence mediate tumour-specific endocytosis through the folate receptor, which is overexpressed in human nasopharyngeal carcinoma, for example. Unlike large nanoparticles, the CPT-Gly-PEG-folate conjugate has low molecular weight and thus less immunogenic.[62]

Another CPT-bioconjugate approach used substance P, an 11-residue neuropeptide, to conjugate to CPT to specifically target tumour cells that overexpress neurokinin-1 receptor. This proof-of-concept approach showed selective and strong cytotoxicity across many malignant tumour types, including breast cancer, astrocytoma and glioblastoma. Since substance P is a short peptide, it can be chemically synthesised. Another feature of substance P is that only C-terminal amino acids are essential for neurokinin-1 receptor-mediated binding and cellular internalization, while the N-terminal could be used for CPT conjugation. Similar to other linkers, cysteine amino acid on the N-terminal allowed for proteolytic release of CPT, and amino-hexanoic acid was used as a spacer to avoid steric hindrance. However, CPT solubility improved only about 38 times, and when loading of CPT increased, the conjugate solubility worsened. Moreover, both cellular internalization and cytotoxicity could improve when a spacer was used. More preclinical studies are needed to verify the pharmacokinetic and pharmacodynamic feature of this novel drug. Moreover, new formulations should be used to improve its stability, solubility, and cytotoxicity.[63]

Vekhoff *et al.* described a new generation of CPT conjugates with sequence-specific DNA ligands that enhanced CPT positioning to its binding site on the DNA of target cells. This could be achieved by conjugating a bromoalkyl analogue of gimatecan to the $3'$ end of triplex-forming oligo-nucleotides (TFO). A TFO–gimatecan conjugate proved to be both stable and potent against HeLa cells *in vitro.*[64]

Other bioconjugate approaches include hyaluronic acid-conjugated CPT-11 (Meditech Research);[65] transferrin–PEG nanoparticles;[66] and camp-tothecin coupled to bombesin analogue.[67]

8.2.7   Non-polymeric Nanoparticles

Non-polymeric nanoparticles are listed and reviewed elsewhere. They include IHL-305 CPT-11 liposomes (Yakult Honsha, Japan);[68] CPX-1 irinotecan-floxuridine liposome;[69] STEALTH liposomal S-CKD-602;[70] SN38 liposomes from Neopharm;[71] and NK012 SN38 micelle.[72]

8.3   Hypoxia Inducible Factor-1 Inhibitors

HIF-1 is a transcription factor that plays an essential role in the cellular hypoxic response. When activated, its two subunits, HIF-1α and HIF-1β, dimerise and enhance the expression of survival genes.[73] Genetic analysis revealed that more than 60 putative genes could be direct targets of HIF-1 activation. These target genes fall largely into four categories: angiogenesis, glycolysis, tumour survival, and invasion. Within hypoxic tumour tissue, HIF-1 activation leads to the transcription of target genes responsible for pathophysiological phenotypes such as immortalization, stem cell-like features, dedifferentiation, neo-angiogenesis, invasion, and metastasis. In cancer therapy, HIF-1α overexpression has been linked to antiangiogenic-associated drug resistance and poor prognosis across several tumour types including breast cancer, ovarian cancer, and gastric cancer.[74–76] Moreover, an important indicator of HIF-1 abnormal activity, carbonic anhydrase-IX (CAIX), is up-regulated in many tumour types and linked to poor prognosis.[77–82]

8.3.1   2ME2

2-methoxyestradiol (2ME2; Panzem), is a microtubule polymerization inhibitor that has been shown to interact with HIF-1α and block angiogenesis.[83] In a phase II safety study of orally administrated 2ME2 in patients with homrmone-refractory prostate cancer, the drug was well tolerated in general. However, since pharmacokinetic results also showed low sustainability,[84] 2ME2 was reformulated as a NanoCrystal® colloidal dispersion (NCD). In a phase II study of 2ME2 NCD in 21 patients with taxane-refractory, metastatic castrate-resistant prostate cancer (ClinicalTrials.org identifier NCT00394810), only 5.35% reached PFS at 6 months, while the mPFS was 56 days and there was no objective response. Grade 3/4 adverse events were observed in 33% of patients. However, compared to the free drug, 2ME2 NCD showed favourable pharmacokinetics with 10–20 times higher levels of 2ME2 and its metabolite 2ME1 in plasma.[85] Additionally, a multicenter phase II study was conducted (ClinicalTrials.org identifier NCT00444314) of 2ME2 NCD alone with its combination with sunitinib (Sutent®; Pfizer) in second-line metastatic RCC. Only 12 patients were evaluable in the trial and the study was terminated due to the drug's minimal antitumour activity and high intolerability. New formulations are currently under development.[86] Other clinical trials with 2ME2 NCD include a phase II study in combination of with bevacizumab (Avastin®; Genentech/Roche) in patients with metastatic carcinoid tumours (ClinicalTrials.org identifier NCT00328497); a phase II study in patients with relapsed or plateau-phase multiple myeloma (ClinicalTrials.org identifier NCT00592579); a phase II study in patients with glioblastoma (ClinicalTrials.org identifier NCT00306618; a phase II study in patients with ovarian cancer (ClinicalTrials.org identifier NCT00400348); and a phase II study in combination with the chemotherapeutic drug temozolomide (Temodar®; Merck) in patients with recurrent glioblastoma multiforme (ClinicalTrials.org identifier NCT00481455).

8.3.2 Camptothecins

CPT and TPT are also known to inhibit HIF-1α expression in addition to the TOP1 inhibitory effects described above. CPT derivatives inhibit HIF-1α expression when administered daily at low nanomolar concentrations.[87] Additionally, the presence of CPT under hypoxic conditions have also been shown to down-regulate VEGF expression.[88]

Little is known about how CPT inhibits HIF-1α. A recent paper by Giovanni Capranico and colleagues at the University of Bologna (Italy), suggested a molecular mechanism by which CPT inhibits HIF-1α activity.[89] Using microarray analysis, two micro-RNAs (miR-17-5p and miR-155) were identified as CPT-associated suppressors of HIF-1α protein expression. Moreover, it has been suggested that TPT acts on HIF-1α through a signalling pathway downstream of TOP1.[87]

Several clinical studies have been conducted to test the therapeutic potential of using CPT derivatives to inhibit HIF-1 in tumour tissue. A phase I study in which oral TPT was administrated in combination with radiotherapy in patients with rectal cancer was conducted in September 2005 (ClinicalTrials.org identifier NCT00215956). A phase I pharmacokinetic study of weekly low doses of TPT administrated intravenously in combination with carboplatin and etoposide in patients with extensive stage SCLC was conducted in August 2006 (ClinicalTrials.org identifier NCT00025272).[90] Another phase I dose-escalation pharmacokinetic study of TPT in combination with erlotinib (Tarceva®; Genentech/Roche) in patients with refractory solid tumours was conducted in January 2008 (ClinicalTrials.org identifier NCT00611468).

8.3.3 siRNA Technologies

Using locked nucleic acid technology, Enzon Pharmaceuticals developed a specific HIF-1α mRNA antagonist (EZN-2968) with durable and potent antitumour effects against human prostate and glioblastoma cell lines.[91] A phase I pilot study of EZN-2968 in patients with advanced solid tumours with liver metastases has been conducted in order to determine the safety and potency of EZN-2968 (ClinicalTrials.org identifier NCT01120288).

Liu *et al.* from the University of Science and Technology of China recently developed a micellar nanoparticle containing HIF-1α short interfering (si)RNA, which could block hypoxic tumour growth in prostate cancer xenograft models.[92]

8.3.4 Endogenous HIF-1α Inhibitors

In addition to HIF-1α, two other family members (HIF-2α and HIF-3α) have been identified which alternatively dimerise with HIF-1β and regulate the hypoxic response.[93,94] Other HIF-1α endogenous regulators include the von Hippel–Lindau (VHL) tumour suppressor protein, which facilitates its degradation through the ubiquitin pathway.[95] Other endogenous inhibitors of HIF-1α are listed in Bao *et al.*[96] They include oxygen-sensitive factor-inhibition HIF (FIH); p53; phosphatase and tensin homologue (PTEN) and liver kinase B1 (LKB1). Recent studies showed the role of miRNAs in regulating HIF-1α expression. miR-21 overexpression is associated with high levels of HIF-1α and VEGF. Hypoxia leads to increased expression of miR-21 in pancreatic and prostate cancer cells. Knocking down miR-21 blocks VEGF expression and the self-renewal capacity of cancer stem cells. miR-107 negatively regulates HIF-1α and VEGF expression under hypoxic conditions. miR-210 up-regulates VEGF and CAIX in pancreatic cancer through HIF-1α. miR-373 is positively associated with HIF-1α in hypoxic cells. miR-20 inhibits VEGF and expression through HIF-1α in hypoxic breast cancer cells. miR-22 has anti-angiogenic effects through down-regulation of HIF-1α and VEGF expression in tumour tissue. miR-101 expression suppresses the self-renewal capacity of cancer stem cells.[96]

8.3.5 Other HIF-1α Inhibitors

Due to the key role that HIF-1 seems to play in tumour progression, several clinical and preclinical approaches have been taken to pharmaceutically modulate HIF-1 pathways or its upstream effectors. Comprehensive reviews of HIF-1 inhibitors can be found in Semenza[74] and Xia *et al.*,[97] but we list some here. Developed by Bayer (Germany), BAY87-2243 is a potent HIF-1α and HIF-2α inhibitor with antitumour activity.[98] A phase I clinical study conducted to evaluate the MTD of BAY87-2243 in patients with advanced malignancies (ClinicalTrials.org identifier NCT01297530) was terminated due to unanticipated high toxicities.[99] Elara Pharmaceuticals (Germany) has developed ELR510444, which is a novel HIF-1α and HIF-2α inhibitor that reduced RCC viability in a VHL-independent manner. ELR510444 induced necrosis, apoptosis, and inhibited angiogenesis in 786-O and A498 RCC xenograft models.[100] Sorafenib (Bay4-9006, Nexavar®; Bayer/Onyx) is a RAF kinase inhibitor that is approved for hepatocellular carcinoma, RCC, and thyroid cancer. Studies also showed that sorafenib had inhibitory effects on HIF-1 activity.[101,102] Temsirolimus (CCI-779, Torisel®; Pfizer) is an mTOR inhibitor that is approved for RCC. Temsirolimus also inhibited HIF-1α in human lung cancer cell lines under hypoxic conditions.[103] A phase I/II clinical trial (Clinical-Trials.gov identifier NCT00329719) is currently under way to evaluate the synergistic effects of combining temsirolimus with sorafenib. Geldanamycin, an HSP-90-specific inhibitor, has been shown to promote HIF-1α degradation under both hypoxic and normoxic conditions.[104] Tanespimycin (17-AAG), a derivative of geldanamycin with similar HIF-1α inhibitory properties is currently being evaluated as both monotherapy and combination therapy with anti-angiogenic agents in patients with multiple myeloma and breast cancer.[105]

Many other small molecules that target HIF-1α are currently tested preclinically: the Trx-1 inhibitor PX-12 was reported to inhibit HIF-1 activity;[106] PD98059 is an MEK inhibitor shown to inhibit HIF-1 activity preclinically;[107] NSC-134754 is a novel small molecule that inhibits HIF-1α activity and thus inhibits tumour progression.[108,109] Other inhibitors include pleurotin (1-methylpropyl 2-imidazolyl disulphide), a thioredoxin-1 inhibitor;[106] YC-1 [3-(5′-hydroxymethyl-2′-furyl)-1-benzyl indazole], a synthetic compound with HIF-1α-inhibitory activity;[110,111] heteroarylsulfonamides (3,4-dimethoxy-*N*-[(2,2-dimethyl-2*H*-chromen-6-yl)methyl]-*N*-phenylbenzenesulfonamide), a novel HIF-1α inhibitor tested against human glioma cells;[112] and cyclic peptide inhibitors of HIF-1.[113]

8.4   Cancer Stem Cells

The importance of targeting HIF-1 signalling pathways partly comes from its role in regulating cancer stem cells and its role during epithelial-mesenchymal transition. Although induced hypoxia has anti-tumour effects, it enhances the self-renewal capacity and maintenance of the dedifferentiated state of cancer stem cells. Surface marker studies confirmed that hypoxia induces stem cell-like phenotypes and thus leads to tumour aggressiveness and poor prognosis.[96] Conversely, cancer stem cells use induced hypoxia to protect from DNA damage by reduction of reactive oxygen species. This can be achieved by activating HIF-1α, which is preferentially expressed in cancer stem cells compared with other cancer cells in the same tissue. *In vivo* knockdown studies showed that both HIF-1α and HIF-2α are required for cancer stem cell survival and tumour propagation. In fact, knocking down these factors led to reduced tumour formation or inhibited VEGF-mediated angiogenesis.[114]

Other evidence of the role of HIF-1 in cancer stem cell regulation is covered in a review by Sarkar and colleagues at Wayne State University (Detroit, MI, USA).[96] In brief, activation of HIF-1 enhances the expression of stem cell-associated genes such as Oct4, Nanog, c-Myc, Notch-1, and CD133 pathways. Moreover, factors downstream of HIF-1 such as CAIX can regulate the acidic microenvironment of surrounding cancer stem cells. Clinically, both HIF-1α and CAIX up-regulation have been associated with poor overall and relapse-free survival.[96]

A paper by Wicha and colleagues at the University of Michigan Comprehensive Cancer Center (Ann Arbor, MI, USA) demonstrated how antiangiogenic therapy with either bevacizumab or sunitinib could increase cancer stem cells in orthotropic breast xenograft models by making tumours more hypoxic, thereby activating HIF-1α. These findings suggest that combining antiangiogenic therapy with inhibitors of HIF-1α will be an improvement over antiangiogenic therapy alone.[115]

Ambasta *et al.* described a novel technology that uses twin nanoparticles to target both cancer stem cells and adjacent endothelial cells at the same time.[116]

8.5 Combination Therapy

VEGF overexpression under hypoxic conditions was first reported in 1992.[117] A few years later, HIF-1 activation was linked to VEGF tumour progression.[118,119] A recent review by Rapisarda and Melillo at the National Cancer Institute (NCI; Bethesda, MD, USA) illustrated how combining antiangiogenic drugs with drugs that target HIF-1α is likely to be mechanistically synergistic.[120] Since anti-angiogenic treatment eventually induces hypoxia, it will up-regulate HIF-1α, which in turn will promote tumour survival by turning on pathways involving angiogenesis, cell metabolism, proliferation, differentiation, and cancer invasion and metastasis. Therefore, VEGF inhibitors undo their own effects, so combining antiangiogenic therapy such as bevacizumab with inhibitors of HIF-1α could have a synergistic effect.

For example, a combination of bevacizumab with metronomic TPT overcame the increase in hypoxia and HIF-1α induced by bevacizumab alone. This strategy provided superior anti-angiogenic effects, which also resulted in superior anti-tumour activity in a glioblastoma xenograft model.[3,121] Another remarkable preclinical study was performed by Kerbel and colleagues at the University of Toronto. They demonstrated that combining pazopanib (Votrient®; GlaxoSmithKline), a VEGF receptor inhibitor, with metronomic TPT, is synergistic and resulted in 100% survival of the drug combination (but not the individual monotherapies) after 6 months of continuous therapy in an aggressive, metastatic, orthotropic xenograft ovarian model.[122]

Two orthotropic ovarian cancer model studies showed that combining pazopanib with metronomic TPT is synergistic and resulted in a significant improvement in survival *vs.* the individual monotherapies.[123] Burkitt *et al.* at the University of Michigan Comprehensive Cancer Center demonstrated that genetic disruption of both HIF-1α and HIF-2α had synergistic anti-tumour activity when combined with sunitinib in a colorectal xenograft model.[124]

Clinically, Kummar *et al.* at the NCI demonstrated that metronomic dosing of TPT inhibited HIF-1α in patients, and that correlated with anti-angiogenics as well, as measured by decreased tumour expression of VEGF and decreased tumour blood flow and permeability. These data validate extensive preclinical data showing that metronomic TOP1 inhibition can inhibit HIF-1α and important downstream targets of HIF-1.[125] Recently, a multicenter phase II trial combining TPT with cisplatin and bevacizumab in recurrent cervical cancer was published (ClinicalTrials.org identifier NCT00548418). Although the study significantly improved mPFS (7.1 months) and mOS (13.2 months), grade 3/4 toxicity was common (thrombocytopaenia 82%, leukopaenia 74%, and anaemia 63%) and most patients (78%) required unanticipated hospital admission.[126]

The SN38-bearing nanoparticle EZN-2208 (described in Section 8.2.4) has been shown to inhibit HIF-1α and its downstream targets such as VEGF1 in a U251 glioma xenograft model.[127] The nanoparticle is currently evaluated for potential HIF-1 inhibition and anti-tumour effects in combination with bevacizumab in patients with refractory tumours (ClinicalTrials.org identifier NCT01251926).

Combining radiotherapy with HIF-1α inhibition is another frontier under investigation. Inhibition and knockdown of both HIF-1α and HIF-2α increased tumour sensitivity to radiotherapy.[96] A meta-analysis study of more than 4800 patients connected hypoxic modification of radiotherapy and improved therapeutic benefit in patients with head and neck cancer.[128]

Janssen *et al.* synthesized an RGD-peptide (described in Section 8.2.2) labeled with radioactive Yttrium-90 (^{90}Y) to target the $\alpha_v\beta_3$ integrins on the endothelium associated with OVCAR-3 ovarian tumour xenografts. The radiolabeled peptide showed effective tumour tissue penetration and delayed tumour growth.[129] A similar approach used RGD-peptide labelled with ^{90}Y and Indium-111 (^{111}In).[130] Other studies demonstrated that HPMA-based nanoparticles (described in Section 8.2.3) could be tagged with ^{90}Y and conjugated to RGD-peptides.[131] A combination of short-range and long-range particle emitters has been suggested. In this way, emitted alpha particles may have a stronger impact on angiogenesis-associated endothelial cells, while emitted beta particles are ideal for solid tumour irradiation.[132]

8.6   CRLX101

CRLX101 (Cerulean Pharma Inc., Cambridge, MA, USA) is a nanopharma-ceutical composed of 20-(S)-CPT conjugated to a linear, cyclodextrin-PEG-based polymer (CDP). The polymer–drug conjugate self-assembles into soluble nanoparticles composed of several conjugate strands when dissolved in aqueous solution. The clinical development of CRLX101 is focused on two unique properties of this nanoparticle, including the ability to increase the circulatory half-life of CPT and the ability to enhance distribution of CPT into tumour tissue for prolonged periods with a low plasma concentration of free CPT.[1] Given these unique properties, it was hypothesised that CRLX101 could enhance antitumour activity while avoiding the toxicity observed with traditional TOP1 inhibitors.

8.6.1 CRLX101 Chemistry

In order to use CRLX101 safely in clinical settings, the *de novo* design included key features addressing biocompatibility and low toxicity. The backbone of this nanoparticle contains β-cyclodextrin (β-CD), which is a cyclic oligosaccharide consisting of (α-1,4)-linked-α-D-glucopyranose. β-CD is copolymerised with difunctionalised polyethylene glycol (PEG_{3400}) in order to give a high molecular weight (>50 kDa) polymer. Copolymerising β-CD with PEG_{3400} serves two main purposes. First, the high molecular weight of PEG_{3400} creates a relatively long distance between reactive groups and thus minimises the steric hindrance between β-CD molecules in the construction of polymer chains. Second, PEG_{3400} plays a key role in avoiding immunorecognition. The large size of CRLX101 also does not allow for first-pass renal clearance, and thus high levels of circulating nanoparticles are maintained. The copolymer–CPT conjugate self-assembles into nanoparticles (30–40 nm) and when dispersed into neutral PBS buffer, their zeta potential is *ca.* −2mV. Importantly, CPT solubility increased 1000-fold after conjugation. Although CDP nanoparticles are not biodegradable, they disassemble into small (∼8 nm) individual CD polymer strands as soon as CPT is released, allowing renal clearance to slowly take place. This unique feature of CRLX101 grants favourable pharmacokinetics with minimal a toxicity profile (Figure 8.2).[1,133,134]

Lactone ring stabilisation is a key feature of CRLX101, since CPT is conjugated to the polymer in such a way as to protect the lactone ring from being opened up. CPT is functionalised to form an ester bond at the 20-hydroxyl group and a glycine linker was used to provide optimal antitumour efficacy with low toxicity. CPT is conjugated to the carboxylate group on the linear CD polymer while the lactone ring is maintained (Figure 8.3).[6,135]

8.6.2 CRLX101 Preclinical Results

Raubitschek *et al.* at the City of Hope Comprehensive Cancer Center (Duarte, CA, USA) demonstrated that intravenous administration of CRLX101 was tolerable and effective against lymphoma models tested, resulting in significantly prolonged survival of animals compared to CPT-11. The superior efficacy of CRLX101 over CPT-11 was demonstrated to result in part in

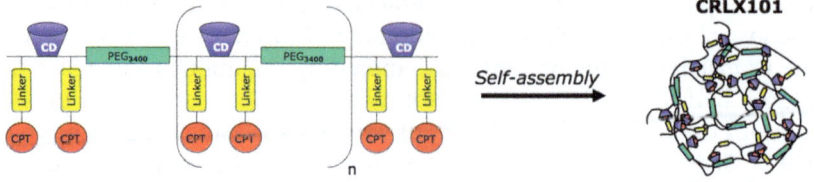

CRLX101

20-30nm in diameter
4-5 strands form one nanoparticle

Camptothecin covalently attached to polymer strands - NCE

Figure 8.2 Polymer self-assembly of CRLX101 nanoparticles. CRLX101 consists of a potent topoisomerase 1 (TOP1) inhibitor, camptothecin (CPT), conjugated to a poly(ethylene) glycol (PEG)–cyclodextrin copolymer *via* an ester bond using a glycine linker. CRLX101 self-assembles into ~30 nm nanoparticles from 4–5 strands of CPT–copolymer conjugate. Strands are held together *via* guest–host complexes between CPT and cyclodextrin. PEG provides "stealth" against immune recognition, resulting in prolonged circulation of CRLX101. The nanoparticles are large enough to avoid rapid kidney filtration, but small enough to penetrate deep into tumours. CPT is released gradually inside tumour cells, resulting in sustained inhibition of TOP1 and hypoxia-inducible factor (HIF)-1α. CPT is released *via* chemical hydrolysis; strands come apart and are excreted primarily *via* the kidney. NCE = new chemical entity.

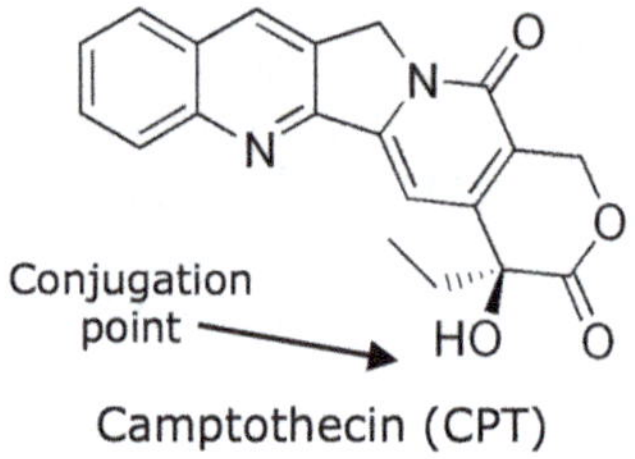

Conjugation
point →

Camptothecin (CPT)

Figure 8.3 Camptothecin chemical structure.

sustained and prolonged concentrations of active drug in lymphoma tumours, which was not achievable with CPT-11.[136]

Davis *et al.* at California Institute of Technology (Pasadena, CA, USA), demonstrated that CRLX101 has excellent tolerability and anticancer effectiveness in various solid tumour xenografts (NSCLC, SCLC, breast cancer, colon cancer, pancreatic cancer and Ewing's sarcoma). In every case, CRLX101 was demonstrated to be superior to the MTD of CPT-11, and in some models resulted in complete regressions. Three out of the seven models tested were resistant to CPT-11, and striking efficacy of CRLX101 was still observed in these models, demonstrating that CRLX101 can work even in resistant models.[137] In a pharmacokinetic and biodistribution study performed by the same group, intravenous administration of CRLX101 in rats and tumour-bearing mice resulted in prolonged plasma half-life and enhanced distribution to tumour tissue when compared to CPT alone. Moreover, the tumour concentration of active CPT was 160-fold higher after administration of polymer-bound CPT compared to the administration CPT alone.

These effects likely play a significant role in the enhanced antitumour activity of CRLX101 when compared to CPT alone or CPT-11.[138]

Notably, CRLX101 tumour growth inhibition of squamous and non-squamous NSCLC was superior to approved therapeutic agents such as TPT, docetaxel (Taxotere®; Aventis), erlotinib (Tarceva®), gemcitabine (Gemzar®; Lilly) and carboplatin (Paraplatin®; Bristol-Myers Squibb). Thanks to its sustained levels within penetrated tumour tissue, CRLX101 inhibited cell proliferation uniformly. Moreover, CRLX101 effectively inhibited tumour growth and increased survival in both KRAS-driven and epithelial growth factor receptor (EGFR)-driven mutation models suggesting that CRLX101 can be effective against drug-resistant NSCLC tumours.[139]

Furthermore, multi-organ pharmacokinetics and accumulation in tumour tissue of CRLX101 were investigated using positron emission tomography in tumour-bearing mice. Tumour vascular permeability was calculated and the data indicated that the majority of nanoparticles stay intact in circulation and do not disassemble into individual polymer strands. Histological measurements using confocal microscopy and transmission electron microscopy (Figure 8.4) showed that CRLX101 localises within tumour cells and provides the sink in the tumour for the nanoparticles.[140] Similar results using serial diffusion magnetic resonance imaging showed a remarkable lymphoma tissue penetration and early treatment response in mice treated with CRLX101 compared to free CPT-11.[141]

In collaboration with Max Wicha at the University of Michigan and others at the University of Pennsylvania and Massachusetts General Hospital,

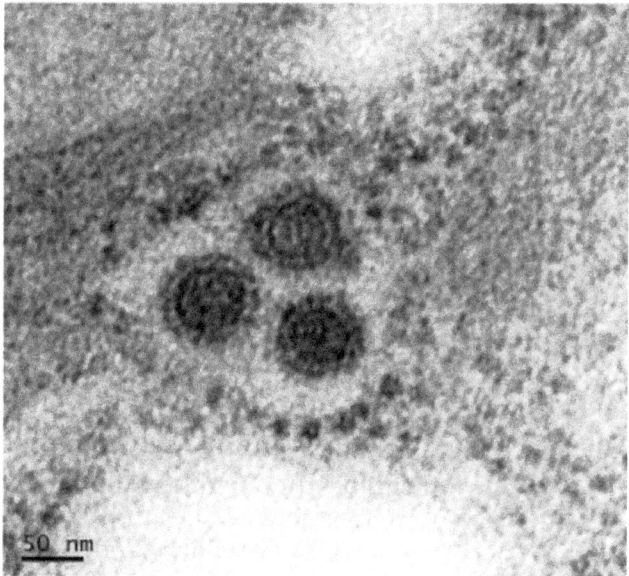

Figure 8.4   Transmission electron microscope image of CRLX101 nanoparticles in a xenograft tumour cell.

Eliasof *et al.* at Cerulean Pharma Inc. demonstrated recently that the combination of CRLX101 with antiangiogenic therapeutics have synergistic antitumour properties. CRLX101, through sustained release of its CPT payload, showed a durable HIF-1α inhibition in multiple human tumour xenograft models including highly metastatic ovarian tumours. These promising results encouraged investigators to study the effects of combining CRLX101 with conventional antiangiogenics. As expected, elevated levels of HIF-1α associated with bevacizumab, aflibercept (Zaltrap®; Sanofi) and pazopanib treatments were significantly suppressed when combined with CRLX101. HIF-1α inhibition by CRLX101 was not only sustained for 2 weeks, but substantial tumour growth inhibition was also achieved. Similar synergistic effects were observed with radiation therapy in a head and neck xenograft tumour model. Moreover, confocal microscopic imaging showed that HIF-1α was translocated into the nucleus after CRLX101 treatment. At the cancer stem cell level, CRLX101 prevented the bevacizumab-induced HIF-1α expression and inhibited tumour regrowth of triple negative breast cancer cells implicating an effective targeting of cancer stem cells.[7]

8.6.3   CRLX101 Clinical Results

In an initial phase I/IIa dose-finding, safety, and pharmacokinetic study of CRLX101 in patients with advanced solid tumours, a total of 62 patients were enrolled and treated (ClinicalTrials.gov identifier NCT00333502). The MTD was determined to be 15 mg m^{-2} on days 1 and 15 of a 28-day cycle. Haematologic toxicity, including neutropaenia, thrombocytopaenia, anaemia, and leucopaenia, was dose-limiting at 18 mg m^{-2} administered on a weekly schedule, and thrombocytopaenia and neutropaenia were also dose-limiting at 18 mg m^{-2} on an every-other-week schedule. In addition, cystitis, which is an expected toxicity for this drug class, was dose-limiting in a weekly dosing schedule evaluated in phase I. Hydration before and following CRLX101 administration was added to the protocol to mitigate the potential for cystitis. In the phase IIa expansion, 38 patients were enrolled and treated at the established MTD and the MTD appeared to be safe and well tolerated for up to six or more treatment cycles, with the majority of adverse events being grade 1 or 2. The most common of these were fatigue, cystitis, nausea, anaemia, dysuria, haematuria, and neutropaenia. There were no treatment-related deaths, and eight patients, including five treated at the MTD, experienced treatment-related serious adverse events. Three patients experienced serious infusion-related hypersensitivity reactions, leading to amendment of the protocol to include pretreatment with a corticosteroid, an antihistamine, and an H2 antagonist to minimise infusion-related hypersensitivity reactions. Following intravenous administration of CRLX101 at 6, 12, 15, and 18 mg m^{-2} in study CRLX-001, systemic plasma exposure to both polymer-conjugated and unconjugated CPT was observed in all subjects. Pharmacokinetic analysis demonstrated exposure that is generally proportional to dose and prolonged plasma exposure that is heavily weighted to nanoparticle-conjugated CPT. CRLX101 plasma concentrations increased sharply following intravenous infusion, whereas unconjugated CPT plasma concentrations increased gradually, consistent with slow release of CPT from the polymer conjugate. Clearance and volume of distribution values for the conjugated CPT were dose-independent and suggest this material was retained within the vasculature and highly perfused tissues. The majority of subjects receiving CRLX101 every other week at 15 mg m^{-2} or 18 mg m^{-2} had measureable levels of unconjugated CPT in plasma at 14 days post-CRLX101 administration, indicating that in order to avoid significant carry-over of unconjugated plasma CPT from one dose to the next, a dosing interval greater than 1 week is required. Therefore, all subsequent clinical studies were performed with an every-other-week schedule.[8,142,143]

A phase II monotherapy evaluation of CRLX101 include a multi-site, multi-country comparison of CRLX101 to best supportive care in NSCLC. Enrolment in this trial is complete and patients remain on study (ClinicalTrials.gov identifier NCT01380769). In March 2013, Cerulean Pharma announced that the phase II NSCLC clinical study did not meet its primary efficacy endpoints and overall survival benefits.[144] Multiple phase II evaluations of CRLX101 are still active to date.

8.7   Conclusion

Preclinical testing of CRLX101 in monotherapy in multiple xenograft models (NSCLC, squamous NSCLC, KRAS mutant NSCLC, EGFR mutant NSCLC, SCLC, ovarian cancer, colorectal cancer, gastric cancer, CPT-11-resistant breast cancer, RCC, pancreatic cancer, and lymphoma) has demonstrated superiority of CRLX101 over every commercial comparator drug tested. In these models, the commercial drugs were administered at their respective optimal dose and dosing schedule and measured for median survival, tumour shrinkage, and tumour growth delay. In many models CRLX101 achieved complete tumour eradication and cure, unattainable with the commercial drugs tested in these models. CRLX101 has also been shown to be a radio-sensitiser in an animal model of head and neck cancer, suggesting that CRLX101 may achieve clinical utility in combination with radiotherapy.[6,134]

It has further been demonstrated in preclinical studies that CRLX101 is a potent and durable inhibitor of HIF-1α.[7,149] The camptothecin family of compounds (including CPT and TPT) have been described as inhibitors of HIF-1α in the past. However, the literature suggests that to achieve durable inhibition of HIF-1α, sustained concentrations of TPT must be delivered to cells. Due to its short half-life, low tumour targeting, and high toxicities, TPT cannot effectively achieve durable HIF-1α suppression.[125] HIF-1α, which is up-regulated under hypoxic conditions in the tumour microenvironment, has recently become a target of increasing interest in cancer research as it appears to be a master regulator for many key cancer cell survival pathways.[74,76,150] It is hypothesised that simultaneous inhibition of TOP1 and HIF-1α by CRLX101 could lead to notable clinical benefit, particularly when the nanopharmaceutical is combined with anti-cancer therapies such as VEGF receptor inhibitors, which are known to create hypoxia and up-regulate HIF-1α. In preclinical models, CRLX101 is demonstrated not only to inhibit HIF-1α, but also to achieve synergistic activity in combination with bevacizumab and other VEGF receptor inhibitors, including the tyrosine kinase inhibitor pazopanib and the VEGF trap molecule aflibercept.[7]

In several animal models, the combination of CRLX101 and leading anti-angiogenesis drugs demonstrate the following: (1) HIF-1α protein expression in each case is significantly up-regulated in the presence of each of the tested anti-angiogenesis drugs; (2) HIF-1α protein expression is significantly down-regulated when exposed to a low dose of CRLX101; (3) when CRLX101 is combined with anti-angiogenesis drugs, HIF-1α protein expression is down-regulated compared to control, confirming that the CRLX101 down-regulation can counteract the up-regulation normally produced by these anti-angiogenesis drugs; and (4) the combination of a low dose of CRLX101 with anti-angiogenesis drugs is synergistic, resulting in markedly longer animal survival than either drug by itself.[7,130]

Similar results have been observed in the preclinical assessment of the nanopharmaceutical's effect on cancer stem cells that cause cancer heterogeneity, drug resistance and metastases. Findings here include: (1) pretreatment with CRLX101 leads to a reduction in cancer stem cells expansion compared to control; (2) pretreatment with bevacizumab leads to enhanced cancer stem cells expansion compared to control; and (3) pretreatment with a combination of CRLX101 and bevacizumab leads to reduced cancer stem cell expansion compared to CRLX101 treatment alone, demonstrating CRLX101's ability to overcome the HIF-1α-associated and undesirable cancer stem cells.[149]

More than 200 human cancer patients have now been treated with CRLX101 in phase I/IIa and phase II clinical trials. In addition to the completed clinical trials, other monotherapy studies of CRLX101 are still running. They include a multi-site randomised phase II clinical trial of 112 patients comparing CRLX101 monotherapy to TPT in advanced small cell lung cancer, or SCLC, conducted at the University of Chicago (Chicago, IL, USA) and at other major medical centres in the USA (ClinicalTrials.gov identifier NCT01803269). The study objectives are to establish feasibility of enrolling this advanced patient population, and differentiate the safety and efficacy of CRLX101 at $15\,mg\,m^{-2}$ dosed every 2 weeks *vs.* TPT, the only approved second-line SCLC agent. The first patient was enrolled in January 2013.[145] A multisite single-arm phase II clinical trial in advanced ovarian cancer patients in monotherapy is conducted at the Massachusetts General Hospital, Brigham and Women's Hospital, Dana–Farber Cancer Center, and Beth Israel Deaconess Medical Center (Boston, MA, USA) (ClinicalTrials.gov identifier NCT01652079). The objectives of this trial are to establish activity worthy of further investigation based on predefined criteria and to confirm safety and tolerability of CRLX101 at $15\,mg\,m^{-2}$ dosed every 2 weeks in ovarian cancer patients. The first patient was enrolled in June 2012 and enrolment is now complete with patients remaining on study at this time.[146] A single-arm phase II pharmacodynamic clinical trial in advanced human epithelial growth receptor 2 (HER2)-negative gastric cancer patients at the City of Hope National Comprehensive Cancer Center (ClinicalTrials.gov identifier NCT01612546). This study's objectives are to establish signals of activity in HER2-negative gastric cancer patients and to utilise sequential tumour biopsies to establish differential nanopharmaceutical tumour penetration. The first patient was enrolled in December 2012 and enrolment continues.

Combination therapy evaluations of CRLX101 include a single-arm phase II clinical trial in advanced RCC patients of CRLX101 at 15 mg m^{-2} dosed every 2 weeks in combination with bevacizumab (10 mg kg^{-1} every 2 weeks) conducted at the University of Pennsylvania (Philadelphia, PA, USA) (ClinicalTrials.gov identifier NCT01625936). After establishing combinability of CRLX101 at the MTD (phase Ib), the objectives for this trial are to fulfil the predefined efficacy criteria of the combination, supporting further evaluation of the combination in RCC. The first patient was enrolled in June 2012 and enrolment is continuing. Very recent results showed that ORR was 30% and mPFS was 7.9 months, which is better than benchmarks such as everolimus (Afinitor®; Novartis) (ORR 4%; mPFS 4.7 months or less).[147,148] A multisite single-arm phase II clinical trial in advanced ovarian cancer patients of CRLX101 at 15 mg m^{-2} dosed every 2 weeks in combination therapy with bevacizumab (10 mg kg^{-1} every 2 weeks) conducted at the Massachusetts General Hospital, Brigham and Women's Hospital, Dana–Farber Cancer Center, and Beth Israel Deaconess Medical Center. The objectives of this trial are to establish activity worthy of further investigation based on pre-defined criteria and to confirm safety and tolerability of CRLX101 in combination with bevacizumab in patients with ovarian cancer. A two-stage design is being employed to enrol a maximum of 43 women. During Stage I, 18 women will be enrolled and if two or fewer achieve PFS through six cycles of therapy (PFS6), the trial will be terminated. Otherwise, a stage II trial will enrol 25 additional women and if, ultimately, eight or more patients achieve PFS6, the combination will be considered worthy of further investigation. CRLX101 is under investigation in two combination treatments for relapsed ovarian cancer. A Phase 2 trial of CRLX101 plus bevacizumab (Avastin®) is enrolling patients, and a Phase 1b trial of CRLX101 plus weekly paclitaxel (Taxol) has enrolled patients.[149]

An additional clinical trial evaluating CRLX101 in combination with chemoradiotherapy (CRT) in the neo-adjuvant treatment of rectal cancer is about to commence enrolment at the University of North Carolina Chapel Hill and affiliated centres in the USA (ClinicalTrials.gov identifier NCT02010567). Radiotherapy causes DNA strand breaks which, if not repaired, lead to desired apoptosis of radiated tumour cells. However, cell repair mechanisms, including TOP1, which re-ligates DNA strand breaks, can undo the radiation damage to tumour cells, thus interfering with the desired effects of the radiation therapy. Since TOP1 is instrumental in repairing radiation-induced DNA single strand breaks, a combinable TOP1 inhibitor is hypothesised to be effective as a radiosensitiser. In fact, combinations of CPT-11 and CRT have shown synergy, raising pathological complete response rates in this setting across various trials. However, the toxicity of CPT-11 prevents its addition to this treatment regimen. CRLX101 in combination with radiotherapy in a head and neck cancer animal model demonstrates notable synergy and radiosensitivity. In addition, hypoxia is a by-product of radiotherapy and there is a well-documented role of hypoxia-mediated HIF-1α induced up-regulation in the induction of resistance to radiotherapy. Accordingly, it is hypothesised that CRLX101, building on both its anti-TOP1 and anti-HIF-1α activity, as well as its favourable safety profile, offers strong potential as an add-on drug to standard of care CRT in neoadjuvant rectal cancer. This open-label, single-arm phase Ib/II study is designed to identify the MTD/recommended phase II dose of CRLX101 administered in combination with capecitabine-based CRT and to detect signals of increased efficacy over CRT alone. The trial is designed to enrol up to 53 patients, with a primary endpoint of pathological complete response and secondary endpoints of disease-free survival and mOS.[7,149]

References

1. M. E. Davis, *Adv. Drug Delivery Rev.*, 2009, **61**, 1189–1192.
2. A. Lorence and C. L. Nessler, *Phytochemistry*, 2004, **65**, 2735–2749.
3. A. Rapisarda, J. Zalek, M. Hollingshead, T. Braunschweig, B. Uranchimeg, C. A. Bonomi, S. D. Borgel, J. P. Carter, S. M. Hewitt, R. H. Shoemaker and G. Melillo, *Cancer Res.*, 2004, **25**, 6845–6848.
4. E. K. Rowinsky, L. B. Grochow, C. B. Hendricks, D. S. Ettinger, A. A. Forastiere, L. A. Hurowitz, W. P. McGuire, S. E. Sartorius, B. G. Lubejko and S. H. Kaufmann, *Phase I and pharmacologic study of topotecan: a novel topoisomerase I inhibitor.*, 1992, vol. 10.
5. H. Hochster, L. Liebes, J. Speyer, J. Sorich, B. Taubes, R. Oratz, J. Wernz, A. Chachoua, B. Raphael and R. Z. Vinci, *Phase I trial of low-dose continuous topotecan infusion in patients with cancer: an active and well-tolerated regimen.*, 1994, vol. 12.
6. S. Svenson, M. Wolfgang, J. Hwang, J. Ryan and S. Eliasof, *J. Controlled Release*, 2011, **153**, 49–55.
7. S. Eliasof, S. Conley, S. M. Keefe, R. Kerbel, N. Carolyn, D. Lazarus, C. Peters, E. Pham, M. S. Wicha, E. G. Garmey and C. Pharma, *Am. Assoc. Cancer Res.*, 2013, **B1**, 95–96.
8. G. J. Weiss, J. Chao, J. D. Neidhart, R. K. Ramanathan, D. Bassett, J. a Neidhart, C. H. J. Choi, W. Chow, V. Chung, S. J. Forman, E. Garmey, J. Hwang, D. L. Kalinoski, M. Koczywas, J. Longmate, R. J. Melton, R. Morgan, J. Oliver, J. J. Peterkin, J. L. Ryan, T. Schluep, T. W. Synold, P. Twardowski, M. E. Davis and Y. Yen, *Invest. New Drugs*, 2013, **31**, 986–1000.
9. M. E. Wall, M. C. Wani, C. E. Cook, K. H. Palmer, A. T. McPhail and G. A. Sim, *J. Am. Chem. Soc.*, 1966, **88**, 3888–3890.

10. Y. H. Hsiang, R. Hertzberg, S. Hecht and L. F. Liu, *J. Biol. Chem.*, 1985, **260**, 14873–14878.
11. C. G. Moertel, a J. Schutt, R. J. Reitemeier and R. G. Hahn, *Cancer Chemother. Rep.*, 1972, **56**, 95–101.
12. R. P. Hertzberg, M. J. Caranfa, K. G. Holden, D. R. Jakas, G. Gallagher, M. R. Mattern, S. Mong, J. O. L. Bartus, R. K. Johnson and W. D. Kingsbury, *J. Med. Chem.*, 1989, **32**, 715–720.
13. Z. Mi and T. G. Burke, *Biochemistry*, 1994, **33**, 10325–10336.
14. R. Garcia-Carbonero and J. G. Supko, *Clin. Cancer Res.*, 2002, **8**, 641–661.
15. G. Pratesi, M. De Cesare, N. Carenini, P. Perego, C. Pisano, S. Penco, P. Carminati, L. Vesci and F. Zunino, *Clin. Cancer Res.*, 2002, 3904–3909.
16. C. Pisano, M. De Cesare, G. L. Beretta, V. Zuco, G. Pratesi, S. Penco, L. Vesci, R. Foderà, F. F. Ferrara, M. B. Guglielmi, P. Carminati, S. Dallavalle, G. Morini, L. Merlini, A. Orlandi and F. Zunino, *Mol. Cancer Ther.*, 2008, **7**, 2051–2059.
17. S. Dallavalle, S. Gattinoni, S. Mazzini, L. Scaglioni, L. Merlini, S. Tinelli, G. L. Beretta and F. Zunino, *Bioorg. Med. Chem. Lett.*, 2008, **18**, 1484–1489.
18. L. Scaglioni, S. Mazzini, R. Mondelli, S. Dallavalle, S. Gattinoni, S. Tinelli, G. L. Beretta, F. Zunino and E. Ragg, *Bioorg. Med. Chem.*, 2009, **17**, 484–491.
19. S. Cananzi, L. Merlini, R. Artali, G. L. Beretta, N. Zaffaroni and S. Dallavalle, *Bioorg. Med. Chem.*, 2011, **19**, 4971–4984.
20. R. Cincinelli, L. Musso, S. Dallavalle, R. Artali, S. Tinelli, D. Colangelo, F. Zunino, M. De Cesare, G. L. Beretta and N. Zaffaroni, *Eur. J. Med. Chem.*, 2013, **63**, 387–400.
21. J. W. Singer, R. Bhatt, J. Tulinsky, K. R. Buhler, E. Heasley, P. Klein and P. de Vries, *J. Controlled Release*, 2001, **74**, 243–247.
22. Y. Matsumura and H. Maeda, *Cancer Res.*, 1986, **46**, 6387–6392.
23. A. V. Yurkovetskiy and R. J. Fram, *Adv. Drug Delivery Rev.*, 2009, **61**, 1193–1202.
24. E. a. Sausville, L. Garbo, G. J. Weiss, D. Shkolny, a. V. Yurkovetskiy, C. Bethune, R. K. Ramanathan and R. J. Fram, *Mol. Cancer Ther.*, 2009, **8**, B52–B52.
25. Y. Ochi, Y. Shiose, H. Kuga and E. Kumazawa, *Cancer Chemother. Pharmacol.*, 2005, **55**, 323–332.
26. Y. Shiose, Y. Ochi, H. Kuga, F. Yamashita and M. Hashida, *Biol. Pharm. Bull.*, 2007, **30**, 2365–2370.
27. M. Harada, H. Sakakibara, T. Yano, T. Suzuki and S. Okuno, *J. Controlled Release*, 2000, **69**, 399–412.
28. M. Harada, J. Imai, S. Okuno and T. Suzuki, *J. Controlled Release*, 2000, **69**, 389–397.
29. S. Okuno, M. Harada, T. Yano, S. Yano, S. Kiuchi, N. Tsuda, Y. Sakamura, J. Imai, T. Kawaguchi and K. Tsujihara, *Cancer Res.*, 2000, **60**, 2988–2995.
30. M. Harada, J. Murata, Y. Sakamura, H. Sakakibara, S. Okuno and T. Suzuki, *J. Controlled Release*, 2001, **71**, 71–86.

31. R. Bhatt, P. de Vries and J. Tulinsky, *J. Med. Chem.*, 2003, **46**, 190–193.
32. Y. Zou, Q. Wu, W. Tansey, D. Chow, M. Hung, C. Charnsangavej, S. Wallace and C. Li, *Int. J. Oncol.*, 2001, **18**, 11172600.
33. J. Homsi, G. R. Simon, C. R. Garrett, G. Springett, R. De Conti, A. A. Chiappori, P. N. Munster, M. K. Burton, S. Stromatt, C. Allievi, P. Angiuli, A. Eisenfeld, D. M. Sullivan and A. I. Daud, *Clin. Cancer Res.*, 2007, **13**, 5855–5861.
34. S. Dallavalle, G. Giannini, D. Alloatti, A. Casati, E. Marastoni, L. Musso, L. Merlini, G. Morini, S. Penco, C. Pisano, S. Tinelli, M. De Cesare, G. Beretta and F. Zunino, *J. Med. Chem.*, 2006, **49**, 5177–5186.
35. A. Dal Pozzo, M.-H. Ni, E. Esposito, S. Dallavalle, L. Musso, A. Bargiotti, C. Pisano, L. Vesci, F. Bucci, M. Castorina, R. Foderà, G. Giannini, C. Aulicino and S. Penco, *Bioorg. Med. Chem.*, 2010, **18**, 64–72.
36. R. Duncan, *Adv. Drug Delivery Rev.*, 2009, **61**, 1131–1148.
37. V. R. Caiolfa, M. Zamai, A. Fiorino, E. Frigerio, C. Pellizzoni, R. D'Argy, A. Ghiglieri, M. G. Castelli, M. Farao, E. Pesenti, M. Gigli, F. Angelucci and A. Suarato, *J. Controlled Release*, 2000, **65**, 105–119.
38. G. Chirico, M. Collini, F. Olivini, M. Zamai, E. Frigerio and V. R. Caiolfa, *Biophys. Chem.*, 2004, **110**, 281–295.
39. D. Bissett, J. Cassidy, J. S. de Bono, F. Muirhead, M. Main, L. Robson, D. Fraier, M. L. Magnè, C. Pellizzoni, M. G. Porro, R. Spinelli, W. Speed and C. Twelves, *Br. J. Cancer*, 2004, **91**, 50–55.
40. F. M. Wachters, H. J. M. Groen, J. G. Maring, J. a Gietema, M. Porro, H. Dumez, E. G. E. de Vries and a T. van Oosterom, *Br. J. Cancer*, 2004, **90**, 2261–2267.
41. F. F. Davis, *Adv. Drug Delivery Rev.*, 2002, **54**, 457–458.
42. F. M. Veronese and J. M. Harris, *Adv. Drug Delivery Rev.*, 2002, **54**, 453–456.
43. R. B. Greenwald, A. Pendri, C. Conover, C. Gilbert and R. Yang, *J. Med. Chem.*, 1996, 1938–1940.
44. R. B. Greenwald, A. Pendri, C. D. Conover, C. Lee, Y. H. Choe, C. Gilbert, A. Martinez, J. Xia, D. Wu and M. Hsue, *Bioorg. Med. Chem.*, 1998, **6**, 551–562.
45. E. K. Rowinsky, J. Rizzo, L. Ochoa, C. H. Takimoto, B. Forouzesh, G. Schwartz, L. A. Hammond, A. Patnaik, J. Kwiatek, A. Goetz, L. Denis, J. McGuire and A. W. Tolcher, *J. Clin. Oncol.*, 2003, **21**, 148–157.
46. D. Yu, P. Peng, S. S. Dharap, Y. Wang, M. Mehlig, P. Chandna, H. Zhao, D. Filpula, K. Yang, V. Borowski, G. Borchard, Z. Zhang and T. Minko, *J. Controlled Release*, 2005, **110**, 90–102.
47. L. C. Scott, J. C. Yao, a B. Benson, a L. Thomas, S. Falk, R. R. Mena, J. Picus, J. Wright, M. F. Mulcahy, J. a Ajani and T. R. J. Evans, *Cancer Chemother. Pharmacol.*, 2009, **63**, 363–370.
48. P. Sapra, H. Zhao, M. Mehlig, J. Malaby, P. Kraft, C. Longley, L. M. Greenberger and I. D. Horak, *Clin. Cancer Res.*, 2008, **14**, 1888–1896.
49. H. Zhao, B. Rubio, P. Sapra, D. Wu, P. Reddy, P. Sai, A. Martinez, Y. Gao, Y. Lozanguiez, C. Longley, L. M. Greenberger and I. D. Horak, *Bioconjugate Chem.*, 2008, **19**, 849–859.

50. S. A. L. Zander, W. Sol, L. Greenberger, Y. Zhang, O. van Tellingen, J. Jonkers, P. Borst and S. Rottenberg, *PLoS One*, 2012, 7, e45248.

51. R. Kurzrock, S. Goel, J. Wheler, D. Hong, S. Fu, K. Rezai, S. K. Morgan-Linnell, S. Urien, S. Mani, I. Chaudhary, M. H. Ghalib, A. Buchbinder, F. Lokiec and M. Mulcahy, *Cancer*, 2012, **118**, 6144–6151.

52. C. R. Garrett, T. S. Bekaii-Saab, T. Ryan, G. a Fisher, S. Clive, P. Kavan, E. Shacham-Shmueli, A. Buchbinder and R. M. Goldberg, *Cancer*, 2013, **119**, 4223–4230.

53. D. L. Nielsen and N. Brünner, *Lancet Oncol.*, 2013, **14**, 1149–1151.

54. A. Awada, A. a Garcia, S. Chan, G. H. M. Jerusalem, R. E. Coleman, M. T. Huizing, A. Mehdi, S. M. O'Reilly, J. T. Hamm, P. J. Barrett-Lee, V. Cocquyt, K. Sideras, D. E. Young, C. Zhao, Y. L. Chia, U. Hoch, A. L. Hannah and E. a Perez, *Lancet Oncol.*, 2013, **14**, 1216–1225.

55. A. Awada, S. Chan, G. H. M. Jerusalem, R. E. Coleman, M. T. Huizing, A. Mehdi, S. M. O'Reilly, J. T. Hamm, P. J. Barrett-Lee, V. Cocquyt, K. Sideras, D. E. Young, M. Brown, C. Zhao, A. L. Hannah, L. K. Masuoka, A. a Garcia and E. a Perez, *Am. Soc. Clin. Oncol.*, 2011, **1**, 102.

56. I. B. Vergote, A. Garcia, J. Micha, C. Pippitt, J. Bendell, D. Spitz, N. Reed, G. Dark, P. M. Fracasso, E. N. Ibrahim, V. a Armenio, L. Duska, C. Poole, C. Gennigens, L. Y. Dirix, A. C. F. Leung, C. Zhao, R. Soufi-Mahjoubi and G. Rustin, *J. Clin. Oncol.*, 2013, **31**, 4060–4066.

57. R. Kunii, H. Onishi and Y. Machida, *Eur. J. Pharm. Biopharm.*, 2007, **67**, 9–17.

58. L. Zhang, Y. Hu, X. Jiang, C. Yang, W. Lu and Y. H. Yang, *J. Controlled Release*, 2004, **96**, 135–148.

59. L. Zhang, M. Yang, Q. Wang, Y. Li, R. Guo, X. Jiang, C. Yang and B. Liu, *J. Controlled Release*, 2007, **119**, 153–162.

60. K. H. Min, J.-H. Kim, S. M. Bae, H. Shin, M. S. Kim, S. Park, H. Lee, R.-W. Park, I.-S. Kim, K. Kim, I. C. Kwon, S. Y. Jeong and D. S. Lee, *J. Controlled Release*, 2010, **144**, 259–266.

61. T. Minko, P. V. Paranjpe, B. Qiu, A. Lalloo, R. Won, S. Stein and P. J. Sinko, *Cancer Chemother. Pharmacol.*, 2002, **50**, 143–150.

62. P. V. Paranjpe, Y. Chen, V. Kholodovych, W. Welsh, S. Stein and P. J. Sinko, *J. Controlled Release*, 2004, **100**, 275–292.

63. W. Zhang, J. Song, L. Mu, B. Zhang, L. Liu, Y. Xing, K. Wang, Z. Li and R. Wang, *Bioorg. Med. Chem. Lett.*, 2011, **21**, 1452–1455.

64. P. Vekhoff, L. Halby, K. Oussedik, S. Dallavalle, L. Merlini, C. Mahieu, A. Lansiaux, C. Bailly, A. Boutorine, C. Pisano, G. Giannini, D. Alloatti and P. B. Arimondo, *Bioconjugate Chem.*, 2009, **20**, 666–672.

65. K. Y. Choi, H. Y. Yoon, J.-H. Kim, S. M. Bae, R.-W. Park, Y. M. Kang, I.-S. Kim, I. C. Kwon, K. Choi, S. Y. Jeong, K. Kim and J. H. Park, *ACS Nano*, 2011, **5**, 8591–8599.

66. M. Hong, S. Zhu, Y. Jiang, G. Tang, C. Sun, C. Fang, B. Shi and Y. Pei, *J. Controlled Release*, 2010, **141**, 22–29.

67. L.-C. Sun, J. Luo, V. L. Mackey, J. a Fuselier and D. H. Coy, *Anticancer. Drugs*, 2007, **18**, 341–348.

68. J. R. Infante, V. L. Keedy, S. F. Jones, W. C. Zamboni, E. Chan, J. C. Bendell, W. Lee, H. Wu, S. Ikeda, H. Kodaira, M. L. Rothenberg and H. A. Burris III, *Cancer Chemother. Pharmacol.*, 2012, **70**, 699–705.

69. G. Batist, K. A. Gelmon, K. N. Chi, W. H. Miller, S. K. L. Chia, L. D. Mayer, C. E. Swenson, A. S. Janoff, and A. C. Louie, *Safety, pharmacokinetics, and efficacy of CPX-1 liposome injection in patients with advanced solid tumors.*, 2009, vol. 15.

70. W. C. Zamboni, S. Strychor, E. Joseph, D. R. Walsh, B. A. Zamboni, R. A. Parise, M. E. Tonda, N. Y. Yu, C. Engbers and J. L. Eiseman, *Clin. Cancer Res.*, 2007, **13**, 7217–7223.

71. J. Williams, R. Lansdown, R. Sweitzer, M. Romanowski, R. LaBell, R. Ramaswami and E. Unger, *J. Controlled Release*, 2003, **91**, 167–172.

72. Y. Matsumura, *Adv. Drug Delivery Rev.*, 2011, **63**, 184–192.

73. G. L. Wang, B. H. Jiang, E. a Rue and G. L. Semenza, *Proc. Natl. Acad. Sci. U. S. A.*, 1995, **92**, 5510–5514.

74. G. Semenza, *Nat. Rev. Cancer*, 2003, **3**, 721.

75. H. Zhong, A. M. De Marzo, E. Laughner, M. Lim, D. A. Hilton, D. Zagzag, P. Buechler, W. B. Isaacs, G. L. Semenza and J. W. Simons, *Cancer Res.*, 1999, **59**, 5830–5835.

76. B. Onnis, A. Rapisarda and G. Melillo, *J. Cell. Mol. Med.*, 2009, **13**, 2780–2786.

77. D. J. Brennan, K. Jirstrom, A. Kronblad, R. C. Millikan, G. Landberg, M. J. Duffy, L. Rydén, W. M. Gallagher and S. L. O'Brien, *Clin. cancer Res.*, 2006, **12**, 6421–6431.

78. S. K. Chia, C. C. Wykoff, P. H. Watson, C. Han, R. D. Leek, J. Pastorek, K. C. Gatter, P. Ratcliffe and A. L. Harris, *J. Clin. Oncol.*, 2001, **19**, 3660–3668.

79. D. Generali, S. B. Fox, A. Berruti, M. P. Brizzi, L. Campo, S. Bonardi, S. M. Wigfield, P. Bruzzi, A. Bersiga, G. Allevi, M. Milani, S. Aguggini, L. Dogliotti, A. Bottini and A. L. Harris, *Endocr.-Relat. Cancer*, 2006, **13**, 921–930.

80. D. Generali, A. Berruti, M. P. Brizzi, L. Campo, S. Bonardi, S. Wigfield, A. Bersiga, G. Allevi, M. Milani, S. Aguggini, V. Gandolfi, L. Dogliotti, A. Bottini, A. L. Harris and S. B. Fox, *Clin. Cancer Res.*, 2006, **12**, 4562–4568.

81. S. A. Hussain, R. Ganesan, G. Reynolds, L. Gross, A. Stevens, J. Pastorek, P. G. Murray, B. Perunovic, M. S. Anwar, L. Billingham, N. D. James, D. Spooner, C. J. Poole, D. W. Rea and D. H. Palmer, *Br. J. Cancer*, 2007, **96**, 104–109.

82. C. Trastour, E. Benizri, F. Ettore, A. Ramaioli, E. Chamorey, J. Pouysségur and E. Berra, *Int. J. Cancer*, 2007, **120**, 1451–1458.

83. N. J. Mabjeesh, D. Escuin, T. M. LaVallee, V. S. Pribluda, G. M. Swartz, M. S. Johnson, M. T. Willard, H. Zhong, J. W. Simons and P. Giannakakou, *Cancer Cell*, 2003, **3**, 363–375.

84. C. Sweeney, G. Liu, C. Yiannoutsos, J. Kolesar, D. Horvath, M. J. Staab, K. Fife, V. Armstrong, A. Treston, C. Sidor and G. Wilding, *Clin. Cancer Res.*, 2005, **11**, 6625–6633.

85. M. Harrison, N. Hahn and R. Pili, *Invest. New Drugs*, 2011, **29**, 1465–1474.
86. J. Y. Bruce and H. Avenue, *Invest. New Drugs*, 2012, **30**, 794–802.
87. A. Rapisarda, R. Shoemaker and G. Melillo, *Cell Cycle*, 2004, **3**, 172–175.
88. M. Xia, K. Bi, R. Huang, M.-H. Cho, S. Sakamuru, S. C. Miller, H. Li, Y. Sun, J. Printen, C. P. Austin and J. Inglese, *Mol. Cancer*, 2009, **8**, 117.
89. D. Bertozzi, J. Marinello, S. G. Manzo, F. Fornari, L. Gramantieri and G. Capranico, *Mol. Cancer Ther.*, 2013, **13**, 239–248.
90. A. a Miller, A. Al Omari, D. J. Murry and D. Case, *Lung Cancer*, 2006, **54**, 379–385.
91. L. M. Greenberger, I. D. Horak, D. Filpula, P. Sapra, M. Westergaard, H. F. Frydenlund, C. Albaek, H. Schrøder and H. Ørum, *Mol. Cancer Ther.*, 2008, **7**, 3598–3608.
92. X. Liu, M.-H. Xiong, X.-T. Shu, R.-Z. Tang and J. Wang, *Mol. Pharm.*, 2012, **9**, 2863–2874.
93. K. Brusselmans, F. Bono, P. Maxwell, Y. Dor, M. Dewerchin, D. Collen, J. M. Herbert and P. Carmeliet, *J. Biol. Chem.*, 2001, **276**, 39192–39196.
94. M. a Maynard, A. J. Evans, T. Hosomi, S. Hara, M. a S. Jewett and M. Ohh, *FASEB J.*, 2005, **19**, 1396–1406.
95. P. Maxwell, M. Wiesener and G. Chang, *Nature*, 1999, **399**, 271–275.
96. B. Bao, A. S. Azmi, S. Ali, A. Ahmad, Y. Li, S. Banerjee, D. Kong and F. H. Sarkar, *Biochim. Biophys. Acta*, 2012, **1826**, 272–296.
97. Y. Xia, H.-K. Choi and K. Lee, *Eur. J. Med. Chem.*, 2012, **49**, 24–40.
98. P. Ellinghaus, I. Heisler, K. Unterschemmann, M. Haerter, H. Beck, S. Greschat, A. Ehrmann, H. Summer, I. Flamme, F. Oehme, K. Thierauch, M. Michels, H. Hess-Stumpp and K. Ziegelbauer, *Cancer Med.*, 2013, 611–624.
99. http://trialfinder.bayerscheringpharma.de/html/pdf/15044_Study_Synopsis_CTP.pdf.
100. J. S. Carew, J. a Esquivel, C. M. Espitia, C. M. Schultes, M. Mülbaier, J. D. Lewis, B. Janssen, F. J. Giles and S. T. Nawrocki, *PLoS One*, 2012, **7**, e31120.
101. Y. S. Chang, J. Adnane, P. a Trail, J. Levy, A. Henderson, D. Xue, E. Bortolon, M. Ichetovkin, C. Chen, A. McNabola, D. Wilkie, C. a Carter, I. C. a Taylor, M. Lynch and S. Wilhelm, *Cancer Chemother. Pharmacol.*, 2007, **59**, 561–574.
102. L. Liu, R. L. K. Ho, G. G. Chen and P. B. S. Lai, *Clin. Cancer Res.*, 2012, **18**, 5662–5671.
103. M. Wangpaichitr and N. Savaraj, *Mol. Cancer*, 2008, **7**, 1506–1513.
104. N. J. Mabjeesh, D. E. Post, M. T. Willard, B. Kaur, E. G. Van Meir and J. W. Simons, *Cancer Res.*, 2002, **62**, 2478–2482.
105. M.-A. Dimopoulos, C. S. Mitsiades, K. C. Anderson and P. G. Richardson, *Clin. Lymphoma, Myeloma Leuk.*, 2011, **11**, 17–22.
106. S. Welsh, R. Williams, A. Birmingham, J. Newman, D. Kirkpatrick and G. Powis, *Mol. Cancer Ther.*, 2003, **2**, 235–243.

107. E. Hur, K. Y. Chang, E. Lee, S. Lee and H. Park, *Mol. Pharmacol.*, 2001, **59**, 1216–1224.
108. L. C. J. Baker, J. K. R. Boult, S. Walker-Samuel, Y.-L. Chung, Y. Jamin, M. Ashcroft and S. P. Robinson, *Br. J. Cancer*, 2012, **106**, 1638–1647.
109. N. Chau, P. Rogers, W. Aherne, I. F. That, D. Block, V. Carroll, I. Collins, E. Mcdonald, P. Workman and M. Ashcroft, *Cancer Res.*, 2005, **65**, 4918–4929.
110. E.-J. Yeo, Y.-S. Chun, Y.-S. Cho, J. Kim, J.-C. Lee, M.-S. Kim and J.-W. Park, *J. Natl. Cancer Inst.*, 2003, **95**, 516–525.
111. S. H. Li, D. H. Shin, Y.-S. Chun, M. K. Lee, M.-S. Kim and J.-W. Park, *Mol. Cancer Ther.*, 2008, 7, 3729–3738.
112. J. Mun, A. A. Jabbar, N. S. Devi, S. Yin, Y. Wang, C. Tan, D. Culver, J. P. Snyder, E. G. Van Meir and M. M. Goodman, *J. Med. Chem.*, 2012, **55**, 6738–6750.
113. E. Miranda, I. K. Nordgren, A. L. Male, C. E. Lawrence, F. Hoakwie, F. Cuda, W. Court, K. R. Fox, P. a Townsend, G. K. Packham, S. a Eccles and A. Tavassoli, *J. Am. Chem. Soc.*, 2013, **135**, 10418–10425.
114. J. M. Heddleston, Z. Li, J. D. Lathia, S. Bao, a B. Hjelmeland and J. N. Rich, *Br. J. Cancer*, 2010, **102**, 789–795.
115. S. J. Conley, E. Gheordunescu, P. Kakarala, B. Newman, H. Korkaya, A. N. Heath, S. G. Clouthier and M. S. Wicha, *Proc. Natl. Acad. Sci.*, 2012, **109**, 2784–2789.
116. R. K. Ambasta, A. Sharma and P. Kumar, *Vasc. Cell*, 2011, **3**, 26.
117. D. Shweiki, A. Itin, D. Soffer and E. Keshet, *Nature*, 1992, **359**, 843–845.
118. J. Forsythe, B. Jiang, N. Iyer, F. Agani, S. W. Leung, R. D. Koos and G. L. Semenza, *Mol. Cell Biol.*, 1996, **16**, 4604–4613.
119. B. Jiang, F. Agani, A. Passaniti and G. Semenza, *Cancer Res.*, 1997, **57**, 5328–5335.
120. A. Rapisarda and G. Melillo, *Nat. Rev. Clin. Oncol.*, 2012, **9**, 378–390.
121. A. Rapisarda, M. Hollingshead, B. Uranchimeg, C. a Bonomi, S. D. Borgel, J. P. Carter, B. Gehrs, M. Raffeld, R. J. Kinders, R. Parchment, M. R. Anver, R. H. Shoemaker and G. Melillo, *Mol. Cancer Ther.*, 2009, **8**, 1867–1877.
122. K. Hashimoto, S. Man, P. Xu, W. Cruz-Munoz, T. Tang, R. Kumar and R. S. Kerbel, *Mol. Cancer Ther.*, 2010, **9**, 996–1006.
123. W. M. Merritt, A. M. Nick, A. R. Carroll, C. Lu, K. Matsuo, N. Jennings, S. Zhang, Y. G. Lin, W. A. Spannuth, A. A. Kamat, R. L. Stone, M. M. K. Shahzad, R. L. Coleman and A. K. Sood, *Mol. Cancer Ther.*, 2010, **9**, 985–995.
124. K. Burkitt, S. Y. Chun, D. T. Dang and L. H. Dang, *Mol. Cancer Ther.*, 2009, **8**, 1148–1156.
125. S. Kummar, M. Raffeld, L. Juwara, Y. Horneffer, A. Strassberger, D. Allen, S. M. Steinberg, A. Rapisarda, S. D. Spencer, W. D. Figg, X. Chen, I. B. Turkbey, P. Choyke, A. J. Murgo, J. H. Doroshow and G. Melillo, *Clin. Cancer Res.*, 2011, **17**, 5123–5131.

126. I. Zighelboim, J. D. Wright, F. Gao, A. S. Case, L. S. Massad, D. G. Mutch, M. a Powell, P. H. Thaker, E. L. Eisenhauer, D. E. Cohn, F. a Valea, A. Alvarez Secord, L. T. Lippmann, F. Dehdashti and J. S. Rader, *Gynecol. Oncol.*, 2013, **130**, 64–68.

127. P. Sapra, P. Kraft, F. Pastorino, D. Ribatti, M. Dumble, M. Mehlig, M. Wang, M. Ponzoni, L. M. Greenberger and I. D. Horak, *Angiogenesis*, 2011, **14**, 245–253.

128. J. Overgaard, *Radiother. Oncol.*, 2011, **100**, 22–32.

129. M. L. Janssen, W. J. Oyen, I. Dijkgraaf, L. F. Massuger, C. Frielink, D. S. Edwards, M. Rajopadhye, H. Boonstra, F. H. Corstens and O. C. Boerman, *Cancer Res.*, 2002, **62**, 6146–6151.

130. M. Yoshimoto, K. Ogawa, K. Washiyama, N. Shikano, H. Mori, R. Amano and K. Kawai, *Int. J. Cancer*, 2008, **123**, 709–715.

131. A. Mitra, A. Nan, J. C. Papadimitriou, H. Ghandehari and B. R. Line, *Nucl. Med. Biol.*, 2006, **33**, 43–52.

132. S. Sofou, *Int. J. Nanomed.*, 2008, **3**, 181–199.

133. J. Cheng, K. T. Khin, G. S. Jensen, A. Liu and M. E. Davis, *Bioconjugate Chem.*, 2003, **14**, 1007–1017.

134. S. Eliasof, D. Lazarus, C. G. Peters, R. I. Case, R. O. Cole, J. Hwang, T. Schluep, J. Chao, J. Lin, Y. Yen, H. Han, D. T. Wiley, J. E. Zuckerman and M. E. Davis, *Proc. Natl. Acad. Sci.*, 2013, **110**, 15127–15132.

135. C. Young, T. Schluep, J. Hwang and S. Eliasof, *Curr. Bioact. Compd.*, 2011, **7**, 8–14.

136. T. Numbenjapon, J. Wang, D. Colcher, T. Schluep, M. E. Davis, J. Duringer, L. Kretzner, Y. Yen, S. J. Forman and A. Raubitschek, *Clin. Cancer Res.*, 2009, **15**, 4365–4373.

137. T. Schluep, J. Hwang, J. Cheng, J. D. Heidel, D. W. Bartlett, B. Hollister and M. E. Davis, *Clin. Cancer Res.*, 2006, **12**, 1606–1614.

138. T. Schluep, J. Cheng, K. T. Khin and M. E. Davis, *Cancer Chemother. Pharmacol.*, 2006, **57**, 654–662.

139. S. R. Kabir, D. Lazarus and S. Eliasof, *Mol. Cancer Ther.*, 2011, **10**, C38.

140. T. Schluep, J. Hwang, I. J. Hildebrandt, J. Czernin, C. Hang, J. Choi, C. A. Alabi, B. C. Mack and M. E. Davis, *Proc. Natl. Acad. Sci. U. S. A.*, 2009, **106**, 11394–11399.

141. T. S. Ng, D. Wert, H. Sohi, D. Procissi, D. M. Colcher, A. a Raubitschek and R. E. Jacobs, *Clin. Cancer Res.*, 2013, **19**, 2518–2527.

142. Y. Yen, L. a. Coerver, E. G. Garmey, D. L. Kalinoski, M. Koczywas, J. a. Neidhart, J. D. Neidhart, J. J. Peterkin, R. K. Ramanathan, J. Ryan and G. J. Weiss, *Mol. Cancer Ther.*, 2011, **10**, A97.

143. Y. Yen, T. Synold, G. J. Weiss, T. Schluep and J. Ryan, *Eur. J. Cancer Suppl.*, 2010, **8**, 134–135.

144. (http://ceruleanrx.com/wp-content/uploads/2013/03/Top-line-Phase-2b-NSCLC-Results-032213-FINAL.pdf).

145. T. A. Hensing, T. Karrison, E. G. Garmey, M. G. Hennessy and R. Salgia, *J. Clin. Oncol.*, 2013, **31**, suppl; abstr TPS7610.

146. D. Lazarus, C. Peters and K. Deotale, *Cancer Res.*, 2013, **73**, 12–14.

147. S. M. Keefe, R. Cohen, S. Eliasof, E. G. Garmey, M. Hennessy, K. M. Mykulowicz, D. Pryma and N. Haas, *Am. Assoc. Cancer Res.*, 2013, **2419**, 1625936.

148. *Cancer Discov.*, 2013, 3, OF3, DOI: 10.1158/2159-8290.CD-NB2013-155.

149. http://ir.ceruleanrx.com/releasedetail.cfm?releaseid=914714.

150. G. L. Semenza, *Drug Discovery Today*, 2007, **12**, 853–859.

151. E. A. Perez, A. Awada, J. O'Shaughnessy, H. S. Rugo, C. Twelves, S.-A. Im, P. Gomez-Pardo, L. S. Schwartzberg, V. Dieras, D. A. Yardley, D. A. Potter, A. Malliez, A. Moreno-Aspitia, J. S. Ahn, C. Zhao, U. Hoch, M. Tagliaferri, A. I. Hannah and J. Cortes, *Lancet Oncol.*, 2015, **16**, 1556–1568.

CHAPTER 9

Nanodelivery Strategies in Breast Cancer Chemotherapy

VUONG TRIEU, OSMOND J. D'CRUZ AND LARN HWANG*

Autotelic Inc., 11100 Warner Ave., Ste 266, Fountain Valley, CA 92708, USA
*Email: lhwang@autotelicinc.com

9.1 Introduction

Breast cancer remains the second leading cause of cancer-related deaths in women, affecting more than 180 000 women in the United States each year. The high mortality observed in breast cancer patients indicates that improvements are needed in the diagnosis, treatment and prevention of this disease. Chemotherapy for patients with breast cancer has been effective in reducing the recurrence of this disease and improving survival in the adjuvant setting as well as in prolonging survival and improving quality of life in the metastatic setting. Approximately 20–30% of women diagnosed with breast cancer develop metastatic breast cancer (MBC). The most successful way of treating MBC is systemic therapy in the form of chemotherapy or immunotherapy. MBC is a heterogeneous disease and hence a variety of treatment options are available to meet different treatment aims in individual patients. The most common drugs used for early breast cancer include the anthracyclines (doxorubicin (DOX)/Adriamycin® and Epirubicin/Ellence®) and the taxanes (such as paclitaxel (PTX)/Taxol® and docetaxel/Taxotere®). These may be used in combination with other drugs (fluorouracil and cyclophosphamide (Cytoxan®).

Anthracyclines are effective as single agents in the treatment of MBC and combination therapy with anthracycline-containing regimens significantly

RSC Drug Discovery Series No. 51
Nanomedicines: Design, Delivery and Detection
Edited by Martin Braddock
© The Royal Society of Chemistry 2016
Published by the Royal Society of Chemistry, www.rsc.org

improves response rates and progression-free survival. The addition of taxanes further increases the efficacy of anthracyclines as first-line therapy. Taxanes are an essential component in the treatment of breast cancer in the neoadjuvant, adjuvant, and metastatic settings. Combination therapy with taxanes, trastuzumab (Herceptin®), and anthracyclines has been shown to produce significant improvements in survival, compared with taxanes and anthracyclines alone; however, use of these drugs is associated with significant side effects. New anthracycline and taxane formulations such as pegylated liposomal DOX (Doxil®), albumin-bound PTX (Abraxane®, ABI-007), and the polymeric micellar nanoparticle (NP) formulation of PTX (Genexol-PM®) have an important role in the treatment of MBC and represent the new advances in nanomedicine. This chapter attempts to review the evolutionary development of nanomedicines against breast cancer using Doxil®, Abraxane®, and Genexol-PM® as examples.

9.2 Nanocarriers for Drug Delivery to Solid Tumors

Nanodelivery strategies have emerged as a viable platform for the development of targeted therapeutic approaches to solid tumors. Lipid-based NPs (liposomes, lipid micelles, microemulsions, niosomes, and solid lipid NPs) and polymer-based nanocarriers (polymer micelles, polymeric NPs, nonogels, nanocapsules, and dendrimers) are being developed for sustained and targeted delivery of anticancer agents.[1,2] The goal of these nanocarriers is to slow drug degradation and loss, minimize harmful adverse effects or increase the bioavailability of the drug at the tumor site in a controlled manner. To be effective, NPs must be able to exist in the circulation for a sufficient period of time and be able to cross the vascular endothelial barrier into the solid tumor interstitium. As a result, NP formulations are able to alter (i) drug delivery, (ii) blood and tissue pharmacokinetic (PK) profiles, (iii) pharmacodynamics, (iv) tissue distribution, and (v) toxicity.

In addition, nanocarriers are being engineered to react to stimuli and to be site specific, either by passive or active targeting mechanisms. Passive targeting of drugs in solid tumors is achieved mainly due to differences in the vascularization of the tumor tissue as a result of increased permeability compared with healthy tissue. NPs are thought to extravasate from the leaky tumor vasculature to a higher degree than in healthy tissue and can remain at the site due to the enhanced permeability and retention (EPR) effect.[3] Drug-containing NPs (50–200 nm) can extravasate from the blood into the tumor interstitial space through these pores. This differential accumulation and penetration of NPs in tumor tissues relative to normal cells is one possible basis for the increased tumor specificity of nanocarrier drugs relative to free drugs. Active targeting involves the chemical "decorating" of the surface of nanocarriers with molecules enabling them to selectively target the cancer cells.[4,5] Tumor cells express many molecules on their surface which are not expressed by normal cells. Some receptors show higher expression in the tumor cells compared to normal cells. Therefore, solid tumors represent a group of malignancies which will get most of the benefits of targeted drug delivery. Various ligands can increase the tumor accumulation of targeted NPs leading to site-specific accumulation of the drug. These targeting ligands include antibodies, small peptides or molecules, lectins, aptamers, engineered proteins, and protein fragments.[6,7]

9.2.1 Liposomal Nanoparticles

Liposomal NPs (LNPs) are simple, self-assembling vesicles with either single or multiple phospholipid bilayers that enclose an aqueous core containing the therapeutic drug.[8] The most important structural components of liposomes are phospholipids and cholesterol. Liposomes can be designed to incorporate hydrophilic or hydrophobic compounds. Unilamellar vesicles comprising one lipid bilayer have diameters of 50–250 nm contain a large aqueous core for the encapsulation of hydrophilic drugs. Multilamellar vesicles composed of several concentric lipid bilayers have diameters of 1–5 μm. The high lipid content allows these multilamellar vesicles to entrap lipid-soluble drugs passively. Drug loading through passive entrapment is less efficient for hydrophilic drugs and active entrapment is a more efficient and preferable strategy.

The clearance of liposomes from the circulation is accomplished by the mononuclear phagocytic system based on the lipid composition, size of the NPs, and the extent of protein binding or opsonization by serum proteins.[9] Larger LNPs (>250 nm) are unable to pass through the fenestrations between the tumor endothelial cells. Smaller LNPs (10 nm) are rapidly filtered out of the circulation by the kidney. Surface charges on tumor-targeting LNPs can affect their ultimate circulation life and the potential for enhanced permeation and retention (EPR). Anionic liposomes have decreased self-aggregation and increased clearance by the kidney. Large amounts of cationic liposomes can cause a tissue inflammatory response. Both positively or negatively charged liposomes can trigger opsonization by fixing complement proteins and increase reticuloendothelial clearance. Neutrally charged liposomes have the longest circulation times and least amount of reticuloendothelial uptake but greatest aggregation, which may limit tumor penetration. The addition of polyethylene glycol (PEG), a synthetic hydrophilic polymer, to the liposomal surface reduces nonspecific interactions between liposomes and cells.[10] PEGylation also decreases the binding of opsonizing serum proteins which can potentiate their clearance by phagocytes of the reticuloendothelial system. PEGylated liposomes have a prolonged circulation time, an increased bioavailability, and a greater potential for tumor targeting. PEGylated liposomes are eventually removed and metabolized by the reticuloendothelial in the liver and spleen. LNPs are pharmaceutically proven delivery vehicles that can encapsulate a drug and also display ligands that target cancer cell-surface receptors.[11] LNPs to fight adult cancers have centered on the delivery of well-known and widely applicable cytotoxic agents, such as DOX, daunorubicin, vinorelbine, cytarabine, and vincristine.

9.3 Doxil®—The First FDA-approved Nano-drug

DOX is a cytotoxic anthracycline antibiotic isolated from cultures of *Streptomyces peucetius* var. *caesius.*[12,13] The anthraquinone part of the molecule is highly lipophilic, while the sugar component is hydrophilic. DOX molecule is both amphiphilic and amphoteric, which results in DOX binding to cell membranes and proteins. In order to increase aqueous solubility, the amino group of the sugar is protonated by forming a DOX hydrochloride (HCl) (Figure 9.1). DOX binds to nucleic acids by intercalating the DNA double helix and stabilizing topoisomerase II cleavage complexes, leading to DNA strand breaks at specific DOX-induced sites.[14]

As a hydrophilic drug, DOX HCl faces physical barriers to gain entry into the tumor. Solid tumors behave differently to normal tissues, displaying several abnormalities, such as leaky blood vessels and a poor lymphatic system.[15] The variations in distribution of blood vessels in a solid tumor can range from well-perfused periphery to semi-necrotic to avascular or necrotic regions at the core of the tumor. In addition, hydrostatic pressure generated by proliferating cancer cells within the tumor can lead to impaired blood flow and lymphatic drainage. The high interstitial pressure in tumors compared to normal tissues results in a net convection flow in the tumor interstitium outwards from the core of the tumor. To reach cancer cells in a solid tumor, therapeutic drugs must make their way to the blood vessels of a tumor, cross the vessel wall into the interstitium, and then transport through the interstitial matrix.[16,17] The mechanisms of transport may involve diffusion and/or convection depending upon the size, charge, and configuration of the therapeutic drug.[17] Transport of hydrophilic small molecules in the interstitial space is mainly by diffusion, which is governed

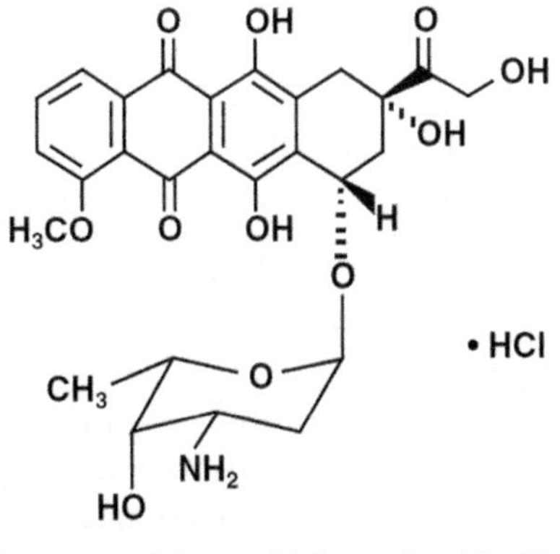

Figure 9.1 Chemical structure of doxorubicin Hydrochloride/Adriamycin. Molecular formula: $C_{27}H_{29}NO_{11} \cdot HCl$. Molecular weight: 579.99. Chemical name: (8S,10S)-10-[(3-amino-2,3,6-trideoxy-α-L-*lyxo*-hexopyranosyl)-oxy]-8-glycoloyl-7,8,9,10-tetrahydro-6,8,11-trihydroxy-1-methoxy-5,12-naphthacenedione hydrochloride. It is supplied in the HCl form as a sterile red-orange lyophilized powder containing lactose and as a sterile parenteral, isotonic solution with sodium chloride for intravenous use.

by diffusivity and the concentration gradient. The transport of larger molecules is by convection, which depends on hydraulic conductivity and pressure difference.[16,18] However, the unique nature of solid tumors with high interstitial pressure and heterogeneity in vasculature pose a formidable barrier to delivery of these therapeutic agents.[17,18]

Because DOX HCl is rapidly cleared from the circulation ($24–73 \, \text{L h}^{-1} \, \text{m}^{-2}$), resulting in slow penetration of tumor tissue *via* passive diffusion, there was a desire to stabilize plasma exposure, allowing time to achieve sufficient tumor penetration. This gave rise to the first US Food and Drug Adminstration (FDA)-approved (1995) nanomedicine: PEGylated liposomal DOX (Doxil®), a NP formulation of DOX-encapsulated in PEG-coated liposomes. Doxil® was designed to enhance the efficacy and reduce the dose-limiting toxicities (DLTs) of DOX by altering the plasma PK and tissue distribution.[19–21] Doxil® has a linear PK profile at lower doses ($10–20 \, \text{mg m}^{-2}$), while in the dose interval of $20–60 \, \text{mg m}^{-2}$, Doxil® is nonlinear. Doxil® enabled $\geq 90\%$ of the drug to remain encapsulated within the liposome during circulation, and unlike DOX, which displays a large volume of distribution (V_d, $809–1214 \, \text{L m}^{-2}$), Doxil® has a small steady-state volume and, therefore, is confined mostly to the vascular compartment (Table 9.1).[19–21]

The persistence of LNPs ($\sim 100 \, \text{nm}$) in the blood allows them to penetrate the altered vasculature of tumors and enter the interstitial space in malignant tissues. The circulation half-life for Doxil® is 2–3 days *vs.* $< 5 \, \text{min}$ for DOX. Doxil® increases DOX bioavailability nearly 90-fold at 1 week from injection of PEGylated liposomes *vs.* free drug. At the tumor sites, the accumulating liposomes gradually break down and release DOX to the surrounding tumor cells. Concentration of drug within the tumor can be up to six times greater with Doxil® than with DOX, likely accounting for the ability of Doxil® to kill DOX-resistant cells.

Doxil® offers an alternative to DOX for women with MBC, especially those who are at increased cardiac risk. In a large randomized phase III trial of Doxil® *vs.* DOX in first-line treatment of women ($n = 509$) with MBC, Doxil® ($50 \, \text{mg m}^{-2}$ every 4 weeks) *vs.* DOX ($60 \, \text{mg m}^{-2}$ every 3 weeks) were broadly comparable in efficacy (progression-free survival 6.9 *vs.* 7.8 months), but

Table 9.1 Pharmacokinetic parameters of Doxil®

Parameter (units)	Doxil® dose[a,b,c]	
	$10 \, \text{mg m}^{-2}$	$20 \, \text{mg m}^{-2}$
C_{max} ($\mu\text{g mL}^{-1}$)μ	4.12 ± 0.215	8.34 ± 0.49
Clearance ($\text{L h}^{-1} \, \text{m}^{-2}$)	0.056 ± 0.01	0.041 ± 0.004
V_{ss} (L m^{-2})	2.83 ± 0.145	2.72 ± 0.120
AUC ($\mu\text{g mL}^{-1} \, \text{h}^{-1}$)	277 ± 32.9	590 ± 58.7
$t_{1/2}$ (h)	52.3 ± 5.6	55.0 ± 4.8

[a]Administered by a 30-minute infusion ($n = 23$). Mean $\pm$ SE.
[b]*vs.* Doxil ($24–35 \, \text{L h}^{-1}/\text{m}^{-2}$).
[c]*vs.* Doxil® ($700–1100 \, \text{L}^{-1} \, \text{m}^{-2}$). AUC: area under the concentration curve; C_{max}: maximum concentration; $t_{1/2}$: terminal half-life; V_{ss}: volume of distribution at steady state.

Doxil® had a different safety profile, with significantly reduced cardiac toxicity compared with DOX [hazard ratio (HR) 3.16, 95% CI 1.58–6.31; $P<0.001$]. With Doxil®, there was less risk of developing cardiotoxicity than with DOX in all patient subgroups, including subgroups at increased risk of developing a cardiac event. Overall survival was similar (21 and 22 months for Doxil® and DOX, respectively; HR 0.94, 95% CI 0.74–1.19). Alopecia (66% *vs.* 20%), nausea (53% *vs.* 37%), vomiting (31% *vs.* 19%), and neutropenia (10% *vs.* 4%) were more often associated with DOX than Doxil®.[22]

Given the lower toxicity, but non-significant improvement in efficacy, it is unclear whether the drug delivery concepts that worked so well in animal experimental models are operational in human clinical tumors. The expected high tumoral accumulation of Doxil® due to EPR did not seem to confer a clinical meaningful advantage in various clinical trials comparing Doxil® to DOX.

9.4 Taxane-based Nanodelivery (Abraxane® and Genexol-PM®)

PTX, a natural product with antitumor activity, is a white to off-white crystalline powder that is highly lipophilic and practically insoluble in water. Taxol® (PTX) is obtained *via* a semi-synthetic process from *Taxus baccata* (Figure 9.2). However, its clinical use is limited by poor aqueous solubility, high toxicity, and low bioavailability. Consequently, a major challenge facing anticancer drug development is the delivery system for hydrophobic drugs. As a result, considerable attention has been given to the development of novel delivery techniques that can provide high bioavailability for insoluble drugs.

The taxanes, such as PTX and docetaxel, are one of the most active and widely used classes of cytotoxic agents in breast cancer treatment.[23,24] PTX is marketed as Taxol® and Abraxane® and is indicated for adjuvant treatment

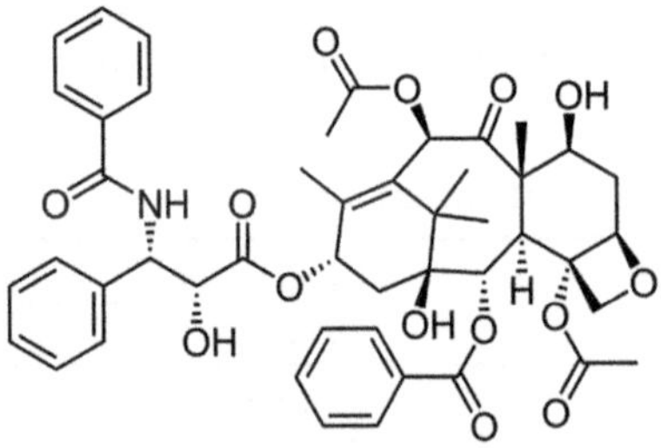

Figure 9.2 Chemical structure of paclitaxel (PTX). Molecular formula: $C_{47}H_{51}NO_{14}$. Molecular weight: 853.9. Chemical name: 5β,20-Epoxy-1,2α,4,7β,10β,13α-hexahydroxytax-1 1-en-9-one 4,10-diacetate 2-benzoate 13-ester with (2R,3S)-*N*-benzoyl-3-phenylisoserine. PTX is a white to off-white crystalline powder that is highly lipophilic, insoluble in water, and melts at ∼216 °C. Taxol® (PTX) is obtained *via* a semi-synthetic process from *Taxus baccata*.

of node-positive breast cancer after standard DOX-containing combination chemotherapy and for treatment of breast cancer after failure of combination chemotherapy for metastatic disease or relapse within 6 months of adjuvant chemotherapy.[25,26] PTX requires polyethylated castor oil [Cremophor-EL (CrEL)] as a solvent, whereas docetaxel is usually delivered in polysorbate 80 plus ethanol. Taxol® is the first-generation formulation of PTX that uses CrEL as an excipient to solubilize PTX. PTX is one of the most effective chemotherapeutic agent used in the treatment of MBCs. PTX promotes the assembly of microtubules from tubulin dimers and stabilizes microtubules by preventing depolymerization leading to subsequent defects in mitotic spindle assembly, chromosome segregation and cell division, resulting in cell death.[24] Recent data suggest that in cell cultures, most mitotic cells in primary human breast cancers contain multipolar spindles after PTX treatment because of chromosome mis-segregation on highly abnormal, multipolar spindles.[27] PTX, like other antimicrotubule agents, activates the intrinsic mitochondrial apoptotic pathway. Loss of the mitochondrial membrane potential by opening of the permeability transition pore results in the release of pro-apoptotic factor cytochrome c, caspase cascade activation, and DNA fragmentation.[28–30] These effects are modulated by the members of Bcl-2 family composed of pro-apoptotic proteins (Bax-like proteins and BH3-only proteins) and anti-apoptotic proteins (Bcl-2-like proteins). PTX turns the balance of pro-apoptotic and anti-apoptotic proteins towards apoptosis by translocation of Bax from cytosol to mitochondria and/or by phosphorylation mediated inhibition of Bcl-2.[31,32]

Higher doses of PTX result in a higher response rate. The relationships between objective disease response and PTX dose intensity in breast cancer and recurrent ovarian cancer were highly statistically significant ($P = 0.004$ and 0.022, respectively). At 135 mg m^{-2} per 21 days, the overall response rate (ORR) was 13.2%, and at 250 mg m^{-2} per 21 days, the ORR was 35.9%. The response rate seen at the intermediate dose of 175 mg m^{-2} was linear, with a correlation coefficient of 0.946.[33] However, systemic administration of solvent-based PTX is associated with serious toxic effects including myelosuppression and peripheral neuropathy, in addition to hypersensitivity reactions due to the formulation components.[34–37] The use of these solvents in the formulation of PTX limits its clinical effectiveness leading to severe anaphylaxis; acute, possibly fatal hypersensitivity reactions and peripheral neuropathy; and can reduce the extent of drug delivery to the tumor.[38,39] CrEL prevents PTX distribution from the vascular compartment into the tissue compartment due to CrEL micelles that are capable of neither interacting with endothelial cell receptors nor diffusing across biological membranes. Failure to maintain adequate concentration at the tumor microenvironment could result in regrowth of the tumor cells and the development of resistance. However, since higher doses of PTX achieve improved clinical responses, albeit with higher toxicity, it is desirable to develop a formulation of PTX which can achieve these doses while minimizing these toxicities.

9.5 CrEL-free Formulations of PTX

Several PTX nanodelivery systems have been developed and clinically confirmed to effectively enhance the efficacy and reduce the DLTs of CrEL-based
PTX in adult solid tumors. An ideal drug delivery system should improve the
therapeutic efficacy of PTX, overcoming the barriers to drug delivery, and
also minimize toxicity and lower the total dose required for the therapy. PTX
nanodelivery systems that have been examined include the use of carrier
protein (albumin), polymeric micelles, polymers, liposomes, low-density
lipoproteins, dendrimers, hydrophilic drug–polymer complexes, and cer-
amic NPs. Typical polymeric materials utilized in polymeric particulate drug
delivery systems include polylactic acid (PLA), poly(D,L-glycolide), and poly-
(lactide-co-glycolide) (PLGA). PLA and PLGA are listed as Generally Recognized as Safe under Sections 201(s) and 409 of the Federal Food, Drug, and
Cosmetic Act (FD&C Act), and are approved for use in medical devices.

Several CrEL-free formulations of PTX have undergone clinical testing, *e.g.*
NP albumin-bound PTX (*nab*-PTX, Abraxane®; Abraxis BioScience/Celgene,
Summit, NJ, USA), liposomal encapsulated PTX (LEP-ETU; NeoPharm,
Petach Tikva, Israel; EndoTAG®-1; Medigene, Planegg/Martinsried,
Germany), PTX polyglumex (Xyotax®; Cell Therapeutics, Seattle, WA, USA),
polymeric-micellar PTX (Genexol-PM®; Samyang Biopharm, Seoul, South
Korea/NantWorks, Los Angeles, CA, USA; and Nanoxel®; Dabur Pharma,
Ghaziabad, India), PTX vitamin E emulsion (Tocosol®; Sonus Pharmaceuticals,
Bothel, MA, USA), and a polymer microsphere formulation of PTX (Paclimer®;
Guilford Pharmaceuticals, Baltimore, MD, USA). Abraxane® and Genexol-PM®
are the only novel PTX formulations approved for clinical use to treat various
solid tumors. Many other novel PTX formulations have failed in advanced
clinical trials. These failures can be attributed to unfavorable PK, poor delivery,
low local concentrations, limited accumulation in the target cells and failure to
optimize dosing schedules in pivotal clinical trials.

9.6 Albumin-bound PTX (*nab*-PTX/Abraxane®)

Abraxane® (*nab*-PTX, ABI-007) is an FDA-approved CrEL-free albumin-bound
formulation of PTX (130–150 nm) for pretreated MBC patients, first-line
treatment of patients with nonsmall cell lung cancer (NSCLC) in combination with carboplatin and in combination with Gemzar® in patients with
pancreatic cancer.[26,40–42] Abraxane® was approved by the FDA in 2005 under
Section 505(b)(2) of the FD&C Act for the treatment of MBC based upon a
direct comparative trial *vs.* Taxol®. Abraxane® is formulated with human
serum albumin from pooled human plasma at a concentration similar to the
concentration of albumin in the blood.[43] Abraxane® has the following
advantages over Taxol®:

(i) Higher dosing of PTX with comparable toxicity (260 mg m^{-2} *vs.*
175 mg m^{-2} for Taxol®). In human cancer xenograft models, the

median lethal doses for Abraxane® and Taxol® were $47 \, \text{mg} \, \text{kg}^{-1}$ day^{-1} and $30 \, \text{mg} \, \text{kg}^{-1} \, \text{day}^{-1}$, respectively.[44] At the $30 \, \text{mg} \, \text{kg}^{-1} \, \text{day}^{-1}$ dose, mortality in the Abraxane® group was considerably lower in comparison to Taxol® (4% *vs.* 49%, respectively). Maximum tolerated doses (MTDs) for Abraxane® and Taxol® were $30 \, \text{mg} \, \text{kg}^{-1} \, \text{day}^{-1}$ and $13.4 \, \text{mg} \, \text{kg}^{-1} \, \text{day}^{-1}$, respectively, and these doses were equitoxic. This is comparable to an increase in MTD from $175 \, \text{mg} \, \text{m}^{-2}$ every 3 weeks for Taxol® to $300 \, \text{mg} \, \text{m}^{-2}$ every 3 weeks for Abraxane.[40]

(ii) Increased intratumoral PTX concentrations. Comparative studies in five human tumor xenografts (lung: H522; breast: MX-1; ovarian: SKOV-3; prostate: PC-3; and colon: HT29), revealed that at equitoxic doses, Abraxane® had significantly better antitumor effects (tumor volume, tumor doubling time, time to tumor recurrence, and tumor-free survival) compared with Taxol®.[45] The extent of endothelial cell binding and transcytosis of PTX was significantly greater for Abraxane® than Taxol® and was abrogated by an inhibitor of caveolae-mediated transcytosis. CrEL also suppressed PTX binding to endothelial cells and albumin.[44] These findings indicated that Abraxane® mediates greater intratumoral accumulation of PTX and associated increased efficacy compared with Taxol®. In the MX-1 human breast cancer xenograft model, the intratumoral PTX was 33% higher for Abraxane® compared to Taxol®. *In vitro* human umbilical vascular and human lung microvessel endotherlial cell binding and transport studies revealed the potential mechanisms for increased intratumoral PTX concentration. Endothelial cell binding of PTX was 9.9-fold higher and transport of PTX across an endothelial cell monolayer was 4.2-fold higher with Abraxane®. Inhibition of caveolae-mediated transcytosis by methyl β-cyclodextrin, completely suppressed endothelial cell transcytosis of Abraxane®.

(iii) More favorable PK and reduced neutropenia. In preclinical studies, Abraxane® achieved a higher plasma clearance and a larger V_d *vs.* Taxol®.[46] It was suggested that micellar entrapment of PTX by CrEL could affect the PK linearity of Taxol®.[45,47] In clinical studies, Abraxane® showed a linear PK. Plasma levels of PTX declined in a biphasic manner after intravenous (IV) administration of Abraxane®, with a rapid first phase representing distribution to the peripheral tissue, and a second slower phase of drug elimination. In a direct comparison of the PK of the formulations of PTX at dose schedules commonly used in the clinic at that time, the maximum concentration of PTX was 6.5-fold higher for Abraxane® than for Taxol®. Despite the differences in administration dose, the area under the curve (AUC) from time zero to infinity (AUC_{∞}) and plasma terminal elimination rate for PTX was not significantly different for the two formulations.[46,48] The total body clearance of PTX from plasma and V_d were both $\sim$50% higher for Abraxane® than for Taxol®. The large V_d for Abraxane® suggests extensive extravascular distribution and/or

Table 9.2 Comparison of total paclitaxel (PTX) pharmacokinetics between Abraxane[®] and Taxol[®,a]

Total PTX pharmacokinetic parameter	Abraxane[®] 260 mg m^{-2}, IV 30 min $(n=14)$		Taxol[®] 175 mg m^{-2}, IV 3 h $(n=14)$	
	Mean	%CV	Mean	%CV
$t_{1/2}$ (h)	20.0	21.3	20.9	21.4
C_{max} (ng mL^{-1})	19 556.4	36.2	5128.0	32.7
AUC$_\infty$ (h*ng mL^{-1})	20 324.5	19.5	20 821.1	25.9
V_z (L m^{-2})	375.4	21.9	270.8	34.2
CL (L h^{-1} m^{-2})	13.2	17.9	8.9	25.7

[a]AUC: area under the concentration curve; AUC$_\infty$: AUC from time zero extrapolated to infinity; CL: clearance; C_{max}: maximum concentration; $t_{1/2}$: terminal half-life; V_z: volume of distribution during terminal phase.

tissue binding of PTX. This seems to confirm the hypothesis that CrEL in Taxol[®] prevents the escape of PTX from the vascular compartment and the distribution to tissues.[38]

The rapid dissociation of Abraxane[®] in circulation is in contrast to Doxil[®]. Rather than stabilizing the PK of PTX, Abraxane[®] actually promote rapid tissue penetration of PTX as shown by decrease in area under the curve (AUC)/dose and large apparent V_d and apparent V_d at steady state (V$_{ss}$) (Table 9.2).[46,48] This is due to the inherent instability of the albumin-encapsulated NPs (Figure 9.3). The NPs break down rapidly below 100 μg mL^{-1}, which is much higher than the maximum concentration (C_{max}) of PTX, and therefore it is not expected to form stable NPs *in vivo*. Because of this breakdown, it is able to rapidly penetrate the underlying tissue. There was only marginal increase in tumor deposition in animals in contrast to Doxil[®]. However, clinical studies have shown that Abraxane[®] is superior to Taxol[®] with a similar safety profile to Taxol[®].

9.6.1 Clinical Efficacy and Safety of Abraxane[®]

In the United States, Abraxane[®] has been approved by the FDA (2005) for the treatment of MBC after failure of combination chemotherapy or relapse within 6 months of adjuvant chemotherapy. Abraxane[®] has been globally approved in more than 50 countries for the treatment of MBC. In the European countries, Abraxane[®] was approved as monotherapy (2008) for the treatment of MBC in adult patients who have failed first-line treatment for metastatic disease and for whom standard, anthracycline containing therapy is not indicated.

In the pivotal phase III MBC trial $(n=454)$, patients received either Abraxane[®] (260 mg m^{-2} every 3 weeks) or Taxol[®] (175 mg m^{-2} every 3 weeks).[40] Abraxane[®] was administered without premedication over 30 min and Taxol[®] was administered over 3 h with corticosteroid and antihistamine premedication. The ORR for patients with MBC treated with Abraxane[®] and Taxol[®] were 33% and 19% $(P<0.001)$, respectively, and in patients receiving

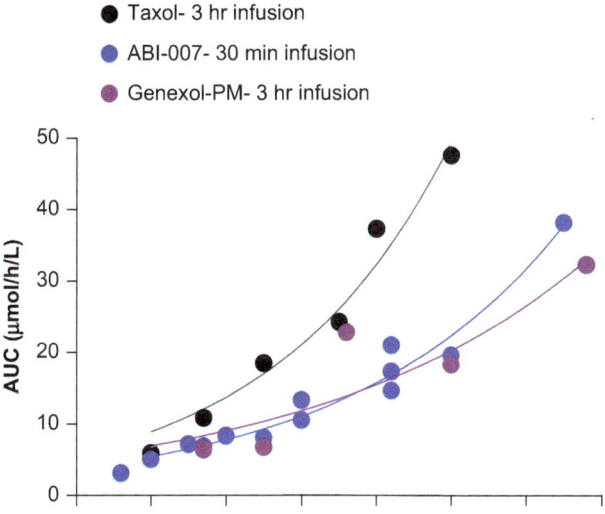

Figure 9.3 Dose proportionality of Genexol-PM®, Abraxane® (ABI-007) and Taxol®. Genexol-PM® and Abraxane® exhibited improved dose proportionality *vs.* Taxol® due to their rapid disintegration in plasma/serum which allows for the utilization of the albumin transport mechanism of paclitaxel unimpeded by Cremophor EL in the Taxol® formulation. AUC: area under the curve.

first-line treatment the response rates were 42% and 27% ($P = 0.029$). Time to progression was also significantly longer with Abraxane® for all patients (23 *vs.* 16.9 weeks, HR 0.75; $P = 0.006$), and among those receiving the drug as second-line or greater therapy (20.9 *vs.* 16.1 weeks, HR 0.73; $P = 0.02$). However, there was no significant difference in median overall survival among all patients between the two groups of treatment. The incidence of grade 4 neutropenia was lower with Abraxane® (9% *vs.* 22%, $P < 0.001$), while a higher rate of grade 3 sensory neuropathy (10% *vs.* 2%, $P < 0.001$) was apparent. No severe hypersensitivity reactions occurred with this agent, even without premedication. However, 8% of the Abraxane®-treated patients received premedication for toxicity other than hypersensitivity.[40]

Given the antitumor activity of Abraxane® in MBC and its favorable toxicity profile, this agent has been evaluated in several other solid tumors. For NSCLC, Abraxane® at a weekly dose of 100 mg m^{-2} was combined with carboplatin at AUC once every 3 weeks and compared to 200 mg m^{-2} Taxol® plus carboplatin (AUC once every 3 weeks). The outcome of these studies showed a 1 month benefit for overall survival for the patients treated with the combination and a higher response rate (33% *vs.* 25%, $P = 0.005$).[41] Consequently, Abraxane® was approved by the FDA (2012) for first-line treatment of locally advanced or metastatic NSCLC, in combination with carboplatin, in patients who are not candidates for curative surgery or radiation therapy. Abraxane® is also approved for the treatment of NSCLC in Argentina, Australia, Chile, Ecuador, Guatemala, Hong Kong, Japan, New Zealand, and Singapore. The most common adverse events ($\geq$20%) of Abraxane® in combination with carboplatin for NSCLC are anemia, neutropenia, thrombocytopenia, alopecia, peripheral neuropathy, nausea, and fatigue.

For advanced pancreatic cancer, Abraxane® (125 mg m^{-2}) was combined with 1000 mg m^{-2} Gemzar® and compared with 1000 mg m^{-2} Gemzar® monotherapy on a 3 week schedule. The outcome of these studies showed $\sim$2 months benefit for overall survival for the chemotherapy-naïve patients ($n = 861$) treated with the combination, and in addition a higher response rate (23% *vs.* 7%, $P = 0.0001$).[42] In the pivotal phase III trial, the addition of Abraxane® increased the median survival by 1.8 months (8.5 *vs.* 6.7 months, HR 0.72; $P < 0.0001$), and 2 year survival was increased from 4% to 9%. As a result, Abraxane® was approved by the FDA (2013) as first-line treatment of patients with metastatic pancreatic cancer, in combination with Gemzar®. Grade 3 and above adverse events that were reported more often with Abraxane® plus Gemzar® *vs.* Gemzar® alone were neutropenia, fatigue (10%), and peripheral neuropathy (16%).[42]

The most frequently reported Abraxane®-related adverse events seen among eight phase I, 14 phase II and four phase III clinical trials in patients with solid tumors (especially MBC, NSCLC, and pancreatic cancer) are neutropenia (74%), neuropathy (58%), various infections (32%), arthralgia (17%), myalgia (14%), and hepatotoxicity (14%), which were dose-related. Hypersensitivity, interstitial pneumonitis, and cardiotoxicity accounted for <1% of adverse events in clinical trials.

9.7 Polymeric Paclitaxel Micelles (Genexol-PM®)

Since the development of Abraxane®, it is widely thought that the albumin in Abraxane® formulation was responsible for its observed clinical effectiveness. However, due to the substantial instability of *nab*-PTX in blood it became apparent that albumin is not critical as a carrier protein since the albumin delivered as Abraxane® in the dosage form is much smaller than the amount of endogenous albumin in circulation (4.4 g *vs.* 175–250 g, respectively).

Genexol-PM® is a lyophilized non-biological, polymeric micellar formulation of PTX which employs a colloidal carrier system that allows IV delivery of a much higher PTX dose without the presence of CrEL.[49] Genexol-PM® utilizes biodegradable amphiphilic di-block copolymer composed of monomethoxy poly(ethylene glycol)-*block*-poly(D,L-lactide) (mPEG-PDLLA) to form NPs with a PTX-containing hydrophobic core and a hydrophilic shell.[49,50] The hydrophobic core of Genexol-PM® entraps free PTX by hydrophobic–hydrophobic interactions.

mPEG-PDLLA micelles have a proven capacity for drug solubilization and have entered phase III clinical trials as a alternative to Taxol® and Abraxane® in the delivery of PTX in breast cancer therapy. Unlike Abraxane®, Genexol-PM® is free of human albumin, benefits from a simple handling, preparation, and storage conditions, and has displayed a 15–30% increase in MTD over Abraxane®.[49] Unlike Abraxane®, the Genexol-PM®-based NP formulation is versatile in that it may be readily modified in multiple ways for incorporating other hydrophobic cancer drugs into the core of the Genexol-PM® NPs to further enhance its broad-spectrum antitumor activity.

Release of drug from the polymeric micelles is instantaneous. Genexol-PM® is not expected to remain as stable NPs *in vivo*. Dissolution studies employing mean particle size measurements, as determined by dynamic light scattering, have demonstrated similar NP dissolution profiles in serum/plasma for both Genexol-PM® and Abraxane®.[50] The NPs are only stable at concentration above 100 μg mL^{-1} for Abraxane® and 1000 μg mL^{-1} for Genexol-PM®—concentrations well above the clinical C_{max} of both formulations.[50] Below these concentrations, they quickly release their PTX cargo.[50] The released PTX is able to form complexes with endogenous circulating albumin.[51] Because of their instability in serum, both formulations are equally capable of utilizing the natural albumin-mediated transport system that is believed to be the major carrier of PTX in blood circulation.[52,53] This transport mechanism is probably functional for all hydrophobic drugs which cannot be transported by diffusion since the majority of the drug would be protein bound. The clinical success achieved with PTX-loaded polymeric micelles over Taxol® have mitigated the importance of albumin in Abraxane® as a drug carrier protein for PTX formulation.[50–53]

The clinical PK of Genexol-PM® displayed marked differences compared to equivalent doses of Taxol®. Genexol-PM® displayed more consistent dose proportionality over the dose range of 85–435 mg m^{-2}, lower AUC$_{\infty}$ and C_{max}, and higher systemic total body clearance.[54,55] The mean apparent V_d of PTX in the terminal elimination phase following infusions of Genexol-PM® was greater than the V_d of Taxol®. The increase in the clearance and the apparent V_d estimates of PTX when administered as Genexol-PM® is likely due to increased distribution from the blood compartment through an albumin-mediated pathway. The PK properties of Genexol-PM® are similar to those of Abraxane®. Genexol-PM® and Abraxane® exhibit similar plasma instability such that NPs are not expected *in vivo* because the C_{max} is below the dissolution concentration of both drugs.[50] In comparison to historical data, Genexol-PM® PK parameters for human, dog, and mouse were similar to Abraxane®.[56] Comparison of historical published data of two phase I clinical trials revealed that Genexol-PM® and Abraxane® display similar PK profiles under the same dosing regimen.[54,55] The dose proportionality of Genexol-PM® is compared to that of Taxol® and Abraxane® (ABI-007) in Figure 9.3. Similarity between Genexol-PM® and Abraxane® demonstrated that albumin as a carrier for PTX can be replaced with a chemical polymer.

9.7.1 Clinical Studies of Genexol-PM®

Genexol-PM® has completed phase I or phase II trials in MBC, NSCLC, pancreatic cancer, ovarian cancer, and bladder cancer in the United States and/or non-US countries. Genexol-PM® has been clinically investigated for the treatment of refractory and metastatic cancers in South Korea, the United States, and Russia. Thus far, Genexol-PM® has been approved in South Korea, marketed as Genexol-PM®, and in India, Vietnam, Philippines, Thailand as Paxus-PM®. The approved indications are (i) first-line therapy for recurrent or MBC; (ii) first-line therapy for recurrent or metastatic NSCLC in combination with cisplatin; (iii) first-line therapy for ovarian cancer in combination with carboplatin; and (iv) second-line therapy for recurrent or MBC after failure of previous therapy. Genexol-PM® is currently being developed in the United States. for approval to treat MBC and NSCLC by the FDA under a 505(b)(2) regulatory pathway with Abraxane® as the reference drug product.

9.7.2 Clinical Efficacy of Genexol-PM®

Phase II trials have been conducted against MBC, anthracycline-resistant breast cancer, NSCLC in combination with cisplatin, and advanced pancreatic cancer.[57–60] In these trials, Genexol-PM® (300 mg m^{-2} *via* a 3 h infusion at each cycle, every 3 weeks), showed high response rates and relatively low incidences of severe myelosuppression in spite of high PTX doses compared to Taxol®.[57–60]

In a single-arm, multicenter phase II trial in patients with MBC, a total of 331 chemotherapy cycles of Genexol-PM® were administered with a median eight cycles per patient.[57] Forty-one women received Genexol-PM® by IV infusion at 300 mg m^{-2} over 3 h every 3 weeks without routine premedication. The ORR for the overall intention-to-treat population was 58.5%, with five complete responses and 19 partial responsess. Thirty-seven patients who received Genexol-PM® as a first-line therapy for their metastatic disease showed a response rate of 59.5%, and two responses were reported in four subjects treated in the second-line setting for their metastatic disease. Grade 3 toxicities included sensory neuropathy (51.2%) and myalgia (2.4%). A hypersensitivity reaction was observed in 19.5% of the patients.

A multicenter phase II trial of Genexol-PM® was conducted in 69 patients with advanced NSCLC. Genexol-PM® was dosed at 230 mg m^{-2} and cisplatin was dosed at 60 mg m^{-2} on day 1 of a 3 week cycle as first-line therapy. Overall response was 37.7%, median time to progression was 5.8 months, and median survival time was 21.7 months. Grade 3 toxicities include sensory neuropathy (13.0%) and arthralgia (7.3%). A hypersensitivity reaction was observed in 5.8% of the patients.[58] A phase IIB trial was subsequently performed comparing the Genexol-PM® (230 mg m^{-2})/cisplatin (60 mg m^{-2}) combination against Taxol® (175 mg m^{-2})/cisplatin (60 mg m^{-2}) combination. The patients were randomized to the

Genexol-PM$^®$ arm ($n = 140$) and the Taxol$^®$ arm ($n = 136$), and the response rate was 43.6% *vs.* 41.9%, respectively.[59] Noninferiority was demonstrated for response, progression-free survival and overall survival. The overall rate of adverse events was similar between the two groups.[59] A phase II trial was also conducted with Genexol-PM$^®$ at 230 mg m^{-2} on day 1 and gemcitabine at 1000 mg m^{-2} on day 1 and day 8 of a 3 week cycle. Forty-three patients were treated with an ORR of 46.5%, median progression-free survival of 4.0 months, and median overall survival of 14.8 months.[61]

For ovarian cancer, a phase I and a phase II trial were conducted to demonstrate noninferiority of Genexol-PM$^®$ to that of Taxol$^®$. For the phase I trial, 18 patients with stage IIB–IV epithelial ovarian cancer were treated with Genexol-PM$^®$ and carboplatin (AUC 5). The Genexol-PM$^®$ was dose-escalated from 220 mg m^{-2} to 300 mg m^{-2}. MTD was not reached and the recommended dose for the phase II trial was 260 mg m^{-2}.[62,63]

9.8 Conclusion

In this chapter, we have reviewed the three NP platforms. The Doxil$^®$ formulation of DOX allows for higher circulating concentrations of the drug and reduces its cardiotoxicity and other toxicities while maintaining clinical efficacy. Abraxane$^®$ is an albumin NP formulation of PTX which improves the drug's efficacy by improving tissue penetration of the drug resulting in large V_{ss} and low AUC/dose. This resulted in less tumor accumulation in preclinical models but superior clinical results. Genexol-PM$^®$, a non-albumin containing NP formulation of PTX which uses a chemical polymer, demonstrated bioequivalence to Abraxane$^®$.

Genexol-PM$^®$ as a CrEL-free, nonbiological micellar NP formulation of PTX showed an improved efficacy and therapeutic index in multiple preclinical tumor models, especially against drug-resistant tumor models. In clinical studies, unlike Taxol$^®$, Genexol-PM$^®$ maintained PK dose-proportionality up to 435 mg m^{-2}, and, unlike Abraxane$^®$ which displayed a rapid increase in toxicity at doses >300 mg m^{-2}, Genexol-PM$^®$ displayed DLT at doses >400 mg m^{-2}. Unlike Abraxane$^®$, Genexol-PM$^®$ is free of human serum albumin, allows for simple handling and preparation and broader storage conditions, and has displayed a 15–30% increase in MTD over Abraxane$^®$. Through therapeutic drug monitoring, Genexol-PM$^®$ has the potential to out-perform Abraxane$^®$ in enhancing the delivery of PTX to tumors while further reducing the toxicity of both single and combination treatment regimens. Both formulations quickly release PTX into the circulation where it binds to endogenous circulating albumin, believed to be the major carrier of PTX in blood circulation. This eliminates the need for albumin as a carrier protein in the PTX formulation. Moreover, Genexol-PM$^®$ exhibits exceptional stability in protein-free fluids ($\sim$10-fold higher stability than Abraxane$^®$) and could potentially be better suited for intraperitoneal administration for ovarian cancer and intravesicle administration for bladder cancer. Unlike Abraxane$^®$, the Genexol-PM$^®$ NP formulation platform is versatile in that it can be readily modified in multiple ways for incorporation of other lipophilic anticancer drugs into the core of the Genexol-PM® to further enhance its broad-spectrum antitumor activity.

References

1. D. Peer, J. M. Karp, S. Hong *et al.*, Nanocarriers as an emerging platform for cancer therapy, *Nat. Nanotechnol.*, 2007, **2**, 751–760.
2. N. Bertrand, J. Wu, X. Xu, N. Kamaly *et al.*, Cancer nanotechnology: The impact of passive and active targeting in the era of modern cancer biology, *Adv. Drug Delivery Rev.*, 2013, pii: S0169-409X(13)00268-8.
3. U. Prabhakar, D. C. Blakey, H. Maeda *et al.*, Challenges and key considerations of the enhanced permeability and retention effect (EPR) for nanomedicine, *Cancer Res.*, 2013, **73**, 2412–2417.
4. H. Maeda, The enhanced permeability and retention (EPR) effect in tumor vasculature: the key role of tumor-selective macromolecular drug targeting, *Adv. Enzyme Regul.*, 2001, **41**, 189–207.
5. L. M. Bareford and P. W. Swaan, Endocytic mechanisms for targeted drug delivery, *Adv. Drug Delivery Rev.*, 2007, **59**, 748–758.
6. R. Weissleder, K. Kelly, E. Yi Sun *et al.*, Cell-specific targeting of nanoparticles by multivalent attachment of small molecules, *Nat. Biotechnol.*, 2005, **23**, 1418–1423.
7. J. Gao, S.-S. Feng and Y. Guo, Antibody engineering promotes nanomedicine for cancer treatment, *Nanomedicine*, 2010, **5**, 1141–1145.
8. V. P. Torchilin, Recent advances with liposomes as pharmaceutical carriers, *Nat. Rev. Drug Discovery*, 2005, **4**, 145–160.
9. A. Gabizon and D. Papahadjopoulos, Liposome formulations with prolonged circulation time in blood and enhanced uptake by tumors, *Proc. Natl. Acad. Sci. U. S. A.*, 1988, **85**, 6949–6953.
10. M. O'Brien, N. Wigler, M. Inbar *et al.*, Reduced cardiotoxicity and comparable efficacy in a phase III trial of PEGylated liposomal doxorubicin HCl (CAELYX™/Doxil®) versus conventional doxorubicin for first-line treatment of metastatic breast cancer, *Annu. Oncol.*, 2004, **15**, 440–449.
11. M. I. Koukourakis, S. Koukouraki, A. Giatromanolaki *et al.*, High intratumoral accumulation of stealth liposomal doxorubicin in sarcomas: rationale for combination with radio therapy, *Acta Oncol.*, 2000, **39**, 207–211.
12. S. H. Kim and J. H. Kim, Lethal effect of adriamycin on the division cycle of HeLa cells, *Cancer Res.*, 1972, **32**, 323–325.
13. R. L. Momparler, M. Karon, S. E. Siegel and F. Avila, Effect of adriamycin on DNA, RNA, and protein synthesis in cell-free systems and intact cells, *Cancer Res.*, 1976, **36**, 2891–2895.
14. G. Capranico, K. W. Kohn and Y. Pommier, Local sequence requirements for DNA cleavage by mammalian topoisomerase II in the presence of doxorubicin, *Nucleic Acids Res.*, 1990, **18**, 6611–6619.

15. R. K. Jain, Normalization of tumor vasculature: an emerging concept in antiangiogenic therapy, *Science*, 2005, **307**, 58–62.
16. R. K. Jain, Delivery of molecular and cellular medicine to solid tumors, *J. Controlled Release*, 1998, **53**, 49–67.
17. J. H. Jang, M. G. Wientjes, D. Lu and J. L. Au, Drug delivery and transport to solid tumors, *Pharm. Res.*, 2003, **20**, 1337–1350.
18. R. K. Jain, Barriers to drug delivery in solid tumors, *Sci. Am.*, 1994, **271**, 58–65.
19. D. W. Northfelt, F. J. Martin, P. K. Working *et al.*, Doxorubicin encapsulated in liposomes containing surface-bound polyethylene glycol: pharmacokinetics, tumor localization, and safety in patients with AIDS-related Kaposi's sarcoma, *J. Clin. Pharmacol.*, 1996, **36**, 55–63.
20. N. M. Marina, D. Cochrane, E. Harney *et al.*, Dose escalation and pharmacokinetics of pegylated liposomal doxorubicin (Doxil) in children with solid tumors: a pediatric oncology group study, *Clin. Cancer Res.*, 2002, **8**, 413–418.
21. A. Gabizon, H. Schmeeda and Y. Barenholz, Pharmacokinetics of pegylated liposomal doxorubicin: review of animal and human studies, *Clin. Pharmacokinet.*, 2003, **42**, 419–426.
22. M. E. R. O'Brien, N. Wigler, M. Inbar, R. Rosso, E. Grischke, A. Santoro *et al.*, Reduced cardiotoxicity and comparable efficacy in a phase III trial of pegylated liposomal doxorubicin HCl (CAELYX™/Doxil®)versus conventional doxorubicin for first-line treatment of metastatic breast cancer, *Annu. Oncol.*, 2004, **15**, 440–449.
23. M. C. Wani, H. L. Taylor, M. E. Wall, P. Coggon and A. T. McPhail, Plant antitumor agents. VI. The isolation and structure of taxol, a novel antileukemic and antitumor agent from *Taxus brevifolia*, *J. Am. Chem. Soc.*, 1971, **93**, 2325–2327.
24. C. M. Spencer and D. Faulds, Paclitaxel. A review of its pharmacodynamic and pharmacokinetic properties and therapeutic potential in the treatment of cancer, *Drugs*, 1994, **48**, 794–847.
25. Taxol® (paclitaxel) injection, April 2011. Bristol-Myers Squibb Company, Princeton, NJ, USA. http://packageinserts.bms.com/pi/pi_taxol.pdf.
26. Abraxane® for Injectable Suspension (paclitaxel protein-bound particles for injectable suspension) (albumin-bound), July 2015. Celgene Corporation, Summit, NJ. http://www.abraxane.com/downloads/Abraxane_PrescribingInformation.pdf.
27. L. M. Zasadil *et al.*, Cytotoxicity of paclitaxel in breast cancer is due to chromosome miiegregation on multipolar spindles, *Sci. Transl. Med.*, 2014, **6**, 229ra43.
28. A. Goncalves, D. Braguer, G. Carles, N. Andre, C. Prevot and C. Briand, Caspase-8 activation independent of CD95/CD95-L interaction during paclitaxel-induced apoptosis in human colon cancer cells (HT29-D4), *Biochem. Pharmacol.*, 2000, **60**, 1579–1584.
29. M. Carre, G. Carles, N. Andre, S. Douillard, J. Ciccolini, C. Briand and D. Braguer, Involvement of microtubules and mitochondria in the

antagonism of arsenic trioxide on paclitaxel-induced apoptosis, *Biochem. Pharmacol.*, 2002, **63**, 1831–1842.

30. N. Andre, M. Carre, G. Brasseur, B. Pourroy, H. Kovacic, C. Briand and D. Braguer, Paclitaxel targets mitochondria upstream of caspase activation in intact human neuroblastoma cells, *FEBS Lett.*, 2002, **532**, 256–260.

31. G. W. Makin, B. M. Corfe, G. J. Griffiths, A. Thistlethwaite, J. A. Hickman and C. Dive, Damage-induced Bax N-terminal change, translocation to mitochondria and formation of Bax dimers/complexes occur regardless of cell fate, *EMBO J.*, 2001, **20**, 6306–6315.

32. S. Haldar, N. Jena and C. M. Croce, Inactivation of Bcl-2 by phosphorylation, *Proc. Natl. Acad. Sci. U. S. A.*, 1995, **92**, 4507–4511.

33. E. Reed, R. Bitton, G. Sarosy and E. Kohn, Paclitaxel dose intensity, *J. Infus. Chemother.*, 1996, **6**, 59–63.

34. R. B. Weiss, R. C. Donehower, P. H. Wiernik, T. Ohnuma, R. J. Gralla, D. L. Trump, J. R. Baker, D. A. Van Echo, D. D. Von Hoff and B. Leyland-Jones, Hypersensitivity reactions from taxol, *J. Clin. Oncol.*, 1990, **8**, 1263–1268.

35. H. Gelderblom, J. Verweij, K. Nooter and A. Sparreboom, Cremophor EL: the drawbacks and advantages of vehicle selection for drug formulation, *Eur. J. Cancer*, 2001, **37**, 1590e8.

36. A. J. ten Tije, J. Verweij, W. J. Loos and A. Sparreboom, Pharmacological effects of formulation vehicles: implications for cancer chemotherapy, *Clin. Pharmacokinet.*, 2003, **42**, 665–685.

37. N. I. Marupudi, J. E. Han, K. W. Li, V. M. Renard, B. M. Tyler and H. Brem, Paclitaxel: a review of adverse toxicities and novel delivery strategies, *Expert Opin. Drug Saf.*, 2007, **6**, 609–621.

38. A. Sparreboom, O. van Tellingen, W. J. Nooijen and J. H. Beijnen, Nonlinear pharmacokinetics of paclitaxel in mice results from the pharmaceutical vehicle Cremophor EL, *Cancer Res.*, 1996, **56**, 2112–2115.

39. A. Sparreboom, L. van Zuylen, E. Brouwer, W. J. Loos, P. de Bruijn, H. Gelderblom, M. Pillay, K. Nooter, G. Stoter and J. Verweij, Cremophor EL mediated alteration of paclitaxel distribution in human blood: clinical pharmacokinetic implications, *Cancer Res.*, 1999, **59**, 1454–1457.

40. W. J. Gradishar, S. Tjulandin, N. Davidson, H. Shaw, N. Desai, P. Bhar, M. Hawkins and J. O'Shaughnessy, Phase III trial of nanoparticle albumin-bound paclitaxel compared with polyethylated castor oil-based paclitaxel in women with breast cancer, *J. Clin. Oncol.*, 2005, **23**, 7794–7803.

41. M. A. Socinski, I. Bondarenko, N. A. Karaseva, A. M. Makhson, I. Vynnychenko, Okamoto *et al.*, Weekly nab-paclitaxel in combination with carboplatin versus solvent-based paclitaxel plus carboplatin as first-line therapy in patients with advanced non-small-cell lung cancer: final results of a phase III trial, *J. Clin. Oncol.*, 2012, **30**, 2055–2062.

42. D. D. Von Hoff, T. Ervin, F. P. Arena, E. G. Chiorean, J. Infante, M. Moore *et al.*, Increased survival in pancreatic cancer with nab-paclitaxel plus gemcitabine, *N. Engl. J. Med.*, 2013, **369**, 1691–1703.

43. Y. Yamamoto, I. Kawano and H. Iwase, Nab-paclitaxel for the treatment of breast cancer: efficacy, safety, and approval, *OncoTargets Ther.*, 2011, **4**, 123–136.

44. N. Desai, V. Trieu, Z. Yao, L. Louie, S. Ci, A. Yang, C. Tao, T. De, B. Beals, D. Dykes, P. Noker, R. Yao, E. Labao, M. Hawkins and P. Soon-Shiong, Increased antitumor activity, intratumor paclitaxel concentrations, and endothelial cell transport of cremophor-free, albumin-bound paclitaxel, ABI-007, compared with cremophor-based paclitaxel, *Clin. Cancer Res.*, 2006, **12**, 1317–1324.

45. M. Roser, D. Fischer and T. Kissel, Surface-modified biodegradable albumin nano- and microspheres. II: effect of surface charges on in vitro phagocytosis and biodistribution in rats, *Eur. J .Pharm. Biopharm.*, 1998, **46**, 255–263.

46. A. Sparreboom, C. D. Scripture, V. Trieu, P. J. Williams, T. De, A. Yang, B. Beals, W. D. Figg, M. Hawkins and N. Desai, Comparative preclinical and clinical pharmacokinetics of a cremophor-free, nanoparticle albumin-bound paclitaxel (ABI-007) and paclitaxel formulated in Cremophor (Taxol), *Clin. Cancer Res.*, 2005, **11**, 4136–4143.

47. M. J. Hawkins, P. Soon-Shiong and N. Desai, Protein nanoparticles as drug carriers in clinical medicine, *Adv. Drug Delivery Rev.*, 2008, **60**, 876–885.

48. N. K. Ibrahim, N. Desai, S. Legha, P. Soon-Shiong, R. L. Theriault, E. Rivera, B. Esmaeli, S. E. Ring, A. Bedikian A, G. N. Hortobagyi and J. A. Ellerhorst, Phase I and pharmacokinetic study of ABI-007, a cremophor-free, protein-stabilized, nanoparticle formulation of paclitaxel, *Clin. Cancer Res.*, 2002, **8**, 1038–1044.

49. S. C. Kim, D. W. Kim, Y. H. Shim, J. S. Bang, H. S. Oh, S. Wan Kim and M. H. Seo, In vivo evaluation of polymeric micellar paclitaxel formulation: toxicity and efficacy, *J. Controlled Release*, 2001, **72**, 191–202.

50. K. Motamed, Y. Goodman, L. Hwang, C. Hsiao and V. Trieu, IG-001 – a non-biologic nanoparticle paclitaxel for the treatment of solid tumors, *J. Nanomater. Mol. Nanotechnol.*, 2014, **3**, 1.

51. K. Paál, J. Müller and L. Hegedûs, *High Affinity Binding of Paclitaxel to Human Serum Albumin.* 2001, vol. 268, pp. 2187–2191.

52. D. A. Yardley, nab-Paclitaxel mechanisms of action and delivery, *J. Controlled Release*, 2013, **170**, 365–372.

53. A. Sethi, M. Sher, M. R. Akram, S. Karim, S. Khiljee, A. Sajjad, S. N. Shah and G. Murtaza, Albumin as a drug delivery and diagnostic tool and its market approved products, *Acta Pol. Pharm.*, 2013, **70**, 597–600.

54. T. Y. Kim, D. W. Kim, J. Y. Chung, S. G. Shin, S. C. Kim, D. S. Heo, N. K. Kim and Y. J. Bang, Phase I and pharmacokinetic study of Genexol-PM, a cremophor-free, polymeric micelle-formulated paclitaxel, in patients with advanced malignancies, *Clin. Cancer Res.*, 2004, **10**, 3708–3716.

55. W. T. Lim, E. H. Tan, C. K. Toh, S. W. Hee, S. S. Leong, P. C. Ang, N. S. Wong and B. Chowbay, Phase I pharmacokinetic study of a weekly

liposomal paclitaxel formulation (Genexol-PM) in patients with solid tumors, *Annu Oncol.*, 2010, **21**, 382–388.

56. J. Hsu, K. Motamed and V. Trieu, IG-001 pharmacokinetics: Bioequivalence to nab-paclitaxel across mouse, dog, monkey and human. Presented at the 31st Annual Miami Breast Cancer Conference, March 6–9, 2014, Miami Beach, FL, USA.

57. K. S. Lee, H. C. Chung, S. A. Im, Y. H. Park, C. S. Kim, S. B. Kim, S. Y. Rha, M. Y. Lee and J. Ro, Multicenter phase II trial of Genexol-PM, a Cremophor-free, polymeric micelle formulation of paclitaxel, in patients with metastatic breast cancer, *Breast Cancer Res. Treat.*, 2008, **108**, 241–250.

58. D. W. Kim, S. Y. Kim, H. K. Kim, S. W. Kim, S. W. Shin, J. S. Kim, K. Park, M. Y. Lee and D. S. Heo, Multicenter phase II trial of Genexol-PM, a novel Cremophor-free, polymeric micelle formulation of paclitaxel, with cisplatin in patients with advanced non-small-cell lung cancer, *Annu. Oncol.*, 2007, **18**, 2009–2014.

59. S. Y. Lee, H. S. Park, K. Y. Lee, H. J. Kim, Y. J. Jeon, T. W. Jang, K. H. Lee, Y. C. Kim, K. S. Kim, J. Oh and S. Y. Kim, Paclitaxel-loaded polymeric micelle ($230\,mg/m^2$) and cisplatin ($60\,mg/m^2$) vs. paclitaxel ($175\,mg/m^2$) and cisplatin ($60\,mg/m^2$) in advanced non-small-cell lung cancer: a multicenter randomized phase IIB trial, *Clin. Lung Cancer*, 2013, **14**, 275–282.

60. M. W. Saif, N. A. Podoltsev, M. S. Rubin, J. A. Figueroa, M. Y. Lee, J. Kwon, E. Rowen, J. Yu and R. O. Kerr, Phase II clinical trial of paclitaxel loaded polymeric micelle in patients with advanced pancreatic cancer, *Cancer Invest.*, 2010, **28**, 186–194.

61. H. K. Ahn, M. Jung, S. J. Sym, D. B. Shin, S. M. Kang, S. Y. Kyung, J. W. Park, S. H. Jeong and E. K. Cho, A phase II trial of Cremophor EL-free paclitaxel (Genexol-PM) and gemcitabine in patients with advanced non-small cell lung cancer, *Cancer Chemother. Pharmacol.*, 2014, **74**, 277–282.

62. Y. Kim, S. W. Lee, S. Kim, Y. Kim, B. Kim and S. Kang, Phase I trial of Cremophor-free, polymeric micelle formulation of paclitaxel with carboplatin in patients with advanced epithelial ovarian cancer, *J. Clin. Oncol.*, 2011, **29**, supplement. Abstract e15533.

63. Y. M. Kim, S. W. Lee, C. H. Cho, S. Y. Hur, B. G. Kim, J. H. Kim, S. C. Kim, S. M. Kim, Y. T. Kim, H. S. Ryu and S. B. Kang, An open label, randomized, parallel, phase II trial to evaluate the efficacy and safety of cremophor-free, polymeric micelle formulation of paclitaxel compared to paclitaxel in subjects with ovarian cancer, *J. Clin. Oncol.*, 2013, **31**, (suppl; abstract 5568).

Developing a Predictable Regulatory Path for Nanomedicines by Accurate and Objective Particle Measurement

AMY J. PHILLIPS

Izon Science Ltd, Burnside, Christchurch 8053, New Zealand
Email: amy.phillips@izon.com

10.1 Introduction

The use of nanotechnology—that is, technologies based on materials and devices 10–100 nm in diameter—is a rapidly expanding area of medicine.[1] Nanomedicine is defined as the monitoring, repair, construction and control of human biological systems at the molecular level, using engineered nanodevices and nanostructure.[2] Compared with current therapeutics (for example, small-molecule drugs), the addition of a carrier material to form a nanoparticle can improve the solubility, release profile and circulation time, as well as the targeted delivery of the drug.[3–5] The promise of technologies such as nanorobots that can enter the body and carry out specific and localised tasks are becoming ever more possible, while current applications are already delivering improved diagnostic abilities and targeted therapeutic effects. For example, a well-known successful nanomedicine is Doxil®, a

RSC Drug Discovery Series No. 51
Nanomedicines: Design, Delivery and Detection
Edited by Martin Braddock
© The Royal Society of Chemistry 2016
Published by the Royal Society of Chemistry, www.rsc.org

~ 90 nm polyethylene glycol (PEG)-ylated liposomal formulation of the anti-cancer drug doxorubicin hydrochloride. The encapsulated form of doxorubicin has a longer circulation time, as well as improved targeting due to the enhanced permeability of the liposomal formulation compared to the solubilised form.[6–8]

The emergence of new therapies or technologies brings the potential to improve human health and quality of life, but any technology with medical application must first be subject to stringent quality control and safety checks. In order to ensure that these are carried out to an appropriate standard, these standards must first be agreed upon and suitably enforced, which are not trivial matters. The European Commission has published a definition of the term "nanomaterials"[9] to be used for specifically for regulatory procedures. However, it is acknowledged that in order to best implement this definition, precise measurements of a material must be carried out in order to verify whether it truly qualifies as a "nanomaterial".[10] Suitable protocols, standardised techniques and reference materials are therefore essential, and must be accessible to manufacturers, in order to be able to provide accurate reports to regulatory authorities.

The many stages of development that need to be regulated are beyond the scope of this chapter, as are the *in situ* measurement of nanoparticles and the exposure and retention of particles in products or in the body. The focus of this chapter is, therefore, the measurement of purified nanoparticles post-production and prior to administration. The current approaches to regulation and standardisation are discussed, as well as considerations for the future of the field, as there are considerable issues which remain to be solved at the present time. The capabilities and limitations of current measurement techniques of nanoparticles in suspension are discussed, particularly in relation to the regulatory requirements. For the purposes of this chapter, the term "particle" primarily pertains to synthesised nanoparticles such as liposomes, metallic nanoparticles and so on, which require regulation. Other nanoparticles that are of medical significance (such as exosomes, aggregates of molecular drug preparations and similar) may also require regulation, but the measurement and analysis of these is beyond the scope of this chapter and therefore is not discussed.

10.2 Regulation of Nanomedicines

10.2.1 The Need to Develop Regulatory Pathways for Nanomedicines

When a material enters the nanoscale, its properties and functions can be altered dramatically and uniquely, due to quantum size effects and the increased surface area to volume ratio. The restriction of electron movement is responsible for the altered electrical and optical properties of nanoscale materials compared to their bulk counterparts, while the increased surface area affects the melting point, reactivity, adhesion and other physical properties of the material.[11–13] A well-known example of a material with altered properties when the particle size is brought to nanoscale dimensions is titanium dioxide (TiO_2). The UV protection that TiO_2 nanoparticles provide is due to the light-scattering properties that are associated with the reduced particle dimensions compared to those of the bulk form.[14] The unique properties of nanoscale materials offer huge potential to improve the standard of patient care; for example, use of materials in this size range can allow increased contrast for diagnostic imaging[15,16] or can allow the material to access to locations that were previously inaccessible.[5] In-depth understanding of the surface properties of nanoparticles—and with that, the ability to modify these properties—has enabled the development of nano-carriers with improved pharmacokinetics through increased circulation times and reduced toxicity.[17] This is highlighted by the design of "stealth" liposomes, which are able to remain in circulation and target specific *in vitro* environments through surface modifications which alter the particle charge[18] (discussed further in later sections).

The relevance of the physical properties of nanomedicines to their biological effects and critical need for improved understanding of these effects have been emphasised by the United States Food and Drug Administration (US FDA), particularly in relation to the development of generics. In 2013, one of the first generic complex drugs—Lipodox™, a generic form of Doxil®—was approved.[19] The approval of Lipodox™ was streamlined in order to meet drug shortages, and approval took only a year, when on average, generic medicines take 3 years to be approved.[20] The effects of small differences remain poorly understood, and, as approval of a generic drug is dependent upon the bio- and physical-equivalence to the branded drug, it is clear that a better understanding of the shape, size and composition of the formulation will provide higher confidence in the equivalence. In view of these knowledge gaps, the speed at which the approval of Lipodox™ progressed led to concerns that the assessments were insufficient, which were reinforced by the lack of regulations.[21] Reliable measurement techniques that are reproducible between operators and different laboratory groups are therefore essential, but these will only be effective once strict guidelines are decided upon.

As with any emerging technology, despite the enormous proposed benefits, there are growing concerns about the potential detrimental effects to human health, and the concept of nanotoxicology is now widely accepted.[22,23] Although nanomedicines have been authorised by licensing authorities for a number of decades, the field is still considered to be in its infancy, and many of the studies of toxicity have yielded inconclusive results—the physicochemical properties are known to correlate to toxicity, but their impacts are not fully understood.[24–26] This has led to concern about the approval of products whose toxicity or safety are not fully understood, as in the case of Lipodox™.[27] The commercial success of the field will depend on consumer acceptance of nanomedicine products. If any product fails, or causes health concerns, the loss of consumer confidence could significantly undermine future developments. This is supported by the findings of the recent European Commission Strategic Nanotechnology Action Plan (SNAP 2010–2015[28]), which revealed that although the possible benefits of nanomedicines are acknowledged, among researchers, industry and public authority there is a strong perception of potential risk. It is clear that appropriate and effective regulatory structures will be essential to the progression of nanomedicines, and governing bodies realise the need for these structures now rather than later, in order to prevent future problems.

10.2.2 Current Status of Nanomedicine Regulation

To date there is no country which has a complete set of laws referring to nanomedicine. While the legislation and regulatory systems that apply to medicine and medical devices can be applicable to nanomedicines, the speed at which new technologies are progressing provides a challenge to the regulators' expertise. The question of whether nanomaterials are considered "new" or already existing products further complicates their regulation. A well-known example of this is the case of carbon nanotubes, which were classified as a new material after their discovery in 1991.[29] Although carbon has been used for centuries in medical applications, the novel characteristics and behaviours of the material in the nanoscale compared to the bulk counterpart led to the decision that carbon nanotubes should be treated as a new material, with the necessary new risk assessments.[30]

Many nanomaterials are used as carriers or encapsulation agents for small-molecule drugs, meaning that the boundaries between medical products and therapies and medicinal devices are becoming increasingly blurred. The European Medicines Agency (EMA) defines a medicinal product as

> *Any substance or combination of substances presented as having properties for treating or preventing disease in human beings. Any substance or combination of substances which may be administered to human beings with a view to making a medical diagnosis or to restoring, correcting or modifying physiological functions in human beings is likewise considered a medicinal product.*[31]

A medical device is defined as

> *any instrument, apparatus, appliance, software, material or other article, whether used alone or in combination, including the software intended by its manufacturer to be used specifically for diagnostic and/or therapeutic purposes and necessary for its proper application, intended by the manufacturer to be used for human beings for the purpose of: (a) diagnosis, prevention, monitoring, treatment or alleviation of disease, (b) diagnosis, monitoring, treatment, alleviation or of compensation for an injury or handicap, (c) investigation, replacement or modification of the anatomy or of a physiological*

process, (d) control of conception, and which does not achieve its principal intended action in or on the human body by pharmacological, immunological or metabolic means, but which may be assisted in its function by such means.[32]

From these definitions it can be seen that many nanomedicines have the characteristics of both a medicinal product and a medical device. In this case the product will have to conform to regulations for medicinal products, while also meeting the standards for medical devices.[33] The blurring of these conventional boundaries is commonly referred to as producing "borderline products", an example of which is the "honeycomb" silicon structure Bio-Silicon, which biodegrades to release drugs encapsulated in the matrix.[34] In order to establish whether existing regulations are adequate, and how to regulate products which are not adequately covered by current legislation, it is essential that the definitions and classifications conventionally applied to medical products are redefined specifically to nanomaterials. It has been suggested that the challenges associated with regulation of borderline products is likely to increase, meaning a complete overhaul of regulatory procedures may be required as each sector will adjust the technology to its own specific needs.[34] It is critical, therefore, that specific sectors are examined, and regulatory policies address the specific challenges posed by each sector. A detailed discussion of the regulatory frameworks currently in place for nanomedicines is beyond the scope of this chapter, but an overview is given below.

10.2.2.1 Regulation of Nanomedicine in the USA

In the USA, FDA approval is essential for clinical applications of new technologies, and it became clear that to harness the potential benefits, improve the use of current drugs and minimise the risks associated with nanoscale materials, revisions of current regulations were necessary. Appropriate regulation remains a challenge, and within the FDA the regulation of nanomaterials intended for medical use falls under the responsibility of committees and groups that govern the general medical legislation. These are primarily the Center for Drug Evaluation and Research, supported by the Center for Biologics Evaluation and Research and the Center for Devices and Radiological Health.

In 2006 the Nanotechnology Task Force was established, with the goal of establishing the regulatory approaches that will allow the development of innovative, safe and effective FDA-regulated products that utilise nanotechnology materials.[35] The task force highlighted the need for "a timely development of transparent, consistent, and predictable regulatory pathway" for these products,[36] and emphasised the importance of characterisation of materials—in particular, particle size—in understanding the biological response. Recent FDA advice acknowledged that the major concern regarding nanoparticles is the uncertainty of the relationship of particle size to their biological interactions.[37] It is clear, therefore, that precise and high-resolution tools are necessary to measure particles. As well as improving the efficacy and potential toxicity of current medicines, accurate characterisation will allow prediction and preparation for the regulation of future products.

10.2.2.2 Regulation of Nanomedicine in the European Union

The primary regulatory body for medicines in Europe is the EMA, whose main task is to evaluate and advise on research and development programmes pertaining to medicinal products, in order to allow market access to new medicines.[38] Within the EMA the key committees involved in nanomedicine regulation are the Committee for Medicinal Products for Human Use (CHMP), the Committee for Medicinal Products for Veterinary Use, and the New and Emerging Technologies Working Group. All chemical substances within the European Union (EU) are covered by the regulation Registration, Evaluation, Authorisation and Restriction of Chemicals (REACH). Although there is no specific reference to nanomaterials, they are currently covered under REACH.

As they emerge, nanomedicine products fall under a complex framework of regulations, directives and guidelines relating to medicinal products for human use, medical devices and their production and use. The current regulatory system was not designed to incorporate nanomedicine, and given the altered properties of nanomaterials compared to their non-nanoscale counterpart, it is questionable whether these regulations will be sufficient. Similarly to the FDA, in 2006 the EMA established their own task force—the Innovation Task Force (ITF)—in order to evaluate the specific challenges posed by nanomedicine regulation, with a particular aim to identify the need for specialised expertise early in the development process.[39] Further objectives of the ITF include to establish a discussion platform for the EMA committees to assess emerging technologies, to provide advice to scientists and committees and to review the suitability of existing regulatory procedures to emerging technologies.

While the EMA is primarily responsible for the approval of pharmaceuticals, the approval of "medical devices", including nanoscale devices, falls to the regulatory authorities of the individual states of the EU. France has been particularly proactive in introducing new legislation pertaining to nanotechnology, but in the rest of the EU nanomaterials are typically assessed on a case-by-case basis. The new legislation in France has been met with optimism from nanomedicine companies, and as the field grows it will be necessary for the rest of the EU to introduce similar legislation to cope with the number of emerging products and ensure transparency in their regulation and production. CHMP comprises an expert panel which provides the EMA with experience gleaned from researchers in different EU states, and the EMA intends to use this knowledge to achieve harmonisation across the EU.

10.2.2.3 Worldwide Regulation of Nanomedicines

As demonstrated by the example of France above, the ability of member states to establish their own regulatory procedures has led to disparities across the EU. The EMA has acknowledged the need for greater consistency across the EU with regards to nanomedicines, and has released a number of reflection papers underlining risk assessments and similar to provide a basis for regulatory guidelines.[40] The task of consistency across the world is an even greater challenge—a recent example that highlights the differences worldwide in the classification of nanomedicines is that of Nanobiotix. The company's lead product—NBTXR3—is a novel cancer nanotherapeutic, which comprises a compound intended to enhance the destruction of tumor tissue through radiotherapy. The product has been classified as a medical device in the EU, but the FDA classify it as a drug in the US.[41] Discrepancies such as this could lead to companies wrongly classifying products either mistakenly, or in order to get the product approved more quickly, or to keep within the guidelines of their country. Of more cause for concern is the somewhat unregulated release of nanomedicines in other countries such as Mexico, India and Argentina.[21] Without strict guidelines, many nanomedicines have entered the markets in these nations without sufficient physicochemical characterisation, leading to questions about their safety.[21]

Currently, the FDA has no regulatory definition for "nanomedicine", and although the EMA provide one, a consensus on the definition has not been reached worldwide.[42] There is, therefore, a need for firm guidelines on the classification of nanomedicine products as well as improved consistency in their regulation, both within regulatory organisations and across the world. Accepted definitions will improve the reliability of product characterisation, and will help to achieve transparency across the field.

10.3 Accurate and Objective Particle Measurement

The definition of a nanomaterial, as provided by the European Commission, is:

> ...*a natural, incidental or manufactured material containing particles, in an unbound state or as an aggregate or as an agglomerate and where, for 50% or more of the particles in the number size distribution, one or more external dimensions is in the size range 1 nm–100 nm.*

Current EU legislation requires manufacturers of materials and regulators to disclose whether raw materials fulfill this definition, and if the final product contains nanoparticles.[43,44] Manufacturers must be able to accurately measure their materials in order to identify whether the materials are in fact nanomaterials; and authorities must be able to measure the materials with equal accuracy to ensure the classifications of nano-/non-nanomaterial provided by manufacturers are correct, and labeled as such. Reliable measurement techniques and guidelines are therefore needed to reduce the potential for error or uncertainty, and improved standardisation would increase the efficiency of regulation.

10.3.1 Specific Challenges of Nanoscale Measurements

The development of nanomedicines is accompanied by an increase in the demand for accurate, reliable characterisation of nanoscale particles. Novel technologies present new challenges, and those associated with the characterisation of nanoparticles is compounded by a lack of knowledge relating to the biological interaction of nanomedicines, and models for *in vitro* and *in vivo* behaviour of nanoparticles. The concerns about this knowledge gap were highlighted in SNAP 2010–2015,[28] in which 90% of respondents involved in the consultation expressed concerns about the toxicity of poorly understood nanomaterials. It is also unclear whether current testing methods of, for example, biodistribution, are applicable and relevant for nanomedicines. The EMA has released a number of reflection papers detailing the agency's scientific guidelines on nanomedicines. A common theme of these documents is that the physicochemical properties (in particular, size) of particles influence their uptake, *in vivo* behaviour and stability. Accurate measurement and reporting of the physicochemical properties of nanoparticles is absolutely fundamental to improving our understanding of nanomaterials, fill the knowledge gaps, relate these properties to the *in vivo* effects, and, not least, achieve agency approval for products. To enable adequate reporting, current regulatory procedures would benefit from standard definitions referring to the size and shape of substances that are considered to be nanoscale, along with standardised means to reliably measure these parameters. For instance, the EU definition of nanomaterial is restricted to materials containing "particles", which are defined as "minute pieces of matter with defined physical boundaries".[10] As well as the ambiguity of "minute pieces", this definition can be taken to include spherical, rod-shaped, and plate-like particles, which are defined specifically as "objects" in the International Organization for Standardization's (ISO) guideline terminology.[10,45]

Several factors complicate the measurements, including the poor dispersibility or aggregation of many nanoparticles, and the variance of size determination when different methods are used. Nanoscale particles are more prone to "sticking" together, forming either weakly or strongly bound masses termed agglomerates or aggregates, respectively.[45] This dynamic tendency means that the "particle size" is not always easy to define.

The effectiveness of current characterisation methods is in question, and the need for the development of new methods and standards is becoming ever clearer. The EMA and FDA have been moving toward new tools, standards and approaches for the characterisation of nanoparticles intended for medical use.[46,47] SNAP 2010–2015[28] identified an overwhelming demand for an inventory of the types and uses of nanomaterials, further establishing the need for reliable and standardised characterisation, in order to effectively document these. The second part of this chapter discusses current methodologies used in characterisation of nanoparticles: their theory, advantages and limitations.

10.3.2 Important Parameters of Nanoscale Materials Used in Bioapplications

The development and quality assurance of nanomedicines rely on precise control of the chemical and physical properties that confer efficacy or even potential toxicity. In order to achieve the desired effects and avoid toxicity, various characteristics including particle material, size, surface charge/coating and concentration must be controlled and measured accurately. These factors play important roles in determining how each nanotechnology will interact with the body and its immune and reticuloendothelial system (RES), and it has been stated that the ultimate fate of nanoparticles within the body is determined by a combination of size and surface properties.[3] Additionally, these properties can play a fundamental role in the stability of the product, which has implications for shelf life. Accurate knowledge of these parameters is therefore essential in the development and regulatory processes.

A number of considerations must be taken into account for any measurements of nanoparticles. Firstly, the sample size that is measured will often represent an extremely small population of the entire sample. In order to achieve a representative sample it must be ensured that enough particles are measured, which can be a limiting factor for many techniques (further discussed under individual techniques). Secondly, the sample preparation can greatly influence the measurement. If the measurement technique requires the dispersion of agglomerates or aggregates, this can influence the nature of the individual particles, and furthermore could be misrepresentative of the sample under storage or usage conditions.

10.3.2.1 Size/Volume

10.3.2.1.1 Size Distribution. The size range of particles has large effects on the clearance, biodistribution and toxicokinetics of nanoformulations, and the need for accurate knowledge of this parameter is a recurrent theme in regulatory guidelines.[28,36] It has been reported that particles smaller than 10 nm are cleared from the body *via* the kidneys while particles 10–150 nm are sequestered in the bone marrow. Particles larger than 200 nm can undergo splenic clearance; those that are 50–500 nm in diameter experience clearance *via* the liver; and particles larger than 500 nm are removed by the macrophages and monocytes of the RES.[48,49] Furthermore, size is of the utmost importance in the cellular processing of particles: a direct relationship between size and uptake route has been reported.[50]

Although the importance of size is well known, the relationship to *in vivo* behaviour is unknown in many cases,[51] and the need for accurate sizing in order to indentify trends or discrepancies is frequently expressed.[14] Despite the critical need, there is a considerable lack of reliable data in this area. One reason for this lack of knowledge is that very few nanoparticle suspensions are truly monodisperse, and more typically, the particle size distribution (PSD) is broad, indicating a number a subpopulations. This poses challenges as few measurement techniques offer reliable measurement of polydisperse samples, which becomes increasingly difficult in complex biological fluids due to the increased presence of background noise.[51] This feature also introduces complexities to the regulation as the exact pharmaceutical identity is difficult to define, in particular the case of generic drug development becomes complicated as the inherent polydispersity adds unavoidable variability, making reproducibility difficult.[21] The monodispersity of manufactured particles is often poorly controlled during production, and therefore must be monitored.[52] The most accurate way to represent the PSD is to present a histogram of the number of particles in different size ranges (Figure 10.1), and typically, a combination of methods is needed for assurance.

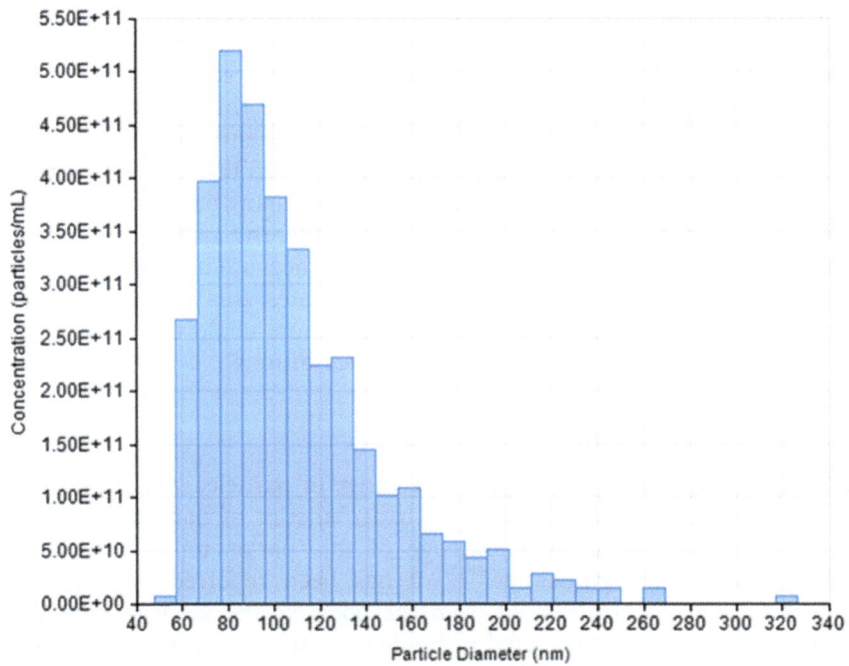

Figure 10.1   Example of a size distribution histogram of a polydisperse sample. The histogram shows the number of particles in each 10 nm bin, allowing the polydispersity to be easily assessed. Depending on the measurement technique the size distribution may be a function of the particle volumes, surface area or scattering intensity. The data shown here was collected by tunable resistive pulse screening analysis of a sample of iron nanoparticles, in-house at Izon Science Ltd (Christchurch, New Zealand).

Given that the nature of real-world materials is to show a size distribution like that in Figure 10.1, rather than a true monomodal or symmetric distribution, the way in which size is reported can introduce significant variation. The modal value for the data in Figure 10.1 will differ from the mean (average) or median values. Following the definition of a nanomaterial, in the case of material classification the most appropriate value to report would be the median, meaning a material would be classified as a "nanomaterial" if the median particle size is between 1 and 100 nm. An intrinsic problem of any measurement technique is the limit of the size range that is captured— no single method will be able to capture a large enough range to be able to calculate the absolute fraction within this range, as the upper limit needs to be several micrometers.[10]

Although particle size is arguably the most important parameter to measure in terms of regulation, a number of reliability issues complicate the matter. First of all is the interpretation of the term itself. For spherical particles a value for the radius or diameter defines the size, but when irregular particles are considered the definition becomes more complex. The Feret diameter is commonly used in microscopy and represents the distance between two parallel tangents on opposite sides of a particle. Typically the maximum and minimum Feret diameters are reported, and this has been suggested as a useful procedure for reporting of nanoparticle size.[10] However, this is only appropriate for visual measurement techniques, and different techniques may report particle diameter, radius of hydration, volume or so on. A useful definition for particle size (which has been discussed at ISO meetings[10]) would include the requirement for method specification and account for the impact of the measurement environment.

10.3.2.1.2 Agglomeration of Particles. In many cases, the "particle" that is seen in colloidal suspensions is in fact an agglomerate, due to the inevitable interaction of particles in solution. The rate of agglomeration and distribution of free and agglomerated particles will, unsurprisingly, influence the pharmacokinetics and potential toxicity of the particle, due to the altered size and exposed surface areas. Given the difficulty in controlling the size distribution, and the consideration that interactions or toxic effects can be focused to a specific size range, accurate measurement of this parameter is crucial in the development of any nanomedicine.

The presence of small amounts of agglomerates can significantly skew the mean particle size, particularly where ensemble or solution-averaged techniques are used for measurement of the particles. The most useful techniques for size determination of nanoparticles will therefore be those that offer single particle resolution rather than a solution averaged result.

10.3.2.1.3 Permeability Effects. The size of nanoparticles is directly related to their permeability. In certain pathological conditions, namely tumors, the permeability of the tissue is increased. Because of this, relatively large particles are able to diffuse into the tumor vasculature while being excluded from healthy tissues, while the lack of lymphatic drainage results in the retention of such particles.[53] This phenomenon is known as the enhanced permeability and retention effect (EPR) and has allowed passive targeting of nanoparticles to tumor environments. It is commonly accepted that most nanoscale materials within the range of 10–200 nm will accumulate in tumor tissue due to the EPR.[54] However, it has been suggested that there is an optimal size for accumulation, which may depend on a difference of as little as 30 nm.[55] These findings highlight the importance of accurate size measurement of nanoparticles to ensure that predictable—and desirable—targeting is achieved.

While permeability changes with size, the solubility of nanoparticles does not.[56] Many drugs have failed due to solubility issues, which could be resolved through the use of appropriately sized, and therefore permeable, nanocarriers. The issue of solubility must be considered differently when regulating nanomedicine, and the factors of size and permeability may be more appropriate parameters to monitor.

10.3.2.2   Concentration

The importance of accurate particle concentration measurements is also often underestimated, even though concentration is critical for determining dosages and diagnostic sensitivity. Conventionally, drug dose is expressed in mass, but for nanoparticles the surface area or particle number might be more appropriate, and therefore particle concentration must be monitored. Recently, the importance of reporting the size range across which concentration has been measured has been realised.[57,58] As with particle size, a single number value for particle concentration is somewhat misleading, given that depending on the technique used and instrument settings for measurement, the measured range can vary significantly.

In the case of nanocarriers being used to administer a soluble drug, accurate size-based concentration measurements are also important. Summing the volume of each particle of a particular size multiplied by the concentration of particles that size will provide the volume fraction. This is an important parameter to measure as the volume fraction of the particles to administered solution volume will determine the amount of drug that is delivered, that is, the payload of the particle.

10.3.2.3   Charge

The *in vivo* fate of nanoparticles also relies on the surface charge of the particles. It is generally accepted that positively charged particles have a higher rate of cell uptake and are more quickly cleared from the circulation than negatively charged particles.[3,49] The reason for the removal of negative particles from circulation is thought to be mediated by the adsorption of proteins onto the particles. It is hypothesised that the adsorption of specific proteins, or combination of proteins, triggers receptor-mediated mechanisms, and while this effect is known to influence the clearance and biodistribution, the exact mechanism is not understood.[49] Understanding of this has allowed the development of stealth nanocarriers and targeted nanoparticles, which carry surface modifications to alter biodistribution and reduce clearance from the body. Coating negatively charged nanoparticles with PEG shields the charge and reduces the adsorption of proteins, which is thought to be the reason for the increased circulation time of stealth nanocarriers, compared to uncoated, negatively charged particles which were quickly cleared after injection.[18] Furthermore, it has been seen that positively charged particles preferentially adsorb proteins with isoelectric points below 5.5, while negatively charged particles preferentially adsorb those with isoelectric points above 5.5.[59] Currently, the accurate measurement of surface charge is challenging. To be able to predict and direct the protein adsorption and, therefore, uptake and targeting of particles, reliable methods of analysing surface charge are needed. This is also important in predicting the stability of nanoparticles within the body, as an alteration of surface charge due to protein adsorption can result in a loss of stabilising interactions.[13]

10.4 Current Techniques for Characterising Nanomaterials

Due to the important and interconnected role that particle size and surface charge play in determining particle circulation time, biological interactions and ultimate biodistribution, precise and accurate means of measuring these characteristics are essential. An understanding of how size and surface properties allow nanoparticles to access novel locations will only become possible by introduction of standardised techniques. In addition to aiding researchers in understanding the biological interactions, improved particle characterisation is required to ensure quality control over manufactured particles and improve current formulations.

A range of techniques are currently used to characterise nanoparticle suspensions, but none have currently been validated for their use in nanomedicine regulation. These can be broadly categorised as ensemble or single-particle techniques,[60] and are summarised in Table 10.1. Ensemble techniques measure the bulk dispersion and report an averaged result of the particle properties. Because of the specific qualities of nanoparticles, a population comprising predominantly small particles with a few large agglomerates could have considerably different properties to a homogenous sample of medium-sized particles. This averaging can lead to low resolution and potential bias or information exclusion. A further limitation of ensemble techniques for nanomedicine applications is the inability to provide a number-based particle size distribution. Single-particle techniques are able to measure and report the properties of individual particles, and the resulting number-based size distributions provide more accurate

Table 10.1 A summarised comparison of techniques used to measure particles in suspension.

Technique	Advantages	Disadvantages
Dynamic light scattering	Easy and quick to run Sample can be recovered Applicable to many particle and solvent types	Not suitable for truly polydisperse samples Accurate knowledge of the refractive index is required Intensity weighted, small subpopulations can skew results
Differential centrifugal sedimentation	High resolution down to 2 nm Reproducible and resistant to sample bias	Best suited for homogenous samples
Electron microscopy	High resolution Single particle analysis Visual technique	Low throughput Labour intensive
Nanoparticle tracking analysis	Single particle technique Increased ability to detect subpopulations Can resolve down to 10 nm	Accurate knowledge of refractive index is required Applicable buffers are limited User-defined settings can bias results
Resistive pulse sensing	Single particle resolution Accurate identification of subpopulation Size, concentration and charge can be measured simultaneously	Highly polydisperse samples can be challenging Instrument set can limit the dynamic range

information about the sample properties. However, single-particle techniques can be low-throughput, sampling only hundreds or a few thousand particles, and require sufficient sample volume to ensure that low concentration populations are accurately represented.[61]

Any method used for regulation of nanomedicines will need to be high-resolution, non-biased and high-throughput to ensure a statistically relevant recording. The most useful will be those that measure and collate the properties of individual particles. It is well known that PSD measurements carried out using different techniques, or even machines from different manufacturers can show considerable variability.[62] Intra- and inter-laboratory measurements are subject to further variation, and institutions such as the National Institute of Standards and Technology (NIST) provide some guidelines to control the potential sources of error (for example, guidelines on the use of photo correlation spectroscopy for the measurement of nanoparticles[63]). Although NIST began producing nanoscale reference materials in 2008,[64] there is still a need for a reliable and reproducible standard method of characterisation, particularly when the requirements for bio- and physical-equivalence are considered (as discussed previously). Particle-by-particle analysis was chosen to ensure the safety of injectables with particle sizes between 10 and 25 μm (USP 788[65]), and it is most likely that a similar approach—that is, single particle size and concentration analysis—will be incorporated into the measurement of sub-micrometer nanomedicines.

10.4.1 Ensemble Measurement Techniques

10.4.1.1 Dynamic Light Scattering

Light scattering-based techniques are commonly used for the measurement of nanoparticle size and charge.[66] Dynamic light scattering (DLS; also known as photon correlation spectroscopy) measures the size of particles in solution by analysing the intensity of light scattered by the sample. The technique relies on the particles being smaller than the wavelength of the incident beam, and measures the diffusion of the particles by detecting intensity fluctuations of the scattered light. The hydrodynamic diameter of the particles in a monodisperse solution can then be calculated from the diffusion coefficient using the Stokes–Einstein equation[67,68] (eqn (10.1)).

$$d_H = \frac{kT}{3D\pi\eta} \tag{10.1}$$

Where d_H is the hydrodynamic diameter of the particle, k is the Boltzmann constant, T is the temperature, is the viscosity of the medium and D is the diffusion coefficient (for full calculations see Anderson *et al.*[62]).

The ease of use, sample recovery and applicability to a wide range of particle and solvent types have made DLS a popular method for PSD determination since its development in the 1960s.[62] However, due to the intensity-weighted approach, DLS does suffer from a number of limitations. High-quality data can only be achieved when the refractive index of the solvent is accurately known, and the concentration of particles and electrolyte are low.[66]

A major limitation of DLS is its sensitivity to skewing of the PSD in the presence of large particles.[69,70] The Rayleigh approximation states that the intensity of scattered light is proportional to the sixth power of the particle diameter ($I \propto d^6$). Therefore, a small number of large particles can significantly bias the result, swamping the scattering of the smaller particles and overestimating the average diameter. A range of algorithms have been developed to account for this limitation (for example, CONTIN[71]), and more recent DLS machines can compensate by incorporating microfluidic flow field fractionation to sort particles prior to measurement.[72] However, the former requires considerable prior knowledge of the sample, and the added cost of the latter mean it is yet to be widely adopted. For these reasons DLS is not suitable for highly polydisperse samples.

10.4.1.2 Differential Centrifugal Sedimentation

Sedimentation is a staple technique for the separation of particles based on size. However, when the particle diameter decreases below ~1 μm, the settling time becomes limiting, and once the gravitational energy becomes commensurate with the thermal energy of the particles, settling will not occur at all. Centrifugation increases the speed of sedimentation of particles and is used widely in colloid science to separate particles based on size and shape.[73] The velocity of the particle away from the axis of rotation is dependent on three forces—centrifugal, buoyant and frictional forces—as well as the density of the particle and fluid, and the volume of the particle. The particles will sediment until the forces balance, resulting in distinct bands. The resolution of this separation is poor, so to overcome this, particles are centrifuged in a tube containing a density gradient created by layering liquids of different concentration. Differential centrifugal sedimentation (DCS) uses the same principle, with the density gradient set up inside a rotating disc, and the sample is introduced to the centre. The Stokes diameter (d_s) can be calculated using eqn (10.2).

$$d_s = \left(\frac{18\eta \ln \frac{R_f}{R_0}}{t(\rho_p - \rho_f)\omega^2} \right)^{0.5} \tag{10.2}$$

Where η is the viscosity of the fluid medium, R_f is the radial position of the detector, R_0 is the radial position of the surface of the density gradient carrier fluid, t is the time taken to reach the detector located at R_f, ρ_p is the particle density, ρ_f is the fluid density and ω is the rotation speed. This allows separation of particles from 2 nm to 80 μm with a resolution of 2% of particle diameter.[74,75] DCS is an ensemble method, which can efficiently produce data that are reproducible and resistant to size bias, although it requires the use of standard calibration particles as well as knowledge of the solution viscosity and density, meaning it is best suited for homogenous samples. DCS can reliably characterise particles with neutral buoyancy and lower densities, which are otherwise difficult to measure with centrifuge technology.[74]

10.4.2 Single-particle Techniques

10.4.2.1 Electron Microscopy

Transmission electron microscopy (TEM) has been an essential tool for the characterisation of nanoparticles. As the TEM uses electron waves rather than light waves to image the sample, the resolving power is far greater than a light microscope. The wavelength of electrons in a TEM operated at 200 kV is 0.025 Å (full calculation can be found in Zou *et al.*[76]); however, in practice the resolution limit is usually ~1 Å due to chromatographic aberrations of the lenses. In the TEM a beam of electrons is produced by a thermionic or field emission electron gun, accelerated by high voltage, and focused onto the sample by magnetic condenser lenses in the same way that light can be focused by glass lenses. The transmitted beam is focused onto either photographic film or a charge-coupled device (CCD) camera by the objective and projection lenses, and the electron density, phase and periodicity of the transmitted electrons are used to produce an image.

Typically, the transmitted beam will be imaged, and the contrast in the image is produced by the loss of electrons due to absorption by the sample. Successful imaging relies on the density of the sample, therefore, many commonly used nanoparticles including metallic or oxide particles are easily imaged on the TEM. Hollow or less electron-dense particles such as liposomes and other biological vesicles show less contrast relative to the background. Negative staining is often used to overcome this, whereby the sample is coated in a heavy metal stain, thus increasing contrast. Although this is commonly used and can provide useful preliminary information, negative-stain TEM brings the potential for artifacts, such as flattening or collapse of hollow particles, giving misleading results. Furthermore, the adsorption of particles onto the carbon or polymeric support film which is required for TEM can cause bias in the results, and further structural artifacts. The use of ''holey'' grids—in which the sample is suspended in a film of stain—can resolve this issue, but the problems associated with negative stain remain.

Given these challenges, cryo-transmission electron microscopy (cryoTEM) has become the leading method of characterisation of particles such as liposomes. In this technique, the sample is rapidly frozen, resulting in the sample embedded in a vitrified layer of buffer, enabling examination of in-solution conditions. A particular advantage of TEM is the ability to collect detailed information about particle shape and composition. However, with non-spherical particles, the measurement of a diameter can give significant variations depending on the orientation of the particle on the grid, or the user-defined measurement parameters.

Despite the benefits of visualisation, TEM is often considered to be too labour intensive to be able to obtain a statistically significant measurement of PSD.[62] A number of semi-automated systems have been developed (for example, Stagg *et al.*[77]) to increase the throughput of cryoEM; however, the technique is still largely qualitative as the analysis of large datasets is prohibitively time consuming, and user-defined parameters for automation can result in information bias. Furthermore, the high-vacuum environment prevents *in situ* analysis, and the high-energy electron beam can damage biological and polymer based samples.[78]

10.4.2.2 Nanoparticle Tracking Analysis

Similarly to DLS, nanoparticle tracking analysis (NTA) calculates the hydrodynamic radius of individual particles *via* the Stokes–Einstein equation.[67,68] The Brownian motion of the particles is tracked using the light scattered from an incident laser light source. The diffraction patterns produced are projected onto a CCD camera and tracked using video processing software.[79] The diffusion constant is therefore directly measured, and particles with diameters below the wavelength of the incident light can be detected due to their Rayleigh scattering. The ability to track individual particles reduces the influence of small populations of impurities, and as it is not an ensemble method it is more capable of detecting individual populations in polydisperse samples.[80]

NTA can accurately characterise particle size as low as 10–35 nm depending on the sample type, and can resolve populations with less than 40 nm difference in their radii.[79] A potential disadvantage of NTA is that the refractive index of the sample must be sufficiently different from the buffer, and some prior knowledge of the refractive index is required.[80] Instrument set up can bias the results—the ultramicroscope must be isolated from mechanical vibration, and the user-defined adjustments to the detection parameters can affect the detection limit, therefore biasing the sample.[14]

10.4.2.3 *Resistive Pulse Sensing*

Resistive pulse sensing (RPS) of nanoparticles using Coulter-type counters has been shown to be a fast and accurate alternative to established sizing methods, and is becoming increasingly popular in the field of nanomedicines.[81,82] This technique provides a direct measure of particle concentration, and high resolution analysis of particle size and surface charge.[83–86] RPS was historically used for measuring microparticles, but recent advances in the fabrication of pores have lead to a renewed interest in the technique for single particle characterisation of nanoparticles.[87] A voltage is applied across a pore which is filled with electrolyte, resulting in an ionic current (Figure 10.2).

Particles traverse the pore with a velocity that is dependent on the zeta potential of the particle, causing a transient blockage in the ionic current (Figure 10.3). The magnitude of this blockade is dependent on the particle volume[83,88] (eqn (10.3)). Measurement of these blockade events allows high-throughput, single-particle analysis of colloidal samples.

$$\frac{\Delta R}{R} = \frac{d^3}{D^2 L} \tag{10.3}$$

Where d is the particle diameter, D is the pore diameter and L is the pore length. This represents the simplest case; for complex particle and pore geometries other factors must be taken into account.[87]

Nanofabrication techniques have enabled the production of pores using materials including glass, silicon, polycarbonate and carbon nanotubes, to name a few.[82,89–91] Although the fixed pore size allows sensitive size measurement, the limited analysis range has implications for the measurement of polydisperse samples.[92] This can be overcome by the use of size-tunable elastomeric pores, which allows optimisation of the resistance pulse magnitude relative to the background current, and is known as tunable resistive pulse sensins (TRPS). An inherent disadvantage of any pore-based technique

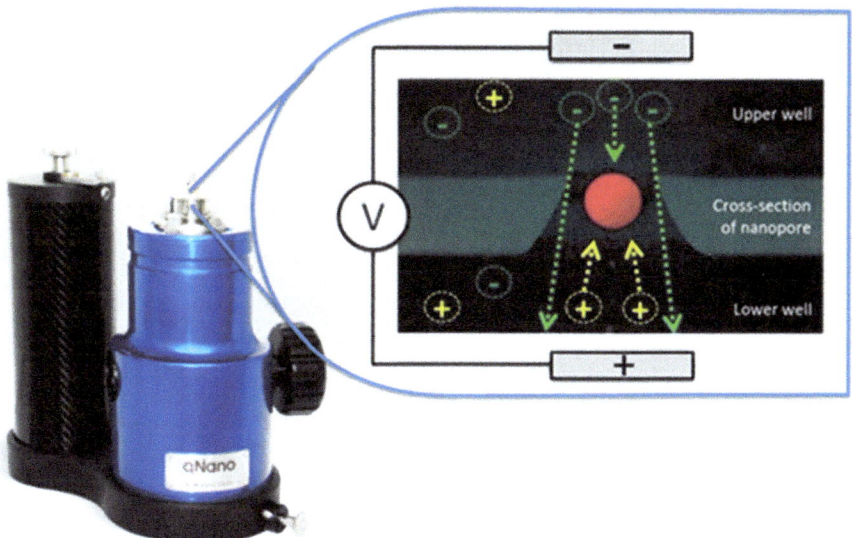

Figure 10.2 Instrument setup and theory of tunable resistive pulse screening (TRPS). The qNano (Izon Science Ltd, Christchurch, New Zealand) is a benchtop instrument for nanoparticle measurement. An elastomeric membrane with a single pore punctured through it is held by four "teeth" which can be adjusted to stretch or relax the nanopore to optimise measurements. The sample is applied in the fluid cell above the pore, and an applied voltage causes particles to move to the lower fluid due to electrophoresis, passing through the pore. The upward force is due to the zeta-potential of the pore, and the balance of these forces determines the particle speed through the pore, which is used to calculate particle zeta-potential (Section 10.4.3.3).

is the potential for pore blockages to occur due to large or adhesive particles, which can be time consuming and require disassembly and cleaning of the instrument.[62] Very small particles may not be detected if their volume does not displace a sufficient volume of electrolyte relative to the pore volume. Although the use of tunable pores does increase the dynamic range of analysis, these issues can mean that highly polydisperse samples may be difficult to analyse using TRPS.

10.4.3 TRPS for Accurate Particle-by-particle Measurement

10.4.3.1 *Size Measurements*

The relationship between particle volume and the magnitude of the resistive pulse that is generated as it traverses the pore is linear, and can therefore be used to calculate this parameter, as defined in eqn (10.4). However, the constantly changing pore diameter makes direct measurement of particle size challenging. Although this is possible when the pore geometry and

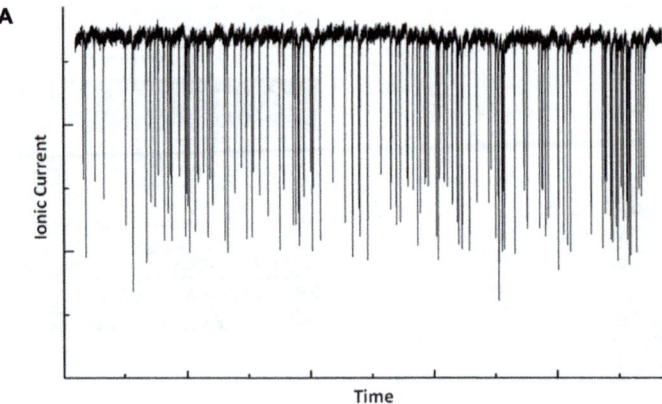

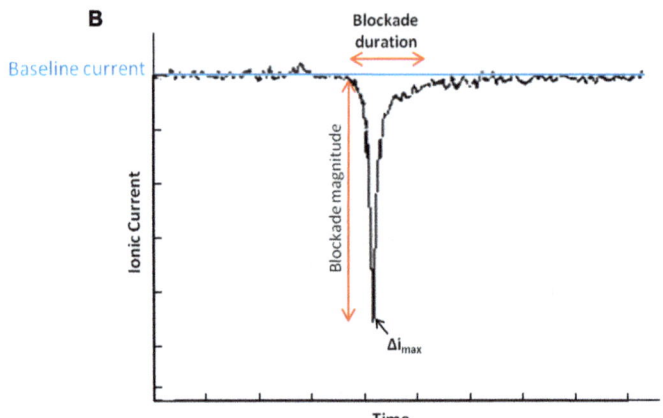

Figure 10.3 Ionic current resistive pulses as seen in tunable resistive pulse screening (TRPS). A: Ionic current "pulses" as seen in real time, generated by
individual particles passing through the pore. B: The close-up view of a
single current pulse shows the characteristic pulse shape. From this,
the blockade magnitude and duration (indicated) can be used to
calculate the particle size and zeta potential, respectively, through
calibration to known standards.

degree of stretch are known, a calibration method has become the preference. Using the relationship outlined in eqn (10.3), a single-point calibration
using particles of known volume has been found to produce highly accurate
size analysis, as long as the pore dimension is not changed for sample
measurement.[83]

$$\frac{\Delta R_S}{V_S} = \frac{\Delta R_C}{V_C} \tag{10.4}$$

Where V_S is the volume of the sample particle, V_C is the volume of the
calibration particle, and ΔR_S and ΔR_C are the resistive pulses generated by

the sample and calibration particles, respectively. The calibration method means that measurement of the pore, buffer or analysis conditions are not required, nor is prior knowledge of the sample. As the size of the resistive pulse is directly proportional to particle volume (eqn (10.4)), doubling the particle diameter results in an eight-fold increase in resistive pulse magnitude, meaning that TRPS is highly sensitive to small differences in particle size. This allows highly sensitive measurement, particularly in the case of samples that contain multimodal or aggregated populations,[62] and can be beneficial in monitoring the further processing of nanoparticles. For example, the reduction in polydispersity of a sample through additional filtering can be monitored with confidence using TRPS to ensure that the desired product is created.[86]

10.4.3.2 Concentration Measurements

As for particle size, particle concentration can be calculated directly if the analysis conditions are well known. Alternatively, a calibration method can be used as the number of particles that traverse the pore in a given time period is linearly proportional to the particle concentration[93] (eqn (10.5)).

$$\frac{C_S}{r_S} = \frac{C_C}{r_C} \tag{10.5}$$

This assumes that the forces driving the sample and calibration particles are equivalent, which is often not the case. To overcome this, pressure is applied to the system, which essentially removes the electrokinetic component of the particle velocity.[93] In the case of highly charged particles where electrokinetic forces become dominant, the concentration is calculated by comparing the slopes of the linear rate *versus* applied pressure of the sample and calibration.

While TRPS can be used to derive a "total" (that is, single number value) concentration measurement, recently the preference has been to report concentration as a function of particle size. For example, the concentration of a nanoparticle suspension may be reported as: $C_{80} - C_{150} = 1.5 \times 10^6$ particles per mL, thus reporting the concentration of particles within the size range 80–150 nm.

10.4.3.3 Charge Measurements

A unique feature of TRPS is its potential to measure individual particle size, based on the duration of the resistive pulse.[86] Kozak *et al.*[86] demonstrated zeta-potential measurements of individual nanoparticles in dispersion using size-tunable pore sensors. This method calculates the distance each particle has travelled within the pore in a specified time from the respective resistive pulse signals. Then electrokinetic and fluidic velocity contributions are modeled by fitting the distance *versus* time curves. The zeta potential of each particle can be calculated from the measured electrophoretic mobility using the Smoluchowski equation (described by Kozak and colleagues.[83]). This is independent of the magnitude of the pulse, meaning that simultaneous size and charge measurements can be performed, given a unique approach for investigating particle properties. A calibration-based zeta-potential method was recently developed (Vogel *et al.*, to be published), based on the measurement of signal durations of translocation events as a function of voltage and applied pressure (Vogel *et al.*, to be published). This provides a very reproducible technique, which is consistent across different instrument settings and users, which has been seen to be a problem with some measurement techniques (see, for example, Section 10.3.2.2). The particle-by-particle nature of TRPS measurements means that it is possible to resolve subpopulations, which is a unique feature of zeta-potential analysis, as most other techniques report an average value.

A recent example of the importance of measuring zeta potential was the characterisation of PEGylated liposomes.[61] In this study zwitterionic liposomes were PEGylated, which led to a change in zeta potential proportional to the degree of PEGylation. The study demonstrated that the zeta-potential change generated by the incorporation of 5% mol per mol PEG lipid into the liposome could be detected by TRPS, which also revealed that the concentration and size distribution was unchanged. The magnitude of the shift in zeta potential was supported by similar studies that used alternative techniques.[61,94] This study is a critical step to validating TRPS as a method for nanomedicine regulation, particularly in cases like that of Lipodox™–a generic version of Doxil®. Lipodox™ is generated by encapsulating doxorubicin hydrochloride in PEGylated liposomes. The approval of generics is dependent on the bio- and physical-equivalence, and given that the targeting and circulation time of these formulations is dependent on the PEGylation, accurate measurement of surface modification is required to ensure equivalence.

The single-particle nature of TRPS means that subpopulations within a sample can be discriminated. Very recently, this was shown to be possible not only for multimodal size distributions, but also for multimodal charge distributions (Vogel, unpublished; Figure 10.4). It has been shown that a mixed sample of nanoparticles with equivalent sizes but different zeta potentials were not resolvable by an ensemble technique (phase analysis light scattering). The same sample, when analysed using TRPS, showed clearly defined populations. Although a number of techniques are able to give useful and accurate data relating to particle size or particle charge, TRPS is unique in its ability to be able to analyse both parameters simultaneously. The potential to link the charge and size data of nanoparticles will be useful for applications such as determining the success of modification (such as liposome PEGylation) and the uniformity of modification within a preparation of nanoparticles.

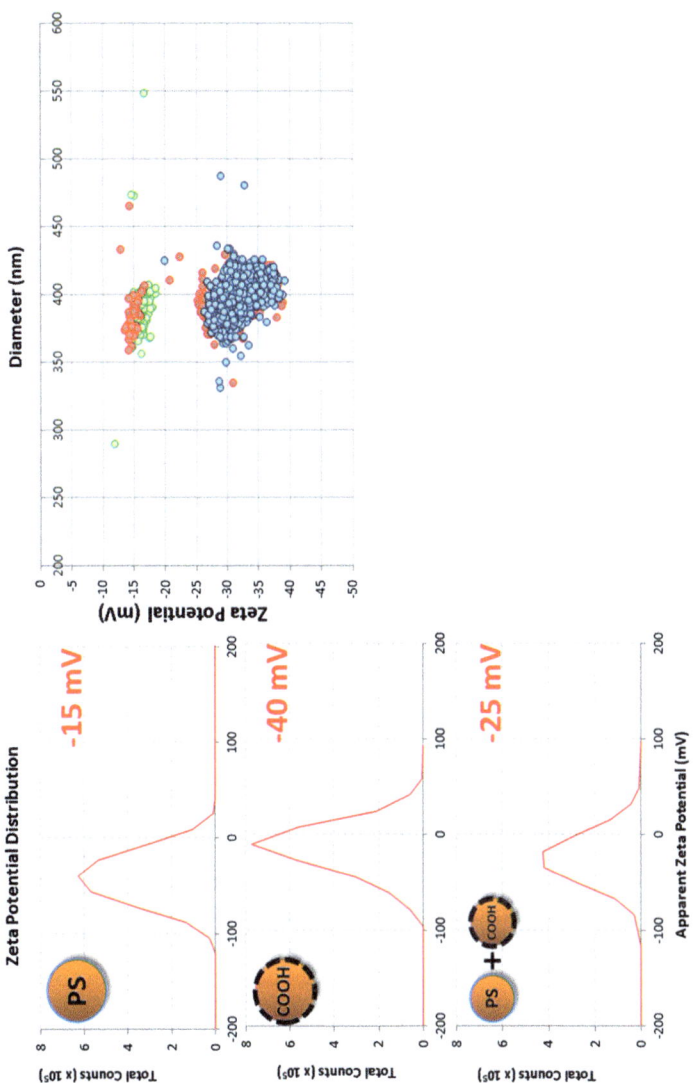

Figure 10.4 Comparison of tunable resistive pulse screening (TRPS) and phase analysis light scattering (PALS) analysis of bimodal charged samples. Analysis of the two samples of neutral polystyrene (PS) particles and negatively charged COOH particles (TRPS data points: green = neutral PS particles, blue = negative COOH particles; traces as labeled for PALS data) show good agreement between the two techniques. However, when the two populations are mixed to give a bimodal sample PALS is unable to resolve the two, giving a solution-averaged zeta-potential value (bottom trace). TRPS shows two distinct populations (red data points), which agree well with the zeta-potential values as determined from unmixed samples.

10.5 Conclusions and Outlook

The regulation of nanomaterials brings a number of specific challenges, which are only recently beginning to be addressed. The measurement of such materials is, at present, not subject to strict controls and lacking in standardised methods and reference materials. As a result, the regulation and guidelines relating to this process are largely incomplete. Compared to the measurement of their bulk counterparts, the difficulties in detecting and counting nanoparticles in statistically significant numbers leads to dramatic variations depending on the technique and settings used. Improvement of current methods, and the validation of accepted methods by regulatory authorities will help (and is necessary) to resolve these issues. Along with this, the provision of guidelines for standardised sample preparation and reporting of results (for example, defining the size range of concentration measurements) will increase the accuracy and therefore confidence levels of the information provided to governing bodies. Given the unique properties of nanomaterials, it is expected that specific guidance documents and regulatory procedures will be required for specific sectors, and as the field is expanding at a rapid rate, the need for these is urgent. As with other materials, a combination of methods of characterisation will be needed for robust assessment of the characteristics, with each of these requiring thorough validation and inter-laboratory comparison checks. This has been approached by NIST, with a set of guidelines produced in order to minimise intra- and inter-laboratory discrepancies,[63] and regulation would benefit from the strict enforcement of these guidelines.

Given the speed at which the field is advancing, regulatory guidelines and definitions need to be reviewed regularly. There is a need for consistency both within regulatory authorities and between those of different countries and states, which is recognised and being moved toward.[36] Fairly recently, a large-scale project was launched to incorporate 24 European partners and six partners from Brazil, Canada, India and the USA (NanoValid 2011[95]). The aim was to develop reliable reference methods and materials, to be standardised throughout the participating groups across the world. It is anticipated that more worldwide projects like this will be employed, suggesting modifications of current definitions, methods and reference standards.[96]

Accurate measurement of nanoparticle characteristics including size, concentration and surface charge is essential to provide the regulatory authorities with sufficient information on which to base their approval of the particles in question. It is likely that multiple techniques will be required to give confidence in the results, and the most useful will those that can give single particle resolution, as has been accepted for the measurement of injectables.[65]

References

1. T. Lammers, F. Kiessling, W. E. Hennink and G. Storm, *J. Controlled Release*, 2012, **161**, 175–187.

2. R. Freitas Jr., in *Nanomedicine Volume I: Basic Capabilities*, Landes Bioscience, Georgetown, TX, 1st edn, 1999.
3. M. E. Davis, Z. G. Chen and D. M. Shin, *Nat. Rev. Drug Discovery*, 2008, 7, 771–782.
4. D. Peer, J. M. Karp, S. Hong, O. C. Farokhzad, R. Margalit and R. Langer, *Nat. Nanotechnol.*, 2007, **2**, 751–760.
5. N. Bertrand and J. C. Leroux, *J. Controlled Release*, 2012, **161**, 152–163.
6. V. P. Torchilin, *Nat. Rev. Drug Discovery*, 2005, **4**, 145–160.
7. V. P. Torchilin, *AAPS J.*, 2007, **9**, E128–47.
8. Y. Barenholz, *J. Controlled Release*, 2012, **160**, 117–134.
9. J. Potocnik, *Commission Recommendation of 18 October 2011 on the Defintion of Nanomaterial*, Brussels, 2011.
10. T. Linsinger, G. Roebben, D. Gilliland, L. Calzolai, F. Rossi, N. Gibson and C. Klein, *Luxemb. Publ. Off. Eur. Union*, 2012.
11. P. Antoniammal and D. Arivuoli, *J. Nanomater.*, 2012, **2012**, 11.
12. B. Gilbert, R. K. Ono, K. A. Ching and C. S. Kim, *J. Colloid Interface Sci.*, 2009, **339**, 285–295.
13. A. A. Shemetov, I. Nabiev and A. Sukhanova, *ACS Nano*, 2012, **6**, 4585–4602.
14. N. C. Bell, C. Minelli, J. Tompkins, M. M. Stevens and A. G. Shard, *Langmuir*, 2012, **29**, 10860–10872.
15. S. Ahn, S. Y. Jung and S. J. Lee, *Molecules*, 2013, **18**, 5858–5890.
16. K. A. White and N. L. Rosi, *Nanomedicine*, 2008, **3**, 543–553.
17. J.-M. Rabanel, P. Hildgen and X. Banquy, *J. Controlled Release*, 2014, **185**, 71–87.
18. M. T. Peracchia, E. Fattal, D. Desmaele, M. Besnard, J. P. Noel, J. M. Gomis, M. Appel, J. d'Angelo and P. Couvreur, *J. Controlled Release*, 1999, **60**, 121–128.
19. FDA, *Press release*, 2013.
20. B. Pollock, *June approval times for ANDAs – a snapshot in time*, 2013, New York.
21. M. Davenpot, *Chem. Eng. News*, 2014, **92**, 10–13.
22. G. Oberdorster, *J. Intern. Med.*, 2010, **267**, 89–105.
23. J. Zhao and V. Castranova, *J. Toxicol. Environ. Health, Part B*, 2011, **14**, 593–632.
24. D. M. Berube, *Nano-Hype: The truth behind the nanotechnology buzz*, Prometheus Books, New York, 2006.
25. A. Dhawan and V. Sharma, *Anal. Bioanal. Chem.*, 2010, **398**, 589–605.
26. K. W. Powers, S. C. Brown, V. B. Krishna, S. C. Wasdo, B. M. Moudgil and S. M. Roberts, *Toxicol. Sci.*, 2006, **90**, 296–303.
27. B. Halford, *Chem. Eng. News*, 2008, **84**, 47.
28. G. Kirmizidis, *Report on the European Commission's Plublic Online Consultation: Towards a Strategic Nanotechnology Action Plan (SNAP) 2010–2015*, 2010.
29. S. Iijima, *Nature*, 1991, **354**, 56–58.
30. TCSA, *Fed. Regist.*, 2008, **73**, 64946–64947.

31. *Directive 2001/83/EC of the European Parliament and of the council of 6 November 2001 on the Community code relating to medicinal products for human use*, 2001.
32. *Council directive 93/42/EEC of 14 June 1993 concerning medical devices*, 1993.
33. P. Spyridoula, Nanomedicine in Europe: Regulating Under Uncertainty, Masters Thesis, Universiteit van Tilburg, 2013.
34. J. D'Silva and G. van Calster, *Eur. J. Health Law*, 2009, **16**, 249–269.
35. FDA Press Off., 2006, *P06–111*.
36. A. C. Von Eschenbach, *Nanotechnology: A Report of the U.S. Food and Drug Administration Nanotechnology Task Force*, 2007.
37. M. N. Duvall, *FDA Regulation of Nanotechnology*, Washington, DC, 2012.
38. EMA, 1995.
39. European Medicines Agency, *Annual Report of the European Medicines Agency 2006*, London, 2006.
40. M. Sean, *BioPharm Int.*, 2012, **26**, 12–13.
41. Nanobiotix Starts Clinical Trial With Lead Product For Soft Tissue Sarcoma, *The Pharma Letter*, http://www.thepharmaletter.com/article/nanobiotix-starts-clinical-trial-with-lead-product-for-soft-tissue-sarcoma, accessed 10 October 2014.
42. S. Tinkle, S. E. McNeil, S. Muhlebach, R. Bawa, G. Borchard, Y. C. Barenholz, L. Tamarkin and N. Desai, *Ann. N. Y. Acad. Sci.*, 2014, **1313**, 35–56.
43. *Regulation No. 1223/2009 of the European Parliament and of the Council on Cosmetic Products*, European Commission, 2009.
44. *Regulation No. 1169/2011 of the European Parliament and of the Council on the Provision of Food Information to Consumers*, European Commission, 2011.
45. *ISO/TS 27687:2008 Nanotechnologies – Terminology and definitions for nano-objects – Nanoparticle, nanofibre and nanoplate*, International Organisation for Standardisation, Geneva, 2008.
46. M. A. Hamburg, *Science*, 2012, **336**, 299–300.
47. F. Ehmann, K. Sakai-Kato, R. Duncan, D. Hernan Perez de la Ossa, R. Pita, J. M. Vidal, A. Kohli, L. Tothfalusi, A. Sanh, S. Tinton, J. L. Robert, B. Silva Lima and M. P. Amati, *Nanomedicine*, 2013, **8**, 849–856.
48. D. E. Owens 3rd and N. A. Peppas, *Int. J. Pharm.*, 2006, **307**, 93–102.
49. F. Alexis, E. Pridgen, L. K. Molnar and O. C. Farokhzad, *Mol. Pharm.*, 2008, **5**, 505–515.
50. J. Rejman, V. Oberle, I. S. Zuhorn and D. Hoekstra, *Biochem. J.*, 2004, **377**, 159–169.
51. E. V. B. van Gaal, G. Spierenburg, W. E. Hennink, D. J. a Crommelin and E. Mastrobattista, *J. Controlled Release*, 2010, **141**, 328–38.
52. M. Gaumet, A. Vargas, R. Gurny and F. Delie, *Eur. J. Pharm. Biopharm.*, 2008, **69**, 1–9.
53. Y. Matsumura and H. Maeda, *Cancer Res.*, 1986, **46**, 6387–6392.

54. H. Kobayashi, R. Watanabe and P. L. Choyke, *Theranostics*, 2013, **4**, 81–89.
55. L. Tang, X. Yang, Q. Yin, K. Cai, H. Wang, I. Chaudhury, C. Yao, Q. Zhou, M. Kwon, J. A. Hartman, I. T. Dobrucki, L. W. Dobrucki, L. B. Borst, S. Lezmi, W. G. Helferich, A. L. Ferguson, T. M. Fan and J. Cheng, *Proc. Natl. Acad. Sci. U. S. A.*, 2014, **111**, 15344–15349.
56. J. Bobe, M. Harada and I. Nakatomi, in *Bio-Nanotechnology: A Revolution in Food, Biomedical and Health Sciences*, ed. D. Bagchi, M. Bagchi, H. Moriyama and F. Shahidi, 2013.
57. A. J. Phillips, in *Australasion Extracellular Vesicles Conferece*, 2014, Cairns.
58. A. J. Phillips, *ASEMV2014*, San Francisco, 2014.
59. A. Gessner, A. Lieske, B. R. Paulke and R. H. Muller, *J. Biomed. Mater. Res., Part A*, 2003, **65**, 319–326.
60. H. G. Merkus, *Particle Size Measurements*, Springer, 17th edn, 2009.
61. D. Kozak, M. F. Broom and R. Vogel, *Accurate Size, Charge & Concentration Analysis of Liposomes using Tunable Resistive Pulse Sensing*, Christchurch, 2012.
62. W. Anderson, D. Kozak, V. A. Coleman, A. K. Jamting and M. Trau, *J. Colloid Interface Sci.*, 2013, **405**, 322–330.
63. ASTM, *ASTM E2490-09 Standard Guide for Measurement of Particle Size Distribution of Nanomaterials in Suspension by Photon Correlation Spectroscopy (PCS)*, West Conshohocken, 2009.
64. M. Baum, www.smalltimes.com, 2008.
65. The United States Pharmacopeial Convention, *<788> Particulate Matter in Injections*, 2012.
66. S. K. Brar and M. Verma, *TrAC, Trends Anal. Chem.*, 2011, **30**, 4–17.
67. A. Einstein, *Ann. Phys.*, 1905, **322**, 549–560.
68. M. von Smoluchowski, *Ann. Phys.*, 1906, **326**, 756–780.
69. O. Elizalde, G. P. Leal and J. R. Leiza, *Part. Part. Syst. Charact.*, 2000, **17**, 236–243.
70. C. Hoo, N. Starostin, P. West and M. Mecartney, *J. Nanoparticle Res.*, 2008, **10**, 89–96.
71. S. W. Provencher, *Comput. Phys. Commun.*, 1982, **27**, 229–242.
72. T. J. Cho and V. A. Hackley, *Anal. Bioanal. Chem.*, 2010, **398**, 2003–2018.
73. F.-K. Liu and G.-T. Wei, *Chromatographia*, 2004, **59**, 115–119.
74. S. T. Fitzpatrick, *US Pat.*, 5 786 898, 1998.
75. H. Vegad, *An introduction to particle size characterization by DCS: Do you know the real size of your nano particles*, 2010.
76. X. Zou, S. Hovmoller and P. Oleynikov, in *Electron Crystallography: Elecetron Microscopy and Electron Diffraction*, ed. X. Zou, Oxford University Press Inc., New York, 1st edn, 2011.
77. S. M. Stagg, G. C. Lander, J. Pulokas, D. Fellmann, A. Cheng, J. D. Quispe, S. P. Mallick, R. M. Avila, B. Carragher and C. S. Potter, *J. Struct. Biol.*, 2006, **155**, 470–481.
78. L. Marton, *Nature*, 1934, 911.

79. B. Carr, A. Malloy and P. Hole, in *Simultaneous sizing of nanoparticles by individually visualizing and separately tracking their Brownian motion within a suspension*, 2007.
80. A. Hawe, V. Filipe and W. Jiskoot, *Pharm. Res.*, 2010, **27**, 314–326.
81. R. W. Deblois, C. P. Bean and R. K. A. Wesley, *J. Colloid Interface Sci.*, 1977, **61**, 323–335.
82. T. Ito, L. Sun, M. A. Bevan and R. M. Crooks, *Langmuir*, 2004, **20**, 6940–6945.
83. R. Vogel, G. Willmott, D. Kozak, G. S. Roberts, W. Anderson, L. Groenewegen, B. Glossop, A. Barnett, A. Turner and M. Trau, *Anal. Chem.*, 2011, **83**, 3499–3506.
84. G. S. Roberts, D. Kozak, W. Anderson, M. F. Broom, R. Vogel and M. Trau, *Small*, 2010, **6**, 2653–2658.
85. T. Ito, L. Sun and R. M. Crooks, *Anal. Chem.*, 2003, **75**, 2399–2406.
86. D. Kozak, W. Anderson, R. Vogel, S. Chen, F. Antaw and M. Trau, *ACS Nano*, 2012, **8**, 6990–6997.
87. D. Kozak, W. Anderson, R. Vogel and M. Trau, *Nano Today*, 2011, **6**, 531–545.
88. R. W. Deblois, *Rev. Sci. Instrum.*, 1970, **41**, 909–914.
89. W. Li, N. A. W. Bell, S. Hernandez-Ainsa, V. V. Thacker, A. M. Thackray, R. Bujdoso and U. F. Keyser, *ACS Nano*, 2013, 7, 4129–4134.
90. C. C. Harrell, Y. Choi, L. P. Horne, L. A. Baker, Z. S. Siwy and C. R. Martin, *Langmuir*, 2006, **22**, 10837–10843.
91. J. Kong, H. Wu, L. Liu, X. Xie, L. Wu, X. Ye and Q. Liu, *J. Nanosci. Nanotechnol.*, 2013, **13**, 4010–4016.
92. D. Kozak, W. Anderson, M. Grevett and M. Trau, *J. Phys. Chem. C*, 2012, **116**, 8554–8561.
93. G. S. Roberts, D. Kozak, W. Anderson, M. F. Broom, R. Vogel and M. Trau, *Small*, 2012, **6**, 2653–2658.
94. M. C. Woodle, M. S. Newman and J. A. Cohen, *J. Drug Targeting*, 1994, **2**, 397–403.
95. R. Ruether, *J. Biomed. Nanotechnol.*, 2011, 7, 8–11.
96. W. Wohlleben, Muller and Philipp, in *Safety of Nanomaterials Along Their Lifecycle: Release, Exposure, and Human Hazards*, ed. W. Wohlleben, T. A. J. Kuhlbusch, J. Schnekenburger and C.-M. Lehr, Boca Raton, 1st edn, 2015, p. 57.

CHAPTER 11

Nanomedicine: Promises and Challenges

R. L. JULIANO

Division of Molecular Pharmaceutics, UNC Eshelman School of Pharmacy, University of North Carolina, Chapel Hill, NC 27599, USA
Email: arjay@med.unc.edu

11.1 The Evolution of Nanomedicine

As this Royal Society of Chemistry volume amply demonstrates, nanomedicine has emerged as a dynamic and exciting area of research. Despite its apparent novelty and timeliness, it is important to realize that this field has strong historical roots. Thus, substantial aspects of contemporary nanoparticle technology derive from the basic principles of classic colloid chemistry. Methods for evaluating particle size and charge and concepts of diffusional or convective transport of nanoparticles all stem from this source. More immediately, the first clinically significant implementation of nanomedicine was the advent of drug-loaded liposomes in the late 1980s. Thus, the histories of liposomal doxorubicin (DOXIL®)[1] for cancer and liposomal amphotericin B (ABELCET® and AMBISOME®)[2] for systemic fungal disease provide many lessons for the development of current nanoparticle delivery systems.

Over the past two decades many exciting new nanotechnologies with potential biomedical applications have come to the fore. Carbon nanotubes, metallic nanoparticles, silica-based particles, novel polymeric and lipid nanoparticles have all attracted substantial interest. In addition to nanoparticles for delivery, a wide range of nanomaterials have been proposed for

RSC Drug Discovery Series No. 51
Nanomedicines: Design, Delivery and Detection
Edited by Martin Braddock
© The Royal Society of Chemistry 2016
Published by the Royal Society of Chemistry, www.rsc.org

use in diagnosis, tissue engineering, genomics, proteomics and other areas of biomedicine. However, much of the core of modern nanomedicine research, and the emphasis of this volume, has dealt with the delivery of therapeutic and imaging agents to desired locations within the body. This is also the emphasis of this chapter.

There is much evidence that nanomedicine is a healthy and dynamic field. Thus, there has been a dramatic growth in publications, patents, clinical trials activity, and commercialization. For example, in late 2013 a search of the US Patent Office database using 'nanoparticle and drug' resulted in over 4250 hits. Perusal of ClinicalTrials.gov using the term 'nanoparticle' found 149 current studies, while many other trials may not be reported on this site. Nanowerk, a leading nanotechnology website, currently lists 305 companies with interests in nanotechnology and life sciences. Grant support for academic research in nanomedicine remains strong. A decade ago the leadership of the US National Institutes of Health (NIH) initiated several major thrusts in nanomedicine. In addition to traditional individual investigator (R01) type grants, various large-scale research programs were established. For example, the National Cancer Institute's Alliance for Nanotechnology in Cancer has invested approximately $150 million, primarily in large research centers. The National Heart, Lung, and Blood Institute's Programs of Excellence in Nanotechnology also support several large research centers, while the NIH Office of the Director funded eight nanomedicine development centers that addressed basic research issues in the field. All of these thrusts utilized monies that were sequestered for these specific projects. While this great expansionary phase may be slowing, overall NIH support for remains solid with more than $300 million in funding for the field in 2012 (NIH RePORTER).

However, amid all these positive developments in nanomedicine, a few voices of concern have been raised.[3–5] They point out the difficulties in translating nanomedicine from the laboratory into the clinic and caution against over-promising the benefits of the technology. Some of these concerns are addressed below. But first, let us touch on some of the potentials and promise of contemporary nanomedicine.

11.2 Unique Capabilities of Nanomaterials: The Promise of Nanomedicine

Working with devices and materials on the nanoscale allows the exploitation of novel physical and chemical properties that do not exist on the macroscale.[6] This fundamental difference has been creatively exploited in many different ways in nanomedicine. Here we briefly mention a few of the most elegant examples in diagnosis, therapy and technology development. For example, metallic nanoparticles have unique optical properties that depend on their state of aggregation. By conjugating oligonucleotides to gold nanoparticles a rapid, ultrasensitive assay for genome analysis was developed[7,8] and eventually commercialized (www.nanosphere.us). Analysis of tumor cells circulating in the blood is an important but challenging aspect of cancer diagnosis. This has been addressed by using a device containing antibody-coated nanoposts to separate circulating tumor cells from other blood cells,[9] while other nano-based approaches have relied exclusively on the physical properties of tumor cells.[10] The fact that carbon nanotubes readily emit electrons in response to a voltage gradient has allowed the development of a radically new design for X-ray equipment that allows rapid and precise imaging, even of organs in motion such as the heart.[11,12] Iron oxide nanoparticles coated with passivating polymers such as dextran have proven to be a boon to magnetic resonance imaging.[13] New technologies for the production of nanoparticles now allow the generation of particles with precisely controlled sizes and shapes and physical properties from a variety of source materials, including proteins.[14,15] Finally, significant progress is being made with nanoparticle systems for the delivery of short interfering (si)RNA and other therapeutic oligonucleotides.[16] All of these examples testify to the present dynamic status of the nanomedicine field.

Going beyond current developments, it is exciting to visualize future potential, especially through the intersections of nanotechnology, information technology and electro-optical technology. Computation is growing ever more powerful. Although photolithography, the technique currently used to produce the features on microchips, is nearing its size limit because of the basic characteristics of light, new approaches to computation including DNA-based computation[17] and quantum computing[18] promise even faster processing capabilities and thus ever more powerful computation. Artificial intelligence (AI) is already being applied to the design and analysis of nanotechnology systems[19] and it is possible to envision nanoparticles with modest AI capabilities built into their structures. Nanoparticles that can respond to external cues from light or magnetic fields are already a reality,[20] and while 'steering' nanoparticles with magnetic fields is challenging in the *in vivo* environment, interesting new approaches are moving toward that goal. Although we are still a long way from the nanobot submarine traversing the bloodstream in the much-parodied 1966 science fiction movie *Fantastic Voyage*, there is increasing interest in developing nanoparticles that have self-motive capabilities.[21] Thus, we can visualize the evolution of highly sophisticated nanoparticles that could do some or all of the following: (a) respond to local environmental cues such as pH or the presence of specific receptors to release drugs; (b) be responsive to external electromagnetic cues; and (c) move through tissues either autonomously (perhaps guided by AI) or in response to external fields. Thus, the promise of the continued exciting evolution of the nanomedicine field seems bright.

11.3 Conceptual Issues in Nanomedicine

Despite its promise, there are also a number of challenges and limitations that will affect the development of nanomedicine, as we highlight in the following sections. In his famous Rede Lecture in 1959 entitled 'The Two Cultures', C. P. Snow deplored the growing gap between scientists and humanists and predicted that this would slow progress in both arenas.[22] On a more limited level there is also a 'culture gap' within nanomedicine that may slow or limit its evolution.[23]

As with many rapidly moving fields of research, nanomedicine has resulted from the intermingling of highly divergent scientific disciplines. From the physical sciences side it draws heavily on chemistry, materials science, and engineering while on the biological side a broad range of biomedical disciplines, including cell and molecular biology, pharmacology, pharmaceutics and medicine contribute to nanomedicine's emphasis on translation to the clinic. To attain true mastery of the field, an investigator in nanomedicine would need broad and deep expertise that encompassed both the quantitative bases of the physical sciences and the diversity and complexity of biological systems. Although a few outstanding investigators have attained this level of mastery, the more typical situation is where physical scientists and biologists, each with limited understanding of the other's field, must collaborate to pursue problems in nanomedicine. This situation presents both opportunities and challenges. Pooling knowledge from different disciplines can be very productive. However, there are also many possibilities for miscommunication; the participants literally do not speak the same language! Often problems occur around the issue of an appropriate match-up between the nanotechnology being developed and the therapeutic or diagnostic application being addressed.

One aspect might well be termed the 'muddle-headed biologist' factor. Because of the nature of their training, many biomedical scientists are poorly equipped to understand the physical and chemical principles that underlie nanotechnology. For example, they sometimes fail to grasp the dramatic differences that prevail between the nanoscale and the macroscale in terms of the behavior of materials. Thus biological scientists often cannot cogently critique the assertions of their physical science colleagues regarding the nanotechnology involved in a particular project. The nanotechnologist may claim that a nanodevice will have certain valuable properties in the intended application; whether that is valid or not, the biologist does not have the intellectual wherewithal to question or rebut the claim. Thus for the biological scientist, nanotechnology assumes a 'magical' aspect and its claims are accepted on faith rather than being challenged and refined through meaningful scientific dialog.

The converse issue is the tendency of physical scientists to try to impose simplicity on the inherent diversity and complexity of biology. Faced with an area of science that is rich in detail but short on broadly applicable generalizations, physical scientists tend to naively seize upon simplistic half-truths about biological systems and then base their design of nanomedicine approaches on these oversimplified generalizations. This usually leads to a great waste of time, energy and money since the technology developed will not accurately deal with the biological basis of the therapeutic or diagnostic problem being addressed. Thus, the nanomedicine literature is rife with articles based on over-simplified and misleading concepts. Typical examples include the idea that all tumors exhibit substantially lower pH than surrounding tissues and thus can be preferentially affected by pH-sensitive reagents, that all tumors have 'leaky' blood vessels and thus readily accumulate nanoparticles, or that there are receptors that are uniquely expressed on one cell type *versus* another, thus easily allowing differential targeting. While each of these assertions may have partial validity in certain instances they are by no means universally true. A recent report[24] provides an instructive example of the need to pay attention to biological details in the design of nanomedicines. Thus much of the effort in the area of nanoparticles for drug delivery has focused on how the physical properties of the particles, such as size, surface charge, and surface coatings will affect their clearance and biodistribution. In contrast, nanotechnologists have given scant attention to the biological status of the recipients of the nanoparticles. However, the observations of Jones *et al.*[24] indicate that the characteristics of the immune system, particularly T-helper (Th) cells, is as important a determinant of nanoparticle clearance as any physical characteristic of the particles; thus mice strains with a Th2 prevalence cleared nanoparticles from the blood far more rapidly than strains with a Th1 prevalence. Certainly there will be many other instances when subtle biological parameters will influence the *in vivo* behavior of nanomedicines.

Thus, a major challenge to nanomedicine is inherent in its highly interdisciplinary nature requiring contributions from both physical scientists and biological scientists. For those interactions to be truly productive, better communication is needed between practitioners of both disciplines. Chemists, physicists and engineers working in the field must develop a deep appreciation for the complexity, redundancy and adaptability of living organisms. Biomedical scientists need better understand the physical and chemical principles that underlie nanotechnology. Ultimately there needs to emerge a new generation of truly interdisciplinary investigators who have a strong grounding in the physical sciences accompanied by deep and broad training in the biomedical sciences.

11.4 Challenges to the Implementation of Nanomedicine

Beyond the scientific and technical challenges of developing nanomedicines into clinically relevant tools, there are economic, social and political considerations that will affect the evolution of the field. Focusing on the area of nanoparticle-mediated drug delivery for example, a number of key problems can be anticipated.

The pharmaceutical development of nanoparticle drugs presents many difficulties. The drug-bearing nanoparticle is a far more complex entity than the parent drug alone and thus formulation and scale-up may be technically difficult and far more costly than for conventional agents. Deploying the resources to overcome these challenges would clearly be justified if major benefits were to accrue to patients. But exactly what therapeutic benefit does nanoformulation provide? One answer to this question might be gleaned from the experience with early-generation nanoparticle drugs that entered the clinic in the 1990s and 2000s. This would include the liposomal agents DOXIL® and AMBISOME®, as well as the protein–drug nanocomplex ABRAXANE®. Although some of these formulations are now off-patent, they remain substantially more costly than their parent drugs (doxorubicin, amphotericin B, and paclitaxel). For the most part, the main clinical advantage of the nanoformulations is based on reduced toxicity rather than on significantly greater efficacy in treating disease. Certainly, experiencing fewer side effects is very important to the quality of life of patients receiving these agents. However, relatively high cost coupled with minimal improvements in effectiveness may become an issue for nanomedicines.

Early-generation nanodrugs distributed passively in the body, reflecting their physical characteristics including size and surface properties. As discussed extensively in this volume, a new generation of nanomedicines is under development incorporating 'targeting' moieties to promote selective delivery of the nanoparticle to particular cells or tissues. A recent review cogently discussed the advantages and problems associated with such targeted nanomedicines.[25] There are clearly technical challenges with this approach; for example, the inclusion of a targeting ligand may reduce the circulation time of the nanoparticle, clearly an undesirable outcome. However, the more important issue is that targeted nanomedicines are likely to be extraordinarily costly. The analysis of nanomedicines from an economic perspective is in its infancy.[26] However, it will be important for such studies to go hand in hand with the development of the technology. National and private healthcare reimbursement entities will increasingly use rigorous economic standards such as cost per quality-adjusted life-year[27] to decide whether to include agents in their formulary. Thus it will be essential for the new generation of targeted nanomedicines to provide improved therapeutic outcomes, such as extension of life in cancer patients, as well as reduced toxicity. As an indication of the importance of this issue, a recent small-scale clinical trial indicating that ABRAXANE® had no therapeutic advantage over paclitaxel in breast cancer caused some US payers to reconsider their reimbursement policies on ABRAXANE®.[28] On a more hopeful note, the recent US Food and Drug Administration approval of Abraxane in pancreatic cancer was based on a small but significant increase in therapeutic effectiveness.[29]

11.5 Conclusion

Nanomedicine represents a premier example of the growing coalescence between the physical and biological sciences. It clearly deserves much of the interest and excitement it has created within the scientific community and among the public. Nonetheless, it is important not to allow this enthusiasm to blind us to the many challenges that must be overcome before nano-technology can find its place in clinical medicine. Leaders in the field should exercise restraint and be cautious about over-promising the extent that nanotechnology will alter the provision of healthcare or the rapidity with which it will be introduced into clinical practice.

There has been great emphasis on rapid 'translation' of nanotechnologies from the laboratory bench to the clinic. Although this is a laudable goal, it is inconsistent with the fact that there is still much basic knowledge lacking concerning the interactions of nanomaterials with living organisms. Furthermore, as discussed above, ineffective communication between nanotechnologists and biomedical scientists often results in 'solutions in search of a problem' rather than nanomedicines that realistically address clinical issues. Finally, nanomedicine will ultimately be tested in the forge of real-world healthcare economics. It will only succeed if it brings benefits to patients that cannot be attained by more conventional approaches and does so at reasonable cost.

References

1. Y. Barenholz, Doxil®–the first FDA-approved nano-drug: lessons learned, *J. Controlled Release*, 2012, **160**, 117–134.
2. R. J. Hamill, Amphotericin B formulations: a comparative review of efficacy and toxicity, *Drugs*, 2013, **73**, 919–934.
3. R. Duncan and R. Gaspar, Nanomedicine(s) under the microscope, *Mol. Pharm.*, 2011, **8**, 2101–2141.
4. V. J. Venditto and F. C. Szoka, Cancer nanomedicines: so many papers and so few drugs!, *Adv. Drug Delivery Rev.*, 2013, **65**, 80–88.
5. R. Juliano, Nanomedicine: is the wave cresting?, *Nat. Rev. Drug Discovery*, 2013, **12**, 171–172.
6. C. A. Mirkin, The beginning of a small revolution, *Small*, 2005, **1**, 14–16.
7. H. M. Azzazy and M. M. Mansour, In vitro diagnostic prospects of nanoparticles, *Clin. Chim. Acta*, 2009, **403**, 1–8.
8. D. A. Giljohann, D. S. Seferos, W. L. Daniel, M. D. Massich, P. C. Patel and C. A. Mirkin, Gold nanoparticles for biology and medicine, *Angew. Chem.*, 2010, **49**, 3280–3294.
9. S. L. Stott, C. H. Hsu, D. I. Tsukrov, M. Yu, D. T. Miyamoto, B. A. Waltman, S. M. Rothenberg, A. M. Shah, M. E. Smas, G. K. Korir *et al.*, Isolation of circulating tumor cells using a microvortex-generating herringbone-chip, *Proc. Natl. Acad. Sci. U. S. A.*, 2010, **107**, 18392–18397.
10. R. A. Harouaka, M. Nisic and S. Y. Zheng, Circulating tumor cell en-richment based on physical properties, *J. Lab. Autom.*, 2013, **18**, 455–468.
11. Y. Z. Lee, L. Burk, K. H. Wang, G. Cao, J. Lu and O. Zhou, Carbon Nanotube based X-ray Sources: Applications in Pre-Clinical

and Medical Imaging, *Nucl. Instrum. Methods Phys. Res., Sect. A*, 2011, **648**(Supplement 1), S281–S283.

12. X. Qian, A. Tucker, E. Gidcumb, J. Shan, G. Yang, X. Calderon-Colon, S. Sultana, J. Lu, O. Zhou, D. Spronk *et al.*, High resolution stationary digital breast tomosynthesis using distributed carbon nanotube x-ray source array, *Med. Phys.*, 2012, **39**, 2090–2099.

13. T. Lam, P. Pouliot, P. K. Avti, F. Lesage and A. K. Kakkar, Superparamagnetic iron oxide based nanoprobes for imaging and theranostics, *Adv. Colloid Interface Sci.*, 2013, **199–200**, 95–113.

14. R. A. Petros and J. M. DeSimone, Strategies in the design of nanoparticles for therapeutic applications, *Nat. Rev. Drug Discovery*, 2010, **9**, 615–627.

15. J. Xu, D. H. Wong, J. D. Byrne, K. Chen, C. Bowerman and J. M. DeSimone, Future of the particle replication in nonwetting templates (PRINT) technology, *Angew. Chem.*, 2013, **52**, 6580–6589.

16. R. Kanasty, J. R. Dorkin, A. Vegas and D. Anderson, Delivery materials for siRNA therapeutics, *Nat. Mater.*, 2013, **12**, 967–977.

17. L. Qian, E. Winfree and J. Bruck, Neural network computation with DNA strand displacement cascades, *Nature*, 2011, **475**, 368–372.

18. J. Stajic, Quantum information processing. The future of quantum information processing. Introduction, *Science*, 2013, **339**, 1163.

19. G. M. Sacha and P. Varona, Artificial intelligence in nanotechnology, *Nanotechnology*, 2013, **24**, 452002.

20. S. Fusco, M. Sakar, S. Kennedy, C. Peter, *et al.*, An integrated microrobotic platform for on demand targeted therapeutic Interventions, *Adv. Mater.*, 2013, ePub ahead of print.

21. D. Patra, S. Sengupta, W. Duan, H. Zhang, R. Pavlick and A. Sen, Intelligent, self-powered, drug delivery systems, *Nanoscale*, 2013, **5**, 1273–1283.

22. C. P. Snow, *The two cultures and the scientific revolution*, Cambridge University Press, New York, 1959.

23. R. L. Juliano, The future of nanomedicine: Promises and limitations, *Sci. Publ. Policy*, 2012, **39**, 99–104.

24. S. W. Jones, R. A. Roberts, G. R. Robbins, J. L. Perry, M. P. Kai, K. Chen, T. Bo, M. E. Napier, J. P. Ting, J. M. Desimone *et al.*, Nanoparticle clearance is governed by Th1/Th2 immunity and strain background, *J. Clin. Invest.*, 2013, **123**, 3061–3073.

25. Z. Cheng, A. Al Zaki, J. Z. Hui, V. R. Muzykantov and A. Tsourkas, Multifunctional nanoparticles: cost versus benefit of adding targeting and imaging capabilities, *Science*, 2012, **338**, 903–910.

26. R. Bosetti, W. Marneffe and L. Vereeck, Assessing the need for quality-adjusted cost-effectiveness studies of nanotechnological cancer therapies, *Nanomedicine*, 2013, **8**, 487–497.

27. S. Simoens and M. Dooms, How much is the life of a cancer patient worth? A pharmaco-economic perspective, *J. Clin. Pharm. Ther.*, 2011, **36**, 249–256.

28. E. Licking, Abraxane: The Canary in the Oncology Reimbursement Coal Mine. http://realendpoints.com/2012/09/abraxane-the-canary-in-the-oncology-reimbursement-coal-mine/, 2012.

29. ASCO, FDA Approves Abraxane Plus Gemcitabine for Metastatic Pancreatic Cancer. http://www.asco.org/advocacy/fda-approves-abraxane-plus-gemcitabine-metastatic-pancreatic-cancer, 2013.

CHAPTER 12

The Challenge of Regulating Nanomedicine: Key Issues

RAJ BAWA,*[a,b] YECHEZKEL BARENHOLZ[c] AND ANDREW OWEN[d]

[a] Patent Law Department, Bawa Biotech LLC, Ashburn, Virginia, USA;
[b] Department of Biological Sciences, Rensselaer Polytechnic Institute, Troy, New York, USA; [c] Laboratory of Membrane and Liposome Research, Department of Biochemistry and Molecular Biology, Institute for Medical Research Israel – Canada (IMRIC), The Hebrew University-Hadassah Medical School, Jerusalem, Israel; [d] Department of Molecular and Clinical Pharmacology, University of Liverpool, UK
*Email: bawa@bawabiotech.com

12.1 Introduction

There is no shortage of excitement, exuberance, hype and misinformation when it comes to anything "nano". Many claim it to be the next industrial revolution, some purport that it has already permeated virtually every sector of our economy while others dismiss it as nothing new but a repackaging of old concepts and technologies with a new label.

In reality, nanotechnology is the natural continuation of the miniaturization of materials and biomedicine. One can now say with relative certainty that research and development (R&D) is in full swing with novel nanomedical products starting to arrive, or at least inching their way, into the marketplace. Everyone agrees that this decade has witnessed relatively more advances and product development in nanomedicine than years prior.

Too often though, companies, academia and policymakers exaggerate basic R&D in nanotechnology as potentially revolutionary advances and

RSC Drug Discovery Series No. 51
Nanomedicines: Design, Delivery and Detection
Edited by Martin Braddock
© The Royal Society of Chemistry 2016
Published by the Royal Society of Chemistry, www.rsc.org

claim these early-stage discoveries as confirmation of downstream novel products and applications to come. Many have desperately tagged or thrown around the "nano" prefix to suit their purpose, whether it is for federal research funding, patent approval of supposedly novel technologies, raising venture capital funds, running for office or seeking publication of their manuscript. All of this is happening while thousands of over-the-counter products containing silver nanoparticles, titanium dioxide and carbon nanoparticles continue to stream into the marketplace without adequate safety testing, labeling or regulatory review.

Nanoproducts and applications do continue to evolve and play a pivotal role in various industry segments, spurring new directions in research, patents, commercialization and technology transfer. In spite of anemic product development in the 1980s and 1990s, patent filings and patent grants have continued unabated.[1–3] It is no secret that nanopatents of dubious scope and breadth continue to be granted by patent offices around the world. In fact, since the early 1980s, "patent prospectors" have been on a global quest for "nanopatent land grabs". Universities have jumped into the fray with a clear intention of patenting as much "nano" as they can. As indicated above, academia, start-ups and established companies often continue to exaggerate basic research or project potential downstream applications based upon early-stage discoveries. Venture funding has mostly shied away in recent years, although industry–university alliances have continued to gel. Wall Street's interest in nanotechnology has been somewhat mixed over the years, from cautionary involvement to generally shying away, partly due to the definitional and regulatory issues.

12.2 Defining "Nano": A Problem for Regulators?[†]

How should regulators define nanotechnology in the context of pharma? What does nanoscale drug delivery mean? Are all engineered nanotherapeutics unique purely due to a specific size range? Is there a regulatory or legally plausible definition of nanotech from a pharma perspective with respect to size? Does a specific or unique size range of nanotherapeutics impart nanotoxicity?

All these questions are related to the definition of the prefix "nano." In fact, this definitional issue, or lack thereof, continues to be one of the most significant problems shared by regulators, policy-makers, drug companies, patent practitioners, and legal professionals.[4–10] In particular, regulatory agencies and governmental entities such as the US Food and Drug Administration (FDA), the European Medicines Agency (EMA), Environmental Protection Agency, Centers for Disease Control and Prevention, National Institute for Occupational Safety and Health and the US Patent and Trademark Office are grappling with this critical issue. Clearly, the need for an internationally agreed definition for key terms like nanotechnology,

---

[†]This section is adapted, with permission, from ref. 16.

nanoscience, nanomedicine, nanobiotechnology, nanodrug, nanotherapeutic, nanopharmaceutical and nanomaterial has gained urgency. This is essential for harmonized regulatory governance, accurate patent searching and prosecution, standardization of procedures, assays and manufacturing, quality control, discussion of ethical issues, safety assessment and much more.

In this chapter, we use the following terms interchangeably: nanomedicine, nanodrug, nanotherapeutic and nanopharmaceutical. Similarly, we use these terms interchangeably: nano, nanotech and nanotechnology.

So, what does the prefix "nano" refer to? Any term with this prefix is broad in scope. Consider the widely used terms nanotechnology, nanomedicine and nanopharmaceutical, all of which are misnomers/misleading because they do not refer to a single technology. The terms nanotechnology and nanomedicine refer to interdisciplinary areas that draw from the interplay among numerous materials, products and applications from several technical and scientific fields. In other words, "nano" is an umbrella term/prefix encompassing several technical/scientific fields, processes and properties at the nano/micro scale.[4-13]

Clearly, there is confusion over the definition of nanotechnology. Moreover, due to a lack of any standard nomenclature, various inconsistent definitions have sprung up over the years.[4-16] For instance, nanotechnology has been *inaccurately* defined by the US National Nanotechnology Initiative (NNI)[‡] since the 1990s as "the understanding and control of matter at the nanoscale, at dimensions between approximately 1 and 100 nanometers, where unique phenomena enable novel applications...". Some definitions increase the upper limit to 200 nm or 300 nm or even 1000 nm. Some definitions omit a lower range. Others refer to sizes in one, two or three dimensions while others require a size plus special/unique property or *vice versa*.[14] Given this backdrop, Dr Bawa proposed the following definition of nanotechnology in 2007, one that is unconstrained by an arbitrary size limitation:[15,16] "The design, characterization, production and application of structures, devices, and systems by controlled manipulation of size and shape at the nanometer scale (atomic, molecular, and macromolecular scale) that produces structures, devices and systems with at least one novel/superior characteristic or property." This definition has three key features. First, the definition recognizes that the properties and performance of the synthetic, engineered *structures, devices and systems* are inherently rooted in their nanoscale dimensions. Second, the focus of this flexible definition is on "technology" that has commercial potential from a consumer perspective, not "nanoscience" or basic R&D in a laboratory setting that may

---

[‡]The NNI is the US government's interagency program for coordinating, planning and managing R&D in nanoscale science, engineering, technology and related efforts across 25 agencies and programs. The NNI has been regularly reviewed by the President's Council of Advisors on Science and Technology since the council was designated in 2004.

lack commercial implication. Third, the *structures, devices and systems* that result from or incorporate nano must be *novel/superior* compared to their bulk, conventional counterparts. Fourth, the concept of *controlled manipulation* (as compared to "self-assembly") is critical here.

One of the major impacts of nanomedicine is taking place in the context of drug delivery. Novel nanodrugs[§] and nanocarriers are being developed that address fundamental problems of traditional drugs, ranging from poor water solubility, unacceptable toxicity profiles, poor bioavailability, solubility issues, physical/chemical instability and a lack of target specificity. Additionally, *via* tagging with targeting ligands, these nanodrug formulations can serve as innovative drug delivery systems for enhanced cellular uptake of therapeutics into tissues of interest (site-specific delivery). As a result, nanodrugs are being developed to allow delivery of active agents more efficaciously to the patient while minimizing side effects, improving drug stability *in vivo* and increasing blood circulation time. There are a number of FDA-approved, commercialized first-generation nanodrugs for intravenous and non-intravenous delivery. Some of these are completely novel while others are redesigned variations of earlier versions (Figure 12.1). The first-generation nanodrugs mainly address single challenges such as targeted delivery while the second and third generations currently in development can offer two or more functions together (*e.g.*, delivery and imaging) or overcome multiple physiological barriers to deliver their therapeutic payloads. However, most nanodrugs are still in their infancy, being at the pre-clinical development stage or in clinical trials. As these nanodrugs move out of the laboratory and into the clinic, governmental regulatory agencies such as the FDA and EMA continue to struggle to encourage their development while imposing some sort of order in light of regulatory and safety concerns. If a size limit must be imposed on the definition of "nano", then an upper limit of 1000 nm (1 μm) may be most appropriate, at least regarding a discussion on nanodrugs. It appears that the FDA in 2014 finally expanded the upper limit to 1000 nm (from 100 nm), albeit unofficially.[¶]

[§]There is no formal definition for a nanoscale drug delivery system (nanodrug product). Dr Bawa[16] defines it as being: (1) a formulation, often colloidal, containing therapeutic particles (nanoparticles) ranging in size from 1 to 1000 nm; and (2) either (a) the carrier(s) is/are the therapeutic (*i.e.*, a conventional therapeutic agent is absent) or (b) the therapeutic is directly coupled (functionalized, solubilized, entrapped, coated, *etc.*) to a carrier. The FDA defines the term "colloid for regulatory purposes as a chemical system composed of a continuous medium (continuous phase) throughout which are distributed small particles, 1 to 1000 nm in size (disperse phase), that do not settle out under the influence of gravity; the particles may be in emulsion or in suspension."

[¶]Even the FDA, which has not adopted any "official" regulatory definition, appears to now be emphasizing a loose definition for products that involve or employ nanotechnology in the context of pharma that either: (1) have at least one dimension in the 1–100 nm range; or (2) are up to 1000 nm, provided the novel/unique properties or phenomenon exhibited are attributable to these dimensions outside of 100 nm.

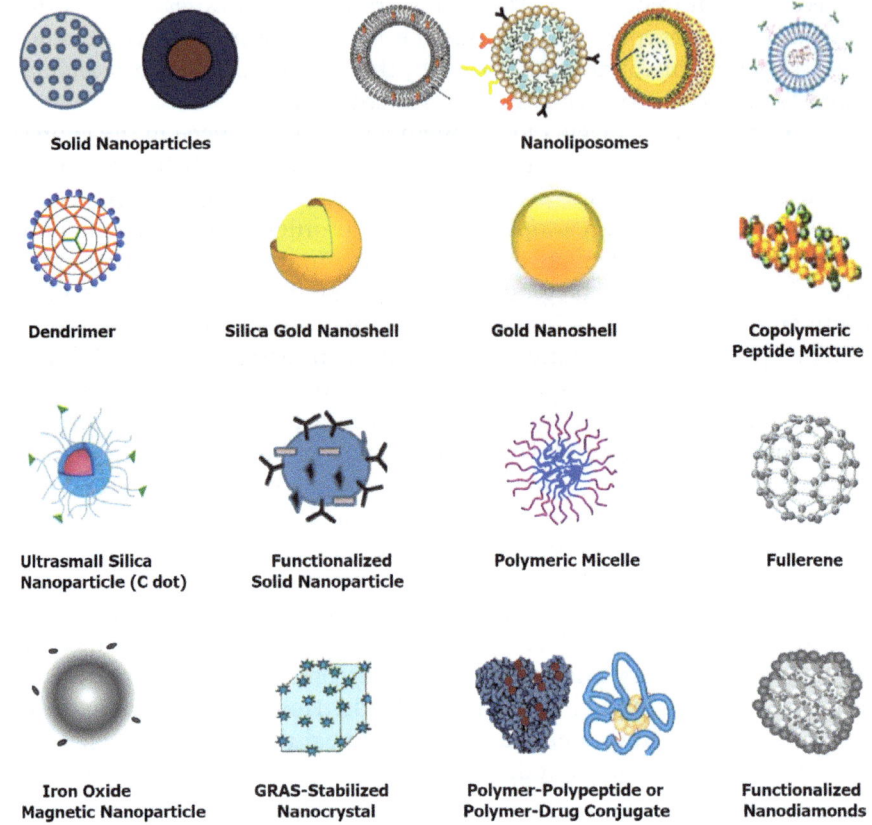

Solid Nanoparticles   **Nanoliposomes**

Dendrimer   **Silica Gold Nanoshell**   **Gold Nanoshell**   **Copolymeric Peptide Mixture**

Ultrasmall Silica Nanoparticle (C dot)   **Functionalized Solid Nanoparticle**   **Polymeric Micelle**   **Fullerene**

Iron Oxide Magnetic Nanoparticle   **GRAS-Stabilized Nanocrystal**   **Polymer-Polypeptide or Polymer-Drug Conjugate**   **Functionalized Nanodiamonds**

Figure 12.1   Schematic illustrations of nanoscale drug delivery system platforms (nanotherapeutics or nanodrug products). Shown are nanoparticles (NPs) used in drug delivery that are either approved, are in preclinical development or are in clinical trials. They are generally considered as first or second generation multifunctional engineered NPs, generally ranging in diameter from a few nanometers to a micron. Active biotargeting is frequently achieved by conjugating ligands (antibodies, peptides, aptamers, folate, hyaluronic acid) tagged to the NP surface *via* spacers or linkers like PEG. NPs such as carbon nanotubes and quantum dots, although extensively advertised for drug delivery, are specifically excluded from the list as this author considers them commercially unfeasible for drug delivery. Non-engineered antibodies and naturally occurring NPs are also excluded. Antibody-drug conjugates (ADCs) are encompassed by the illustration labelled "Polymer-Polypeptide or Polymer-Drug Conjugate." This list of NPs is not meant to be exhaustive, the illustrations are not meant to reflect three dimensional shape or configuration and the NPs are not drawn to scale. Abbreviations: NPs: nanoparticles; PEG: polyethylene glycol; GRAS: generally recognized as safe; C dot: Cornell dot; ADCs: antibody-drug conjugates. Copyright © 2016 Raj Bawa. All rights reserved. The copyright holder permits unrestricted use, distribution and reproduction of this figure (plus legend) in any medium, provided the original author and source are clearly and properly credited. Reproduction without proper attribution constitutes copyright infringement.

12.3 Lessons Learned from Doxil®: The First FDA-approved Nanodrug

Doxil® was the first FDA-approved nanodrug. The success of Doxil® largely stemmed from it being a nanoliposome. The development of Doxil® is reviewed elsewhere[17,18] and also in this book. In this chapter, we deal with two aspects of the development of Doxil®: can it be classified as a nanodrug and what are the special requirements for approval of generic Doxil® by the FDA and/or EMA? To answer the first question we have to go back to 1959, when the new discipline of nanotechnology was first mentioned, although not by name, by Nobelist Richard Feynman. He highlighted the importance of scaling issues, suggesting that certain very small assemblies have different physical behavior from large ones, to a large extent due to their very large surface-to-volume ratios and their small sizes.|| This imparts unique properties to most nano-assemblies that low molecular weight molecules and larger assemblies generally lack. Given this backdrop, we suggest that nanodrugs based on nanoliposomes (such as Doxil®) meet the above criteria and can be classified as nanodrugs primarily for the following two reasons:

(1) Biological reason (nano-anatomy): nanoscale particulates take advantage of the unique porous vasculature that is the Achilles' heel of cancer and inflammation, as it enables "passive targeting" of ≤100 nm particulates to the tumor and inflamed tissues. This effect is termed the enhanced permeability and retention (EPR) effect.[19–21] This selection is a decision process made by the biology/physiology/anatomy related to the disease. Neither low molecular weight drugs, nor drugs associated with particles in the micro range can take advantage of the EPR effect.
(2) Nano-related reasons:
  (a) For liposome fabrication, the bottom-up approach is used. In the presence of water, there is a self-assembly from single lipid molecules to particulate assemblies. The lipid bilayer width is 5 nm and many of the assemblies are <100 nm in diameter, structures that fall under the nano classification.
  (b) Nanovolume provides the nanoliposomes unique properties of highly efficient, stable drug loading and controlled release profile. The nanovolume enables the efficient remote loading driven by transmembrane ion and/or pH gradients, thereby achieving very high drug levels per liposome. Therefore, these nanoliposomes can bring sufficient drug load to the tumor or inflamed tissue in spite of their nanovolume. This nanovolume also enhances drug release by the ammonia present in the tumor due to the unique metabolic pathway of glutaminolysis.[18] This is demonstrated by the fact that in such a nanoliposome, the shift between internal pH 5.0

|| As a drug particle is granulated (nanonized) into smaller particles, the total surface area of the smaller particles become much greater relative to their volume (*i.e.*, an enormously increased surface-area-to-volume ratio).

and pH 6.0 requires the movement of >10 protons. Finally, the high concentration of active pharmaceutical ingredient (API) at the intraliposome aqueous phase in the presence of a suitable counter ion results in intraliposome reversible precipitation and/or crystallization which helps to stabilize drug loading during storage and circulation in the blood of the animal or human. This effect enables liposomes to reach the diseased tissue loaded with drug.

Doxil® is a successful anticancer drug, especially given its worldwide use for the treatment of ovarian cancer. So far, more than 500 000 patients have been treated with Doxil® worldwide. However, in spite of its extended use in the more than 4 years since the last Doxil®-related patent expired[22] and in spite of major efforts by many companies around the globe to try to produce it and obtain regulatory approval, there is only one FDA-approved "generic Doxil®" available. This formulation is Lipodox®, which was approved by the FDA in 2014** (not approved yet by the EMA). Why are there not many generic Doxil®-like products approved by the FDA, despite enormous efforts by many companies to get approval? The answer lies in the FDA requirements for the approval of generic Doxil®. In fact, these requirements can also serve as a good example to the tough and broad spectrum of requirements of FDA for the approval of similar nanodrugs.

The scientific, technological, and regulatory difficulties in getting Doxil® generics approved can be better understood after evaluating carefully the example of FDA requirements for approval of "generic Doxil®". This information is available in a 2013 FDA document.[23] The development of a generic Doxil® requires that certain criteria be met. Specifically, *in vitro* characterization and chemistry, manufacturing and control (CMC) levels include: (1) the same drug product composition; (2) manufactured by an active (remote) liposome drug-loading process with an ammonium sulfate gradient; and (3) on the CMC level, this means having liposomal characteristics equivalent to Doxil®, including liposome composition, state of encapsulated drug, internal environment of the liposome, liposome size distribution, number of lamellae, grafted polyethylene glycol (PEG) at the liposome surface, electrical surface potential or charge and *in vitro* leakage rates.

The clinical requirements are deceptively simple in appearance, and include:

- a single dose,
- a two-way crossover study in ovarian cancer patients whose disease has recurred or progressed after platinum-based chemotherapy,

---

**In October 2011, drug shortages were such a pressing issue in the US that an executive order was issued directing the FDA to streamline the approval process for new therapeutics that could fill the voids. One of the major drugs whose supply was deficient in the US was Doxil®, and to curb this shortage, the FDA authorized the temporary importation of Lipodox® in February 2012. Following this, the FDA evaluated and approved the drug formulation within a year, roughly one-third of the time it takes for an average generic to receive premarket regulatory approval. As a result, Lipodox® became the first generic nanodrug approved in the US.

- dose of 50 mg m^{-2},
- pharmacokinetic variables,
- free doxorubicin and liposome-encapsulated doxorubicin,
- bioequivalence based on (90% confidence interval) area under the curve (AUC) and maximum concentration (C_{max}).

The pivotal bioequivalence study should be conducted using a test product produced by the proposed commercial-scale manufacturing process. Due to low plasma levels of free doxorubicin there is a potential for large patient-to-patient variability. As a result, overcoming this issue in the clinical trials may require a large number of patients. Other aspects of the comparison studies that cannot be ignored today are the already recognized side effects during Doxil's® first clinical trial:[24] hand–foot syndrome[17] and complement activation.[25,26]

The development of Doxil® required the development and validation of a large repertoire of quality control methods to deal with the chemical, biological and physicochemical characterization of a liposomal drug.[27] The FDA draft guidelines on liposomal drugs and on generic Doxil® further increases this repertoire. While the chemical aspect of the liposomal drug quality control is familiar to the drug industry, this is not the case for the unique biological assays (such as quantification of complement activation) and the unique physico-chemical methods that are a new topic to the pharma industry. The latter includes use of sophisticated equipment and methodologies such as electron microscopy (especially cryo-transmission electron microscopy or cryo-TEM), X-ray diffraction (small- and wide-angle X-ray scattering), differential scanning calorimetry and methods to determine size distribution and morphology distribution. Size and size distribution are accepted as critical attributes and can be approached using various methods.[27] These methods are divided into two major groups:

(1) Methods based on measuring the diffusion coefficient from which the particle hydrodynamic radius is determined using the Stokes–Einstein equation[27,28] The, radius obtained *via* dynamic light scattering (DLS) data includes the PEG layer as part of the hydrodynamic radius.
(2) The second group includes methods based on image analysis of cryoTEM images, determining directly the shape and size distribution irrespective of the liposome shape. Due to lack of contrast, the radii determined by this approach do not include the liposome PEG layer. Therefore, the dimension of the PEG layer obtained by other methods such as X-ray diffraction must be added to the radii determined by the image analysis.[28,29] For an ellipsoid such as Doxil®, this image analysis enables the determination of the axial ratio and thereby the volume of the liposomes. When combined with X-ray diffraction data, whether the drug inside the liposome is soluble, or appears as a precipitate or as crystals can be determined.[17,18,30]

Among the first group of methods DLS is the most commonly used and widely accepted approach.[27] Our recent results[28] provide more information on PEGylated liposomal doxorubicin (PLD) size measurements and size distribution based on both DLS and image analysis using cryo-TEM.

12.4   Baby Steps Lead to Regulatory Uncertainty: The FDA as an Example

The consequences of the overregulation of nanomedicine are potentially severe, ranging from unnecessary barriers to commercialization to inadvertent harm to the public and the environment. However, clearly there exists a need for creating regulatory guidelines that follow a science-based approach and are responsive to shifting risks. In other words, a fine balance needs to be struck, one that is flexible enough to be modified as technological innovation proceeds. A balance is always desired between promoting the development of nanomedicine and exercising effective and timely oversight when warranted. After all, laws and regulations always try to keep up with scientific advances in emerging technologies and if they lag too far, as we believe in this case, that is when things become murky, commercialization stifles, public health is negatively impacted and venture recedes.

The "baby steps" that the FDA has undertaken over the past decade or so have generally contributed to regulatory uncertainty.[5] Although this federal agency has held public hearings, posted information on its website, presented at various conferences, created a task force to address nanotech, published a position paper in *Science* and released draft guidance documents, in our opinion it has failed to provide anything concrete with respect to assays, testing, data requirements or practical guidelines. As a result, the pharma and device industries, drug and patent lawyers, business and venture communities, policymakers and the public are left to dig up whatever information they can find on nanoregulation or on marketed products that may incorporate nanomaterials or involve nanotech. Currently, there are few reliable means to identify marketed "nano-containing" products, and consumers are unable to pinpoint which ones may be toxic. This is obviously not very comforting for the consumer, given that the FDA is an agency tasked with protecting public health. In fact, statements that summarized the state of affairs at the FDA few years ago are still applicable today in 2016.[5]

Nanoproducts, whether they are a drug, device, biologic, or combination of any of these, are creating challenges for the FDA regulators as they struggle to accumulate data and formulate testing criteria to ensure the development of safe and efficacious nanoproducts. Emerging technologies are particularly problematic for regulatory agencies like the FDA, given the independent nature, slow response rate, significant inertia of governmental agencies. Currently, major global regulatory systems, bodies and regimes regarding nanomedicines are not fully mature, hampered in part by a lack of specific protocols for preclinical development and characterization. Additionally, in spite of numerous harmonization talks and meetings, there is lack of consensus on procedures, assays and protocols to be employed during pre-clinical development and characterization of nanomedicines. On the other hand, there is a rise of diverse nano-specific regulatory arrangements and systems, contributing to a dense global nanotechnology regulatory landscape, full of gaps and devoid of central coordination. Moreover, governmental regulators often lack technical and scientific knowledge to support risk-based regulation, thereby leaving a significant regulatory void. Therefore, guidance is critically needed to provide clarity and legal certainty to all stakeholders: manufacturers, policymakers, healthcare providers, and the consumer.

There are hundreds, if not thousands, of nanoproducts in the market for human use, but little is known of their health risks, safety data or toxicity profiles. Even less is known of nanoproducts that are released into the environment that can potentially contact humans. Then, there are nanoproducts such as cosmetics containing metallic nanoparticles and other nanomaterials flooding the market yet not subject to premarket review by the FDA. Adding to safety concerns is the fact that manufacturers are neither required to obtain any special premarket approval from the FDA (on a case-by-case basis) nor required to list nano-ingredients on product labels at this time. Meanwhile, evidence continues to mount that many (if not most) nanoproducts inherently possess novel size-based properties and toxicity profiles. FDA continues to adopt a precautionary approach to this issue in hopes of countering negative publicity. A one-size-fits-all approach to nanogovernance may not be ideal.

It is well established that certain nanoproducts marketed for direct and indirect human consumption may be unsafe.[31,32] These products could present unexpected human toxicity effects due to various factors, such as an increased reactivity compared to their ''bulk'' counterparts and an increased potential to traverse biological barriers/membranes to reach or accumulate in tissues.[33,34] Furthermore, there are concerns about the occupational and environmental risks associated with the manufacture and disposal of nanoproducts.[35,36] Not all nanoscale materials are created equal: some may be toxic and their toxicities dependent upon various factors that are material-specific (charge, polarity and chemical residues) and/or geometry-specific (size, shape and nanoscale features). Although nanotoxicity is complex, it is universally recognized that nanoscale products and particles *often* have fundamentally properties distinct from their larger bulk counterparts.[3–7,11,15,16] In other words, "nano" does not just mean that a nanoproduct is merely smaller, but that there is a possibility that the product's "nanocharacter" may render it unsafe. Basically, it cannot be presumed that a nanoproduct will be safe or even "bioequivalent" to its larger bulk counterpart. Since the FDA position is that it only regulates nanoproducts and not nanotechnology *per se*,[37] overlooking potential toxicological risks of a technology poses serious public health concerns.

In 2012, the FDA commissioner summarized in general terms the FDA's so-called "broadly inclusive initial approach" with respect to nanogovernance in a policy paper published in *Science*[38]:

> [The] FDA does not categorically judge all products containing nanomaterials or otherwise involving the application of nanotechnology as intrinsically benign or harmful. As with other emerging technologies, advances in both basic and applied nanotechnology science may be unpredictable, rapid, and unevenly distributed across product applications and risk management tools. Therefore, the optimal regulatory approach is iterative, adaptive, and flexible …. It is iterative by developing and delivering incremental components of a regulatory system, such as guidances specific to product areas, each as warranted and when ready. It is adaptive by providing a mechanism, within statutory constraints, to change the rules, presumptions, or pathways for these regulatory components, in light of new information gained from research or from experience in regulating earlier products. And it is flexible by using all available means, ranging from workshops to consultations to guidances to rules, in order to match the burden of regulation to its need.

However, in spite of public and industry concerns, as of March 2016, no clear guidelines or regulations have been proposed by or expected from the FDA. The "broadly inclusive initial approach" referenced above by the FDA Commissioner needs to be expanded into practical, science-based regulatory guidelines that can be depended upon by industry and consumers alike. Obviously, not every nanoproduct should to be regulated, but more is clearly expected from the FDA than guidance documents that are in draft format, broad lectures that fail to identify key regulatory issues or policy papers that are often short on specifics. In this regard, although the FDA has made important strides, especially in the Hamburg era, numerous challenges continue to confront the agency as important unanswered questions linger regarding nanoregulation (Box 12.1).

Size changes within the nanoscale range and the potential unpredictability arising therefrom are likely to add complexity to the FDA review process. The traditional product-by-product regulatory model that the FDA currently employs may not be effective for all nanoproducts, because it may be difficult to put them into one of the available traditional classifications (*i.e.*, drug, device, biological or combination product). However, in many cases, the FDA may view nanoproducts as technologically overlapping

Box 12.1   Key questions for regulators: balancing public health and encouraging nanomedicine[5].

- Many products that the FDA and EMA see during their review utilize nanotechnologies or contain nanomaterials, some lack disclosure of this while other identify them as such. Should these products be regulated? If so, how and to what degree?
- It is likely that some marketed nanoproducts (*e.g.*, sunscreens containing zinc oxide and titanium dioxide) warrant safety labeling to alert the unsuspecting consumer. Are most nanomaterials used in nanoproducts inherently toxic?
- Are advances in nanoscience and nanomedicine proceeding too fast for meaningful review to take place? Can regulations truly tame the vastness encompassed by "nano"? For example, should a different set of principles apply for regulating cosmetics versus cancer nanomedicines?
- How much harmonization of drug regulation can reasonably be expected between the US and the rest of the world? Should pharmacoeconomic data be required prior to commercialization of nanomedicines to demonstrate both social and economic added value in comparison to "conventional" established treatments?
- It appears that regulatory agencies are pushing industry to provide product-specific data for areas like cosmetics, where the FDA lacks statutory premarket review authority. Are such voluntary industry measures transparent and meaningful?
- Is the FDA's so-called "broadly inclusive approach" of considering whether products contain nanomaterials or involve nanotechnology sufficient?
- Can the safety and efficacy of complex follow-on nanotherapeutics (nanosimilar products) ever be assured without a full slate of clinical trials?
- Have the FDA and EMA kept pace with emerging advances in nanotech R&D with respect to predicting, defining, measuring and monitoring potential "nanotoxicities"?
- Globally, should there be a wider coordinated effort on the part of regulatory agencies to review, amend or create nanoregulations where appropriate and warranted? Who in addition to regulatory agencies like the FDA and EMA be given the key responsibility to regulate nanomedical products for human use?
- Can nanomedicine, as applied to public health, be solely regulated under existing regulations and laws?
- Are new regulations needed for all regulated products containing nanomaterials or involving nanotechnology or should they be limited to only a subset of products containing nanomaterials?

- Have delayed and uncoordinated efforts hurt venture and commercialization activities in the US and Europe?
- The FDA has unofficially embraced the inaccurate definition of nanotechnology proposed by the NNI. What is the "official" position of the FDA regarding the definition of nanotechnology, nanoscale, nanotherapeutic, nanodrug, nanomaterial and nanomedicine?
- Industry and stakeholders fully understand that regulatory agencies cannot formulate generalized guidelines, assay protocols or tests for all nanomedicines or nanoproducts. Nevertheless, regulatory agencies should provide selective guidelines, at least on a case-by-case basis (instead of nonbinding, unofficial "draft guidelines" or "position papers"), for industry to rely upon in determining whether their nanoformulations or nanodrug products might be subject to regulatory examination beyond what is typical for small-molecule drugs?

(miniaturization will blur distinctions between different categories) from a review perspective, and therefore consider them as highly integrated nanomedical combination products. These complexities are likely to pose additional challenges and review issues for the FDA.[5,39]

The FDA's ability to regulate nanoproducts effectively will depend largely on the category into which the product seeking approval falls. However, as alluded to above, certain therapeutics are "combination products", which consist of two or more regulated components (drug, biological or device) that are physically, chemically or otherwise combined or mixed to produce a single entity.[5,39,40] In such cases, the FDA determines the "primary mode of action" (PMOA) of the product, which is "the single mode of action of a combination product that provides the most important therapeutic action". This process is frequently imprecise, as it is not always possible to clearly elucidate a combination product's PMOA. Determining which framework will apply to any combination product is the task of the Office of Combination Products (OCP).

Obviously, the OCP will be the first office within the FDA to review many nanoproducts. The OCP makes its assignments on a case-by-case basis depending on the PMOA. But again, this process is frequently imprecise as it is not always possible to clearly elucidate a combination product's PMOA, often because at the time of an investigational application it is not clear which mode of action provides the most important therapeutic action, or the product has two different equally critical modes of action. It is quite possible that some nanoproducts will blur the distinction between mechanical and chemical action at the nanoscale or that they may be both therapeutic and diagnostic in operation. In fact, such spanning of regulatory boundaries between the various categories often results in inconsistency[5,39,40]

Another key issue is the limited information currently available correlating the physicochemical properties of nanoscale materials with risks, and lack of validated preclinical screens and animal models for the assessment of nanomaterials. In addition, the toxicity issues surrounding many nanoscale materials may not be fully apparent until they are widely used and their exposure fully felt by a diverse population. Hence, some sort of post-market tracking or a surveillance system must be adopted or legislated to assist in product recalls. None of this has happened yet, in spite of serious signals that toxicity may be more widespread than is apparent. Although toxicological testing for health risks of nanoparticles and nano-enabled products is not currently a complete science,[41] it is crucial to monitor their unique properties (if any) that may lead to serious adverse effects and toxicity. It is essential that long-term testing of nanoscale materials and nano-products is in place to allow safety testing. Box 12.2 lists recommendations for regulatory agencies to consider as they tackle the regulatory framework for nanomedicine.

Another critical issue is an important one: How does the FDA currently approve nanoscale therapeutics? Are unique nano-enabled properties assessed? Are these products reviewed on a case-by-case basis? It is clear that some nanomaterial-containing formulations ("nanoformulations") are indeed new chemical entities (NCEs). When warranted, nanoversions of active ingredients should be treated by the FDA as NCEs. This will ensure that drugs and biologicals that have been previously approved by the FDA but later modified as nanoversions will undergo a new and rigorous round of safety testing/clinical trials in order to obtain regulatory approval.

The FDA on its website highlights certain critical issues in its nanoregulatory approach that, at face value appear most appropriate.[42] However, underneath this veneer remains the real point for nanoregulation: "attributes" of materials at the nanoscale (nanocharacter) may require unique assays/testing and premarket approval beyond what the laws on the books can currently tackle to provide safe products to the consumer. There is urgency in this regard and clear guidelines are needed, not nonbinding draft reports that are meaningless from an industry perspective. Real progress in nanoregulation is urged, not mere regulatory discussions or a listing of what will be done in future. Nanoregulation with respect to pharmaceutical products, no matter how challenging or complex, needs to be addressed transparently with proper industry and public feedback. Regulation of nanomedicine must be conducted with the overarching principle that it is science based, not politically motivated or policy based.

Sadly, the conclusions drawn a few years ago regarding the regulation of nanomedicine are no different from those faced by the industry and the public in 2016[5]:

For now, nanoproducts submitted for FDA review will continue to be subjected to an uncertain regulatory pathway. This could negatively impact venture funding, stifle research and development in nanomedicine,

Box 12.2   Recommendations for regulatory agencies regarding nanomedicine regulation.[5]

Safety and risk

- On a case-by-case basis and in conjunction with industry, identify unique safety issues associated with nanoparticles and nanomedical products. The FDA should meet its regulatory and statutory obligations by offering technical advice and guidance to industry beyond what its track record currently reflects.
- Actively seek product safety data from industry where FDA statutory authority exists for pre-market review.
- Incentivize and encourage voluntary industry submissions of safety data on nanomaterials or products that incorporate nanotechnology prior to market launch, especially in cases (*e.g.*, cosmetics) where the FDA lacks statutory pre-market review authority.
- Correlate physiochemical properties with *in vivo* biological behavior and therapeutic outcome.
- Since there are few protocols to characterize nanomedicines at the physicochemical, biological and physiological levels, it is essential to develop a research strategy that involves adsorption, distribution, metabolism, and excretion (ADME) studies. A holistic approach to understanding ADME can be realized through the integration of mechanistic ADME data through the mathematical algorithms that underpin physiologically based pharmacokinetic (PBPK) modelling, routinely utilized to support regulatory submissions for conventional medicines in the US by the FDA and in Europe by the EMA.
- Develop toxicology tests and conduct physico-chemical characterization (PCC) studies for nanomaterials. Although complexity and diversity of nanomedicines and nanomaterials poses a problem, biocompatibility and immunotoxicity must be taken into consideration during preclinical assessment.
- Understand mass transport across biological membranes and body compartments as well as biodistribution profiles following administration via a specific delivery route.
- Develop standards that correlate the biodistribution of various nanomaterials with safety/efficacy by using parameters such as size, surface charge, stability, surface characteristics, solubility, crystallinity and density.
- With industry input, create a comprehensive public databank relating to the bio-interactions and toxicity profiles of engineered nanomaterials (ENMs).

Data

- Adapt existing methodologies, as well as develop new paradigms for evaluating *in vivo* animal and clinical data pertaining to safety and efficacy of nanomedical products before and during the product life cycle.
- Develop guidance that provides specifics as to what kind of data is required at each step of the nanomedical translational process.
- Share data in a transparent and harmonized manner. Seek additional data on safety or effectiveness during premarket review process when warranted. The FDA's excessive reliance on publicly available or voluntarily submitted information, adverse-event reporting and on post-market surveillance activities may not be ideal in the case of ENMs for human use.

Standardization and Nomenclature

- Create reference classes for ENMs that are synthesized and characterized.
- Develop consensus-testing protocols to provide benchmarks for the creation of classes of nanoscale materials, both ENMs and native.
- Create uniform nomenclature for and/or working definitions of nanomaterials. Refine the current definitions of nanomaterial, nanotechnology, nanodrug, nanopharmaceutical, nanoscale and nanomedicine for regulatory purposes.
- Further explore international regulatory harmonization efforts and formal treaties with relevant stakeholders.
- In addition to governmental bodies, involve various standard-setting organizations such as the ISO and ASTM International.
- Consult and collaborate with other federal agencies in a more effective, transparent and science-based manner. The FDA's current engagement in policy dialogue with other federal agencies (*via* the Emerging Technologies Interagency Policy Coordination Committee and other forums) has not produced any important guidelines for industry. The FDA should limit the number of non-binding draft guidance and policy papers that it periodically issues.

Tools and Techniques

- Assist in developing unique tools and techniques to characterize nanoscale materials.
- Develop imaging modalities for visualizing tissue biodistribution.
- Develop mathematical and computer models for risk/benefit analysis that can monitor quality, safety and effectiveness vis-à-vis standard ENMs.

Classification scheme

- Reevaluate the current FDA classification scheme, including the Primary Mode of Action (PMOA) criteria for combination products
- Develop a classification system that is based on (a) function or (b) risk of potential harm.
- Reevaluate the system of differing legal standards for different product classes that may result in divergent regulatory outcomes for different product classes.
- Place more effort in tailoring relevant guidance governing various product-classes and address interpretation of relevant statutory/regulatory standards relative to these classes.

and erode public acceptance of nanoproducts. The end-result of this could be a delay in or loss of commercialized nanoproducts. Whether the FDA eventually creates new regulations, tweaks existing ones, or establishes a new regulatory center to handle nanoproducts, for the time being it should at least look at nanoproducts on a case-by-case basis. The FDA should not attempt regulation of nanomedicine by applying existing statutes alone, especially where scientific evidence suggests otherwise. Incorporating nanomedicine regulation into the current regulatory scheme is unwise. Regulation of nanotech must balance innovation and R&D with the principle of ensuring maximum public health protection and safety. Regulatory oversight must evolve in concert with newer generations of nanomedical products. It is hoped that the "baby steps" that the FDA has taken in the past decade regarding nanogovernance will translate into more meaningful, flexible and science-based guidance in the near future. In the end, the long-term prognosis of nanomedicine will hinge on effective, valid nanogovernance requiring the full commitment of regulatory agencies such as the FDA, as well as the regulated community such as the manufacturing sector.

12.5   Importance of Understanding Pharmacokinetics and Distribution in Development and Regulatory Submission

The drivers for the application of nanotechnology to drug delivery are numerous and can be categorized into delivering either process-specific advantages or pharmacological benefits. Insolubility of APIs is a continuing problem for the pharmaceutical industry[43] and negatively impacts upon the difficulty and cost associated with drug development. As discussed earlier, nanodrugs are broad in terms of their complexity and in the benefits that have been explicitly demonstrated to date. Examples of pharmacokinetic benefits include improved oral bioavailability,[44,45] modified delivery profiles,[44] long-acting delivery[46,47] and overcoming fed/fasted variation.[48] Advanced nanodrugs have also been developed that offer the potential to specifically target diseased cells and tissues by either passive or active mechanisms.[49] While nanoparticles have attracted much attention with respect to understanding whether there are new safety assessments required, it is important to recognize that for many such nanomedicines, adverse drug reactions for encapsulated APIs have been mitigated through altered distribution of the drug.[50] While it is true that certain nanomaterials trigger cellular events that may lead to unexpected safety issues (*e.g.*, within the immune system[51,52]), the real question relates to whether existing regulatory processes are sufficiently robust as to capture them. As for any medicine, nanotechnology-enabled products require a robust assessment prior to clinical application. Pharmacokinetic and targeting benefits may have important related benefits such as the ability to overcome sub-optimal patient compliance in chronic disease or reduction in doses or pill burden required for effective management of the disease. However, by their very nature, nanomedicines behave differently to dissolved molecules and this presents a challenge for accurate determination of the likely *in vivo* behavior during the development process.

Physiologically based pharmacokinetic (PBPK) modelling is now extremely well integrated into the development process for conventional medicines. PBPK modeling is increasingly used to support decision making on whether, when and how to conduct clinical pharmacology studies in humans and frequently form part of the investigational new drug and new drug application review process for the FDA and the EMA. The FDA has a PBPK program nested within the Division of Pharmacometrics that reviews the adequacy of submitted PBPK models, facilitates the review process through *de novo* analyses, supports regulatory policy and harmonises regulatory recommendations on the use of PBPK with non-US regulatory bodies and the wider scientific community.[52] In 2014, the EMA produced a concept paper (EMA/CHMP/211243/2014) on the qualification and reporting of PBPK modeling that highlighted that PBPK explicitly features in numerous guidelines. These include those for pharmacokinetic assessment during hepatic impairment (CPMP/EWP/2339/02), pharmacogenetic methods as applied to pharmacokinetics (EMA/CHMP/37646/2009), drug–drug interactions (CPMP/EWP/560/95/Rev. 1 Corr.*), as well as upcoming new guidelines and guideline revisions.

In basic terms, PBPK modeling involves the integration of mathematical descriptors of medicine-specific and patient-specific factors in order to simulate the pharmacokinetic properties in specific scenarios. Patient-specific factors usually take the form of physiological or anatomical descriptors such as those for regional blood flow or organ mass for example, and typically include known variation in such parameters within the human population.[53] Traditionally, medicine-specific factors have related specifically to the physicochemical properties of the API (*e.g.*, Log *P*, pK_a and fraction unbound, *etc.*) and/or *in vitro* experimental data that provide a surrogate for a pharmacological process. Examples for such *in vitro* data include the transcellular permeation across Caco-2 cells to provide a description of oral absorption of the drug, or metabolism studies in hepatocytes or microsomes to provide a measure of intrinsic clearance of the API. Accordingly, integrating such *in vitro* data with system data allows PBPK modeling to be utilized for *in vitro–in vivo* extrapolation. When the relationship between molecule physicochemical properties and pharmacological processes is understood, it is also possible to model pharmacokinetic behavior directly from the molecular descriptors of the API. However, as with any modeling approach, the robustness of the final models are only as good as the quality of the input data, and the degree of understanding of the processes regulating ADME of the medicine. While PBPK modeling for conventional small-molecule drugs (and environmental toxins) has improved remarkably over the past two decades, there are comparatively few examples of PBPK modeling for nanomedicines.[53–59]

Transcellular permeation across Caco-2 cells is a good example to illustrate the process by which an *in vitro* observation can be validated for inclusion as a parameter in PBPK modeling. In 2002, Sun *et al.* assessed the apparent permeability (P_{app}) across Caco-2 cell monolayers for 20 commercially available drugs for which human pharmacokinetic data were available.[60] Using these data, they were able to define the mathematical relationship between Caco-2 permeability at pH 7.4 and the fraction absorbed (%) when these drugs were dosed to humans. In doing so, they provided the equation necessary for using Caco-2 cell data to model the absorption of drugs. Whether these same equations can accurately inform PBPK models for nanotechnology-enabled products remains uncertain, but the approach was recently applied to a novel solid drug nanoparticle formulation with apparent success, as qualified by observations in pre-clinical species.[57] In this case, the strategy was to improve the oral bioavailability of the API, and as such, augment a process known to occur for any orally delivered medicine be it in aqueous or nanoparticulate form. However, the specific mechanisms involved in oral bioavailability of nanoparticles may differ from those of dissolved molecules (Figure 12.2).

A conventional orally delivered API goes through the process of dissolution within the intestine, and thereafter the oral bioavailability is determined by the lipophilicity of the molecule along with its affinity for drug transporters[61] and drug metabolism enzymes[62] that may adversely impact upon its absorption. However, for various nanotechnologies, other processes have been implicated in absorption that include rate of dissolution (through high surface area to mass ratio[63]), cellular interactions due to inherently reactive nanoparticle surfaces,[64] physical interaction of the nanoparticle within mucous,[65] paracellular permeation across the epithelium,[66] endocytic processes in M-cells and absorption through Peyer's patches.[67,68] There are also good data to suggest that certain lipid-based nanomaterials may favor the delivery

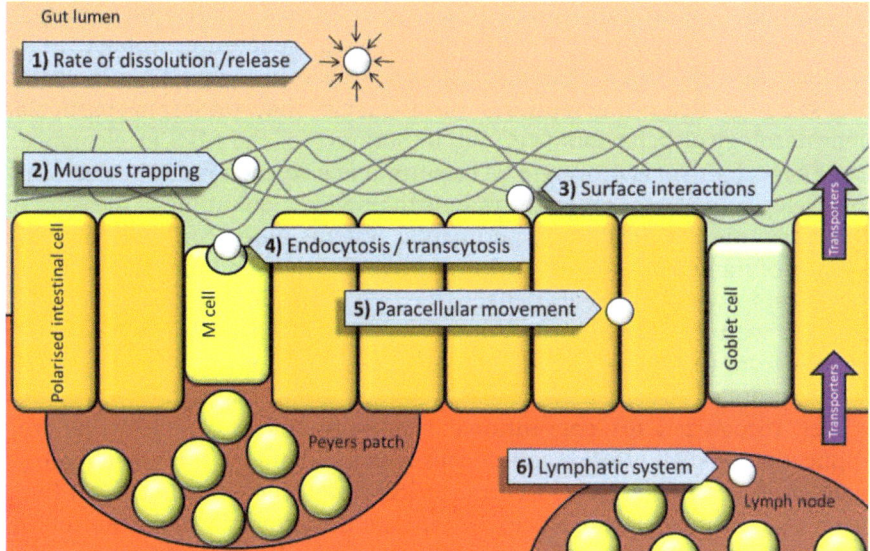

Figure 12.2 Potential mechanisms for oral drug delivery from nanotechnology-enabled products. The unique properties of nanomaterials may (1) impact the rate of dissolution; (2) result in entrapment within the mucous pores; (3) affect interactions mediated by the reactivity of the surfaces; (4) impact endocytic or transytic passage; (5) affect paracellular passage; or (6) entry to the systemic circulation *via* the lymphatics.

of drugs *via* the lymphatics.[69] Clearly, the mechanisms of oral absorption of many nanomedicines are different from those of conventional dissolved molecules and many of these nano-specific processes are unlikely to be captured using conventional Caco-2 cell monolayers. As such, there has been recent interest in developing advanced *in vitro* systems, typically involving co-culture with goblet cells (mucus-producing) and M-cells.[70,71] While the initial validation of these new models has shown early success, they have not yet been utilized to determine the mathematical descriptors of how these processes relate to absorption in humans, which will be needed for robust PBPK modeling of oral nanomedicines. Nonetheless, one can speculate that physical descriptors of the nanomedicine such as size, surface charge or carrier material composition may have utility for informing nanomedicine PBPK alongside the physicochemical properties of the API.

For nanomedicines that enter the systemic circulation as intact particles (either directly or indirectly), many other processes dictate their distribution and clearance, and the fundamental mechanisms are still being explored in a technology-specific manner. For example, the mononuclear phagocyte system (MPS) is known to impact nanoparticle clearance and it seems likely that a better understanding of its net effect on elimination of nanomaterials will enable the development of *in vitro* tools to generate data for informing robust PBPK modelling.[72] These are complex mechanisms to understand, since the overall effect is likely to be influenced by several processes (*e.g.*, protein corona formation can influence MPS uptake). Nonetheless, a schematic representation of the pathway to generating robust PBPK modeling approaches for nanomaterials is shown in Figure 12.3. First it is necessary to determine specific mechanisms that underpin particular pharmacological behaviors (*e.g.*, the importance of the MPS in clearance). Once a mechanism is proven to be important *in vivo*, then it will be necessary to develop robust experimental systems to provide a quantitative measure of its contribution. Once these experimental systems have been applied across a series of nanomaterials it may be possible to determine the mathematical relationship with the *in vivo* effect (for *in vitro–in vivo* extrapolation) and possibly with the physical descriptors of the nanoparticle (for modeling directly from the physical attributes of the formulation).

The utility of PBPK modeling in development and regulatory submission is well recognized for conventional small-molecule drugs. However, the application of such *in silica* approaches is in its infancy for nanotechnology-enabled drug development. Central to improved modeling is the confirmation of pharmacological processes that are important in definition of the *in vivo* pharmacokinetics. As for any PBPK model, ultimate performance will need to be qualified through a series of "predict–learn–confirm" cycles that map to supporting data generated *in vitro* and in preclinical species.

12.6   Conclusions

Nanomedical innovations are in full swing at both academic and industrial sectors as they relate to research, patents, commercial opportunities and technology transfer. However, in the nanomedicine arena, currently there are knowledge gaps relating to understanding the mechanisms that impact benefits for many technologies in terms of pharmacokinetics and ADME. Changes in both of these biological parameters are central to the achieved changes in efficacy but may also underpin new safety considerations that require additional data, assays or research tools to clarify their mechanistic basis. Importantly, a fundamental basis for the application of nanodrugs is the improvement of efficacy through a reduction in toxicity. However, as for any drug, nano-enabled products for human use require a robust, evidence-based assessment that evolves as gaps in knowledge are addressed. The evolution of the regulatory framework for nanodrugs is required to integrate a firm consideration of the needs for commercialization that is, after all, a prerequisite to patients realizing the step-change in their care. Equally, however, it is imperative that science-based regulatory guidelines mitigate the risk of inadvertent harm to patients or the environment.

As we rapidly enter the era of nanosimilars and theranostics, these are further likely to test the limits of regulatory authority, placing further strain on over-burdened regulatory agencies. Therefore, it is critical that the regulatory gaps that currently exist in nanomedicine are handled now; this will alleviate some of the burden later. The guiding principle here should reflect a balance of innovation and R&D with public health protection.

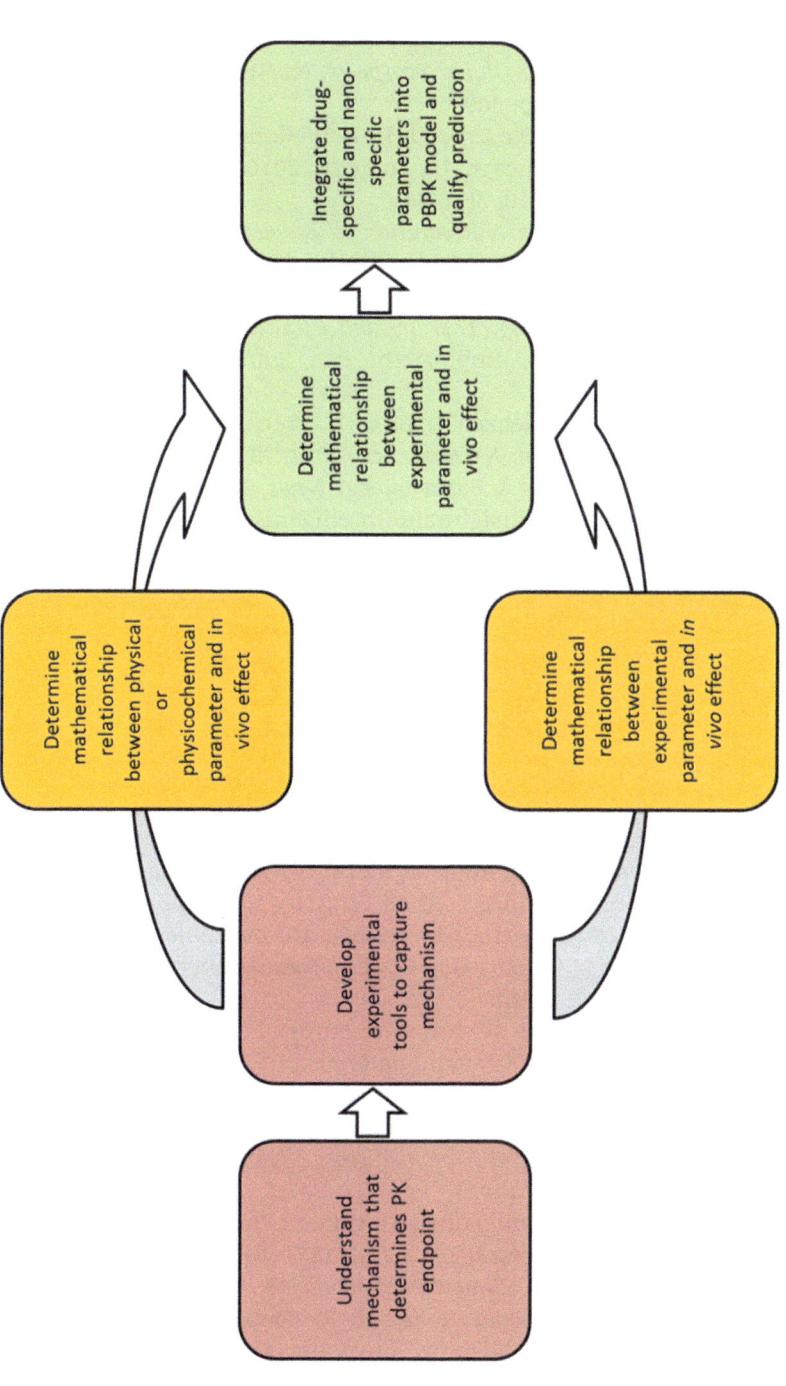

Figure 12.3 Schematic representation of a possible pathway to improved physiologically based pharmacokinetic (PBPK) modeling of nanomedicines. The fundamental processes regulating absorption, distribution, metabolism and elimination of nanomaterials first require confirmation. Development of robust *in vitro* systems may then enable clarification of how nanoparticle properties influence the pharmacological processes. Definition of the mathematical relationships will result in improved models, but additional qualification will be required through a series of "predict–learn–confirm" cycles, as is the case for conventional small-molecule drugs.

References

1. R. Bawa, *Alb. L. J. Sci. Tech.*, 2007, **17**(3), 699.
2. R. Bawa in *Bionanotechnology: Global Prospects*, ed. D. E. Reisner, CRC Press, Boca Raton, FL, 2009, p. 309.
3. R. Bawa in *Biotechnology and the Law*, ed. H. B. Wellons and E. S. Ewing, American Bar Association, Chicago, IL, 2nd edn, 2016.
4. R. Bawa, *Eur. J. Nanomed.*, 2010, **3**(1), 34.
5. R. Bawa in *Bionanotechnology: A Revolution in Biomedical Sciences and Human Health*, ed. D. Bagchi *et al.*, Wiley Blackwell, UK, 2013, p. 720.
6. H. M. Mansour, C.-W. Park and R. Bawa in *Handbook of Clinical Nanomedicine: Nanoparticles, Imaging, Therapy and Clinical Applications*, ed. R. Bawa, G. Audette and I. Rubinstein, Pan Stanford Publishing, Singapore, 2016, ch. 9, p. 233.
7. S. Tinkle, S. E. McNeil, S. Mühlebach, R. Bawa, G. Borchard, Y. Barenholz, L. Tamarkin and N. Desai, *Ann. NY Acad. Sci.*, 2014, **1313**, 35.
8. R. A. Stein, *Genetic Engineering & Biotechnology News*, 2014, Available at: http://www.genengnews.com/insight-and-intelligence/nanotechnology-is39the-magic-bullet-becoming-reality/77900016/ (Accessed on February 14, 2016).
9. S. Fischer, *IEEE Pulse, 2014*, Available at: http://pulse.embs.org/marc2014/regulating-nanomedicine/ (Accessed on February 14, 2016).
10. D. Johnson, 2014, Available at: http://spectrum.ieee.org/nanoclast/at-work/innovation/nanomedicines-will-be-bigonce-they-emergefrom-regulatory-troubles (accessed on February 14, 2016).
11. R. Bawa, S. R. Bawa and R. Mehra, in *Handbook of Clinical Nanomedicine: Law, Business, Regulation, Safety and Risk*, ed. R. Bawa, Pan Stanford Publishing, Singapore, ch. 58, p. 1291.
12. A. Siew, *PharmTech Europe*, 2014, Available at: http://www.pharmtech.com/current-issues-nanomedicines (Accessed on February 15, 2016).
13. M. Davenport, *Chem. Eng. News*, 2014, **92**(45), 10. Available at: http://cen.acs.org/articles/92/i45/Closing-Gap-Generic-Nanomedicines.html (Accessed on February 14, 2016).
14. G. Lovestam, 2010, Available at: https://ec.europa.eu/jrc/sites/default/files/jrc_reference_report_201007_nanomaterials.pdf (Accessed on February 15, 2016).
15. R. Bawa, *Nanomedicine*, 2007, **2**(3), 351.
16. R. Bawa in *Handbook of Clinical Nanomedicine: Nanoparticles, Imaging, Therapy and Clinical Applications*, ed. R. Bawa, G. Audette and I. Rubinstein, Pan Stanford Publishing, Singapore, 2016, ch. 6, p. 127.
17. Y. Barenholz, *J. Controlled Release*, 2012, **160**, 117.
18. Y. Barenholz in *Handbook of Clinical Nanomedicine: Nanoparticles, Imaging, Therapy and Clinical Applications*, ed. R. Bawa, G. Audette and I. Rubinstein, Pan Stanford Publishing, Singapore, 2016, ch. 29, p. 827.
19. Y. Matsumura and H. Maeda, *Can. Res.*, 1986, **46**, 6387.

20. R. K. Jain, *Can. Res.*, 1986, **47**, 3039.
21. H. Maeda, G. Y. Bharate and J. Daruwalla, *Eur. J. Pharm. Biopharm.*, 2009, **71**, 409.
22. Y. Barenholz and G. Haran, 1994. *U. S. Pat.* No. 5316771.
23. Draft guidance on Doxorubicin hydrochloride. Available at: http://www.fda.gov/downloads/Drugs/.../Guidances/UCM199635.pdf (accessed on February 15, 2016).
24. A. Gabizon, R. Catane, B. Uziely, B. Kaufman, T. Safra, R. Cohen, F. Martin, A. Huan and Y. Barenholz, *Cancer Res.*, 1994, **54**, 987.
25. J. Szebeni, F. Muggia, A. Gabizon and Y. Barenholz, *Adv. Drug Delivery Rev.*, 2011, **63**, 1020.
26. J. Szebeni, P. Bedocs, Z. Rozsnyay, Z. Weiszhar, R. Urbanics, L. Rosivall, R. Cohen, O. Garbuzenko, G. Bathori, M. Toth, R. Bunger and Y. Barenholz, *Nanomed. Nanotechnol. Biol. Med.*, 2012, **8**, 176.
27. Y. Barenholz and S. Amselem in *Liposome Preparation and Related Technique*, ed. G. Gregoriadis, CRC Press, Boca Raton, FL, 2nd edn, 1993, vol. I, p. 527.
28. S. P. Damari, D. Shamrakov, Y. Barenholz and O. Regev, submitted.
29. Y. Schilt, T. Berman, X. Wei, Y. Barenholz and U. Raviv, 2015, *Biochim. Biophys. Acta.*, 2016, **1860**, 108.
30. D. D. Lasic, P. M. Frederik, M. C. A. Stuart, Y. Barenholz and T. J. McIntosh, *FEBS Lett.*, 1992, **312**, 255.
31. W. H. De Jong and P. J. A. Borm, *Int. J. Nanomed.*, 2008, **3**, 133.
32. N. Sanvicens and M. P. Marco, *Trends Biotechnol.*, 2008, **26**, 425.
33. W. E. Bawarski, E. Chidlowsky, D. J. Bharali and S. A. Mousa, *Nanomed. Nanotechnol. Biol. Med.*, 2008, **4**, 273.
34. M. E. Davis, Z. G. Chen and D. M. Shin, *Nat. Rev. Drug Discovery*, 2008, **7**, 771.
35. I. Linkov, F. K. Satterstrom and L. M. Corey, *Nanomed. Nanotechnol. Biol. Med.*, 2008, **4**, 167.
36. R. Gaspar, *Nanomedicine*, 2007, **2**, 143.
37. G. Mendel, *Alb. L. Rev.*, 2008, **59**, 1323.
38. M. A. Hamburg, *Science*, 2012, **336**, 299.
39. R. Bawa, *Curr. Drug Delivery*, 2011, **8**, 227.
40. S. B. Foote and R. J. Berlin, *Minn. J. L. Sci. Technol.*, 2005, **6**, 619.
41. R. Bawa, S. Melethil, W. J. Simmons and D. Harris, *SciTech. Lawyer*, 2008, **5**(2), 10.
42. FDA's approach to regulation of nanotechnology products. Available at: http://www.fda.gov/ScienceResearch/SpecialTopics/Nanotechnology/ucm301114.htm (accessed on February 10, 2016).
43. B. E. Rabinow, *Nat. Rev. Drug Discovery*, 2004, **3**(9), 785.
44. F. Haessler, F. Tracik, H. Dietrich, H. Stammer and J. Klatt, *Int. J. Clinical Pharmacol. Ther.*, 2008, **46**(9), 466.
45. S. Weir, R. Torkin and H. R. Henney, *Curr. Med. Res. Opin.*, 2013, **29**(12), 1627.

46. M. N. Samtani, A. Vermeulen and K. Stuyckens, *Clin. Pharmacokinet.*, 2009, **48**(9), 585.
47. G. van't Klooster *et al.*, *Antimicrob. Agents Chemother*, 2010, **54**(5), 2042.
48. B. Deschamps, N. Musaji and J. A. Gillespie, *Int. J. Nanomed.*, 2009, **4**, 185.
49. I. F. Uchegbu and A. Siew, *J. Pharm. Sci.*, 2013, **102**(2), 305.
50. M. E. O'Brien *et al.*, *Ann. Oncol.*, 2004, **15**(3), 440.
51. A. N. Ilinskaya and M. A. Dobrovolskaia, *Brit. J. Pharmacol.*, 2014, **171**(17), 3988.
52. Available at: http://www.fda.gov/AboutFDA/CentersOffices/OfficeofMedicalProductsandTobacco/CDER/ucm365118.htm (Accessed on February 14, 2016).
53. D. M. Moss and M. Siccardi, *Br. J. Pharmacol.*, 2010, **171**(17), 3963.
54. G. Bachler, N. von Goetz and K. Hungerbuhler, *Int. J. Nanomed.*, 2013, **8**, 3365.
55. G. Bachler, N. von Goetz and K. Hungerbuhler, *Nanotoxicol*, 2014, 1.
56. M. Li, K. T. Al-Jamal, K. Kostarelos and J. Reineke, *ACS Nano*, 2010, **4**(11), 6303.
57. T. O. McDonald *et al.*, *Adv. Healthcare Mater.*, 2014, **3**(3), 400.
58. R. K. Rajoli, D. J. Back, S. Rannard, C. L. Freel Meyers, C. Flexner, A. Owen and M. Siccardi, *Clin. Pharmacokinet.*, 2015, **54**(6), 639–650.
59. R. S. Yang, L. W. Chang, C. S. Yang and P. Lin, *J. Nanosci. Nanotechnol.*, 2010, **10**(12), 8482.
60. D. Sun, H. Lennernas, L. S. Welage, J. L. Barnett, C. P. Landowski, D. Foster *et al.*, *Pharma. Res.*, 2002, **19**(10), 1400.
61. M. F. Fromm, *Int. J. Clin. Pharmacol. Ther.*, 2000, **38**(2), 69.
62. A. Bruyere, X. Decleves, F. Bouzom, K. Ball, C. Marques, X. Treton *et al.*, *Mol. Pharm.*, 2010, **7**(5), 1596.
63. W. Li *et al.*, *Int. J. Pharm.*, 2014, **460**(1–2), 13.
64. S. Hirsjarvi, S. Dufort, J. Gravier, I. Texier, Q. Yan, J. Bibette *et al.*, *Nanomedicine*, 2013, **9**(3), 375.
65. S. K. Lai, Y. Y. Wang and J. Hanes, *Adv. Drug Delivery Rev.*, 2009, **61**(2), 158.
66. S. Sadekar and H. Ghandehari, *Adv. Drug Delivery Rev.*, 2012, **64**(6), 571.
67. E. Gullberg *et al.*, *J. Pharm. Exper. Ther.*, 2006, **319**(2), 632.
68. J. Wang, L. Li, Y. Du, J. Sun, X. Han, C. Luo *et al.*, *Mol. Pharm.*, 2015, **12**(2), 463.
69. K. Kohli, S. Chopra, D. Dhar, S. Arora and R. K. Khar, *Drug Discovery Today*, 2010, **15**(21–22), 958.
70. A. Beduneau *et al.*, *Eur. J. Pharm. Biopharm.*, 2014, **87**(2), 290.
71. C. Schimpel *et al.*, *Mol. Pharm.*, 2014, **11**(3), 808.
72. M. A. Dobrovolskaia, P. Aggarwal, J. B. Hall and S. E. McNeil, *Mol. Pharm.*, 2008, **5**(4), 487.

CHAPTER 13

Doxil® – the First FDA-approved Nano-drug: from Basics via CMC, Cell Culture and Animal Studies to Clinical Use

YECHEZKEL (CHEZY) BARENHOLZ

Laboratory of Membrane and Liposome Research, Department of Biochemistry and Molecular Biology, Insitute for Medical Research Israel – Canada (IMRIC), Hebrew University-Hadassah Medical School, Jerusalem, Israel
Email: chezyb@gmail.com; chezyb@ekmd.huji.ac.il

13.1 Introduction

Doxil® (or, given its European brand name, Caelyx®) (Figure 13.1) is the first US Food and Drug Administration (FDA)-approved nano-drug (1995) and was developed through 7.5 years of major efforts in research, research and development, and development work as well as clinical studies by many people located in Jerusalem, Israel (Barenholz laboratory and Gabizon laboratory), in Menlo Park, CA, USA (Liposome Technology Inc. (LTI) team), and in many hospitals worldwide that were involved in the various stages of Doxil® clinical trials. Hadassah University Hospital, Jerusalem, was the site of the first Doxil® clinical trial.[1]

The history of the development of Doxil® is reviewed in brief below and in more detail elsewhere.[2,3] In order not to repeat what was discussed in

RSC Drug Discovery Series No. 51
Nanomedicines: Design, Delivery and Detection
Edited by Martin Braddock
© The Royal Society of Chemistry 2016
Published by the Royal Society of Chemistry, www.rsc.org

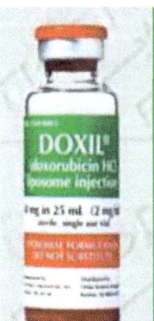

--------------------------INDICATIONS AND USAGE--------------------------
DOXIL is an anthracycline topoisomerase inhibitor indicated for:
- **Ovarian cancer (1.1)**
After failure of platinum-based chemotherapy.
- **AIDS-related Kaposi's Sarcoma (1.2)**
After failure of prior combination chemotherapy or intolerance to such therapy. Results are based on objective response rate; no results are available from controlled trials that demonstrate clinical benefit.
- **Multiple Myeloma (1.3)**
In combination with bortezomib in patients who have not previously received bortezomib and have received at least one prior therapy.

Figure 13.1   The original Doxil was produced by Ben Venue Laboratories for Sequus (which developed the nano-drug when still called Liposome Technology Inc.), reviewed elsewhere.[2,3] After US Food and Drug Administration and European Medicines Agency approvals, Sequus Inc. distributed the drug in the United States, while Schering Plough distributed it in Europe under the brand name Caelyx®. Doxil/Caelyx is now distributed worldwide by Janssen Products LP.

previous recent reviews, in this review I describe only the scientific highlights of Doxil® development and the lessons learned from its 20 years of use, and I will concentrate on the importance and relevance of cell culture studies. These *in vitro* experiments, while being fundamental to the understanding of the mechanism of action of Doxil®, may be misleading if they are interpreted wrongly. Finally, I describe in brief some of the new surprising findings of work with Doxil®, which have not yet been published.

Overall, the development of Doxil® demonstrates the obligatory need for applying a multidisciplinary approach. This relies on a combined understanding of the optimal use of cross-talk between physicochemical, nano-technological, and biological principles. The experience gained in the discovery and development of Doxil® now enables the development of other novel improved anticancer, anti-inflammatory and antibiotic nano-drugs.

13.2   Introducing Doxil®

The brand name Doxil® came from the abbreviation of *DOX*orubicin *I*n *L*iposomes). Doxil®/Caelyx is a PEGylated liposomal doxorubicin (PLD).

In terminology accepted by the FDA, Doxil® and its generics are defined as "doxorubicin HCl in liposomes". However, this definition is questionable, as in the intra-liposome aqueous phase the doxorubicin is in the form of (doxorubicin)2 sulfate (and not the chloride) salt, although if the drug is released from the liposome it is as a free base. Once it is released its counter anion will be dependent on the medium where it is released, and in most cases it will be the chloride anion.

Doxil®, the first FDA-approved nano-drug[2,3] as well as the first FDA-approved liposomal drug, is in extensive clinical use as an anti-cancer drug since its approval in November 1995.[2–5] Doxil® is also recognized as the gold standard of injectable nano-drugs and drug delivery systems. Doxil® by itself or in combination with other low molecular weight drugs, or with various biologicals such as monoclonal antibodies was, and is still, in evaluation in many clinical trials.[6]

The patents that protect Doxil® expired during 2010 (see Section 13.5). However, the first generic form of Doxil®, the Lipodox produced by Sun-Pharma, was approved by the FDA only in February 2013. An unusually long period of 3 years passed between the expiration of the Doxil® patents and the approval of Lipodox®. Doxil® and its generic Lipodox are both PLD, in which the drug is loaded remotely *via* the chemical engine of a transmembrane ammonium gradient having the bi-valent sulfate as ammonium and doxorubicin counter anion.[2,3,7–11] The remote drug loading results in the formation of rod-like long crystals of doxorubicin (dox)-sulfate in the length of the intra-liposome space forcing the nano-liposome to change from spherical to an ellipsoid, "coffee-bean"-like shape (Figure 13.2 and ref. 2, 3, and 12).

The Doxil® membrane is composed of hydrogenated soybean phosphatidylcholine (HSPC), which by itself can form a stable liposome and therefore it can be classified as a liposome-forming lipid.[13–15] The packing parameter (PP) is defined as the ratio between the cross-section of the amphiphile hydrophobic part (such as hydrocarbon chains of phospholipids or the

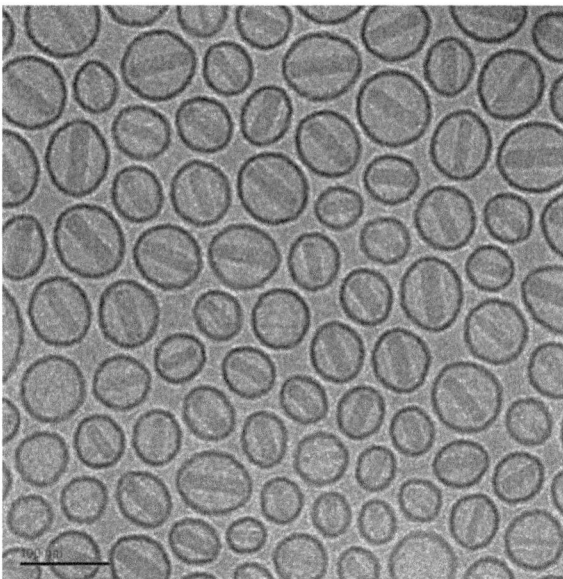

Figure 13.2 Cryo-transmission electron microscope image of Caelyx® batch 101371803. Scale bar = 100 nm.

hydrocarbon rings of sterols) and the cross-section of the hydrophilic part (such as the amphiphile head group). When present in aqueous phase, the hydrophobic part of the amphiphiles is driven away from the water. The type of assemblies formed due to amphiphile self-association is determined by their PP. Amphiphiles having a PP in the range of 0.74–1.0 are defined as liposome-forming lipids.[13] HSPC is a liposome-forming lipid having a PP of 0.8.[16] It is actually a mixture of two phosphatidylcholines (PCs), distearoyl PC (DSPC) and palmitoyl stearoyl PC (PSPC), both having high melting temperature (T_m). The mole ratio between the two PCs determines the exact T_m of HSPC in the range of 50–53 °C. Due to this high T_m, at 37 °C, lipid bilayers of HSPC will be at the solid ordered phase, which is a highly rigid phase, similar to that observed before for liposomes made of PCs having two saturated acyl chains such as DMPC, DPPC and DSPC.[14,15] Cholesterol has a high PP of 1.21; it is neither a liposome nor a micelle-forming lipid.

1,2-Distearoyl-sn-glycero-3-phosphoethanolamine-N-[methoxy(polyethylene glycol)-2000 Da] (sodium salt, DSPE–PEG) at the mole percentage used in Doxil® has a PP of 0.487.[16] This means that by itself it is a "micelle-forming lipid". The small PP in the case of DSPE–PEG means that this molecule has a very large polar head group due to the large (45-*mer*) polyethylene glycol (PEG) moiety. This head group is highly flexible and highly hydrated as each oxygen atom of the 45 ethylene oxide groups binds three or four water molecules, therefore the head group will be highly hydrated and very bulky[17] (Figure 13.3).

Until 2014 almost no studies were undertaken on the thermotropic behavior of the HSPC/cholesterol/DSPE–PEG ternary system, and the few that were performed focused on the effect of the DSPE–PEG mole percentage on the liposomal membrane.[16]

A detailed study including phase diagrams describing the thermotropic behavior of the liposomes composed of HSPC/cholesterol/DSPE–PEG has been recently published.[18] The authors studied the effect of cholesterol mole percentage on the differential scanning calorimetry (DSC) thermograms using the highly sensitive MicroCal DSC calorimeter. Their results resembled the well-established dogma that cholesterol, when present in a lipid bilayer, interacts favorably with the ordered saturated lipid acyl chains, and therefore, it interferes with the chain–chain interaction leading to the disappearance of the solid ordered to liquid disordered phase transition by introducing a new liquid ordered phase. The level of this phase is dependent on cholesterol mole percentage, reaching a complete transformation to liquid ordered phase at 35–40 mol% cholesterol.[14,15,19] Such high mole percentage of cholesterol also ensures high membrane resistance to changes in temperature, including resistance to leakage of liposome content.[14,15] Our recent DSC studies[20] on Doxil® and liposomes having lipid composition and size identical to Doxil® confirm these previous studies.[18]

These unique properties of the PEG head group covalently attached to DSPE underlie steric stabilization.[2,17] The hydrophobic part of this molecule consisting of two stearoyl acyl chains is responsible for the excellent

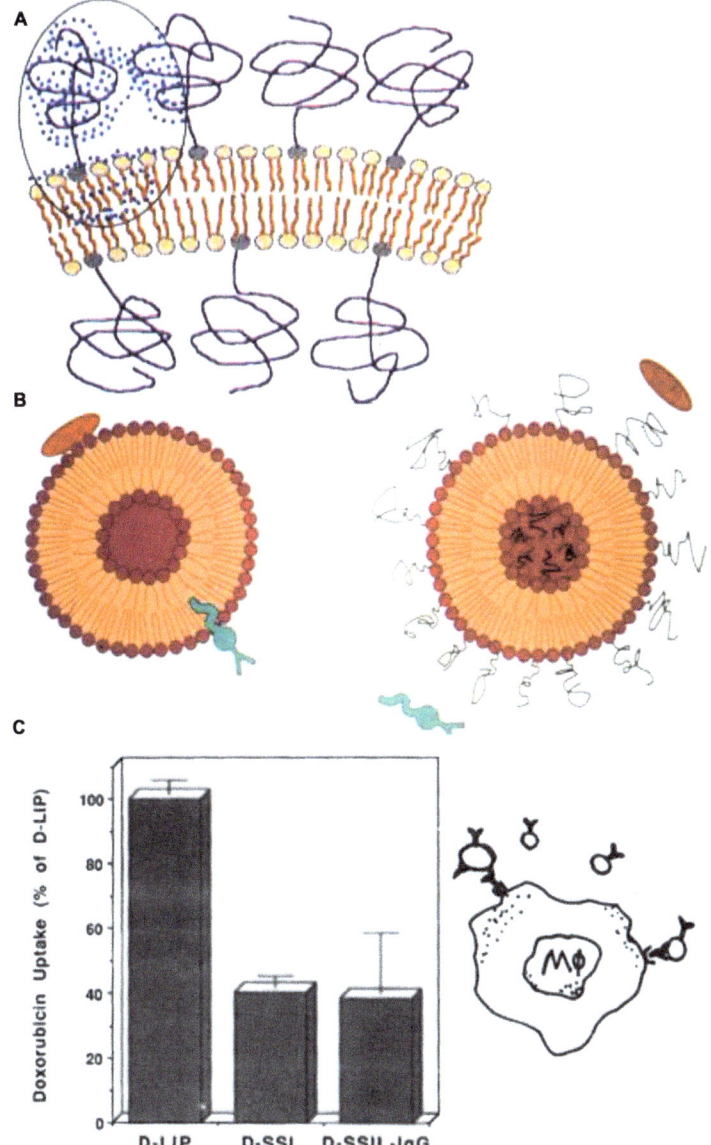

Figure 13.3 A. Structure cartoon pointing to the large size, bulkiness due to flexibility and high level of hydration (three to four water molecules per ethylene oxide moiety) of the 45-*mer* polyethylene glycol (PEG) 2000 Da chains of the liposome "coating" derived from the liposome DSPE–PEG lipopolymer. The blue dots represent water molecules bound to the PEG and to the DSPE head group. At about 4–9 mol% PEG-DSPE water bound to the lipid bilayer is at a minimum and liposome stability is at maximum (based on ultrasound and differential scanning calorimetry studies[14,16,17]). B. The concept of steric stabilization. C. The significance of steric stabilization: uptake of liposomes by macrophages is dramatically inhibited by the inclusion of DSPE–PEG (2000 Da). D-SSL: Doxil-like stericaly stabilized liposomes; D-LIP: similar liposomes loaded with doxorubicin but lacking the DSPE–PEG in the liposome membrane; D-SSIL: Doxil-like liposomes targeted by antibodies.[67,68]

retention of this lipo-polymer in the liposomal membrane during the prolonged circulation in the blood. Similar lipo-polymers having shorter acyl-
chains are desorbed from the liposome membrane during the circulation in
the blood, and therefore steric stabilization is lost with circulation time. The
DSPE–PEG is negatively charged due to the phosphate-monoester group of
the DSPE.[71] This negative charge, when measured as zeta potential, is hindered, which explains the low negative zeta potential of Doxil®; however,
this low negative zeta potential is still enough to induce some complement
activation.[21,22]

The mole ratio of the Doxil® liposome lipids is HSPC:cholesterol:DSPE–
PEG 56.6:38.1:5.3, which is equivalent to a weight ratio of 3:1:1, respectively.[23,24] Liposome stability can be assessed from calculation of the
"additive packing parameter", which describes the molar percentage
weighted sum of PP of the liposome components. In the case of nano-
liposomes composed of HSPC, cholesterol, and DSPE–PEG, good stability
is achieved in the range of 4–9 mol% PEG–DSPE (Figure 13.4).

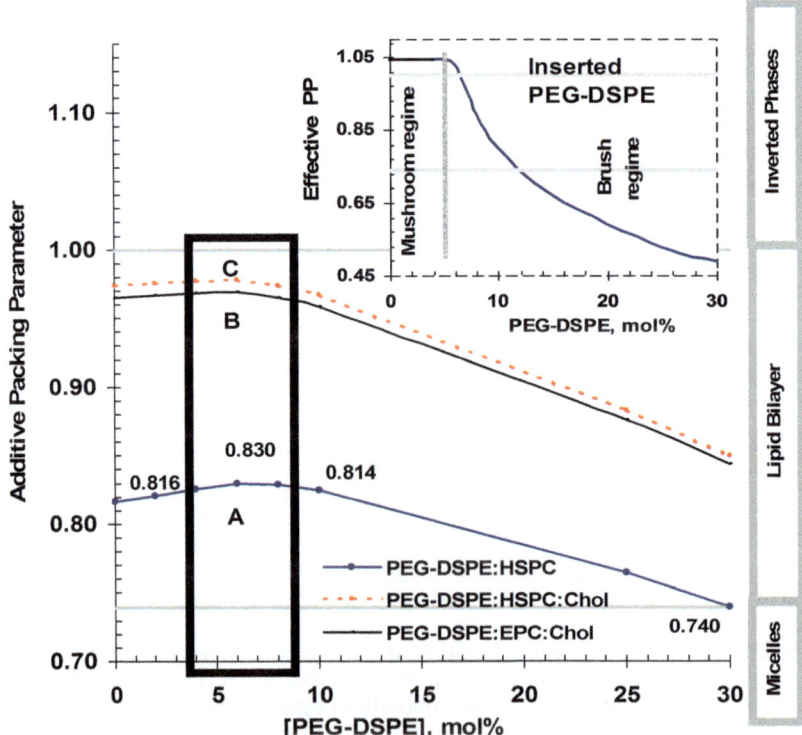

Figure 13.4 Effect of PEG–DSPE on the additive packing parameter (APP) of nano-
liposome lipid bilayer composed of HSPC alone and HSPC/cholesterol
or egg phosphatidylcholine (EPC); both using 56.6/38.1 mol/mol of
EPC/cholesterol. Inset: Changes in effective packing parameter (PP)
when PEG–DSPE is inserted into lipid bilayer with increasing
concentration.[16]

The HSPC is the liposome's matrix; cholesterol at high mole percentage improves the resistance of the liposome membrane to temperature changes and to content leakage from the liposome (Figure 13.4). At such high cholesterol mole percentage the lipid bilayer should be in a rigid liquid-ordered phase.[14,15,18] Such a rigid liquid-ordered phase means that the level of membrane "free volume" is low[16] which enables the achievement of zero-order slow drug release.[25] The DSPE–PEG acts as the membrane steric stabilizer, which enables extension of liposome circulation time.[14,16,17,26] Together with the transmembrane ammonium-sulfate-driven loading, the lipid composition enables the Doxil® to achieve a long circulation time without loss of drug in the blood. A calculation based on the lipid molar concentration of the components and on liposome size estimates that 1 mL of the Doxil® dispersion contains 2.3×10^{14} liposomes and each liposome contains $\sim 10\,000$–$20\,000$ molecules of doxorubicin, of which $>95\%$ are in the dox-sulfate intraliposome nano-crystal.

Doxil® performance is related to a number of unrelated principles:

1. Prolonged drug circulation time and avoidance of the reticuloendothelial system (RES) due to steric stabilization resulting from the use of PEGylated nano-liposomes (see Figure 13.3 and ref. 1–3 and 27).
2. High and stable remote loading of doxorubicin driven by a transmembrane ammonium sulfate gradient.[2,3,7,9,10]

   This unique remote loading allows for almost 100% drug loading and results in the formation of intra-liposomal nano-crystals of doxorubicin-sulfate having the length of the liposome diameter. It forces the nano-liposome to change its shape to a prolate ellipsoid, referred to as a "coffee-bean shape". A cryogenic transmission electron micrograph of Doxil® is shown in Figure 13.2.[2,3,7,12] This remote loading also enables selective drug release at the tumor. Gabizon[28] demonstrated a much faster release of doxorubicin by Doxil® in tumor interstitial fluid than in plasma. This beneficial outcome led us to conclude that the transmembrane ammonium gradient-driven remote loading may also allow for the observed selective drug release at the tumor. We ascribe this to a release inducer, a unique tumor metabolite present in the tumor interstitial fluid and formed due to unique tumor metabolism. This metabolite is practically not present in the blood or in normal healthy tissues. Therefore, a selective and significant release occurs mainly at the tumor tissue and not in plasma or in healthy tissues. We recently demonstrated that ammonia produced at the tumor site due to the metabolic pathway of glutaminolysis,[29–31] may be the main inducer of doxorubicin release from Doxil® at the tumor site. In glutaminolysis the amino acid glutamine serves as the main precursor to override inhibition of part of the Krebs cycle. In this pathway glutamine is converted *via* two steps releasing ammonia and glutamate, which is further metabolized to α-keto-glutarate. See ref. 32 and references listed therein. More details on new findings for Doxil® in 2012–2015 are discussed below.

3. Having the liposome lipid bilayer in a liquid-ordered phase composed of the high-T_{m} (53 °C) fully hydrogenated soy phosphatidylcholine (mainly DSPC) together with cholesterol, and 2000 Da PEG–DSPE. See ref. 2, 9 and 10 and recent X-ray diffraction results of Schilt *et al.* (manuscript in preparation).

   Such a high-T_{m}-based liquid-ordered phase explains the robustness of the liposomes, which is expressed as a minimal drug release during ~ 2 years storage at ~ 40 °C, as well as during prolonged circulation time ($t_{1/2} \sim 90$ h[33]). This liquid-ordered physical state of the lipid bilayer is also an important factor in preventing burst release in blood. The low rate of drug release in the blood circulation of patients was demonstrated for Doxil® in the first clinical trial[1] and was confirmed more recently in the clinical study conducted by Sun Pharma to demonstrate the bioequivalence between Lipodox and Doxil®/Caelyx www.fda/gov/downloads/Forindustry/UserFees/GenericDrugUserFees/UCM450791. pdf. In the Sun Pharma clinical trial, plasma levels of total doxorubicin as well as encapsulated and free doxorubicin were determined. For both PLDs, Doxil® and Lipodox, the area under the curve from zero to infinity ($\mathrm{AUC}_{0-\mathrm{inf}}$) of free doxorubicin was <1% of the $\mathrm{AUC}_{0-\mathrm{inf}}$ of the encapsulated doxorubicin, the latter being very similar to the results of total doxorubicin (for more details see EMA/588790/2011 by the Committee for Medicinal Products for Human Use (CHMP), Committee assessment report of the European Medicines Agency). The remote loading driven by the trans-membrane ammonium ion gradient also enables the desired controlled release at the tumor site, which is induced by tumor metabolites, especially ammonia (and therefore it does not occur in plasma or in most normal tissues where there is almost no ammonia) (Figure 13.5).

4. Being in the nano-size range will allow the liposomes to benefit from the enhanced permeability and retention (EPR) effect[34–36] and to accumulate at the tumor site.[1] Figure 13.2 describes the unique structural aspects of Doxil®.

The above four attributes also help us to understand why Doxil®'s therapeutic activity can be further improved by combining it with local heating achieved by radiofrequency ablation. It was demonstrated that such a combination may modify the tumor micro-environment in a way that improves Doxil®'s therapeutic effect against extra-hepatic tumors in mice.[37,69] In these experiments Doxil® plus radiofrequency ablation showed significantly better therapeutic activity than ThermoDox-like thermo-sensitive liposomes plus radiofrequency (Figure 13.6).

In summary, the experience gained from Doxil®'s development demonstrates that each detail of its composition and morphology (Figure 13.7) is important for Doxil®'s performance.

However, it is worth noting that a certain degree of freedom is allowed. For example, regarding size distribution, there is a certain preferred size

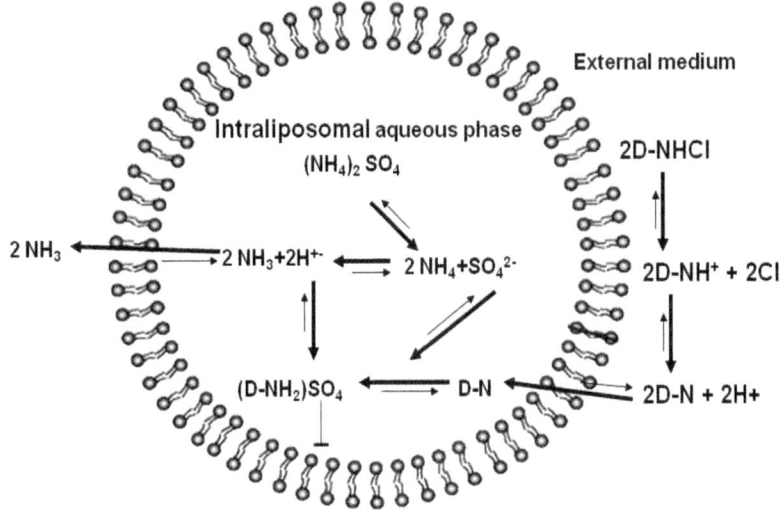

Figure 13.5 Remote loading of amphipathic weak base into liposomes using an ammonium sulfate gradient. Concentration of $(NH_4)_2SO_4$ in the liposomes is much higher than in the extra-liposomal medium. This drives small amounts of neutral ammonia to diffuse out of the liposomes, reducing the intra-liposomal pH, as H_2SO_4, un-ionized drug base (D-N) from the external medium is then drawn in through the liposomal membrane and is "trapped" in the intra-liposome aqueous phase by its ionization and creation of insoluble (doxorubicin)$_2$-sulfate. This process can continue until all the ammonium ions are exchanged (as neutral ammonia) with doxorubicin. However, the process is stopped to retain some of the ammonium gradient. This stabilizes the drug retention inside the liposome. To a large extent this remote loading works well due to the large differences in the permeability coefficients of ammonia ($0.13\ cm\ s^{-1}$), which is very fast, and sulfate ($<10^{-12}\ cm\ s^{-1}$), which is very slow.[8,9] Active loading benefits from the nano-volume of nano-liposomes as the movement of a small number of ions can produce large effects.[7,11,70] US patents 5 192 549[37] and 5 316 771.[38]

range (rather than a specific size) that will result in similar pharmacokinetics and biodistribution and therefore therapeutic performance. This is why an in-depth physicochemical characterization for liposome-based drugs is a must. This may also explain why the FDA produced such detailed requirements (see FDA Draft Guidance on Doxorubicin Hydrochloride, Recommended Feb 2010; Revised Nov–Dec 2014, http://www.fda.gov/downloads/Drugs/. . ./Guidances/UCM199635.pdf).

Our first Doxil® clinical trial demonstrated that Doxil® is indeed passively targeted to tumors (Figure 13.8 and Gabizon *et al.*[1]), probably due to the EPR effect. Doxil® is metabolized intracellularly,[2] therefore the findings that doxorubicin metabolites were found in urine and tumor biopsies is strong evidence that the doxorubicin is released from the liposomes and becomes

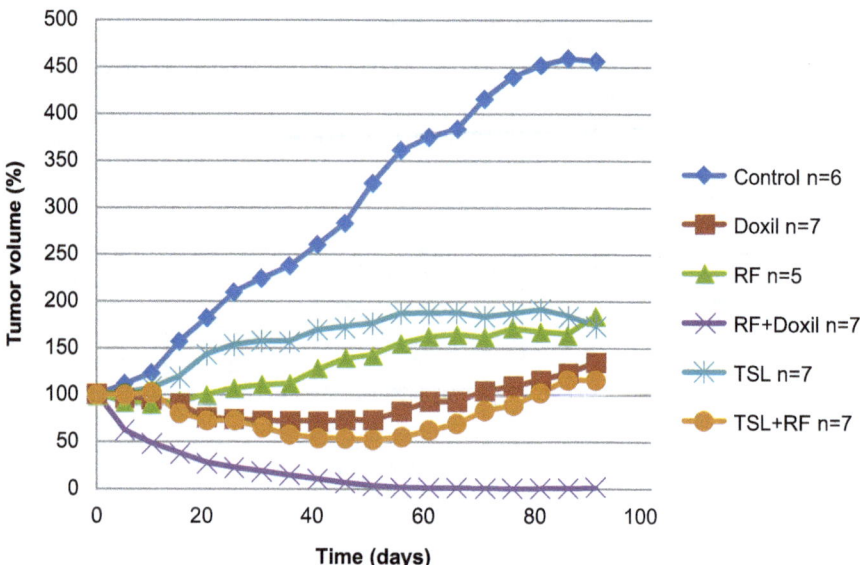

Figure 13.6 Mouse tumor model therapeutic efficacy studies using PEGylated liposomal doxorubicin (PLD; Doxil) and thermosensitive PEGylated liposomal doxorubicin (TSL) with and without radiofrequency (RF) ablation. The effect of the various treatments on tumor volume over 90 days was studied. The following experimental protocol used the human medulloblastoma cell line (Daoy) as the mouse tumor model. ∼4 million Daoy cells were inoculated subcutaneously in the back of 4–5-week-old nude-Hsd:athymic mice.

On day 0, 31 mice with a tumor size of $12 \times 9 \pm 1$ mm were randomized to six groups and treated as follows. Monotherapy: Doxil injection; Monotherapy: TSL injection; RF ablation; Combined therapy: Doxil injection 15 min after RF treatment; Combined therapy: TSL injection 15 min before RF application; Control: non-treated mice.

available to the tumor cells.[1,2,28] This selective drug release at the tumor site is mediated by factors related to the unique tumor micro-environment and metabolism.[2,3,28,32]

13.3   Doxil®: Historical Perspective in Short

In order to better understand why and how Doxil® was invented and developed, it is important to review Doxil®'s historical perspective, which are described in short below and has been reviewed elsewhere.[2,3]

1979–1987 Pre-Doxil® era (liver passively targeted by liposomal doxorubicin)

1979 Gabizon and Barenholz start their basic research on liposomal doxorubicin

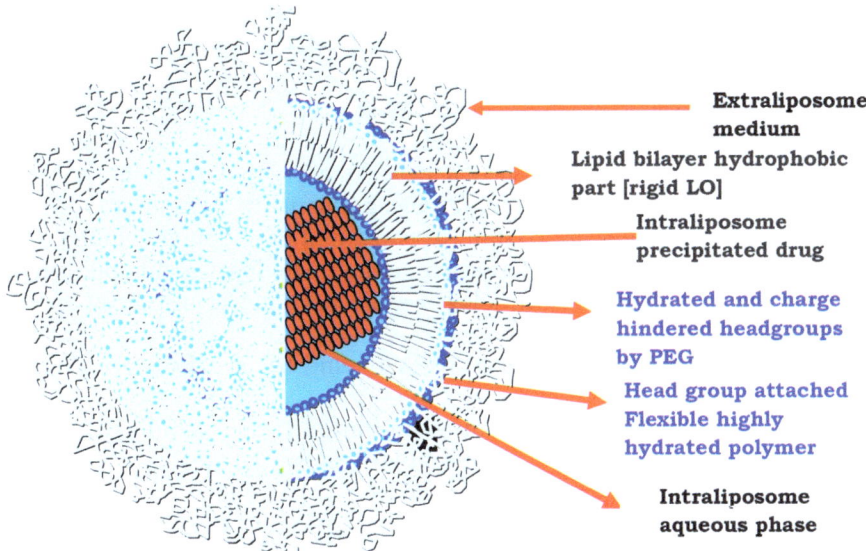

Figure 13.7 Doxil® = PEGylated nano (<100 nm) unilamellar liposome. The cartoon is based on cryo-transmission electron microscopy, short- and wide-angle X-ray scattering, dynamic light scattering, compressibility, and doxorubicin absorbance and fluorescence.[2,3] LO: liquid ordered; PEG: polyethylene glycol.

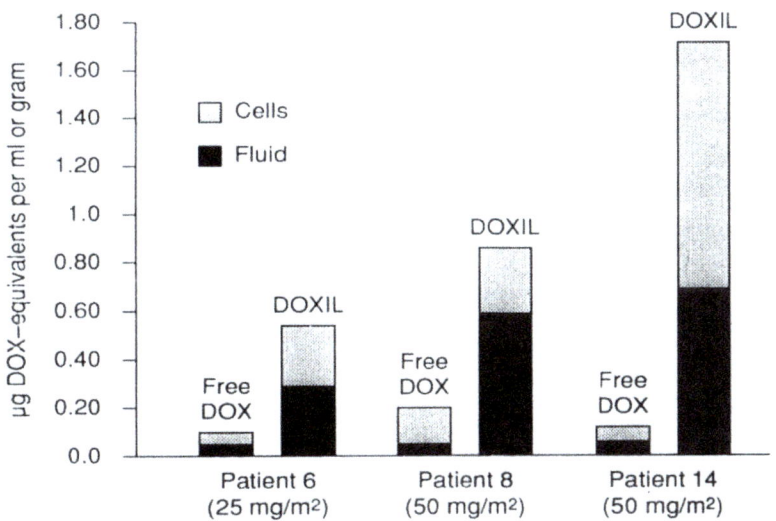

Figure 13.8 Doxorubicin levels in three patients' tumor biopsies, comparing administration of free doxorubicin (DOX) and Doxil.[1] This experiment strongly supports the enhanced permeabliliy and retention (EPR) effect in cancer patients.

1984 The first clinical trial with liposomal doxorubicin based on conventional non-PEGylated large negatively charged oligo-lamellar liposomes (DOX-OLV)

1985 LTI licensed the OLV-DOX technology

1987 Clinical trial of OLV-DOX failed

1987/1988 Barenholz developed a new concept of doxorubicin remote loading driven by a transmembrane gradient of ammonium sulfate, which was patented by Yissum (research and development company of the Hebrew University of Jerusalem, Israel). The first patent application was filed in 1988. This technology was licensed to LTI and serves as one of the "legs" of Doxil®[37–39]

1988 Gabizon and Papahadjopoulos, Allen and LTI studied sterically stabilized (stealth) liposomes

1988 LTI, Gabizon, and Barenholz each started Doxil® development

1989 LTI patented the stealth concept and registered Stealth®

1991–1992 Doxil® first-in-man clinical trial in Jerusalem

1994 Gabizon and Barenholz's major publication on Doxil® clinical trial (*Cancer Research*)

1995 (November 17) FDA approved Doxil®

1996 First Doxil® sales in USA and Europe

2010 (March) US patent expired

2011 Ben Venue Laboratories, Inc., the only site where Doxil®/Caelyx was produced, was shut down

February 2013 FDA approved Lipodox as the first Doxil® generic

13.4 Clinical Indications for Doxil®

Doxil® was approved by the FDA (1995) for the following indications.

Ovarian cancer: Doxil® (doxorubicin HCl liposome injection) is indicated for the treatment of patients with ovarian cancer whose disease has progressed or recurred after platinum-based chemotherapy.

AIDS-related Kaposi's sarcoma: Doxil® is indicated for the treatment of AIDS-related Kaposi's sarcoma in patients with disease that has progressed on prior combination chemotherapy or in patients who are intolerant to such therapy. The treatment of patients with AIDS-related Kaposi's sarcoma is based on objective tumor response rates. No results are available from controlled trials that demonstrate a clinical benefit resulting from this treatment, such as improvement in disease-related symptoms or increased survival.

Multiple myeloma: Doxil® in combination with bortezomib is indicated for the treatment of patients with multiple myeloma who have not previously received bortezomib and have received at least one prior therapy.

This information is taken from the Doxil® label
(www.accessdata.fda.gov/drugsatfda_docs/label/2007/050718s029lbl.pdf).

Caelyx was approved by the European Medicines Agency (EMA) in 1996 for the treatment of the following types of cancer in adults:

Metastatic breast cancer in patients at risk of heart problems. "Metastatic" means the cancer has spread to other parts of the body. Caelyx is used on its own for this disease.

Advanced ovarian cancer in women whose previous treatment, including a platinum-based anti-cancer medicine, has stopped working.

AIDS-related Kaposi's sarcoma multiple myeloma (a cancer of the cells in the bone marrow), in patients with progressive disease who have received at least one other treatment in the past and have already undergone or are unsuitable for a bone marrow transplant. Caelyx is used in combination with bortezomib. This information is taken from EMA/626038/2010 and EMEA/H/C/000089.

The side effect profile of Doxil® is superior to free doxorubicin in all typical doxorubicin side effects. This is especially relevant to cardiotoxicity, which is the most serious side effect of doxorubicin, and this is much lower for Doxil®.[2,3] However, Doxil® presents two side effects which are not typical to doxorubicin but related to the drug being a particle, which results in complement activation pseudo-allergy,[21,22] and to very long circulation time of the nano-drug, which results in its accumulation in the skin, leading to hand and foot syndrome.[1,2,23,24]

Various ways to overcome these two types of toxicity are now under development,[2,3] and more detailed information on the current use of Doxil® has been published.[2–5] Details of more recent clinical trials which involve Doxil® as a monotherapy and mainly on Doxil® combinations with other low molecular weight drugs and/or biologicals can also be found.[6] The latter clinical trials suggest that the use of Doxil® and its generic is expected to expand. This explains the high motivation and major efforts of many companies to develop generic versions of Doxil®.

13.5 Doxil®-related Intellectual Property

Doxil® intellectual property is contained within two families of patents.

One family covers the effect of the lipopolymer PEG-DSPE as a liposome lipid component on liposome circulation time and RES avoidance (US patent 5 013 556[40]).

The second patent family is on the drug loading method. It is based on transmembrane-driven remote loading of amphipathic weak bases such as doxorubicin (US patents 5 192 549 and 5 316 771[37,38]).

It took 7.5 years from the filing of these two families of patent applications in 1988/1989 until Doxil® was approval by the US FDA in November 1995. As remote-loading patents were extended to March 9, 2010, Doxil® enjoyed 14 years of patent protection (see Doxil® historical perspective, Section 13.3).

13.6 From the Failure of DOX-OLV to the Success of Doxil®

Doxil®'s success stems to a large extent from the lessons learned from the lack of success in clinical trials of a previous liposomal doxorubicin formulation developed by Gabizon and Barenholz, referred to as DOX-OLV. DOX-OLV is based on 200–500 nm oligo-lamellar negatively charged liposomes, in which the positively charged doxorubicin was passively loaded by electrostatic interaction.[3] DOX-OLV was aimed at the treatment of liver cancer, and in spite of being efficacious in murine tumor models, the DOX-OLV formulation failed to show efficacy in human clinical trials. To a large extent, this was due to a "dilution-induced release" resulting in relatively poor doxorubicin partitioning between the liposome (DOX-OLV) membrane to the medium, leading to a fast drug release immediately after its infusion.[3,41,42] This phenomenon was not observed in mice where the dilution upon intravenous administration is minimal and >300-fold smaller than in humans. (For a comparison of DOX-OLV and Doxil® see Table 13.1.)

In this case, murine studies can be very misleading due to the very large differences in blood volumes and volume of distribution between mice and humans. In addition, imaging studies revealed that in humans, the DOX-OLV, after losing most of its drug, reached the RES rather than the liver tumor target.[42] In retrospect, such a failure could have been be prevented by applying a very simple *in vitro* dilution release assay, which is now routinely employed.[41] The experience gained from the DOX-OLV studies in murine models and humans led to the application of a better understanding of the cross-talk between physicochemical, nano-technological and biological principles, which ultimately resulted in the success and approval of Doxil®.[2,3,9,10] Table 13.1 describes the requirements to get a viable liposomal doxorubicin formulation in order to meet the lessons from the DOX-OLV failure.

In order to save time, all four aspects described in Table 13.1 were investigated in parallel at four different locations: LTI laboratories at Menlo Park, by LTI scientists; Dimitri Papahadjopoulos' laboratory at University of California San Francisco by Alberto Gabizon; Terry Allen's laboratory at the University of Alberta in Canada; and Chezy Barenholz's laboratory in Jerusalem.

LTI, Terry Allen, Alberto Gabizon and Dimitri Papahadjopoulos worked on generating liposomes of extended circulation and RES avoidance, which, due to being at the nano-range size can take advantage of the EPR effect.[34,35] The EPR effect was expected to result in selective nano-particulate extravasation from the tumor capillaries into the tumor tissue. The liposomes with prolonged circulation time and RES avoidance were termed by Frank Martin of LTI as "stealth" liposomes, and this unique property of such liposomes was referred to as "stealthness", which means particulates which are unseen or unrecognized by the RES.

At the same time, Chezy Barenholz and his student Gilad Haran (now a professor at the Weizmann Institute of Science, Rehovot, Israel) were working on the development of a remote and stable loading method for doxorubicin, which should enable the formulation to reach sufficient level of drug to achieve therapeutic efficacy in humans under conditions that the drug, when reaching the tumor will be bioavailable to the tumor cells.[2,3,7,37,38,68] Table 13.1 describes a comparison between our failed DOX-OLV and the successful Doxil®.

Table 13.1 Comparison between the clinical trials for failed DOX-OLV and FDA-approved Doxil®.

Property	DOX-OLV[67]	Doxil®[2]	Comments
Liposome type by size and lamellarity	Oligolamellar large vesicles (OLV)	Small unilamellar vesicles (SUV)	
Liposome type by function	RES-directed	Stabilized, avoiding RES, stealth	
Lipid composition (mole ratio) and phase	EPC : EPG : cholesterol 7.0 : 3.0 : 4.0 Fluid liquid ordered	HSPC : DSPE–PEG (2000) : cholesterol 55 : 5 : 40 Solid liquid ordered	The negatively charged EPG In DOX-OLV is responsible for the electrostatic association to the liposome membrane
Storing medium	Buffered saline (high ionic strength)	Buffered sucrose histidine (low ionic strength)	
T_m of liposome-forming PC	$< 0\,°C$	50–53 °C (dependent on exact HSPC acyl chain composition)	Fluid *versus* solid PC
Electrical charge (as zeta potential) in saline	Negative due to the EPG in the membrane	DSPE–PEG is negatively charged but the zeta potential is almost neutral	In Doxil® negative charge of DSPE–PEG charge is hindered by the ∼5 nm 200 Da PEG layer
Liposome size distribution	200–500 nm heterogeneous	< 100 nm Uni-modal	The size distribution of Doxil® is much more homogeneous than of DOX-OLV
Location of drug	Membrane	Liposome aqueous phase mostly in a crystalline form	Doxil® is a good example of the advantages of remote loading which results in a high drug-to-lipid ratio
Loading method	Passive	Active (remote)	
Drug/ phospholipid (mole ratio)	0.05–0.07	0.2–0.3	
Release of drug upon large dilution *in vitro* and *in vivo*	Fast and dependent mainly on medium to membrane partition coefficient and dilution	Almost none without an external inducer (either ultrasound, radiofrequency or tumor metabolites)	
$t_{1/2}$ of PK in small rodent	<0.2 hours	>10 hours	
$t_{1/2}$ of PK in humans	<0.5 hours	>30 hours	

13.7 The Obligatory Need for Animal Studies and the Issue of the Relevance of Studies Using Cells in Culture (*in vitro*) to Doxil® Development

13.7.1 The Issue of Animal Studies

Animal use in biomedical and drug development for human and veterinary applications is and has been a very "hot" and highly controversial topic for a long time.[43] It is obvious that so far the contributions of animal studies to the development of many fields in biology, medicine, veterinary, and even agriculture are enormous, and it is difficult to see how all the current knowledge on metabolism, behavior, physiology, and health in humans and animals under normal and various pathological conditions could be achieved without the use of animal studies. Many mammalian species and, recently, other species, including birds and fish are in use. Current practice is changing and animal studies can be reduced as some biological and pharmaceutical research and development can be based more on extensive theoretical (computer modeling), molecular, and cell-culture studies. These other disciplines dramatically reduce the requirement of animal studies for drug development. However, there is still an absolute need for some animal studies, as the intact organism is different in many major aspects from studies based purely on two-dimensional and even three-dimensional co-culture or organ systems. An understanding of metabolism can be benefited extensively from cell and organ culture studies; however, understanding of overall metabolism is highly dependent on the interaction between different organs and tissues, which at this point in time can only be studied using animals. In the case of drug delivery, the drugs need to reach the target tissue and target cells and there is, therefore, a crucial need for an understanding of the various barriers to be overcome, as exemplified by the blood–brain barrier, or the issue of extravasation from the blood vessels into the diseased tissues. These require pharmacokinetic and biodistribution studies in animals. Finally, the "bottom line" is to prove efficacy and lack of (or acceptable) toxicity and these parameters also require the use of animal studies.

The process of drug development from concept to approval by the FDA and/or the EMA is a long one, and can take more than 10 years. It involves many stages before clinical trials (investigational new drugs applications in USA) are approved by the institutional review board committees. A Nuffield Council publication from 2005[44] describes and summarizes animal use during the different stages of drug (new chemical entity) development in terms of percentages of the total use per drug. The Nuffield report divides drug development into the stages required to achieve drug approval. The aim of the discovery stage is to screen to potential candidates, starting from as many as 1 000 000 molecules, requiring 5–15% of the total animal use; identification of lead candidates may involve up to 1000 molecules, this together with the stage of optimizing the performance of lead candidates which may involve up to 200 molecules, requires 60–80% of the animal use; selection based on efficacy and toxicity among up to 20 preferred candidates requires 10–20% of the animal use per drug.

As described above, the development of Doxil®, from the time the first patent application was submitted until its approval, took 7.5 years. Doxil® is a nano-drug based on a well-characterized active pharmaceutical ingredient (API), doxorubicin. In such a case therefore, although the approval still requires similar stages as for new chemical entities, the first stages of development should be shorter in time and may need significantly fewer animals. Animal studies with nano-drugs such as Doxil® require many pharmacokinetic and biodistribution studies in addition to efficacy and toxicity studies. For example, the optimization of liposome lipid composition was conducted based on pharmacokinetic and biodistribution studies as described elsewhere.[26,33,45]

In this part of the review I will try to demonstrate that animal studies in the case of Doxil® development were a must, and a decision based on tissue culture experiments alone would be highly misleading. However, it is also clear that a lot can be learned from *in vitro* studies and *in vitro* cell culture studies, and it is clearly desirable to have data from both systems.

13.7.2 *In vitro–In vivo* Correlation

Methods that can serve to demonstrate a real *in vivo–in vitro* correlation (IVIVC) are highly useful in order to reduce the use of animals. But this correlation has to be well proven.

The FDA defines IVIVC as a "predictive mathematical model describing the relationships between an *in vitro* property of a dosage form and *in vivo* response". Generally the *in vitro* property is the rate or extent of drug dissolution or release, while the *in vivo* response is the plasma drug concentration.[46]

IVIVC has been a highly important tool in the pharmaceutical development of drugs for many years.[47] Recently, IVIVC was also introduced to the pharmaceutical development of generic drugs.

The meaningful and productive use of IVIVC in the pharmaceutical development of drugs and generic drugs requires in-depth understanding of composition/structure/function relationships of the drug. Understanding these relationships enables the design and the development of a drug and generic drug products, thus allowing a reduction in the number of animal studies, and informs decisions to enable the right clinical studies to be conducted. It also allows the development of a reproducible manufacturing process of the drug product, which is another major requirement of new drug application and abbreviated new drug application approval by the FDA.

However, IVIVC of liposomal drugs are not straightforward, and are still at an early stage of evolution. It is well accepted that an assay which deals with the different aspects of API release of the liposomes at various physiologically relevant conditions can serve as the basis for IVIVC. These studies are similar to what is referred by the FDA as "dissolution" studies. A good example is the ammonia-induced doxorubicin release from Doxil®, which is relevant to what happens to Doxil® at the tumor site where there is a significant level of ammonia due to the unique-to-tumor metabolic pathway of glutaminolysis, and at the same time the ammonia level in the plasma is minimal (Section 13.2 and ref. 3, 29 and 32).

13.7.3 Studies Based on Cells in Culture

What role do cell culture studies play in liposomal drug development? Such studies can demonstrate the translation of drug release into the desired cytotoxic effects. They can also be used to study the potential toxicities which are not related to the therapeutic effect, such as lytic effects, effect on platelet aggregation and the like.[48]

The development of Doxil® can be used as a test case to evaluate the issue of if, and to what extent, studies of cells in culture can replace animal studies. The experience gained during Doxil® development clearly demonstrates that studies involving cell culture can help in understanding of the mechanism of action of liposomal formulations. However, they cannot replace animal studies. Cell culture studies can suggest if the liposomes are taken up by the cells followed by drug release intra-cellularly, or whether the mechanism of action involves mainly release of the drug from the liposomes to the medium (or interstitial fluid) from where the drug is then taken up in free form by the cells. This is a very important aspect to understand in order to optimize the performance of liposomal drugs.

During the development of our first-generation OLV-DOX with a formulation that failed in the clinic, we studied liposomal drug cytotoxicity against tumor cell lines in culture. These studies indicated that the encapsulation in the OLV has almost no effect on the median inhibitory concentration (IC_{50}) (compare 1.1×10^{-8} M for the free drug with 1.8×10^{-8} M for the liposomal drug) in the two cell lines studied (J-6456 lymphoma and P388 leukemia).[49] This suggested that encapsulation of the drug did not affect drug availability to the tumor cells. At the time the experiment was performed, we believed that this was a "good" sign, as it should have promised good biological activity. Now it is clear that in 1988 and until the analysis of the clinical trial pharmacokinetics was made available, we did not understand that this actually should be a warning signal as to the "viability" of the DOX-OLV formulation. The reason is that this high cytotoxicity of the liposomal drug suggested unwanted and relatively fast dilution-dependent major drug leakage from the liposomes into the cell growth medium. At a later stage when we evaluated this formulation in humans,[2,3,41,42] we found that fast drug release occurred in the blood almost immediately after intravenous DOX-OLV administration. We also found that the cytotoxicity of DOX-OLV could be totally eliminated when the cation exchanger Dowex-50 was added to the cell growth medium. Dowex-50 also totally eliminated the cytotoxicity of free doxorubicin and very efficiently bound free drug but did not bind liposomal drug at all.[1,41,50] Therefore, the lack of activity indicated that the free drug which is quickly released from the OLV-DOX is responsible for cell killing and not the intact drug-loaded liposomes. In summary, the DOX-OLV mechanism of action is a release of the drug to the medium, and the released drug is responsible for inhibition of cell growth.

13.7.4 Lessons Learned from *In vitro* Cell Culture Studies during Doxil[®] Development

What we learned from the failure of OLV-DOX in the clinic is that rates of drug release in the plasma will be determined by the way liposomes are loaded with the drug; liposome lipid composition and structure; and blood and disease-site microenvironment (such as tumor) conditions and composition.[3,9,32,51–53] In order to achieve a beneficial and superior (to the conventional drug) therapeutic index, two conditions have to be met. The first is that enough drug-loaded liposomes have to reach the tumor, and the second condition is that the API that reaches the tumor encapsulated in liposomes has to find its way to the tumor cells. In order to meet the first condition, for intravenous-administered liposomal drugs such as Doxil[®], only when drug release in the blood (determined by k_{off}, which includes all rates contributing to drug release) is significantly slower than the liposome clearance (k_c) will the liposome determine and control drug pharmacokinetics and biodistribution, to the level that enough drug loaded-liposomes will reach the tumor. When $k_{off} > k_c$, then the ratio k_{off}/k_c is a measure of the rate of drug release *in vivo*. Controlling this ratio is obligatory to achieve controlled drug release in blood or in the tissues reached by the liposomes. Therefore, this ratio also affects therapeutic efficacy of the liposomal drug. For drugs of fast clearance, when $k_{off} \gg k_c$ the benefits of using liposomes for drug delivery will be minimal or none, as the performance of the liposomal drug will be similar to that of the free drug. This is exemplified by our first-generation failed OLV-DOX formulation (see Table 13.2). An efficient and functional way to test the release rate is a cytotoxicity test measuring doxorubicin IC_{50} in cell culture. Horowitz *et al.*[54] showed that Doxil[®] has about a 20-fold higher IC_{50} (lower cytotoxic activity) than free doxorubicin. These are very different results from those achieved with DOX-OLV that show almost identical cytotoxicity effects to free doxorubicin (see above, and Gabizon and Barenholz[49]). At that time the PLD used (referred to as L-DOX) was prepared by hydration of the lipids in 155 mM ammonium sulfate[7] and, therefore, the transmembrane ammonium gradient was smaller than the current Doxil[®] gradient, for which lipid hydration is performed in 250 mM ammonium sulfate. The fact that cytotoxicity of Doxil[®]-like PLDs is much lower than that of free doxorubicin, was recently confirmed with other cell lines.[32] We demonstrated that the differences in cytotoxicity between DOX-OLV and Doxil[®]-like PLDs are related to the fact that when comparing the release rate of the two diffrent liposomal doxorubicins, only DOX-OLV

Table 13.2 The requirements to achieve therapeutically efficacious passively-targeted drug-loaded liposomes and the means to fulfill them.[a]

Main requirements to achieve therapeutically efficacious passive targeting of liposomes into cancer tissues	Physicochemical and biophysical solutions used to meet the requirements
Requirements	Solutions
1. Extended circulation time of drug-loaded liposomes while circulating in the animal/human blood	Development of sterically stabilized liposomes (SSL) composed of high T_m phospholipids, cholesterol, and a lipo-polymer such as ^{2000}PEG–DSPE
2. Sufficient levels and stable loading of drug in order for long-circulating nano-liposomes to reach the disease site with liposomes loaded with drug at a level needed to achieve therapeutic efficacy ($t_{1/2}$ of drug release in blood should be longer than circulation $t_{1/2}$)	Use of pH or transmembrane ammonium ion gradients for remote (active) loading of amphipathic weak bases or acids into long-circulating nano-liposomes
3. Extravasation into tumor tissues	Using small enough (<120 nm, preferably <100 nm) nano-liposomes in order to efficiently extravasate through the gaps in the tumor vasculature (taking advantage of the EPR effect)
4. Getting biologically active drug into the tumor cells	Releasing drug from liposomes through selective drug leakage at site due to diseased tissue properties, or using: collapsible ion gradient, or liposomes sensitive to secretory phospholipases, or by applying physical means such as heat (thermo-sensitive liposomes), or using high intensity ultrasound or radiofrequency, or by internalization as a result of active targeting

[a]For the historical perspective of Doxil® and its development, see Section 13.3 and Barenholz.[2,3]

undergoes a very rapid large (>500-fold) dilution-induced doxorubicin release, while the Doxil®-like PLD lacks such a release.[41] As discussed earlier, and in retrospect, the similarity in the growth inhibition curves when free doxorubicin was compared with DOX-OLV may serve as an indication of fast and large dilution- induced drug release *in vivo*,[42] which was also confirmed *in vitro*.[41] Cytotoxicity of doxorubicin to cells in culture requires that the doxorubicin will be internalized by the cells. This can be achieved in two different ways, either by cell uptake of intact liposomes loaded with drug or by release of drug from the liposomes to the cell growth medium followed by free drug uptake from this medium. To differentiate between these two modes of action we added the cation exchanger Dowex-50 to the cell culture growth medium under conditions that ensure it has no effect on cell viability or cell well-being. The Dowex-50 cation exchanger resin binds free doxorubicin very efficiently, but does not bind liposome-associated doxorubicin at all, either in the form of DOX-SUV or in Doxil®-like PLDs.[1,50,55] The presence of this cation exchanger inhibits almost completely the cytotoxicity of free doxorubicin and of DOX-OLV to cells in culture, suggesting that in both the active species is free doxorubicin. These differences demonstrate nicely that the large dilution-induced release that limited the performance of OLV-DOX in humans does not occur with Doxil®.[1,41,72]

To make sure that our interpretation was correct we studied the effect of nigericin on doxorubicin release and derived IC_{50} values. Nigericin is an ionophore that collapses the pH and ammonium ion gradients by exchanging the intra-liposome proton with extra-liposome medium potassium ions. Exposure of Doxil® to nigericin caused complete drug release due to the collapse of proton and ammonium ion gradients,[10] which was paralleled by reduction in the IC_{50} to the low level of free doxorubicin, proving that Doxil®'s excellent drug retention is the reason for Doxil®'s poor IC_{50} in this assay (Figure 13.9, and ref. 54).

Further strong support that the free drug released and not the liposome-encapsulated drug is the active species comes from comparing growth inhibition of OV-163 cells in culture by PLDs based on different liposome-forming PCs (Figure 13.10).

Figure 13.10 clearly demonstrates that growth inhibition is directly related to the T_m of the liposome-forming PC. The higher the T_m, the lower the growth inhibition. PLD based on the fluid egg PC (T_m $-5\,°C$) has similar growth inhibition and IC_{50} to the free drug, while PLD based on HSPC (T_m $52\,°C$) has a much lower growth-inhibition effect, and PLD based on DPPC (T_m of $41\,°C$) growth inhibition is in between. These growth inhibition curves are strongly and directly related to an effect of PLD PC on doxorubicin release rate.[7]

However, the opposite (very low k_{off}) is also disfavourable. Namely, when k_{off} is too slow and there is no liposome uptake by the target cells, there will be no or poor therapeutic efficacy, even if the loaded liposomes may reach the target very efficiently, as the free drug concentration at the target tissue will be too low, as exemplified by sterically stabilized cisplatin liposomes.[2,9,56–58]

13.8 The Time is Ripe for Generic Doxil®-like PLDs

With such large sales, extensive use (>500 000 patients so far), a growing market, and potentially new indications which may lead to large revenues, it was expected to have generic versions of Doxil®-like PLDs available, given that it is 6 years since the last Doxil®-related patent[38] expired. However, to date, there is only one FDA- and EMA-approved formulation of generic Doxil®-like PLDs on the market. Since the end of 2011, there has been a shortage of Doxil®/Caelyx, and this was a further drive for the development of a generic Doxil®-like PLD. However, in spite of all the above motivation, we still have only one generic formulation, the doxorubicin-HCl

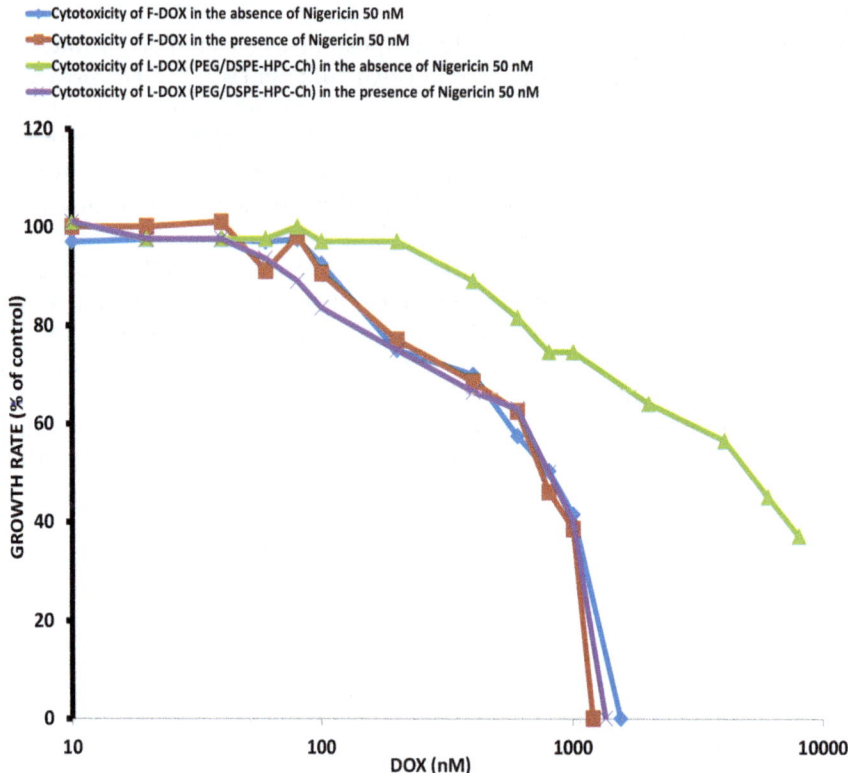

Figure 13.9   Effect of nigericin on Doxil cell growth inhibition. Growth inhibition curves of OV-1063 cells (5000 cells per well) growing for 72 hours comparing the effect of free doxorubicin (F-DOX), PEGylated liposomal doxorubicins (PLDs) based on HSPC (L-DOX) with and without the ionophore nigericin.[54]

liposome injection (Lipodox), approved by the FDA (in February 2013), but not yet approved by the EMA. The question is, why, in spite of these large efforts by many companies worldwide over many years do we not see the approval of more generic Doxil®-like PLD products?

The answer may be found in the FDA requirements for the approval of generic Doxil®, which show a tough, extensive broad spectrum of requirements needed to achieve approval. Special focus is on chemistry, manufacturing, and control (CMC) issues, including the comparability with the reference listed drug and IVIVC. These include many scientific, technological, and regulatory difficulties (for more details see the FDA *Draft Guidance on Doxorubicin Hydrochloride* (2010) updated in 2013 and 2014[59] and ref. 60–62).

13.9   Doxil®: New Findings (2012–2015)

In 2012, I published a major review on Doxil®,[2] which described various aspects of Doxil® development, Doxil® characterization and its clinical

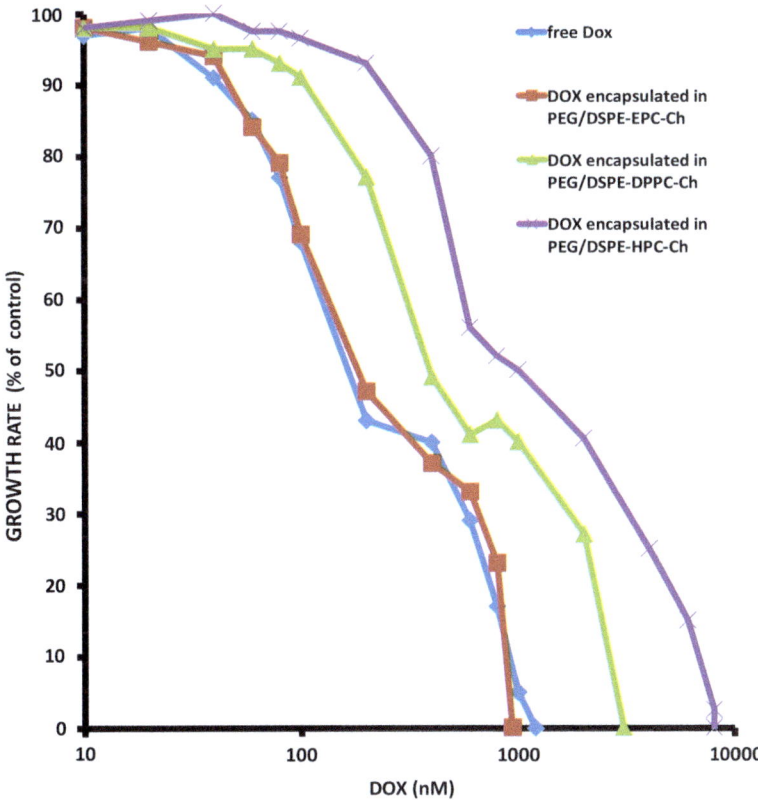

Figure 13.10 Effect of phosphatidylcholine acyl chain composition of PLD on cell growth inhibition. Growth inhibition curves of OV-1063 cells (5000 cells per well) growing for 72 hours in the presence of PEGylated liposomal doxorubicins (PLDs) of three different liposome-forming phosphatiduylcholines (PCs) (egg PC, DPPC and HSPC).[54] DOX: doxorubicin.

applications, including Doxil® historical perspectives. To our surprise, although it was 23 years that had passed from 1989 when the major Doxil®-related patent applications were filed (US patents 5 192 549[37] and 5 316 771,[38] and US patent 5 013 556,[40] Section 13.5), much important new data on Doxil® was obtained and published, including structural findings as well as functional-related findings. In collaboration with other teams,[63,64] we completed X-ray diffraction studies and cryo-transmission electron microscopy (TEM) studies on Doxil® initiated by us in 1992.[12,65] Our recent X-ray diffraction studies describe the high-resolution structures and morphology of PEGylated nanoliposomes before and after remote loading of doxorubicin. It was found that the structure of the liposomal membranes was the same before and after drug remote loading and was independent of the ammonium counter anion (comparing the bi-valent sulfate as in Doxil® to the mono-valent counter anion methanesulfonate). We confirmed the Lasic

1992 data that the drug formed crystals inside PLDs loaded by ammonium sulfate gradient and we obtained quantitative data on the dimensions of the intra-liposome doxorubicin-sulfate nano-crystals. We found that the distance head to head (phosphate to phosphate) across the bilayer of the lipid membrane is 4.80 ± 0.10 nm. The PEG layer thickness as the distance between the end of the hydrophilic layer and the position where the PEG density is at 10% of its effective maximum value differs for the external and internal PEG layers, $L_{out} = 4.8 \pm 1.2$ nm and $L_{in} = 3.6 \pm 1.2$ nm. This difference between the two PEG thicknesses is consistent with the curvature differences and the asymmetric DSPE-PEG distribution between the two leaflets.[63] The same study also demonstrates that the doxorubicin-sulfate intra-liposome nano-crystal, which occupies the full length of the liposome and actually forces the liposome to assume the Doxil® typical coffee-bean shape, has a radius of 8.6 ± 0.4 nm. We also found that in replacing the bivalent sulfate counter anion with the mono-valent counter anion methanesulfonate, the intra-liposome doxorubicin-methanesulfonate salt has an amorphous morphology. These recent studies showed that the drug hardly affects the structure and dimensions of the membrane. They also demonstrate that the lipids in the membrane bilayer were non-uniformly distributed between the two leaflets that form the Doxil® liposome bilayer. In the inner monolayer, the PEG chains assumed mainly mushroom conformations, whereas in the outer layer, the PEG chains assumed more extended conformations of a short polymer brush, or the lipopolymer molecules are in a transition between the mushroom and the brush phases. The data were consistent with little or no interaction between the drug and the lipid bilayer.

Our X-ray diffraction studies[63] demonstrate that solution X-ray scattering, when combined with powerful analytical tools, can determine the high-resolution structure of complex non-crystallized nanoparticle dispersions used in nanomedicine such as Doxil®, hence it satisfies the characterization requirements of the regulatory agencies (FDA and EMA) for nano-drug approval.

We have also dedicated major efforts to analyze size and morphology distributions. These are considered critical attributes in the performance of nano-drugs including PLDs, as they contribute to pharmacokinetics and drug biodistribution including passive targeting related to the EPR effect. Therefore, the FDA and EMA both require a comprehensive analysis of these two variables. We compared detailed dynamic light scattering (DLS) analyses (Z-average, polydispersity index, number average and D10-50-90) with the automated, MATLAB-based quantitative analysis of cryo-TEM images developed by us. This analysis provides direct model-independent structural information, in liquid dispersion. While DLS analysis is performed on a large number of particles ($\sim 10^{13}$), it neither deals with particle morphology nor with their deviation from spherical shape. The cryo-TEM analysis, although based on a much smaller number (10^3) of particles, supplies morphology parameters. Doxil® and its first generic, Lipodox, show the "coffee bean" ellipsoid shape with an average axial ratio of 1.17 ± 0.03, compared with the spherical shape with axial ratio of 1.05 ± 0.06 for the same liposomes but without the remotely loaded drug. We combined cryo-TEM and DLS to meet the FDA requirements: cryo-TEM characterizes the morphology of non-spherical shapes and provides values of trapped volume and trapped drug (*e.g.* crystallinity). For spherical particles, DLS provides size distribution in a more accurate way, with cryo-TEM being confirmatory.[64]

The third physicochemical method, which we added recently to the repertoire of methods used to characterize nano-drugs that are supramolecular-assemblies such as Doxil®, is solution (DSC), which enabled the so-far missing thermodynamic characterization of Doxil®.

The membrane of Doxil® and its generic Lipodox both show the same small and reversible non-leaky phase transition (despite having >35 mol% cholesterol). The lipopolymer DSPE–PEG (2 kDa) increases the cooperativity of this transition. In both, the intraliposomal doxorubicin-sulfate nanocrystals melt at $\sim 70\,^{\circ}$C, almost identical to the melting of the crystals formed in bulk solution of 250 mM ammonium sulfate at high (>60 mg mL^{-1}) doxorubicin. However, the nanocrystals endotherm demonstrates a $\sim$2.0–2.5-fold larger cooperative unit (narrower $\Delta t_{1/2}$), of the transition, suggesting that the liposome nanovolume restricts size and increases the crystallinity of the doxorubicin-sulfate crystals. Although the FDA guidelines for generic Doxil® recommend DSC characterization, no DSC data or referral to such information was published. The reason is that at such high level of membrane cholesterol it was not expected to observe a membrane-related phase transition, and the routine conditions used for liposome characterization by DSC did not show a phase transition. However, when Doxil® and Lipodox "as is" were analyzed the two clear endotherms were identified, one related to the membrane lipids and the second (at higher temperature) to the intra-liposomal doxorubicin-sulfate nano-crystals (Figure 13.11).

DSC analysis, combined with cryo-TEM was performed on PLDs of the same size and doxorubicin/lipid ratio, which were prepared at various ammonium sulfate concentrations. The combined analyses show that crystallization of doxorubicin-sulfate occurs only when the ammonium sulfate concentration used for lipid hydration $\geq$150 mM. It also demonstrates the different roles that ammonium and sulfate ions play in the remote loading process.[20] DSC studies also demonstrated that PLDs in which the doxorubicin was remote loaded using transmembrane ammonium methanesulfonate show the same lipid phase transition but do not show any drug melting, confirming the X-ray results described above.[20,63]

The use of these three physicochemical methods (X-ray diffraction; size and morphology distributions obtained by DLS and cryo-TEM image analyses; and thermodynamic characterization based on DSC) demonstrates that this combination of methods enables us to determine the high-resolution structure of complex supramolecular-assembly nano-drugs, which is important to satisfy the characterization requirements of regulatory agencies for nano-drug approval.

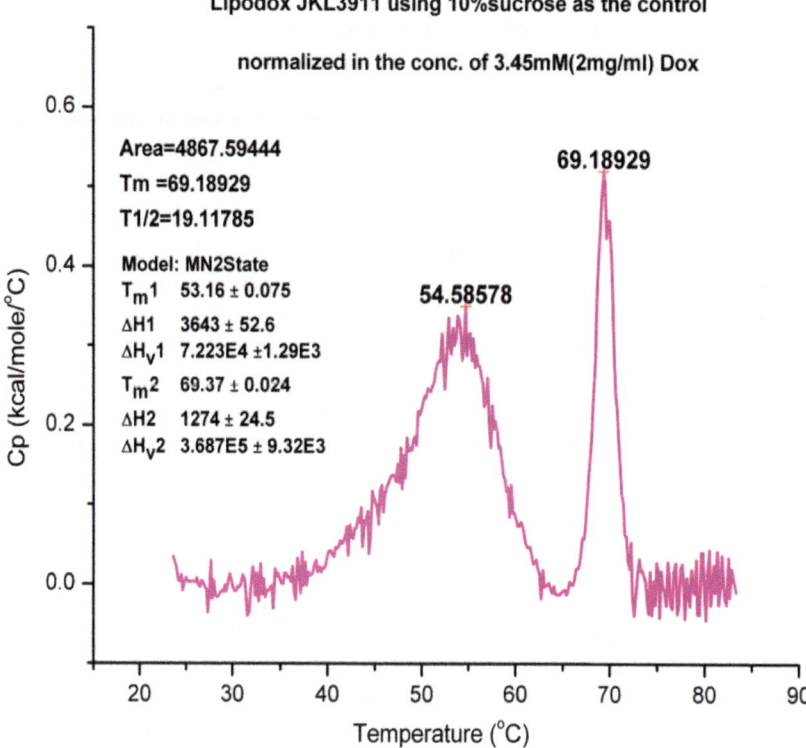

Figure 13.11 Differential scanning colirimetry (DSC) of Lipodox "as is" against 10% sucrose in 10 mM histidine buffer pH 6.5. Scanning was performed using MicroCal VP-DSC differential scanning calorimeter (GE Health-care, US, now Malvern, UK) at a scanning speed of $1\,^{\circ}$C min^{-1}. As demonstrated by Wei *et al.*,[20] the endotherm with a T_m at $54.6\,^{\circ}$C represents a membrane lipid phase transition, while the endotherm at $69.2\,^{\circ}$C represents the melting of the doxorubicin-sulfate nano-crystal.

Recently, new highly relevant information on the tumor microenvironment demonstrated its critical role in determining the therapeutic efficacy of Doxil®-like PLDs. Various anatomical, structural, and immunological aspects related to the microenvironment were studied and were shown to explain differences between different tumors in EPR effect and in Doxil® nano-drug therapeutic efficacy variability (see recent articles by Zamboni and coworkers[52] and Jain and his team[53]). In addition, we demonstrated that the unique tumor metabolome may also be highly relevant. Our results showing that tumor ammonia may be an important factor in doxorubicin release[3,32] adds a new dimension to current knowledge by relating the unique tumor metabolism to the therapeutic efficacy of PLD. The significant levels of ammonia/ammonium in the tumor microenvironment enable the PLD to act as stimulus-responsive liposomes, with the ammonia produced continuously during the unique-to-tumor glutaminolysis acting as the tumor-specific stimulus. This study may help in understanding interpatient variability in the therapeutic efficacy of Doxil®-like nano-drugs. Namely it is possible that, in addition to the previously known differences in the EPR effect,[52,53] there is also variability in the level of ammonia at the tumor site. It may be possible that in the future, *in vivo* imaging may be used to demonstrate the production and level of ammonia/ammonium in tumors. Thus, in addition to the ability to evaluate if the tumors show an EPR effect using *in vivo* imaging, as demonstrated by Andresen and co-workers,[66] it may be possible to predict in individual patients whether and to what extent treatment with a nano-drug such as Doxil® will be therapeutically efficacious.

13.10 Doxil® Still Keeps Some Secrets

Twenty-eight years have passed since the first provisional patent on Doxil® was filed (see Section 13.5), and >500 000 patients have received Doxil® since 1991 when the first patient in the first clinical trial was injected with Doxil®.[1] Research with Doxil® formulations is still revealing new major information on its physicochemical and biological features. These new data enable a better understanding of Doxil®'s mechanism of action. Many clinical studies which involve Doxil® introduce new combinations of Doxil® with other drugs and/or biologicals, as well as the use of other means which may improve its therapeutic index by improving its efficacy and/or reducing its side effects. However, we are still missing important information on the optimal rate of doxorubicin release at tumors and how it is affected by the type of the tumor. An answer to this question may lead to a further improvement of the clinical performance of nano-drugs such as Doxil®.

The 35 years of basic research, pre-clinical and clinical experience with liposomal doxorubicin[2,3] have enabled us to initiate research on how to develop novel nano-drugs in general and especially how to improve the performance of PLD and similar drugs by optimizing drug release rate at the tumor, as well as reducing the major side effects (of hand and foot syndrome and complement activation), thereby further improving the therapeutic index of nano-drugs.

Special Acknowledgements

This, like all my other reviews on Doxil® is dedicated to my wife, Dr Hanna Barenholz. The development of Doxil® was a wonderful "adventure", and a very large highlight of my career. Without Hanna's encouragement, advice, patience, dedication, and continuous moral and actual support throughout our 56 years of life together I would not have been able to accomplish my part in the development of Doxil®.

I also want to thank my four daughters: Chagit, Ayelet, Tamar, and Avigail, who grew up during the years of my work on liposomal doxorubicin including Doxil®; their husbands, Uri, Perri, Ron, and Assaf; and last but not least our 12 grandchildren, Yael, Yuval, Amit, Omri, Inbar, Mika, Rotem, Guy, Eyal, Gal, Dror, and Kfir, who give us so much joy. Our grandchildren were my escape at periods of despair during the development of Doxil®.

Professionally, I would like to name a few of the many people who deserve my gratitude: firstly, Alberto Gabizon, a friend and partner of 36 years in exciting research including Doxil® development (the many shared papers and patents with him are excellent evidence of our highly productive collaborative interaction); the late Demetrios (Dimitri) Papahadjopoulos, for his friendship and intellectual stimulus; and Nick Arvanidis, from whom I learned about priorities in applied research and for the many excited discussions and "hot" disputes we had through the years of Doxil®'s development.

I also want to thank all my many laboratory people and my collaborators who have participated in these exciting Doxil®-related "adventures".

Last but not least I also want to thank Martin Braddock, the editor of this book, for his help in editing this chapter.

References

1. A. Gabizon, R. Catane, B. Uziely, B. Kaufman, T. Safra, R. Cohen, F. Martin, A. Huang and Y. Barenholz, *Cancer Res.*, 1994, **54**, 987.
2. Y. Barenholz, *J. Controlled Release*, 2012, **160**, 117.
3. Y. Barenholz, in *Handbook of Clinical Nanomedicine*, ed. R. Bawa, G. Audette and I. Rubinstein, Pan Stanford Publishing, Singapore, 2016, pp. 827–906.
4. R. Solomon and A. A. Gabizon, *Clin. Lymphoma Myeloma*, 2008, **8**, 21.
5. G. Ferrandina, A. Licameli, D. Lorusso, G. Fuoco, S. Pisconti and G. Scambia, *Ther. Clin. Risk Manage.*, 2010, **6**, 463.
6. www.clinicaltrials.gov.
7. G. Haran, R. Cohen, L. K. Bar and Y. Barenholz, *Biochim. Biophys. Acta*, 1993, **1151**, 2015.
8. E. M. Bolotin, R. Cohen, L. K. Bar, N. Emanuel, S. Ninio, D. D. Lasic and Y. Barenholz, *J. Liposome Res.*, 1994, **4**, 455.
9. Y. Barenholz, *Curr. Opin. Colloid Interface Sci.*, 2001, **6**, 66.
10. Y. Barenholz, in *Liposome Technology*, ed. G. Gregoriadis, Informa Healthcare, New York, NY, 3rd edn, 2007, vol. 2, pp. 1–26.
11. D. Zucker, D. Marcus, Y. Barenholz and A. Goldblum, *J. Controlled Release*, 2009, **139**, 73.
12. D. D. Lasic, P. M. Frederik, M. C. A. Stuart, Y. Barenholz and T. J. Mcintosh, *FEBS Lett.*, 1992, **312**, 255.
13. J. J. Israelachvili, *Intermolecular and Surface Forces*, Academic Press, New York, NY, 2nd edn, 1992, Chapters 16 and 17.
14. Y. Barenholz and G. Cevc, in Physical chemistry of biological surfaces, ed. A. Baszkin and W. Norde, Marcell Dekker, New York, NY, 2000, pp. 171–241.

15. O. G. Mouritsen, *Life–as a Matter of Fat: The Emerging Science of Lipidomics*, Springer, Berlin, 2005.
16. O. Garbzenko, Y. Barenholz and A. Priev, *Chem. Phys. Lipids*, 2005, **135**, 117.
17. O. Tirosh, Y. Barenholz, Y. Katzhendler and A. Priev, *Biophys. J.*, 1998, **74**, 1371.
18. H. Kitayama, Y. Takechi, N. Tamai, H. Matsuki, C. Yomota and H. Saito, *Chem. Pharm. Bull.*, 2014, **62**, 58.
19. J. Hjort Ipsen, G. Karlström, O. Mouritsen, H. Wennerström and M. Zuckermann, *Biochim. Biophys. Acta, Biomembr.*, 1987, **905**, 162.
20. X. Wei, R. Cohen and Y. Barenholz, submitted, 2016.
21. J. Szebeni, F. Muggia, A. Gabizon and Y. Barenholz, *Adv. Drug Delivery Rev.*, 2011, **63**, 1020.
22. J. Szebeni, P. Bedocs, Z. Rozsynyay, Z. Weiszhar, R. Urbanics, L. Rosivall, R. Cohen, O. Garbuzenko, G. Bathori, M. Toth, R. Bunger and Y. Barenholz, *Nanomedicine*, 2012, **8**, 176.
23. www.drugs.com/pro/doxil®/html.
24. www.doxil®.com.
25. Y. Avnir, K. Barhum-Turjeman, D. Tolchinski, A. Sigal, P. Kizelstein, D. Tzemach, A. Gabizon and Y. Barenholz, *PLoS One*, 2011, **6**(10), e25725.
26. M. C. Woodle and D. D. Lasic, *Biochem. Biophys. Acta*, 1992, **113**, 171.
27. A. Gabizon, H. Shmeeda and Y. Barenholz, *Clin. Pharmacokinet.*, 2003, **42**, 419.
28. A. Gabizon, *Adv. Drug Delivery Rev.*, 1995, **16**, 285.
29. C. H. Eng, K. Yu, J. Lucas, E. White and R. T. Abraham, *Sci. Signaling*, 2010, **3**, 31.
30. L. Wang, H. Zhou, Y. Wang, G. Cui and L. J. Di, *Cell Death Dis.*, 2015, **6**, e1620.
31. R. W. Moreadith and A. L. Lehninger, *J. Biol. Chem.*, 1984, **259**, 6215.
32. L. Silverman and Y. Barenholz. *Nanomed.: Nanotechnol. Biol. Med.*, 2015, **11**, 1841–1850.
33. A. Gabizon, Y. Barenholz and M. Bialer, *Pharm. Res.*, 1993, **10**, 703.
34. Y. Matsumura and H. Maeda, *Cancer Res.*, 1986, **46**, 6387.
35. R. K. Jain, Barriers to drug delivery in solid tumors, *Sci. Am.*, 1994, **271**, 58–65.
36. H. Maeda, G. Y. Bharate and J. Daruwalla, *Eur. J. Pharm. Biopharm.*, 2009, **71**, 409.
37. Y. Barenholz and G. Haran. (March 9, 1993). Method of amphipathic drug loading in liposomes by pH gradient, *U.S. Pat.* 5192549.
38. Y. Barenholz and G. Haran. (May 31, 1994). Liposomes: Efficient Loading and Controlled Release of Amphipathic Molecules. *U.S. Pat.* 5316771.
39. Y. Barenholz and S. Amselem, Liposome Technology, in *Liposome Preparation and Related Techniques*, ed. G. Gregoriadis, CRC Press, Boca Raton, FL, 2nd edn, 1993, vol. I, pp. 527–616.
40. M. C. Woodle, F. J. Martin, A. Yau-Yang, C. T. Redmann. Liposomes with enhanced circulation time, *U.S. Pat.* 5013556, 1991.

41. S. Amselem, R. Cohen and Y. Barenholz, *Chem. Phys. Lipids*, 1993, **64**, 219.

42. A. Gabizon, R. Chisin, S. Amselem, S. Druckmann, R. Cohen, D. Goren, I. Fromer, T. Peretz, A. Sulkes and Y. Barenholz, *Br. J. Cancer*, 1991, **64**, 1125.

43. A. Tsafriry, Use of animals in clinical research, in *The Man and the Animal: Scientific Experimentation in Animals*, Haifa University Publisher, Haifa, 2013 (in Hebrew).

44. Nuffield Council on Bioethics, 2005. The ethics of research involving animals, Table 8.1, London, 2005. (www.nuffieldbioethics.org/animal-research).

45. A. Gabizon and D. Papahadjopoulos, *Proc. Natl. Acad. Sci. U. S. A.*, 1988, **85**, 6949.

46. S. Sakore and B. Chakraborty, *J. Bioequivalence Bioavailability*, 2011, **S3**, DOI: 10.4172/jbb. S3-001.

47. D. M. Chilukuri, G. Sunkara and D. Young, *Pharmacutical Product Development: In Vitro - In Vivo Corolation*, Informa Healthcare, New York, NY, 2007.

48. M. J. Parnham and H. Wetzig, *Chem. Phys. Lipids*, 1993, **64**, 263.

49. A. Gabizon and Y. Barenholz, Adriamycin - containing liposomes in cancer chemotherapy, in *Liposomes as Drug Cariers*, ed. G. Gregoriadis, 1988, Wiley, New York, pp. 365–379.

50. S. Druckmann, A. Gabizon and Y. Barenholz, *Biochim. Biophys. Acta*, 1989, **980**, 381.

51. Y. Barenholz and R. Cohen, *J. Liposome Res.*, 1995, **5**, 905.

52. G. Song, D. B. Darr, C. M. Santos, M. Ross, A. Valdivia, J. L. Jordan, B. R. Midkiff, S. Cohen, N. Nikolaishvili-Feinberg, C. R. Miller, T. K. Tarrant, A. B. Rogers, A. C. Dudley, C. M. Perou and W. C. Zamboni, *Clin. Cancer Res.*, 2014, **20**, 6083.

53. V. P. Chauhan, T. Stylianopoulos, J. D. Martin, Z. Popovic, O. Chen, W. S. Kamoun, M. G. Bawendi, D. Fukumura and R. K. Jain, *Nat. Nanotechnol.*, 2012, **7**, 383.

54. A. T. Horowitz, Y. Barenholz and A. A. Gabizon, *Biochim. Biophys. Acta*, 1992, **1109**, 203.

55. S. Amselem, A. Gabizon and Y. Barenholz, *J. Pharm. Sci.*, 1990, **79**, 1045.

56. D. D. Lasic, J. J. Vainer and P. K. Working, *Curr. Opin. Mol. Ther.*, 1999, **1**, 177.

57. T. Peleg-Shulman, D. Gibson, R. Cohen, R. Abra and Y. Barenholz, *Biochim. Biophys. Acta*, 2001, **1510**, 278.

58. W. C. Zamboni, A. C. Gervais, M. J. Egorin, J. H. Schellens, E. G. Zuhowski, D. Pluim, E. Joseph, D. R. Hamburger, P. K. Working, G. Colbern, M. E. Tonda, D. M. Potter and J. L. Eiseman, *Cancer Chemother. Pharmacol.*, 2004, **53**, 329.

59. http://www.fda.gov./downloads/Drugs/GuidanceComplianceRegulatory Information/Guidences/UCM199635.pdf.

60. W. Jiang, R. Lionberger and L. X. Yu, *Bioanalysis*, 2011, **3**, 333.
61. S. Tinkle, S. E. McNeil, S. Mühlebach, R. Bawa, G. Borchard, Y. Barenholz, L. Tamarkin and N. Desai, *Ann. N. Y. Acad. Sci.*, 2014, **1313**, 35.
62. R. Bawa, Y. Barenholz and A. Owen, The challenge of regulating nano-medicines: key issues, in *Nanomedicine: Design, Delivery and Detection*, ed. M. Braddock, Royal Society of Chemistry, Cambridge, UK, 2016.
63. Y. Schilt, T. Berman, X. Wei, Y. Barenholz and U. Raviv, *Biochim. Biophys. Acta*, 2016, **1860**, 108–119.
64. S. Peretz Damari, D. Shamrakov, Y. Barenholz and O. Regev, Submitted for publication, 2015.
65. D. D. Lasic, B. Ceh, M. C. A. Stuart, L. Guo, P. M. Frederik and Y. Barenholz, *Biochim. Biophys. Acta*, 1995, **1239**, 145.
66. A. L. Petersen, T. Binderup, P. Rasmussen, J. R. Henriksen, D. R. Elema, A. Kjærand and T. L. Andresen, *Biomaterials*, 2011, **32**, 2334.
67. N. Emanuel, E. Kedar, E. M. Bolotin, N. I. Smorodinsky and Y. Barenholz, *Pharm. Res.*, 1996, **13**, 352.
68. N. Emanuel, E. Kedar, E. M. Bolotin, N. I. Smorodinsky and Y. Barenholz, *Pharm. Res.*, 1996, **13**, 861.
69. A. V. Andriyanov, E. Koren, Y. Barenholz and S. N. Goldberg, *PLoS One*, 2014, **9**(1), e92555.
70. A. Cern, A. Golbraikh, A. Sedykh, A. Tropsha, Y. Barenholz and A. Goldblum, *J. Controlled Release*, 2012, **160**, 147.
71. O. Garbuzenko, S. Zalipsky, M. Qazen and Y. Barenholz, *Langmuir*, 2005, **21**, 2560.
72. European Medecines Agency CHMP assessment report, Doxorubicin, Sun Pharma, EMA/588790/2011.

Subject Index

2-methoxyestradiol (2ME2), 211–212
N-(2-hydroxypropyl) methacrylamide (HPMA) copolymeric nanoparticles, 205–206

Abraxane®, 238–239
  clinical efficacy and safety of, 242–244
activatable nanomedicines, 171–175
active targeting
  and nanomedicine imaging, 163–170
  and siRNA, 72–73
amphiphilic polymeric nanoparticles, 209–210
animal studies, and MSCs, 114–118
  bone regeneration, 115–116
  cartilage regeneration, 114–115
  muscle regeneration, 116–118
antibodies, 129–130
anticancer agents
  NC-6004 (cisplatin-incorporating micelle), 187–189
    clinical studies, 189
    preclinical studies, 188–189
    preparation and characterization of, 188
  NC-6300 (epirubicin-incorporating micelle), 189–192
    clinical studies, 191–192
    preclinical studies, 191
    preparation and characterization of, 189–190

NK105 (paclitaxel-incorporating micelle), 184–187
  clinical studies, 186–187
  preclinical studies, 186
  preparation and characterization of, 184–186
  overview, 182–184

bioconjugates, 210–211
biodistribution assessment
  imaging as non-invasive method for, 153–157
  rationale for, 152–153
bioluminescence imaging, 112–114
bone regeneration, 115–116

camptothecins, 212–213
cancer, and polymeric nanoparticles
  combination therapy, 215–216
  and CRLX101, 210–221
    chemistry, 217
    clinical results, 220–221
    preclinical results, 217–220
  hypoxia inducible factor-1 inhibitors (HIF-1), 211–214
    camptothecins, 212–213
    endogenous HIF-1α inhibitors, 213
    2-methoxyestradiol (2ME2), 211–212
    and siRNA, 213
  overview, 199–200

stem cells, 214–215
and topoisomerase 1
    inhibitors, 200–211
        amphiphilic, 209–210
        bioconjugates, 210–211
        carbohydrate-based,
            203–204
        HPMA copolymeric
            nanoparticles, 205–206
        non-polymeric, 211
        overview, 200–203
        polyamine, 204–205
        polyethylene glycol (PEG),
            206–209
carbohydrate-based polymeric
    nanoparticles, 203–204
carbon-based nanoparticles, 8
carbon nanotubes (CNT), 85–87
carrier materials, of theranostics,
    124–133
        antibodies, 129–130
        liposomes, 129
        metal nanoparticles, 130–131
            gold nanoparticles,
                130–131
            iron oxide nanoparticles,
                130
            silver nanoparticles, 131
        microbubbles, 132–133
        nanocarbons, 131–132
        polymeric nanoparticles,
            124–129
            dendrimers, 128–129
            micelles, 126–127
            nanocapsules, 127–128
            nanogels, 126
            nanospheres, 127
            polymersomes, 128
cartilage regeneration, 114–115
cationic surface, and siRNA, 75
CDKs. *See* cyclin-dependent kinases
    (CDKs)
cell adhesion rate, and MSCs, 112
cell cycle
    deregulation of, 28–29
    and miRNA, 36

overview, 23–26
proteins, 30–36
regulation of, 27–28
    CDK inhibitors, 27–28
    cyclins and CDKs, 27
restriction points and
    checkpoints, 26–27
and RNA interference, 29–38
    action mechanism, 29–30
    overcoming regulation,
        36–38
    RNAi and cell cycle
        proteins, 30–36
    siRNA expression for RNAi,
        34–36
cell distribution (bioluminescence
    imaging), 112–114
cell membrane, and siRNA, 74–77
cell penetrating peptides (CPPs), 76
cell proliferation, 109–110
cellular barriers, and siRNA, 74–80
    binding to cell membrane,
        74–77
    cationic surface, 75
    cell binding ligands, 76–77
    cell penetrating peptides
        (CPPs), 76
    entering cell, 77
    hydrophobicity, 76
    intra-cytoplasmic trafficking,
        79–80
    lipid bilayer of endosomes,
        78–79
chondrogenic differentiation,
    110–112
circulation degradation, and siRNA,
    68–69
cisplatin-incorporating micelle
    (NC-6004), 187–189
    clinical studies, 189
    preclinical studies, 188–189
    preparation and
        characterization of, 188
combination therapy, for cancer,
    215–216
computed tomography (CT), 134–136

conventional drug therapy, and
    CDKs, 29
CPPs. *See* cell penetrating peptides
    (CPPs)
CrEL-free formulations of PTX, 240
CRLX101, and cancer, 210–221
    chemistry, 217
    clinical results, 220–221
    preclinical results, 217–220
CT. *See* computed tomography (CT)
cyclin-dependent kinases (CDKs)
    and conventional drug therapy,
        29
    and cyclins, 27
    inhibitors, 27–28
    overview, 24
cyclins
    and CDKs, 27
    and conventional drug therapy,
        29

delivery systems, of siRNA, 30–34,
    80–90
    and carbon nanotubes (CNT),
        85–87
    and exosomes, 87
    lipid-based, 81–82
    peptides, 84–85
    polymer-based, 82–84
    polysiRNA, 85
    and silica, 87–88
dendrimers
    nanoparticles, 5–6
    polymeric nanoparticles,
        128–129
deregulation, of cell cycle, 28–29
double-stranded RNA (dsRNA), 47
Doxil®
    and animal studies, 330–331
    cell culture studies, 332–333
    clinical indications for,
        326–327
    development of, 316–324
    and DOX-OLV, 327–329
    historical perspective,
        324–326

-like PLDs, 335–336
    and nanodelivery, 236–238
    and nanomedicine regulations,
        295–298
    new developments, 336–341
    overview, 315–316
    physicochemical and
        biological features of, 341
    -related intellectual property,
        327
    in vivo–in vitro correlation
        (IVIVC), 331–332
dsRNA. *See* double-stranded RNA
    (dsRNA)

ECM. *See* extracellular matrix (ECM)
endogenous HIF-1α inhibitors, 213
epirubicin-incorporating micelle
    (NC-6300), 189–192
    clinical studies, 191–192
    preclinical studies, 191
    preparation and
        characterization of, 189–190
exosomes, and siRNA, 87
external magnetic device, and MSCs,
    107
extracellular barriers, and siRNA,
    66–74
    active targeting, 72–73
    exiting systemic circulation,
        70–73
    extracellular matrix (ECM),
        73–74
    and immune system, 69–70
    passive targeting, 70–72
    surviving circulation
        degradation, 68–69
extracellular matrix (ECM), 73–74

Genexol-PM®, 238–239
    clinical efficacy of, 246–247
    clinical studies of, 246
gold nanoparticles, 130–131

HIF-1. *See* hypoxia inducible factor-1
    inhibitors (HIF-1)

HPMA copolymeric nanoparticles. *See N*-(2-hydroxypropyl)-methacrylamide (HPMA) copolymeric nanoparticles

hydrophobicity, and siRNA, 76

hypoxia inducible factor-1 inhibitors (HIF-1), 211–214
  camptothecins, 212–213
  endogenous HIF-1α inhibitors, 213
  2-methoxyestradiol (2ME2), 211–212
  and siRNA, 213

imaging
  bioluminescence, 112–114
  and nanomedicines
    clinical experience, 175–178
    overview, 151–152
    performance optimization, 157–175
    spatio-temporal distribution characteristics of, 152–157
  and theranostics, 133–146
    computed tomography (CT), 134–136
    magnetic resonance imaging (MRI), 136–139
    nuclear imaging, 133–134
    optical imaging, 141–146
    ultrasound, 139–141

immune system, and siRNA, 69–70

intra-cytoplasmic trafficking, 79–80

in vitro siRNA delivery system, 48–66
  characterization and development of, 53–64
  evaluating specific targets, 64–66
  as investigational tool, 50–53

in vivo siRNA delivery system, 66

iron oxide nanoparticles, 130

lesion targeting ability, and nanomedicine imaging, 157–163

ligands, 12–13

lipid-based siRNA delivery systems, 81–82

lipid bilayer, of endosomes, 78–79

liposomal NPs (LNPs), 235

liposomes, 129
  and micelles, 2–4

LNPs. *See* liposomal NPs (LNPs)

magnetic resonance imaging (MRI), and theranostics, 136–139

mesenchymal stem cells (MSCs)
  adhesion to tissue injured site, 112–114
    cell adhesion rate in *ex vivo* studies, 112
    cell distribution (bioluminescence imaging), 112–114
  animal studies, 114–118
    bone regeneration, 115–116
    cartilage regeneration, 114–115
    muscle regeneration, 116–118
  external magnetic device, 107
  overview, 106–107
  and supraparamagnetic iron oxide, 107–112
    cell proliferation, 109–110
    multipotential differentiation capacity, 110–112

metal nanoparticles, 7–8, 130–131
  gold nanoparticles, 130–131
  iron oxide nanoparticles, 130
  silver nanoparticles, 131

micelles
  cisplatin-incorporating, 187–189
    clinical studies, 189
    preclinical studies, 188–189

micelles (*continued*)
   preparation and characterization of, 188
   epirubicin-incorporating, 189–192
      clinical studies, 191–192
      preclinical studies, 191
      preparation and characterization of, 189–190
   and liposomes, 2–4
   paclitaxel-incorporating, 184–187
      clinical studies, 186–187
      preclinical studies, 186
      preparation and characterization of, 184–186
   polymeric nanoparticles, 126–127
   polymeric paclitaxel, 244–247
microbubbles, and theranostics, 132–133
micro-ribonucleoproteins (microRNP), 47
microRNAs (miRNA), 36, 47
microRNP. *See* micro-ribonucleoproteins (microRNP)
miRNA. *See* microRNAs (miRNA)
mononuclear phagocyte system (MPS), 69
MPF. *See* M-phase promoting factor (MPF)
M-phase promoting factor (MPF), 26
MPS. *See* mononuclear phagocyte system (MPS)
MRI. *See* magnetic resonance imaging (MRI)
MSCs. *See* mesenchymal stem cells (MSCs)
multipotential differentiation capacity, 110–112
   chondrogenic differentiation, 110–112
   osteogenic differentiation, 112
muscle regeneration, 116–118

nanocapsules, 127–128
nanocarbons, and theranostics, 131–132
nanocarriers/nanoparticles
   liposomal NPs (LNPs), 235
   overview, 1–2
   physicochemical factors, 8–13
      ligands, 12–13
      shape, 11
      size, 10–11
      surface charge, 11–12
   types of, 2–8
      carbon-based, 8
      dendritic, 5–6
      metallic, 7–8
      micelles and liposomes, 2–4
      nanocrystals, 7
      peptidic, 6–7
      polymeric, 4–5
      silica, 8
      viral nanoparticles (VNPs), 2
nanocrystals, 7
nanodelivery
   albumin-bound PTX, 240–244
   CrEL-free formulations of PTX, 240
   and Doxil, 236–238
   liposomal NPs (LNPs), 235
   nanocarriers for drug delivery, 234–235
   overview, 233–234
   and stroma modification, 170–171
   taxane-based, 238–239
nanogels, 126
nanomaterials
   capabilities of, 282–283
   parameters for bioapplications, 261–265
      concentration, 264
      size/volume, 261–264
      surface charge, 264–265

techniques for characterising,
265–275
differential centrifugal
sedimentation,
267–268
dynamic light scattering,
267
nanoparticle tracking
analysis (NTA), 269–270
resistive pulse sensing
(RPS), 270–271
transmission electron
microscopy (TEM),
268–269
nanomedicines
accurate and objective particle
measurement, 259–265
challenges of, 260–261
parameters of, 261–265
techniques for
characterising,
265–275
tunable resistive pulse
sensing (TRPS) for,
271–275
activatable, 171–175
capabilities of, 282–283
challenges to implementation
of, 285–286
conceptual issues in, 283–285
evolution of, 281–282
and imaging
clinical experience,
175–178
overview, 151–152
performance
optimization, 157–175
spatio-temporal
distribution
characteristics of,
152–157
regulations of
and Doxil®, 295–298
in the European Union,
258
overview, 290–294

pharmacokinetics and
distribution, 306–310
purpose of, 254–256
status of, 256–259
uncertainty, 298–306
in the USA, 257–258
worldwide, 259
nanospheres, 127
NC-6004. *See* cisplatin-incorporating
micelle (NC-6004)
NC-6300. *See* epirubicin-
incorporating micelle (NC-6300)
non-polymeric nanoparticles, 211
NP. *See* nanocarriers/nanoparticles
nuclear imaging, and theranostics,
133–134

optical imaging, and theranostics,
141–146
osteogenic differentiation, 112

paclitaxel (PTX)
CrEL-free formulations of, 240
-incorporating micelle
(NK105), 184–187
clinical studies, 186–187
preclinical studies, 186
preparation and
characterization of,
184–186
passive targeting, and nanomedicine
imaging, 163–170
passive targeting, and siRNA, 70–72
PCNA. *See* proliferating cell nuclear
antigen (PCNA)
PEG polymeric nanoparticles. *See*
polyethylene glycol (PEG)
polymeric nanoparticles
peptides
nanoparticles, 6–7
and siRNA, 84–85
performance optimization
imaging and nanomedicines,
157–175
activatable nanomedicines,
171–175

performance optimization
(*continued*)
  active *versus* passive
    targeting, 163–170
  size-dependence and
    lesion targeting ability,
    157–163
  and stroma modification,
    170–171
physicochemical factors, of NP,
  8–13
  ligands, 12–13
  shape, 11
  size, 10–11
  surface charge, 11–12
polyamine polymeric nanoparticles,
  204–205
polyethylene glycol (PEG) polymeric
  nanoparticles, 206–209
polymer-based siRNA delivery
  systems, 82–84
polymeric nanoparticles, 4–5
  amphiphilic, 209–210
  bioconjugates, 210–211
  and cancer
    combination therapy,
      215–216
    CRLX101, 210–221
    hypoxia inducible factor-1
      inhibitors (HIF-1),
      211–214
    overview, 199–200
    stem cells, 214–215
    topoisomerase 1
      inhibitors, 200–211
  carbohydrate-based,
    203–204
  dendrimers, 128–129
  HPMA copolymeric
    nanoparticles, 205–206
  micelles, 126–127
  nanocapsules, 127–128
  nanogels, 126
  nanospheres, 127
  non-polymeric, 211
  overview, 200–203
  polyamine, 204–205

polyethylene glycol (PEG),
  206–209
polymersomes, 128
  and theranostics, 124–129
polymeric paclitaxel micelle, 244–247
polymersomes, 128
PolysiRNA, 85
primary miRNA (pri-miRNA), 47
pri-miRNA. *See* primary miRNA
  (pri-miRNA)
proliferating cell nuclear antigen
  (PCNA), 28
PTX. *See* paclitaxel (PTX)

regulations
  of cell cycle, 27–28
    CDK inhibitors, 27–28
    cyclins and CDKs, 27
    and overcoming RNAi,
      36–38
  of nanomedicines
    and Doxil®, 295–298
    in the European Union,
      258
    overview, 290–294
    pharmacokinetics and
      distribution, 306–310
    purpose of, 254–256
    status of, 256–259
    uncertainty, 298–306
    in USA, 257–258
    worldwide, 259
reticuloendothelial system. *See*
  mononuclear phagocyte system
  (MPS)
RISC. *See* RNA-induced silencing
  complex (RISC)
RNAi. *See* RNA interference (RNAi)
RNA-induced silencing complex
  (RISC), 47
RNA interference (RNAi), 47
  and cell cycle, 29–38
    action mechanism, 29–30
    cell cycle proteins, 30–36
    overcoming regulation,
      36–38
    siRNA expression for, 34–36

sentinel lymph node (SLN), 163
shape, of NP, 11
short hairpin RNAs (shRNA). *See*
small interfering RNA (siRNA)
silica, and siRNA, 87–88
silica nanoparticles, 8
silver nanoparticles, 131
siRNA. *See* small interfering RNA
(siRNA)
size, of NP, 10–11
size-dependence, and nanomedicine
imaging, 157–163
SLN. *See* sentinel lymph node (SLN)
small interfering RNA (siRNA)
cellular barriers, 74–80
binding to cell
membrane, 74–77
cationic surface, 75
cell binding ligands,
76–77
cell penetrating peptides
(CPPs), 76
entering cell, 77
hydrophobicity, 76
intra-cytoplasmic
trafficking, 79–80
lipid bilayer of
endosomes, 78–79
delivery systems, 30–34,
80–90
and carbon nanotubes
(CNT), 85–87
and exosomes, 87
lipid-based, 81–82
peptides, 84–85
polymer-based, 82–84
polysiRNA, 85
and silica, 87–88
expression for RNAi, 34–36
extracellular barriers, 66–74
active targeting, 72–73
exiting systemic
circulation, 70–73
extracellular matrix
(ECM), 73–74
and immune system,
69–70

passive targeting, 70–72
surviving circulation
degradation, 68–69
and hypoxia inducible factor-1
inhibitors (HIF-1), 213
overview, 46–48
in vitro delivery, 48–66
characterization and
development of, 53–64
evaluating specific
targets, 64–66
as investigational tool,
50–53
in vivo delivery, 66
spatio-temporal distribution
characteristics
of imaging and
nanomedicines, 152–157
biodistribution
assessment, 152–153
as non-invasive method,
153–157
stem cells, cancer, 214–215
stroma modification, and
nanomedicine imaging,
170–171
supraparamagnetic iron oxide, and
MSCs, 107–112
cell proliferation, 109–110
multipotential differentiation
capacity, 110–112
surface charge, of NP, 11–12
systemic circulation, and siRNA,
70–73

taxane-based nanodelivery,
238–239
theranostics
carrier materials, 124–133
antibodies, 129–130
liposomes, 129
metal nanoparticles,
130–131
microbubbles, 132–133
nanocarbons, 131–132
polymeric nanoparticles,
124–129

theranostics (*continued*)
  and imaging, 133–146
    computed tomography
      (CT), 134–136
    magnetic resonance
      imaging (MRI),
      136–139
    nuclear imaging,
      133–134
    optical imaging,
      141–146
    ultrasound, 139–141
  overview, 120–124
tissue injured site adhesion, and
  MSCs, 112–114
    cell adhesion rate in *ex vivo*
      studies, 112
    cell distribution
      (bioluminescence imaging),
      112–114

topoisomerase 1 inhibitors
  and polymeric nanoparticles,
    200–211
      amphiphilic, 209–210
      bioconjugates, 210–211
      carbohydrate-based,
        203–204
      HPMA copolymeric
        nanoparticles, 205–206
      non-polymeric, 211
      overview, 200–203
      polyamine, 204–205
      polyethylene glycol (PEG),
        206–209

ultrasound, and theranostics,
  139–141

viral nanoparticles (VNPs), 2
VNPs. *See* viral nanoparticles (VNPs)